SEMICONDUCTOR SPINTRONICS

PREFACE

This book introduces the developing history of semiconductor spintronics, basic concepts, and research fruits and prospected future development. It includes Introduction and 10 chapters. The introduction introduces the developing history of semiconductor spintronics. Chapter 1 introduces the properties of magnetic ions in semiconductors, energy-level splitting of magnetic ions in the crystal field, and energy-level characteristics of the basic state and low excited states. Chapter 2 introduces the properties of dilute magnetic semiconductors, giant Zeeman splitting effect, and optical properties. Chapter 3 introduces ferromagnetic semiconductors, ferromagnetic interaction theory, and factors influencing the Curie temperature. Chapter 4 introduces the injection of spin electron, Rashba effect, coherent transport of spin through hetero-interface, experiments and theories of injection of spin-polarized electrons. Chapter 5 introduces spin relaxation, three main mechanisms of spin flip — EY, DP, and BAP mechanisms — and experimental studies of spin relaxation. Chapters 6–10 are special research topics, introducing some recent research fruits. Chapter 6 introduces the theoretical basis and experimental measurements of Rashba and Dresselhaus effects. Chapter 7 is on the optical responsibility of spin, including spin photocurrent induced by optically injected electron spin and electron spin polarization derived by the electric field, in spin-splitting systems. Chapter 8 is on the control of spin coherent electrons, including electron spin coherence, their spatial movement, spin Hall effect, production of spin current, and spin dynamics in semiconductors. Chapter 9 is about transport of spin-polarized electrons and magnetic domains, including the spin transport of two-dimensional electron gas and quantum dot of magnetic semiconductors, and magnetic domain transport in magnetic semiconductors. Chapter 10 deals with spin property control in semiconductor quantum dots and wires, including the control of g-factors and Rashba coefficient by changing the shape and size of dots and wires, N doping, and applied electric or magnetic fields. This book is suitable for high-level students in university, postgraduates, and professors and researchers.

CONTENTS

LIST OF ACRONYMS

1D	one-dimensional
2D	two-dimensional
2DEG	two-dimensional electron gas
BIA	bulk inverse asymmetry
CB	conduction band
CFR	crystal-field resonance
CISP	current induced spin polarization
CP	circularly polarized
CPGE	circular photo-galvanic effect
DBH	dangling band hybrid
DMS	dilute magnetic semiconductor
DTS	differential transmission spectrum
DW	domain-wall
EMA	effective mass approximation
ESR	electron spin resonance
FET	field effect transistor
FFT	fast Fourier transform
FMS	ferromagnetic semiconductor
FT	Fourier transform
FR	Faraday rotation
FWHM	full-width at half-maximum
GMR	giant magnetic resistance
HRTEM	high resolution transmission electron microscopy
KR	Kerr rotation
LED	light emitting diode
LH	left hand
MBE	molecular beam epitaxy
MCD	magnetic circular dichroism
MOKE	magneto-optical Kerr effect
MP	magnetic polaron
MPGE	magneto-photogalvanic effect

MQW	multiple quantum well
MR	magnetoresistance
MRAM	magnetic random access memory
OS	oxidation state
PEM	photoelectric modulator
PL	photoluminescence
QD	quantum dot
QDQW	QD quantum well
QW	quantum well, quantum wire
QUIC	quantum interference control
RF	radio frequency
RH	right hand
RHEED	reflection high-energy electron diffraction
RSA	resonant spin amplification
RSHE	reversed spin Hall effect
RSS	Rashba spin splitting
RT	room temperature
SdH	Shubnikov-de Haas
SEM	scanned electron microscopy
SGE	spin-galvanic effect
SIA	structure inverse asymmetry
SL	square loop
SOC	spin-orbit coupling
SOI	spin-orbit interaction
SPC	spin polarized current
SQUID	superconducting quantum interference device
SRT	spin relaxation time
TE	transverse electric (field)
TEM	transmission electron microscopy
TM	transition metal
TM	transverse magnetic (field)
TMR	tunneling magnetoresistance
TRFR	time-resolved Fadaday rotation
VCSEL	vertical-cavity surface emitting laser
XRD	X-ray diffraction

INTRODUCTION

Spintronics studies the use of spin freedom of electrons. It originated from the giant magnetic resistance (GMR) effect found by Fert and Gruenberg, respectively, in 1988. This discovery produced highly sensitive magnetic sensors, used in the read-out head of the magnetic hard disk. Just after 9 years as the GMR was discovered, in November 1997 IBM Co. declared that they fabricated the commercial read-out head of the magnetic hard disk. This product accomplished an enterprise of several billions of US dollars. The spin electronic devices under study include magnetic random access memory (MRAM), spin field effect transistor (FET), spin-polarized laser, etc. These devices depend on their ability to control spin in the solids and they can be used to decrease the power consumption, to overcome the velocity limit connected with the charge electrons, and in the quantum information treatment and quantum computation in future.

The study contents include: production, transport, tunneling of spin-polarized electrons, optical phenomena, lifetime, decoherence mechanism connected with them, etc. A semiconductor is the best material to study spintronics, because: (1) the number of carriers in a semiconductor is relatively few, and their behavior can be looked as that of a single particle, excluding the many-particle effect; (2) the qualities of semiconductor single crystals, heterostructures, quantum wells, or quantum dots (QDs) can be made to be very perfect, so that the lattice defects and impurities can be decreased to the least degree, and the relaxation of electron spins can be decreased; (3) semiconductors are "transparent" to most part of the light; thus, one can inject and detect spin electrons by circularly polarized (CP) light; (4) the semiconductor device technology is advanced, and it is easier to make devices, integrate devices, or integrate with other devices. But a semiconductor has a shortcoming: it is nonmagnetic. Its magnetism has to dope magnetic ions from outside; while the solubility of the magnetic ions in the semiconductor is smaller, the concentration of magnetic impurity in semiconductors is generally several percent higher. Therefore, the

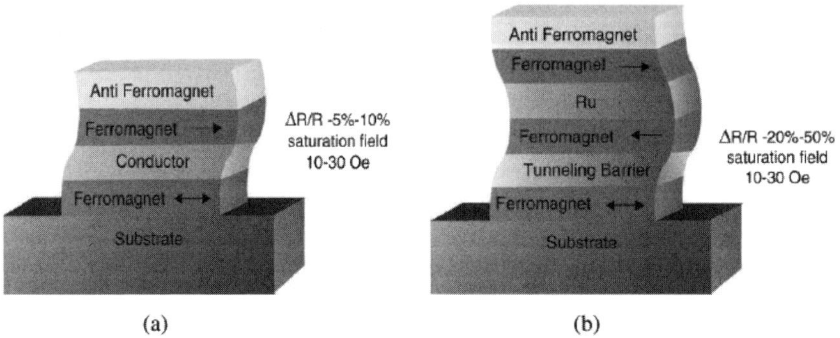

Fig. 0.1. Schematic diagram of (a) spin valve and (b) MTJ proposed on the basis of the GMR effect (Wolf *et al.*, 2001).

formation of the spin-polarized electron in the semiconductor is a difficult problem.

0.1. Origin of Spintronics — GMR Effect Device

In 1988, the GMR effect was discovered in a three-layer thin film structure, as shown in Fig. 0.1(a). In this structure, the above and underneath layers are ferromagnetic materials, for example, alloys of Fe, Co, and Ni, and the middle layer is a nonmagnetic material, for example, Cu. When the magnetic moments of above and underneath layers are parallel, the resistance of the material is smallest; when the magnetic moments are antiparallel, the resistance is largest. The current can be perpendicular to the interface (CPP) and also can be parallel to the interface (CIP). This effect exists at room temperature (RT). When a rather small magnetic field (~100 Oe) changes the orientation of one ferromagnetic layer, one can observe obvious variations in the resistivity (~10%).

On the basis of the GMR effect, the spin valve was proposed, as shown in Fig. 0.1(a). An antiferromagnetic layer is added on the above ferromagnetic layer. The antiferromagnetic layer makes the magnetic moment of the above ferromagnetic layer hardly to change the direction in the external magnetic field, i.e. it plays a "nail-up" role of the magnetic moment. The underneath ferromagnetic layer is free, and its magnetic moment direction can be changed in the external small magnetic field. When the magnetic moments of the above and underneath layers become antiparallel, the resistance generally increases 5–10%.

Magnetic tunnel junction (MTJ), as shown in Fig. 0.1(b), consists of above nail-up layer (two ferromagnetic layers, and a thin layer of Ru placed in between them, form a strong antiferromagnetic coupling), underneath free ferromagnetic layer, and middle thin insulator layer (generally Al_2O_3). The insulator layer acts as a potential barrier layer, and the tunneling current is perpendicular to the interface. When the moments of the above and underneath ferromagnetic layers change from parallel to antiparallel, the tunneling resistance changes 20–30%, similar to the spin valve, but the modulation range is large. Because the tunneling current density is generally small, the MTJ device has a high resistance.

The application of the spin valve and MTJ device is wide, for example, magnetic field sensors, read-out head of the magnetic disk, galvanic isolators, MRAM, etc. GMR spin valve read-out head is the main part of the magnetic disk driver; nearly all commercial GMR heads are the spin valve forms originally proposed by IBM. After improvement, for example, using the antiferromagnetic layer as the nail-up layer (as shown in Fig. 0.1(b)), the increase in the magnetic resistance rises from 5% to today's 20%. Now the memory density of the hard disk has approached 100 Gbits per square inch. The stripe width of the sensors approaches 0.1 μm, the current density becomes very large, and the demand for the sensitivity of the spin valve becomes more high and high.

MRAM uses the property of the magnetic hysteresis loop to store up data and uses the magnetic resistance to read out data. When the GMR-based MTJ and the memory unit of spin valve are integrated into an integrated circuit chip, its function is like a general semiconductor random access memory. But it has an advantage, the data can be retained with power off. Compared with the electrically erasable programmable read-only memory (EEPROM) and the flash memory, the write time of the MRAM is 1000 times faster; besides, there is no wearout in the write cycling and low power consumption for writing, while the EEPROM and flash memory will wear out after one million write circles. The MRAM data access times are about 1/10,000 that of hard disk drives. MRAM is not yet available commercially.

0.2. New Materials for Spintronics Applications

The search for material-combining properties of the ferromagnet and the semiconductor has been a long-standing and elusive goal, due to the large

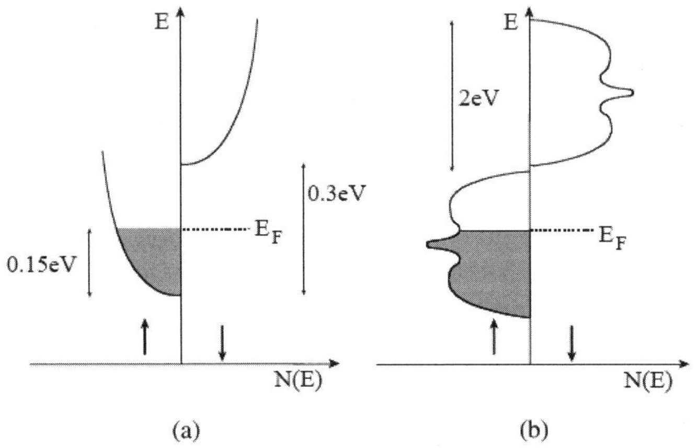

Fig. 0.2. (a) Schematic of density of states of the magnetic semiconductor below T_C and (b) schematic of density of states of half-metallic ferromagnet CrO_2 (Wolf *et al.*, 2001).

differences in the crystal structure and chemical bonding of two materials. The ferromagnetic semiconductor (FMS) is an ideal material, which is a semiconductor, meanwhile, has a higher Curie temperature. EuS, the most early studied magnetic semiconductor, in which the magnetic species (Eu^{2+}) resides on every lattice site, has no practical applications, because its Curie temperature is much lower than RT. Recently, it was found that in $Ga_{1-x}Mn_xAs$, the Curie temperature T_C is as high as 110 K (Ohno *et al.*, 1996). It is an important step in putting forth strength to raise the Curie temperature. The density of states of the magnetic semiconductor is shown in Fig. 0.2(a). Assuming that the magnetic interaction occurred through the free holes in the semiconductor, the theory predicted that the Curie temperatures of many semiconductors can be higher than RT (Dietl *et al.*, 2000). There have been some progresses in search for the FMSs with a high Curie temperature, but the experimental results are not too definite.

Experimentally, another material was found — ferromagnetic oxides and related compounds, which have large spin-polarized carriers, for example, CrO_2, Fe_3O_4, $La_{0.7}Si_{0.3}MnO_3$, etc. Their densities of states of conduction band (CB) for spin-up and spin-down electrons are seperated completely, as shown in Fig. 0.2(b). Thus, they are "half-metallic"; the Fermi level intersects only one of the two spin bands. Many of these oxides have been investigated through Andreev reflection spectroscopy (Soulen *et al.*, 1998), and it was found that they have high carrier spin polarization values (above 70%) at low temperatures. At liquid He temperature, the spin polarization of CrO_2 is as high as 90%. But at RT, the spin polarization decreases to zero rapidly.

Other materials, for example, the laser-ablated 6–8% Co-doped TiO_2, have the RT ferromagnetism (Matsumoto *et al.*, 2001). These materials are transparent to visible light and may be of particular importance in optoelectronic applications. Zincblende CrSb/GaAs/CrSb epilayers have been confirmed to be ferromagnetic at RT (Zhao *et al.*, 2001).

0.3. Spin Injection and Spin Transport of Electrons

The most important key for the fabrication of spin-related devices (spin electronic devices) is the injection of spin-polarized electrons, production of spin current, and electric detection at RT and without magnetic field. The following are several possible production methods of spin current.

(1) *Ohmic injection.* Through a ferromagnetic metal, the spin-polarized electric current is introduced. If the Ohmic contact is formed between the ferromagnetic metal and the semiconductor, it is expected that the spin-polarized current can be derived from the ferromagnetic metal. But generally, the Ohmic contact formed by heavy doping leads to spin-flip scattering, resulting in the loss of spin polarization. At $T < 10\,\text{K}$, the 4.5% spin-polarized electron injection was obtained by the FM-InAs Ohmic contact (Hu *et al.*, 2001), and the 2% spin-polarized electron injection was obtained by the FM-GaAs Ohmic contact (Zhu *et al.*, 2001).

(2) *Tunneling injection.* With a scanning tunneling microscope with a ferromagnetic tip, it is proved that the spin-polarized electrons can be effectively injected into the semiconductor in a vacuum tunneling process (Alvorado and Renand, 1992). The FM–insulator–FM tunneling junction with a high magnetic resistance (Fig. 0.1(b)) has also demonstrated that the tunneling barrier can conserve the spin polarization in the tunneling process. Thus, it is suggested that the tunneling may be a more effective method to achieve spin injection than the diffusive transport.

The above two methods depend on the epitaxial growth of a ferromagnetic thin film on semiconductors, and the formation of sharp interface and high-quality Schottky barrier is demanded.

(3) *Ballistic electron injection.* Besides difussive transport, ballistic transport can also be used. From the energy difference between two spin conduction subbands of the ferromagnetic material and the CB of the

semiconductor, the transmission and reflection probabilities of the spin-dependent interfacial ballistic electron can be determined. With a point contact formed between the ferromagnetic and nonferromagnetic metals, as high as 40% spin-polarized current was obtained experimentally in the nonferromagnetic metal (Upadhyay *et al.*, 1998).

(4) *Using the giant Zeeman splitting effect of the dilute magnetic semiconductor (DMS) in the magnetic field.* The DMS $Be_{0.07}Mn_{0.03}Zn_{0.9}Se$ is used as a spin aligner. At a low temperature (~10 K) and an applied magnetic field (>1 T), when electrons move through 300 nm BeMnZnSe layer, spin-polarized electrons with 90% injection efficiency were obtained (Fiederling *et al.*, 1999). It is because in the magnetic field, Mn ions of low concentration produce giant Zeeman splitting in the semiconductor by the sp–d exchange interaction, and the effective Landè g-factor of electron can reach 100. At low temperatures, the electrons occupy only the lowest $S = -1/2$ subband, and the spin-polarized electrons are obtained.

Spin polarization is measured from the circular polarization of the light-emitted from a GaAs light-emitting diode (LED) injected by the spin-polarized electrons. This experiment also demonstrated that the spin decoherent time is long enough, so that the spin polarization was still conserved in the 100 μm diffusive length. Similarly, in the case of not completely matched DMS $Zn_{0.94}Mn_{0.06}Se$ and GaAs, the effective injection of spin-polarized electrons was also observed (Jonker *et al.*, 2000), and the injection efficiency of spin polarization reached 50%.

(5) *Using the FMS as a spin aligner.* When the p-type FMS $Ga_{0.955}Mn_{0.045}As$ is at a temperature lower than the Curie temperature T_C (≈105 K), the spontaneous magnetic moments are formed in the semiconductor. Under the forward bias voltage, the spin-polarized holes from the p-type GaMnAs and the nonpolarized electrons from the n-type GaAs are injected simultaneously into an InGaAs quantum well, respectively. The holes and electrons combine in the quantum well, resulting in the emission of CP light. The spin polarization of the hole can be derived from the circular polarization of the light; at $T = 6$ K, the spin polarization of the hole reached 10% (Ohno *et al.*, 1999). This experiment demonstrated that the FMS can produce spin-polarized holes, which move through the interface, and conserve polarized diffusive distance larger than 200 nm. And the spin orientation can be modulated by applying a small magnetic field (~40 Oe).

(6) *Using the optical method.* The sample is a GaAs quantum well, and the growth direction is the z-direction. In the GaAs quantum well,

the subbands of the heavy and light holes are separated; the heavy one is the basic state, and the light one is the excited state. When a CP light (σ^+) with the energy of a light hole exciton is irradiated on the sample, it can only excite the electron from the $L_z = -1/2$ light hole subband to the $S_z = 1/2$ electron subband ($\Delta L_z + 1$) due to the transition selection rule between the heavy hole, light hole, and the electron. Therefore, in the CB, the number of $S_z = 1/2$ electrons is much larger than that of $S_z = -1/2$ electrons, resulting in spin-polarized electrons. After the laser pulse is incident on the sample, electrons in this part will recombine with the heavy holes and produce σ^+ and σ^- polarized light, corresponding to the electron–heavy hole transitions ($S_z = -1/2 \rightarrow L_z = -3/2$) and ($S_z = +1/2 \rightarrow L_z = 3/2$), respectively. The intensity of σ^- CP light I_- will be much larger than that of σ^+ CP light I_+, I_-/I_+ equals the ratio of the $S_z = 1/2$ and $S_z = -1/2$ electron numbers. From the time-resolved optical spectroscopy, the spin relaxation time (SRT) obtained was about several hundreds of picoseconds (Heberle *et al.*, 1994). When a magnetic field is applied in the x-direction, the oscillation of the I_- and I_+ decay curves with time can be observed due to the precession of the electron spins around the applied magnetic field. The oscillation period is the Larmor angular frequency $g_e\mu_B B/\hbar$, where g_e is the Landè factor of the electron.

0.4. Optical Modulation of Spin Coherence in Semiconductors and Nanostructures

From the last section, we see that the spin-polarized electrons in semiconductors and nanostructures can be produced, detected, and modulated by optical methods (pulsed laser, time-resolved optical spectroscopy, etc.). One recent experiment (Kikkawa *et al.*, 1997) showed that in the narrow range of doping concentration in the semiconductor, the spin life increases about several orders of magnitude, longer than several nanoseconds. In the heterojunctions and QDs the nanosecond spin life is conserved until RT. This fact opens the road for the practical applications of the coherent quantum spintronics.

(1) *Prolonging spin lifetime.* Assume that the spin direction of an electron is in the z-direction and the applied magnetic field is in the y-direction, then the components a and b of the electron spin wave function change with time,

$$a(t) \sim \cos \omega_L t, \quad b(t) \sim \sin \omega_L t, \tag{0.1}$$

where $\omega_{\rm L} = \mu_{\rm B} g_{\rm e} B/\hbar$ is the Larmor frequency. If we look a and b as two components of a vector in the plane perpendicular to the magnetic field (z–x plane), then the vector rotates with time in the plane, similar to the Larmor precession of a top in the gravity field.

Spin coherent state is the state where all electrons of same spin precess with one frequency, but because of the spin–spin, spin–impurity, or spin–phonon interactions, the total spin of the system (magnetization) will decay exponentially with time

$$M(t) = M(0)\exp\left(-\frac{t}{T_2}\right)\cos\left(\frac{\mu_{\rm B} g_{\rm e} B t}{\hbar}\right), \tag{0.2}$$

where T_2 is called the dephasing time or spin lifetime. Generally T_2 is several hundreds of picoseconds.

The optical methods to measure spin dephasing time are mainly Faraday rotation and Kerr rotation (KR), as shown in Figs. 0.3(a) and 0.3(b), respectively. The light beam is incident normal to the sample, along the z-direction, and the magnetic field is applied in the sample plane, along the y-direction. The CP pump laser pulse excites spin coherent electron states, which make Larmor precession around the magnetic field B. Another probe beam is linearly polarized. The Faraday rotation method is to measure the rotation angle of the linear polarization of the beam transmitting from the sample. The rotation angle is proportional to the projection of the electron total spin in the beam direction. The KR method is to measure the rotation angle of the linear polarization of the beam reflecting from the sample, as shown in Fig. 0.3(b). Applying the Faraday method needs to cut the non-transparent substrate layer, while applying the Kerr method does not need.

The sample is the modulation-doped $Zn_{0.8}Cd_{0.2}Se$ quantum well, and the impurity density is $n \sim 10^{17}\,{\rm cm}^{-3}$, which produces the two-dimensional electron gas (2DEG). At $T = 5\,{\rm K}$, $B = 4\,{\rm T}$, the measured Kerr angle oscillation curves show that they decay with time very slowly, with the spin lifetime $T_2 \sim 10\,{\rm ns}$ (Kikkawa *et al.*, 1997). For the ordinary quantum well sample, the oscillation decays quickly, and T_2 is smaller by three orders of magnitude. This is because the 2DEG sample has a shorter hole lifetime and a weaker electron–hole exchange, so that spin relaxation is incomplete, and remanent spin polarization precesses long after the holes have recombined.

(2) *Spin coherent transport through the heterostructure interface.* For a GaAs/ZnSe heterojunction sample, the CP pump laser pulse excites electron spin coherent states in the GaAs layer. The KR angles in the ZnSe layer were measured by the probe beam. It is found that though only

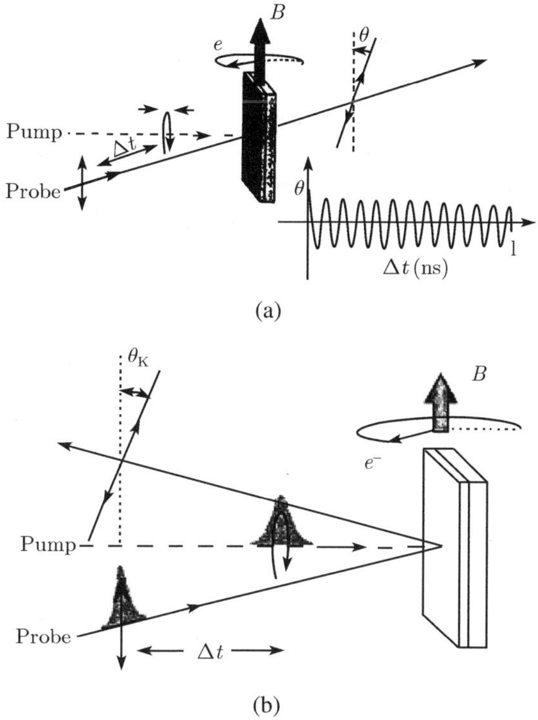

Fig. 0.3. Schematic figures of (a) Faraday rotation and (b) KR experiments.

2.5–10% spin electrons enter into the ZnSe layer from the GaAs layer, these coherent states still conserve several nanoseconds in the ZnSe layer (Malajovich *et al.*, 2000). Besides, the GaAs/ZnSe interface also modifies the amplitude and phase of the electron spin through an associated discontinuity in the electron g-factor.

Further experiments (Malajovich *et al.*, 2001) show that the applied bias electric field will increase the injection efficiency of spin electrons through the interface to five times. The sourcing of coherent spin transfer lasts at least one to two orders of magnitude than in unbiased structures, and the spin signal survives for longer times, longer than the interval between two laser pulses. These experiments reveal promising opportunities for multifunctional spin electronic devices, in which the amplitude and phase of the net spin current are controlled by either electrical or magnetic fields.

(3) *Space resolution of spin coherent states.* The sample is GaAs-doped with donor impurity Si, where the electron can recombine with the excited

hole and increase the spin polarization lifetime. A laser pulse of time interval Δt smaller than the spin polarization lifetime is incident on the sample and excites the spin-polarized electrons in the sample. Adjusting the magnetic field intensity, so that the precession angle of the spin electrons in Δt time equals 2π, i.e. the Larmor precession is driven into resonance with the optical excitation repetition interval, the Faraday rotation intensity will increase greatly (Kikkawa and Awschalom, 1999). Taking $\Delta t = 50\,\text{ps}$, and scanning the magnetic field in the range of -0.1 to $0.1\,\text{T}$, the authors measured the Faraday rotation as a function of the magnetic field. There are several equal-distance resonant peaks in the curve; the intensity of the peaks is larger than that produced by a single pulse by one order of magnitude. It is estimated from the width of the resonant peak that the spin lifetime at RT exceeds 100 ns.

Making spin devices needs to realize effective spin electron transport, without destroying related spin information. On the basis of above experiments, Kikkawa and Awschalom (1999) did the space-resolution Faraday rotation experiments and obtained that in the drag of an electric field the transport distance of coherent precessing spin electrons exceeds $100\,\mu\text{m}$. When spin electrons are excited by the laser pulse, they will move in the electric field direction. The wave pockets of spin electrons produced by 10 sequence laser pulses are distributed in the electric field direction, and the distance of the farthest wave pocket from the excited point is about $100\,\mu\text{m}$. The illuminant intensity is measured simultaneously, and it is found that the spin dephasing effect is medium, with the dephasing time about 29 ns.

0.5. Spin Electronic Devices

On the basis of special properties of DMSs and FMSs, some spin electronic prototype devices have been proposed and fabricated. Either adding the spin degree of freedom to conventional charge-based electronic devices or using the spin alone has the potential advantages of nonvolatility, increased data processing speed, decreased electric power consumption, and increased integration densities compared with conventional semiconductor devices.

(1) *Spin LED.* As mentioned in Sec. 0.3.4, in the n-electrode of GaAs LED, there is a layer of DMS BeMnZnSe. An applied magnetic field (several tesla) makes the electron energy of BeMnZnSe to produce giant Zeeman splitting and the electrons through this layer to be spin-polarized. The

spin-polarized electrons recombine with the spin randomly oriented holes in the GaAs LED, resulting in the CP light. Measuring the polarization of the emitting light, one can obtain the polarization of the electron spin (Fiederling *et al.*, 1999). On the other hand, as a device application, the magnetic field can control the device from emitting or not emitting CP light.

(2) *Ferromagnetic FET.* As mentioned in Sec. 0.2, GaMnAs is a ferromagnetic material, whose Curie temperature reaches as high as 110 K (Ohno *et al.*, 1996). The theoretical study found (Dietl *et al.*, 2000) that the magnetic interaction between magnetic ions is through the free holes in the semiconductor; the more the hole density, the higher the Curie temperature. If the FET is constructed by the ferromagnetic material InMnAs, the metal gate can control the hole concentration in the FMS channel, then changes its ferromagnetic property (Ohno *et al.*, 2000). At 22.5 K, the sheet Hall resistivity R_{Hall}, which is proportional to the magnetization of the magnetic semiconductor layer, is dependent on the magnetic field. When the gate voltage is zero, the R_{Hall}–B curve shows a weak magnetic hysteresis. When $V_G = -125$ V, the hole concentration in the channel increases, and the R_{Hall}–B curve shows a strong magnetic hysteresis, which means that the InMnAs material has strong ferromagnetism. When $V_G = 125$ V, the holes in the channel are exhausted, and the R_{Hall}–B curve shows no magnetic hysteresis, only a linear relation with the magnetic field B, which means that the InMnAs becomes a paramagnetic material. Thus, this device can be used to control the ferromagnetic property of the material by a small gate voltage.

(3) *FMS tunneling junction.* As mentioned in Sec. 0.1, the MTJ is composed of a pinned layer and a free layer separated by a very thin insulating layer, generally aluminum oxide (see Fig. 0.1(b)). Now that the MTJ has been applied to the semiconductor, the up and down layers are FMSs and the middle layer is a nonmagnetic semiconductor; this is called the FMS tunneling junction. It has the property similar to that of MTJ, when the magnetizations of the up and down layers are parallel, the tunneling resistance R_P is small. When the magnetizations are antiparallel, the tunneling resistance R_A is large. Define the tunnel magnetoresistance (TMR) as

$$\text{TMR} = \frac{R_A - R_P}{R_P} = \frac{2P_1P_2}{(1 - P_1P_2)}, \tag{0.3}$$

where P_1 and P_2 are the spin polarizations of the up and down layers, respectively. At low temperatures, the TMR of the $Ga_{1-x}Mn_xAs$/AlAs/

$Ga_{1-x}Mn_xAs$ tunneling junction can reach 75% (Tanaka and Higo, 2001). This TMR was obtained in junctions with a very thin ($\leq$1.6 nm) AlAs tunnel barrier. The TMR decreases rapidly with increasing barrier thickness.

There is another double tunneling junction structure: GaMnAs/AlAs/GaAs/AlAs/GaMnAs (Mattana *et al.*, 2003), where the first junction plays the role of the injector of the ballistic spin electrons and the second junction is used to detect the resulting spin polarization in GaAs. Because the ferromagnetism in GaMnAs is produced by the holes, the tunneling current is the hole current. From the experiment, it is found that at 4 K the TMR of the single tunneling junction is 38%, while the TMR of the double tunneling junction also has the same value. In the double tunneling junction, the width of the barrier layer is two times that of the single junction, then the tunneling resistance would increase by three orders of magnitude. But the experimental resistance of the double junction equals nearly that of the single junction ($\sim 10^{-2}\,\Omega\,\mathrm{cm}^2$); it is concluded that the process is a sequential tunneling process: the holes first enter the GaAs quantum well through the first junction and stay in there. Then the holes tunnel to the GaMnAs layer through the second junction. Because there is still large TMR ($\sim$35%), which means the spin of the hole in the quantum well is conserved for a long time, its SRT τ_{sf} is much larger than the staying time in the well τ_{n}. This is the first example to detect the spin polarization in semiconductors by the electric method.

References

Alvorado SF, Renand Ph. (1992) *Phys Rev Lett* **68**: 1387.
Dietl T *et al.* (2000) *Science* **287**: 1019.
Fiederling R, Keim M, Reuscher G, Ossan W, Schmidt G, Wang A, MolenKamp LW. (1999) *Nature* **402**: 787.
Heberle AP, Ruhle WW, Ploog K. (1994) *Phys Rev Lett* **72**: 3887.
Hu CM *et al.* (2001) *Phys Rev B* **63**: 125333.
Jonker BT, Park YD, Bennett BR, Cheong HD, Kioseeglou G, Petron A. (2000) *Phys Rev B* **62**: 8180.
Kikkawa JM, Smorchkova IP, Samarth N, Awschalom DD. (1997) *Science* **277**: 1284.
Kikkawa JM, Awschalom DD. (1999) *Nature* **397**: 139.
Malajovich I, Kikkawa JM, Awschalom DD, Berry JJ, Samarth N. (2000) *Phys Rev Lett* **84**: 1015.
Malajovich I, Berry JJ, Samarth N, Awschalom DD. (2001) *Nature* **411**: 770.
Matsumoto Y *et al.* (2001) *Science* **291**: 854.
Mattana R, George JM, Jattres H *et al.* (2003) *Phys Rev Lett* **90**: 166601.

Ohno H *et al.* (1996) *Appl Phys Lett* **69**: 363.
Ohno Y, Young DK, Beschton B, Matsukura F, Ohno H, Awschalon DD. (1999) *Nature* **402**: 790.
Ohno H *et al.* (2000) *Nature* **408**: 944.
Soulen RJ *et al.* (1998) *Science* **282**: 85.
Tanaka M, Higo Y. (2001) *Phys Rev Lett* **87**: 026602.
Upadhyay SK *et al.* (1998) *Phys Rev Lett* **81**: 3247.
Wolf SA, Awschalom DD, Buhrman BA *et al.* (2001) *Science* **294**: 1488.
Zhao J *et al.* (2001) *Appl Phys Lett* **76**: 2776.
Zhu HJ *et al.* (2001) *Phys Rev Lett* **87**: 016601.

Chapter 1

PROPERTIES OF MAGNETIC IONS IN SEMICONDUCTORS

1.1. Electron Configuration of Magnetic Ions

Ferromagnetic atoms Ti, V, Cr, Mn, Fe, Co, and Ni are in the fourth row of the Element Periodic Table. The electrons of the outmost shell are 3d electrons, and there are 4–10 3d electrons from Ti to Ni, respectively. Because there are many electrons, we should first classify the electron configurations of the ferromagnetic ions.

There are two basic rules for classification of the electron configuration:

1. Pauli principle: there are no electrons whose four quantum numbers n, l, m_l, and m_s are equal, where n is the principal quantum number, l is the orbital angular momentum quantum number, m_l is the projection of the angular momentum in the z-direction, and m_s is the projection of the spin in the z-direction.
2. The Hunde rule determining the basis term of the atom:
    a. In the given electron configurations, the spectrum term that the spin has the maximum value $S_{\max}$ and L takes the maximum possible value at this $S_{\max}$ is the basis term, i.e. has the lowest energy.
    b. If the filled electrons in the outmost shell have not exceeded the half, then the basis term is $J = |L - S|$; if the fill exceeds the half, the basis term is $J = L + S$.

Now we determine the electron configurations and basis term of the often used Mn^{2+} ion in the DMSs. There are seven 3d electrons in the outmost shell of the Mn atom, so there are five 3d electrons in the Mn^{2+} ion. The method of determining the electron configurations is: under the Pauli principle, first write the distribution table of electrons according to the quantum numbers, as shown in Table 1.1. In Table 1.1, the

Table 1.1. Possible electron configurations of Mn^{2+} ion.

m_l	m_s	×5	×10										
+2	↑	↑	↑	↑↓	↑↓	↑↓	↑↓	↑	↑	↑	↑	↑	↑
+1	↑	↑	↑	↑	↑	↑		↑↓	↑↓	↑↓	↑	↑	↑
0	↑	↑	↑	↑	↑		↑	↑	↑		↑↓	↑↓	↑
−1	↑	↑	↓	↑		↑	↑	↑		↑	↑		↑↓
−2	↑	↓	↓		↑	↑	↑		↑	↑		↑	
M_L	0	0	0	4	3	2	1	3	2	1	2	1	1
M_S	5/2	3/2	1/2	3/2	3/2	3/2	3/2	3/2	3/2	3/2	3/2	3/2	3/2

m_l	m_s	×3	×3	×3	×3	×3	×3	×3	×3	×3
+2	↑↓	↑↓	↑↓	↑↓	↑	↑	↑	↑	↑	↑
+1	↑	↑	↑		↑↓	↑↓	↑↓	↑	↑	↑
0	↑	↑		↑	↑	↑		↑↓	↑↓	↓
−1	↓		↑	↑	↓		↑	↓		↑↓
−2		↓	↓	↓		↓	↓		↓	
M_L	4	3	2	1	3	2	1	2	1	1
M_S	1/2	1/2	1/2	1/2	1/2	1/2	1/2	1/2	1/2	1/2

m_l	m_s																	
+2	↑↓	↑↓	↑↓	↑↓	↑↓	↑↓	↑↓	↑↓	↑↓	↑↓	↑↓	↑			↑		↑	↑
+1	↑↓	↑↓	↑↓	↑			↑			↑		↑↓	↑↓	↑↓	↑↓	↑↓		↑↓
0	↑			↑↓	↑↓	↑↓		↑			↑	↑↓	↑↓	↑↓		↑	↑↓	
−1		↑			↑		↑↓	↑↓	↑↓				↑		↑↓	↑↓	↑↓	
−2			↑			↑			↑	↑↓	↑↓			↑				↑↓
M_L	6	5	4	5	3	2	3	2	0	1	0	4	1	0	2	0	0	0
M_S	1/2	1/2	1/2	1/2	1/2	1/2	1/2	1/2	1/2	1/2	1/2	1/2	1/2	1/2	1/2	1/2	1/2	1/2

arrow represents the projection of spin (m_s), the arrow pointing up represents $m_s = 1/2$, and that pointing down represents $m_s = -1/2$. In writing table we do not need to consider the electrons whose summation of the M_L and M_S is negative, because they do not give any new information. The distribution table of five 3d electrons according to m_l and m_s is Table 1.1.

In Table 1.1, the symbols (×5), (×10), and (×3) represent that there are 5, 10, and 3 selections for the below configurations, respectively. For example, in the (×5) configuration, the m_l of the electron of spin down can be any one of +2, +1, 0, −1, −2. From Table 1.1, the electron configurations of ^6S, ^4G, ^4F, ^4D, ^4P, ^1S, ^2P, $3 \times {}^2D$, $2 \times {}^2F$, $2 \times {}^2G$, ^2H, and ^2I can be made, where S, P, D, F, G, H, and I represent the total angular momentum

$L = 0, 1, 2, \ldots$ states, and the number at left top corner represents spin multiple states $n_s = 2S+1$, for example, $n_s = 6$, then $S = 5/2$. Each electron configuration (L, S) is composed of states $m_l \leq L$ and $m_s \leq S$, for example ^{4}G electron configuration, $L = 4$, $S = 3/2$, is composed of the following states and their corresponding $-m_l$ and $-m_s$ states.

m_l	4	3	2	1	0	4	3	2	1	0
m_s	3/2	3/2	3/2	3/2	3/2	1/2	1/2	1/2	1/2	1/2

Therefore, the ^{4}G configuration is composed of 10 positive (m_l, m_s) states. Similarly, the ^{6}S, ^{4}G, ^{4}F, ^{4}D, ^{4}P, ^{2}S, ^{2}P, $3 \times {}^2$D, $2 \times {}^2$F, $2 \times {}^2$G, ^{2}H, and ^{2}I configurations are composed of 3, 10, 8, 6, 4, 1, 2, 3×3, 2×4, 2×5, 6, and 7 positive (m_l, m_s) states. There are 74 positive (m_l, m_s) states, in agreement with the number of positive (m_l, m_s) states in Table 1.1. If we check the number of each (m_l, m_s) states, they are also in agreement.

In so many electron configurations, which is the basis term? From Table 1.1, according to the Hunde rule (a), it is the spin maximum ^{6}S $(S = 5/2)$, and simultaneously the L maximum term. There is only one term $L = 0$. Then according to the Hunde rule (b), in the Mn^{2+} ion, the 3d electrons fill just half of the shell, and $L = 0$, so the total angular momentum $J = S = 5/2$. The basis term of the Mn^{2+} ion is then $^6S_{5/2}$, where the number at the right bottom corner represents the total angular momentum J.

A law has been discovered in the writing table process that the configuration of d^n $(n < 5)$ electrons equals that of d^{10-n} electrons; therefore, only the configurations of d^n $(n < 5)$ electrons are needed to calculate. The basis state terms of various ferromagnetic ions calculated by the above method are listed in Table 1.2, where S, D, F, etc., represent the orbital states (L), the left top index represents the spin multiple state $(2S + 1)$, and the right bottom index represents the total angular momentum.

Table 1.2. Basis state terms of ferromagnetic ions.

Free ions	Ti^{3+}, V^{4+}	Ti^{2+}, V^{3+}	V^{2+}, Cr^{3+}	Cr^{2+}, Mn^{3+}	Mn^{2+}, Fe^{3+}	Fe^{2+}	Co^{2+}	Ni^{2+}
Number of d electrons	1	2	3	4	5	6	7	8
Basis term	$^2D_{3/2}$	$^3F_2$	$^4F_{3/2}$	$^5D_0$	$^6S_{5/2}$	$^5D_4$	$^4F_{9/2}$	$^3F_4$

1.2. Splitting of the Basis State of Free Ions in the Crystal Field

The magnetic ions in the semiconductor are in the crystal field formed by the surrounding atoms. For semiconductors with zincblende structure (generally III–V and part of II–VI semiconductors), the crystal field has the tetrahedral symmetry (T_d) with respect to the central magnetic ion. The basis state energy levels of the free ion will split in the crystal field. For example, the basis state term of the Cr^{2+} ion is $^5D_0$ (Table 1.2), if the spin degeneracy ($S = 2$) is not considered, the orbital degeneracy of the orbit D ($L = 2$) is 5-fold, which will split in the crystal field of the T_d group symmetry.

According to group theory, the splitting of energy levels of free ion in a specific crystal field can be obtained by the resolution of the characteristic values. Let $\chi^l(\alpha)$ be the characteristic value in the rotation group rotating α angle of the wave function with the angular momentum l, then

$$\chi^l(\alpha) = \sum_{m=-l}^{l} e^{-im\alpha} = \frac{\sin(l + 1/2)\,\alpha}{\sin(\alpha/2)}, \quad (\alpha \neq 0). \tag{1.1}$$

Let $\chi_i(\alpha)$ be the characteristic value of the ith irreducible representation in a point group rotating α angle, which can be got by consulting the group theory books. By using the characteristic value resolution theorem,

$$R_l = \sum_i a_i \Gamma_i, \tag{1.2}$$

$$a_i = \frac{1}{n} \sum_k N_k \chi_i^*(C_k) \chi_l(C_k), \tag{1.3}$$

where R_l is the characteristic value of angular momentum l of the rotation group, Γ_i is the characteristic value of the ith representation of the point group, and a_i is the number of the ith representation of the point group after resolution, which is determined by Eq. (1.3), where the summation is over all rotation elements of the point group, C_k is the rotating angle of the kth element, N_k is the number of the kth element, and n is the total number of all rotation elements.

For example, for the T_d group, there are three kinds of rotation elements: C_2, C_3, C_4, whose rotating angles are $\pi, 2\pi/3$ and $\pi/2$, respectively.

Table 1.3. Resolution of the rotation group representation in the T_d group.

Rotation group representation	Resolved T_d group representations	Number of T_d group representations
S	A_1	1
P	T_1	1
D	$E+T_2$	2
F	$A_2+T_1+T_2$	3
G	$A_1+E+T_1+T_2$	4
H	$E+2T_1+T_2$	4
I	$A_1+A_2+E+T_1+2T_2$	6

From Eq. (1.1), the characteristic values of the rotation group are obtained,

$$\begin{aligned} &\chi(C_2) = \chi(\pi) = (-1)^l, \\ &\chi(C_3) = \chi\left(\frac{2\pi}{3}\right) = \begin{cases} 1 & l = 0, 3, 6, \ldots \\ 0 & l = 1, 4, 7, \ldots \\ -1 & l = 2, 5, 8, \ldots \end{cases} \\ &\chi(C_4) = \chi\left(\frac{\pi}{2}\right) = \begin{cases} 1 & l = 0, 1, 4, 5, \ldots \\ -1 & l = 2, 3, 6, 7, \ldots \end{cases} \end{aligned} \tag{1.4}$$

By using Eq. (1.3) and the characteristic value $\chi_i(C_k)$ tables of the T_d group in the group theory books, we obtained the resolution of the rotation group representation, as shown in Table 1.3.

In the T_d group, the degeneracy of the A_1 and A_2 representations is single, that of the E representation is double, and that of the T_1 and T_2 representations is 3-fold. From the last column of Table 1.3, we see that originally one orbital energy level in the free space splits into several energy levels in the crystal field of the T_d group. As for which level among the splitting levels is the basis state level, i.e. has the lowest energy, this problem cannot be solved by the group theory analysis. It can only be solved by the calculation based on the property of the crystal field or by the experimental observation of the absorption spectra or the magnetic properties.

The above discussion is for the case that the SOC energy is rather small, which is smaller than the crystal field splitting energy, so that the orbit degenerate state splits first in the crystal field. Another case is the contrast one, the spin orbital coupling energy is larger than the crystal field splitting energy. In this case, the free ion state is denoted by the total angular momentum J, which then splits in the crystal field. This time the double group representations should be used, the characteristic value resolution theorem is still Eqs. (1.2) and (1.3), but the $\chi(C_k)$ in the equation is the characteristic value of the double group representations. For the rotation group,

$$\chi_J(\alpha) = \frac{\sin(J+1/2)\alpha}{\sin(\alpha/2)}, \tag{1.5}$$

where J can be half-integer or integer. For example, the ^{2}D state ($L = 2$, $S = 1/2$) splits into 6-fold $^2D_{5/2}$ state and 4-fold $^2D_{3/2}$ state (according to the vector coupling theorem $L + S \leq J \leq |L - S|$). Their characteristic values of the rotation group can be calculated from Eq. (1.5) taking $J = 5/2$ and $3/2$. Then using the characteristic values of the T_d double group representations, we obtain

$$R\left(J = \frac{5}{2}\right) = \Gamma_7 + \Gamma_8,$$

$$R\left(J = \frac{3}{2}\right) = \Gamma_8,$$

where the Γ_8 representation is 4-fold degenerate and the Γ_7 is 2-fold degenerate.

1.3. Crystal Field Theory

In order to calculate the splitting of the magnetic ionic energy levels in the crystal field, including the order of energy levels and the magnitude of splitting energies, we need to inspect the concrete crystal field. The crystal field is formed by the surrounding atoms around the magnetic ion, which are called ligands; thus, the crystal field is also called the ligand field. If the ligand atoms are looked as the point charges with charge $z_k e$, where the index k denotes the kth atom, then the crystal field potential

$$V = \sum_k V(\mathbf{r}_k) = \sum_{m,n} A_n^m r^n P_n^m(\cos\ \theta)e^{im\phi}, \tag{1.6}$$

where the expansion coefficient

$$A_n^m = \sum_k \frac{e^2 z_k}{r_k^{n+1}} \frac{(n-|m|)!}{(n+|m|)!} P_n^m(\cos\ \theta_k)e^{-im\phi_k}, \tag{1.7}$$

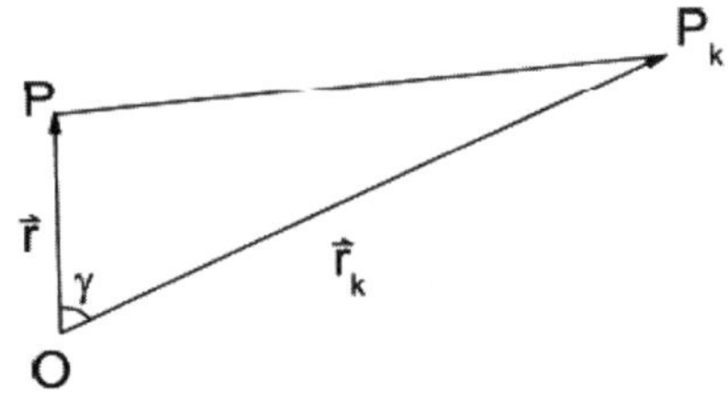

Fig. 1.1. Coordinates of the magnetic ion and surrounding atoms.

θ_k, ϕ_k are the polar angle and the azimuth angle of the kth atom coordinate $\boldsymbol{r}_k$ (see Fig. 1.1).

It is assumed that the magnetic ion is at coordinate origin O, the kth surrounding atom P_k is at $\boldsymbol{r}_k$, carrying charge $z_k e$, then the static electric potential produced by the atom P_k at an arbitrary point P is proportional to $1/R_k$, where R_k is the distance between the P_k and P,

$$R_k^2 = r^2 + r_k^2 - 2rr_k \cos\gamma,$$
$$\frac{1}{R_k} = \frac{1}{r_k}\left[1 + \frac{r}{r_k}\left(\frac{r}{r_k} - 2\cos\gamma\right)\right]^{-1/2}. \tag{1.8}$$

If $r_k > r$, then $1/R_k$ can be expanded to be series of $(r/r_k)^n$,

$$\begin{aligned}\frac{1}{R_k} &= \frac{1}{r_k}\left\{1 - \frac{1}{2}\frac{r^2}{r_k^2} + \frac{r}{r_k}\cos\gamma + \frac{3}{4}\cdot\frac{4}{2}\frac{r^2}{r_k^2}\cos^2\gamma + \cdots\right\}\\ &= \frac{1}{r_k}\left\{1 + \frac{r}{r_k}\cos\gamma + \frac{1}{2}\frac{r^2}{r_k^2}(3\cos^2\gamma - 1) + \cdots\right\}\\ &= \frac{1}{r_k}\sum_{n=0}^{\infty}\frac{r^n}{r_k^{n+1}}P_n(\cos\gamma).\end{aligned} \tag{1.9}$$

From the addition theorem,

$$P_n(\cos\gamma) = \sum_{m=-n}^{n}\frac{(n-|m|)!}{(n+|m|)!}P_n^m(\cos\theta)P_n^m(\cos\theta_k)e^{im}(\phi - \phi_k), \tag{1.10}$$

where $P_n(\cos\gamma)$ is the Legendre polynomial, (θ, ϕ) and (θ_k, and ϕ_k) are the polar and azimuth angles of the $\boldsymbol{r}$ and $\boldsymbol{r}_k$, respectively, and the spherical harmonics

$$Y_n^m(\theta,\phi) = (-1)^n\frac{(2n+1)(n-|m|)!}{4\pi(n+|m|)!}P_n^m(\cos\theta)e^{im\phi}, \tag{1.11}$$

we obtain

$$\begin{aligned} V(\boldsymbol{r}) &= \sum_k \sum_{n=0}^{\infty} \frac{4\pi}{2n+1} \frac{e^2 z_k}{r_k^{n+1}} r^n \sum_{m=-n}^{n} Y_n^m(\theta,\phi) Y_n^{m*}(\theta_k,\phi_k) \\ &= \sum_{n=0}^{\infty} \sum_{m=-n}^{n} A_n^m r^n Y_n^m(\theta,\phi), \end{aligned} \tag{1.12}$$

where

$$A_n^m = \sum_k \frac{4\pi}{2n+1} \frac{e^2 z_k}{r_k^{n+1}} Y_n^{m*}(\theta_k,\phi_k). \tag{1.13}$$

They are just Eqs. (1.6) and (1.7).

For the cubical symmetry crystal O_h, there are six surrounding atoms, whose polar coordinates are $(r,\theta,\phi) = (a,0,0,)$, $(a,\pi/2,0)$, $(a,\pi/2,\pi/2)$, $(a,\pi/2,\pi)$, $(a,\pi/2,3\pi/2)$, and $(a,\pi,0)$, respectively. Nonzero A_n^m are

$$\begin{aligned} A_0^0 &= 6\sqrt{4\pi}\frac{ze^2}{a}, \\ A_4^0 &= \frac{7}{6}\sqrt{4\pi}\frac{ze^2}{a^5}, \\ A_4^4 &= A_4^{-4} = \frac{1}{6}\sqrt{\frac{35}{2}}\sqrt{4\pi}\frac{ze^2}{a^5}. \end{aligned} \tag{1.14}$$

Hence,

$$V_{\mathrm{c}}(\boldsymbol{r}) = A_4^0 r^4 \left[Y_4^0 + \sqrt{\frac{5}{14}}(Y_4^4 + Y_4^{-4}) \right]. \tag{1.15}$$

Let

$$\begin{aligned} D &= \frac{5}{2}\sqrt{\frac{9}{4\pi}}A_4^0 = \frac{35}{4}\frac{ze^2}{a^5}, \\ q &= \frac{2}{105}\langle r^4 \rangle = \frac{2}{105}\int_0^{\infty} R^2(r) r^4 \cdot r^2 \,\mathrm{d}r, \end{aligned} \tag{1.16}$$

where $R(r)$ is the radial wave function of the d state of the magnetic ion. It can be proved that

$$\begin{aligned} \langle \pm 2 | V_{\mathrm{c}} | \pm 2 \rangle &= Dq, \\ \langle \pm 2 | V_{\mathrm{c}} | \mp 2 \rangle &= 5Dq, \\ \langle \pm 1 | V_{\mathrm{c}} | \pm 1 \rangle &= -4Dq, \\ \langle 0 | V_{\mathrm{c}} | 0 \rangle &= 6Dq, \end{aligned} \tag{1.17}$$

where $|m\rangle$ denotes the state whose component of the orbit angular momentum in the z direction is m. The calculation of Eq. (1.17) needs to use the coupling theorem of angular momentum and coupling coefficients of angular momentum (Edmonds, 1957). From the coupling theorem of angular momentum and the formula of $V_c(\boldsymbol{r})$, Eq. (1.15), we obtain that only in the case of $\Delta m = m - m' = 0$ or ± 4 the matrix element of the crystal field $\langle m|V_c|m'\rangle$ does not equal zero.

Solving the secular equation, we obtain the following splitting energy levels and corresponding wave functions of the d orbit energy levels in the crystal field,

$$
\begin{array}{lll}
\text{Energy} & \text{Wave function} & \text{Symmetry} \\
 & \xi = \dfrac{i}{\sqrt{2}}\left(|1\rangle + |-1\rangle\right) & \sqrt{\dfrac{15}{2}}zy \\
\mathrm{T}_2 \quad -4Dq & \eta = -\dfrac{1}{\sqrt{2}}\left(|1\rangle - |-1\rangle\right) & \sqrt{\dfrac{15}{2}}zx \\
 & \varsigma = -\dfrac{i}{\sqrt{2}}\left(|2\rangle - 2\,|\rangle\right) & \sqrt{\dfrac{15}{2}}xy \\
 & \theta = |0\rangle & \dfrac{1}{2}\sqrt{\dfrac{5}{2}}\left(3z^2 - r^2\right) \\
\mathrm{E} \quad 6Dq & \varepsilon = \dfrac{1}{\sqrt{2}}\left(|2\rangle + |-2\rangle\right) & \dfrac{1}{2}\sqrt{\dfrac{5}{2}}\left(x^2 - y^2\right).
\end{array}
\tag{1.18}
$$

It can be proved that for the cubical symmetry crystal O with eight surrounding atoms, and the tetrahedral symmetry crystal T_d with four surrounding atoms,

$$
\begin{aligned}
A_4^0 &= \frac{28}{27}\sqrt{4\pi}\frac{ze^2}{a^5}, \quad (O) \\
A_4^0 &= \frac{14}{27}\sqrt{4\pi}\frac{ze^2}{a^5}. \quad (T_d)
\end{aligned}
\tag{1.19}
$$

Comparing Eqs. (1.19) and (1.14), we obtain

$$
(Dq)_{O_h} : (Dq)_{T_d} : (Dq)_O = 1 : \frac{4}{9} : \frac{8}{9}. \tag{1.20}
$$

1.4. Wave Functions of Many-electron States

There are many d electrons in the outmost shell of the magnetic ion; thus, their wave function is a determinant formed by the wave functions of single electrons. For example, the $3d^3$ configuration (V^{2+}, Cr^{3+}), its basis state is ^4F (28-fold degenerate) from Table 1.2. According to the theorems of maximum component of angular momentum and spin, if there is no state $M_L > M'_L$, or $M_S > M'_S$, then (M'_L, M'_S) are the eigenstate of L^2 and S^2. Similar to Table 1.1, if 3d electrons fill in the $M_L = 2, 1, 0$, respectively, their spin are all up ($\uparrow$), then the component of the total angular momentum $M_L = 3$ and the component of the total spin $M_S = 3/2$. There is no state whose M_S can exceed 3/2: hence, the determinant formed by this state is the eigenstate (Abragam and Bleaney, 1970):

$$\left|{}^4\mathrm{F}; \frac{3}{2}3\right\rangle = (2^+1^+0^+). \tag{1.21}$$

By using the properties of arising and descending operators of angular momentum,

$$L^{\pm}\Big|L, M\ \rangle = \hbar\sqrt{(L \mp M)(L \pm M + 1)}\Big|l, M \pm 1\Big\rangle, \tag{1.22}$$

we obtain the eigenstates of other M_L and M_S values,

$$\begin{aligned} L^-\left|{}^4\mathrm{F}; \frac{3}{2}3\right\rangle &= \sqrt{6}\left|{}^4\mathrm{F}; \frac{3}{2}2\right\rangle \\ &= \sqrt{4}(1^+1^+0^+) + \sqrt{6}(2^+0^+0^+) + \sqrt{6}(2^+1^+-1^+), \end{aligned} \tag{1.23}$$

where the second formula is obtained by the operation of the L^- to each single electron wave function. The former two determinants equal zero because two states in them are the same. Therefore,

$$\left|{}^4\mathrm{F}; \frac{3}{2}2\right\rangle = (2^+1^+-1^+). \tag{1.24}$$

Operating the descending operator L^- again, we obtain

$$\begin{aligned} L^-\left|{}^4\mathrm{F}; \frac{3}{2}2\right\rangle &= \sqrt{10}\left|{}^4\mathrm{F}; \frac{3}{2}1\right\rangle, \\ &= \sqrt{4}(1^+1^+-1^+) + \sqrt{6}(2^+0^+-1^+) + \sqrt{4}(2^+1^+-2^+), \\ \left|{}^4\mathrm{F}; \frac{3}{2}1\right\rangle &= \sqrt{\frac{3}{5}}(2^+0^+-1^+) + \sqrt{\frac{2}{5}}(2^+1^+-2^+). \end{aligned} \tag{1.25}$$

Using the orthogonal relation of wave functions, we obtain the eigenstate of $L = 1$, $M_L = 3/2$,

$$\left|{}^4\mathrm{P}; \frac{3}{2}1\right\rangle = \sqrt{\frac{2}{5}}(2^+0^+-1^+) - \sqrt{\frac{3}{5}}(2^+1^+-2^+). \tag{1.26}$$

Operating the S^- operator,

$$S^-\left|{}^4\mathrm{F};\frac{3}{2}3\right\rangle=\sqrt{3}\left|{}^4\mathrm{F};\frac{1}{2}3\right\rangle$$
$$=(2^-1^+0^+)+(2^+1^-0^+)+(2^+1^+0^-),$$
$$\left|{}^4\mathrm{F};\frac{1}{2}3\right\rangle=\sqrt{\frac{1}{3}}(2^-1^+0^+)+\sqrt{\frac{1}{3}}(2^+1^-0^+)+\sqrt{\frac{1}{3}}(2^+1^+0^-) \quad (1.27)$$

If 3d electrons fill in the 2^+, 2, and 1^+ states, respectively, then $M_L=5$, $M_S=1/2$, there is no state whose M_L can exceeds 5; thus, the state formed by these three states is the state with the maximum component of the angular momentum M_L,

$$\left|{}^2\mathrm{H};\frac{1}{2}5\right\rangle=(2^+2^-1^+), \quad (1.28)$$

$$\left|{}^2\mathrm{H};\frac{1}{2}4\right\rangle=\sqrt{\frac{2}{5}}(2^+1^-1^+)+\sqrt{\frac{3}{5}}(2^+2^-0^+). \quad (1.29)$$

Using the descending operator and the orthogonal relation, we obtain

$$\left|{}^2\mathrm{G};\frac{1}{2}4\right\rangle=-\sqrt{\frac{3}{5}}(2^+1^-1^+)+\sqrt{\frac{2}{5}}(2^+2^-0^+),$$
$$\left|{}^2\mathrm{H};\frac{1}{2}3\right\rangle=\sqrt{\frac{8}{15}}(2^+1^-0^+)-\sqrt{\frac{2}{5}}(2^+1^+0^+)-\sqrt{\frac{2}{15}}(2^-1^+0^+)$$
$$+\sqrt{\frac{1}{5}}(2^+2^--1^+),$$
$$\left|{}^2\mathrm{G};\frac{1}{2}3\right\rangle=-\sqrt{\frac{1}{60}}(2^+1^-0^+)-\sqrt{\frac{3}{20}}(2^+1^+0^+)-\sqrt{\frac{1}{5}}(2^-1^+0^+)$$
$$+\sqrt{\frac{3}{10}}(2^+2^--1^+),$$
$$\left|{}^2\mathrm{F};\frac{1}{2}3\right\rangle=\sqrt{\frac{1}{12}}(2^+1^-0^+)+\sqrt{\frac{1}{12}}(2^+1^+0^+)-\sqrt{\frac{1}{3}}(2^-1^+0^+)$$
$$-\sqrt{\frac{1}{2}}(2^+2^--1^+). \quad (1.30)$$

In summary, using the L^- and S^- operators, and the orthogonal and normalized conditions, we can obtain all wave functions of the many-electron state.

By using the property of the determinant wave function, we can transfer the matrix element of many-electron wave functions in the crystal field into that of single-electron wave functions. Only when $M_L' - M_L = 0$ or ± 4, the matrix element of the many-electron wave functions does not equal zero, otherwise equals zero. For example, the basis state of the 3d^3 configuration ^{4}F (orbital 7-fold degenerate),

$$\begin{aligned}\left\langle {}^4\mathrm{F}; \frac{3}{2}3\,|V_\mathrm{c}|\,{}^4\mathrm{F}; \frac{3}{2}3 \right\rangle &= \langle 2^+1^+0^+\,|V_\mathrm{c}|\,2^+1^+0^+\rangle \\ &= \langle 2|V_\mathrm{c}|2\rangle + \langle 1|V_\mathrm{c}|1\rangle + \langle 0|V_\mathrm{c}|0\rangle \\ &= Dq - 4Dq + 6Dq = 3Dq, \\ \left\langle {}^4\mathrm{F}; \frac{3}{2}3|V_\mathrm{c}|{}^4\mathrm{F}; \frac{3}{2} - 1 \right\rangle &= \left\langle 2^+1^+0^+|V_\mathrm{c}| - \sqrt{\frac{3}{5}}(-2^+0^+1^+) \right. \\ &\quad \left. + \sqrt{\frac{2}{5}}(-1^+ - 2^+2^+) \right\rangle \\ &= \sqrt{\frac{3}{5}}\,\langle 2|V_\mathrm{c}| - 2\rangle = \sqrt{15}Dq. \qquad (1.31)\end{aligned}$$

Similarly, we can calculate other nonzero matrix elements. Solving the secular equations, we obtain the splitting energies of the ^{4}F basis state in the crystal field

$$E_\mathrm{c} = 6Dq \text{ (3-fold)}, -2Dq \text{ (3-fold)}, \quad 12Dq \text{ (1-fold)} \qquad (1.32)$$

and corresponding wave functions.

1.5. Equivalent Operator Method

In the last section, in order to calculate the crystal field matrix elements of many-electron states, we need to know every wave functions, of the many-electron state; the calculation quantities are large. Stevens (1952) proposed the equivalent operator method. According to the Wigner–Eckart theorem of the group theory, in a subspace of multiple states with same L due to the same symmetry, the coordinate operator r_i $(i = x, y, z)$ in the crystal field potential $V_\mathrm{c}(\boldsymbol{r})$ can be replaced by the operators of angular momentum L_i, but the noncommutation of the L_i should be considered. For example, the crystal field of cubical symmetry or tetrahedral symmetry (Eq. (1.15))

$$V_{T_d}(\boldsymbol{r}) = \frac{\sqrt{8}}{15} r^4 \left[Y_4^0 + \sqrt{\frac{5}{14}} \left(Y_4^4 + Y_4^{-4} \right) \right]$$
$$= \frac{1}{20}\left(35z^4 - 30r^2z^2 + 3r^4\right) + \frac{1}{8}\left[(x+iy)^4 + (x-iy)^4\right], \quad (1.33)$$

where

$$z^4 \sim L_z^4, \quad (x+iy)^4 \sim L_+^4, \quad (x-iy)^4 \sim L_-^4. \quad (1.34)$$

We calculate the equivalent operators corresponding to r^2z^2 and r^4. Using the commutation relation of angular momentum operators $[L_x, L_y] = i\hbar L_z$ (taking $\hbar = 1$), we prove,

$$\begin{aligned} z^2r^2 &\sim L_z^2L^2 + \frac{1}{6}L_x^2 + \frac{1}{6}L_y^2 - \frac{2}{3}L_z^2, \\ y^2r^2 &\sim L_y^2L^2 + \frac{1}{6}L_z^2 + \frac{1}{6}L_x^2 - \frac{2}{3}L_y^2, \\ x^2r^2 &\sim L_x^2L^2 + \frac{1}{6}L_y^2 + \frac{1}{6}L_z^2 - \frac{2}{3}L_x^2, \\ r^4 &= \left(x^2+y^2+z^2\right)r^2 \sim L^2L^2 - \frac{1}{3}L^2 = \bar{L}^2\left(\bar{L}+1\right)^2 - \frac{1}{3}\bar{L}\left(\bar{L}+1\right), \end{aligned} \quad (1.35)$$

where $\bar{L}$ is a number, representing the eigenvalue of the operator L. Thus,

$$V_{T_d}(\boldsymbol{r}) \sim \frac{1}{20}\left[35L_z^4 - 30L_z^2\bar{L}(\bar{L}+1) + 25L_z^2 - 6\bar{L}(\bar{L}+1) + 3\bar{L}^2(\bar{L}+1)^2\right]$$
$$+ \frac{1}{8}(L_+^4 + L_-^4). \quad (1.36)$$

By using the equivalent operators, we can easily calculate the crystal field matrix elements of many-electron states within a group of multiple states with same eigenvalues of angular momentum $\bar{L}$. For example, in the Co^{2+} ion, the basis state of the $3d^3$ configuration is ^4F multiple states, $\bar{L} = 3$. From Eq. (1.36), we obtain the matrix elements of the V_c field,

$$\langle M_L | V_c | M_L' \rangle = a \begin{vmatrix} 9 & 0 & 0 & 0 & 3\sqrt{15} & 0 & 0 \\ 0 & -21 & 0 & 0 & 0 & 15 & 0 \\ 0 & 0 & 3 & 0 & 0 & 0 & 3\sqrt{15} \\ 0 & 0 & 0 & 18 & 0 & 0 & 0 \\ 3\sqrt{15} & 0 & 0 & 0 & 3 & 0 & 0 \\ 0 & 15 & 0 & 0 & 0 & -21 & 0 \\ 0 & 0 & 3\sqrt{15} & 0 & 0 & 0 & 9 \end{vmatrix}. \quad (1.37)$$

Table 1.4. Crystal field splitting constant a of magnetic ions in the semiconductor.

Free ion	V^{2+}	Cr^{2+}	Fe^{2+}	Co^{2+}
Configuration	$3d^3$	$3d^4$	$3d^6$	$3d^7$
Constant a of basis state	$a < 0$	$a > 0$ (T_2)	$a > 0$ (E)	$a > 0$ (A_2)

Solving the secular equation, we obtain the crystal field splitting energies,

$$E_c = 18a \text{ (3-fold)}, -6a \text{ (3-fold)}, -36a \text{ (1-fold)}, \tag{1.38}$$

where a is the proportional constant. Comparing with the result of direct calculation Eq. (1.32), we found that both results are equal, if let $a = Dq/3$.

The matrix elements of the equivalent operators between wave functions of multiple states similar to Eq. (1.37) are listed in tables of Stevens (1952) and Hatchings (1964). We need not calculate from the beginning. The constant a in Eq. (1.38) or Dq in Eq. (1.32) needs to be determined by experiments. For example, for the T_d symmetric crystal with four ligand atoms (semiconductor), the constants of magnetic ions a are listed in Table 1.4.

1.6. Magnetic Ion Energy Levels in Semiconductors

The Hamiltonian of the magnetic ion in the semiconductor can be written as

$$H = H_{\text{ion}} + H_c + \lambda \boldsymbol{L} \cdot \boldsymbol{S} + \mu_B \boldsymbol{B} \cdot (\boldsymbol{L} + 2\boldsymbol{S}), \tag{1.39}$$

where the terms on the right side are the free ion term, the crystal field term, the spin–orbit coupling (SOC) term, and the interaction term with the magnetic field, respectively. λ is the SOC constant and μ_B is the Bohr magneton. In general cases, the crystal field splitting energy is larger than the spin orbital coupling energy, so we first consider the crystal field splitting.

The Mn^{2+} ion is more simple, whose basis state is ^{6}S, $L = 0$; thus, there is no crystal field splitting of orbital states. The spin $S = 5/2$ 6-fold degenerate state does not split in the zero magnetic field; and splits in the magnetic field; the splitting energy equals $2\mu_B BS_z$, where S_z is the component of the spin in the direction of the magnetic field.

The Cr^{2+} ion is also more simple, whose basis state is $^6D_0$, $S = 5/2$, $L = 2$. In the T_d crystal field, the orbital D state splits into E and T_2

states. From Table 1.4, the basis state is $^5T_2$ state, and the spin orbital splitting is very small.

Next we consider the energy levels of Fe^{2+} and Co^{2+} ions in the semiconductor of zinc-blende structure (Slack *et al.*, 1969). From Table 1.2, the basis state of the Fe^{2+} $3d^6$ configuration is ^5D. The orbital D state splits into E and T_2 energy levels in the T_d crystal field. From Table 1.4, the basis state is ^5E state. The spin $S = 2$ state is similar to the orbital $L = 2$ state, which also splits into E and T_2 energy levels in the T_d crystal field. According to the multiplication table of group theory, the orbital E state, including the spin, splits into five levels as shown in Eq. (1.40).

$$^5\mathrm{E}\text{: } \mathrm{E} \times (\mathrm{E} + \mathrm{T}_2) = \Gamma_1 + \Gamma_2 + \Gamma_3 + \Gamma_4 + \Gamma_5. \tag{1.40}$$

Similarly, for the orbital T_2 state,

$$^5\mathrm{T}_2\text{: } \mathrm{T}_2 \times (\mathrm{E} + \mathrm{T}_2) = \Gamma_1 + \Gamma_2 + \Gamma_3 + 2\Gamma_4 + 2\Gamma_5. \tag{1.41}$$

By use of the tables of coupling coefficients (Koster *et al.*, 1963), we can write down spin–orbit product wave functions of all irreducible representations. For example, the singlet Γ_2 wave function for the orbital ^5E state (Eq. (1.40)) is given by

$$\psi_{20} = \frac{1}{2}\left[\left(Y_2^2 + Y_2^{-2}\right) S_2^0 - Y_2^0 \left(S_2^2 + S_2^{-2}\right)\right], \tag{1.42}$$

where Y_L^m is the orbital spherical harmonics and S_S^m is the spin spherical harmonics. This is abbreviated by using the kets $|M_L M_S\rangle$ as

$$\psi_{20} = \frac{1}{2}\left[|20\rangle - |02\rangle - |0\bar{2}\rangle + |\bar{2}0\rangle\right], \tag{1.43}$$

where the $|02\rangle$ means $M_L = 0$, $M_S = 2$. The 25 wave functions in Eqs. (1.40) and (1.41) can be obtained with the similar method, which are given in Koster *et al.*, (1963).

Including the SOC term $\lambda \boldsymbol{L}\cdot\boldsymbol{S}$ in Eq. (1.39) makes the matrix $\langle\psi_i|H|\psi_j\rangle$ a 25×25 matrix. Because the Hamiltonian (1.39) does not give any coupling between wave functions belonging to different irreducible representations, the 25×25 matrix will break up into one 1×1 (Γ_2), one 2×2 (Γ_1), two 2×2 (Γ_3), three 3×3 (Γ_4), and three 3×3 (Γ_5) matrices. The observed and calculated energy levels of the ^5E state of Fe^{2+} in CdTe are shown in Fig. 1.2 (Slack *et al.*, 1969). The calculation is done using $\lambda = -100\,\mathrm{cm}^{-1}$, $Dq = -248\,\mathrm{cm}^{-1}$ for Fe^{2+} in CdTe. From Fig. 1.2, we see that the calculated of the order of energy levels is Γ_1, Γ_4, Γ_3, Γ_5, and Γ_2, and the spacings between neighboring levels are nearly equal, $\sim 20\,\mathrm{cm}^{-1}$.

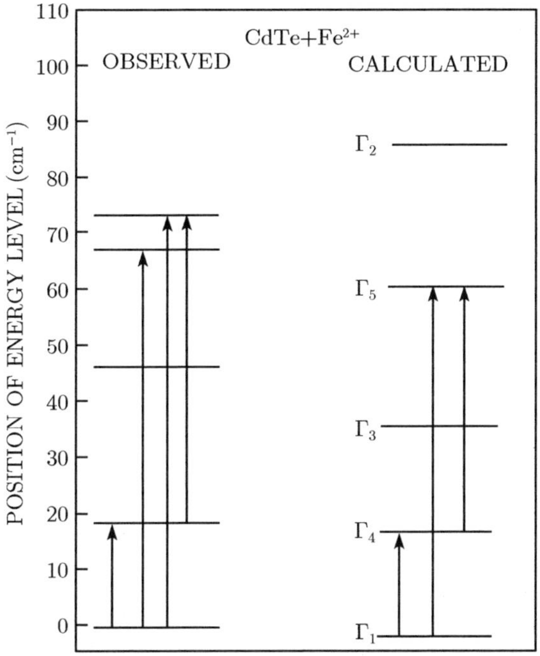

Fig. 1.2. Observed (left) and calculated (right) energy levels of the ^{5}E state of Fe^{2+} in CdTe. The vertical arrows denote the allowed transitions.

Next we consider the semiconductors of the wurtzite structure, whose lattice has C_{3v} symmetry, lower than the T_d symmetry. The symmetric axis is along the *c*-axis, equivalent to the [111] direction of the zincblende structure. The crystal field of the semiconductor of the wurtzite structure can be looked as the field of zero order with T_d symmetry adding the perturbative field with C_{3v} symmetry,

$$V_c(\boldsymbol{r}) = V_{T_d}(\boldsymbol{r}) + V_{C_{3v}}(\boldsymbol{r}). \tag{1.44}$$

The ^{5}D energy levels of the free ion Fe^{2+} split in the V_{T_d} field into ^{5}E and $^5T_2$ levels. Because the symmetry of the C_{3v} group is lower than that of the T_d group, the ^{5}E and $^5T_2$ levels will split further in the C_{3v} field. According to group theory, if we know the characteristic values $\chi(\mathrm{T})$ of all representations of the T_d and C_{3v} groups, then we can resolve the representations of group of high symmetry into representations of group of

Table 1.5. Characteristic values of representations of the T_d and C_{3v} groups.

Group	Operation	E	3	m_x
	A_1	1	1	1
	A_2	1	1	−1
T_d	E	2	−1	0
	T_1	3	0	−1
	T_2	3	0	1
	A_1	1	1	1
C_{3v}	A_2	1	1	−1
	E	2	−1	0

low symmetry (Heine, 1960).

$$\begin{aligned}\chi(\mathrm{T}) &= \chi^{(\alpha)}(\mathrm{T}) + \chi^{(\beta)}(\mathrm{T}) + \cdots + \chi^{(\varepsilon)}(\mathrm{T}) \\ &= \sum_{\lambda} C_\lambda \chi^{(\lambda)}(\mathrm{T}),\end{aligned} \tag{1.45}$$

where $\chi(\mathrm{T})$ is the characteristic value of one representation of the high symmetry group, and $\chi^{(\alpha)}(\mathrm{T})$ is that of the representation α of the low symmetry group, T is a symmetric operation, for example, rotation of an angle, inversion, etc. The characteristic values of representations of the T_d and C_{3v} groups are shown in Table 1.5 (Heine, 1960).

From Table 1.5 and Eq. (1.45), we easily get that the original E level does not split in the C_{3v} field, while the T_2 level splits into $A_1 + E$ levels.

In order to calculate the splitting energy in the crystal field, we need to write the crystal potential as the form of the equivalent operators, and calculate the matrix elements of the crystal field. For the semiconductors of wurtzite structure, which is hexagonal symmetric, the symmetric axis is along the c-axis. To effect the transition to a trigonal crystal, we transform to a system of coordinates whose axes are parallel to ξ, η, ζ, with ζ along [111]. They are defined by

$$\begin{aligned}\hat{x} &= \frac{1}{\sqrt{6}}\left(\hat{\xi} - \sqrt{3}\hat{\eta} + \sqrt{2}\hat{\zeta}\right), \\ \hat{y} &= \frac{1}{\sqrt{6}}\left(\hat{\xi} + \sqrt{3}\hat{\eta} + \sqrt{2}\hat{\zeta}\right), \\ \hat{z} &= \frac{1}{\sqrt{3}}\left(-\sqrt{2}\hat{\xi} + \hat{\zeta}\right).\end{aligned} \tag{1.46}$$

where $\hat{x}, \hat{y}, \hat{z}$ and $\hat{\xi}, \hat{\eta}, \hat{\zeta}$ are unit vectors along the coordinate axes in two coordinate systems. It can be proved that in the new coordinate system (Villeret *et al.*, 1990),

$$\begin{aligned} V_c(T_d) &= a\sum_i\left[-\frac{1}{30}(35\varsigma_i^4 - 30\varsigma_i^2 r_i^2 + 3r_i^4) - \frac{2\sqrt{2}}{3}\varsigma_i\xi_i(\xi_i^2 - 3\eta_i^2)\right], \\ V_c(C_{3v}) &= \frac{b}{3}\sum_i(\xi_i^2 + \eta_i^2 - 2\varsigma_i^2) + \frac{c}{60}\sum_i(35\varsigma_i^4 - 30\varsigma_i^2 r_i^2 + 3r_i^4). \end{aligned} \tag{1.47}$$

Then we transfer the special coordinates ξ, η, and ζ into the equivalent operators and obtain ($L = 2$)

$$\begin{aligned} V_c(T_d) &= a\left(\frac{7}{6}L_\varsigma^4 - \frac{31}{6}L_\varsigma^2 + \frac{12}{5}\right) + \frac{\sqrt{2}a}{6}\left\{L_+^3 + L_-^3, L_\varsigma\right\}, \\ V_c(C_{3v}) &= -b(2 - L_\varsigma^2) - c\left(\frac{7}{12}L_\varsigma^4 - \frac{31}{12}L_\varsigma^2 + \frac{6}{5}\right). \end{aligned} \tag{1.48}$$

where $L_\pm = L_\xi \pm iL_\eta$, $\{u, v\} = uv + vu$.

Figure 1.3 is the schematic diagram of the energy levels of Fe^{2+} in T_d and C_{3v} fields taking into account the spin–orbit interaction (SOI) (Villeret *et al.*, 1990), where the Γ_3 and Γ_5 levels at the left side are the E and T_2 levels (T_d group), and the Γ_1 and Γ_3 levels at the right side are the A_1 and E levels (C_{3v} group), respectively. In Fig. 1.3,

$$\begin{aligned} \Delta &= 6a, \\ \Delta_1 &= 3a - \frac{5}{2}b - \frac{3}{2}c + \frac{1}{2}\Delta_3, \\ \Delta_3 &= \left[32a^2 + (2a + 3b - c)^2\right]^{1/2}, \end{aligned} \tag{1.49}$$

where a, b, and c are all positive. The splitting levels in the T_d crystal field are the same as those in Fig. 1.2, which split further in the C_{3v} field,

Last we consider the Co^{2+} ion. From Table 1.2, its basis state is $^4F(L = 3, S = 3/2)$. From Table 1.3, the F state splits into $A_2 + T_2 + T_1$ levels in the T_d group crystal field. From Table 1.5 in the C_{3v} field, $A_2 \to A_2, T_1 \to A_2 + E, T_2 \to A_1 + E$, all levels split into $A_2 + A_1 + E + A_2 + E$ levels, as shown in Fig. 1.4 (Villeret *et al.*, 1990).

For $L = 3$, the crystal field potential can be written as

$$V_c = \frac{c - 2a}{12}\left(7L_\varsigma^4 - 67L_\varsigma^2 + 72\right) - \frac{\sqrt{2}a}{6}\left\{L_+^3 + L_-^3, L\varsigma\right\} + b\left(4 - L_\varsigma^2\right). \tag{1.50}$$

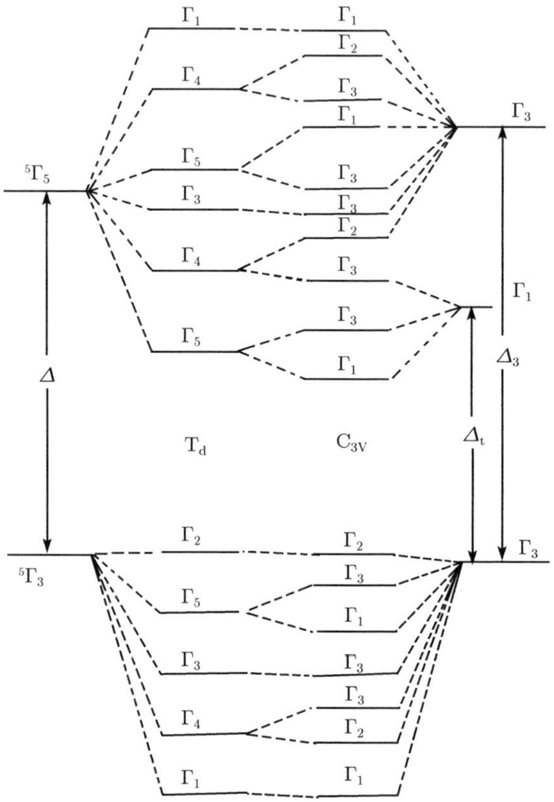

Fig. 1.3. Schematic diagram of the energy levels of Fe^{2+} in T_d and C_{3v} fields.

If the potential has T_d symmetry, the coefficients $b = c = 0$. If the symmetry is C_{3v}, all coefficients will, in general, be nonvanishing. The calculated eigenenergies of energy levels in the C_{3v} crystal field as in Fig. 1.4 are as follows (from down to up),

$$
\begin{aligned}
E(\Gamma_2) &= -9a - \frac{1}{2}b + \frac{9}{2}c - \frac{1}{2}\Delta_4, \\
E(\Gamma_3) &= 6a + \frac{3}{2}b - 3c - \frac{1}{2}(\Delta_3 - \Delta_1), \\
E(\Gamma_1) &= -6a - 5b + 3c, \\
E(\Gamma_3) &= 6a + \frac{3}{2}b - 3c + \frac{1}{2}(\Delta_3 - \Delta_1), \\
E(\Gamma_2) &= -9a - \frac{1}{2}b + \frac{9}{2}c + \frac{1}{2}\Delta_4,
\end{aligned}
\tag{1.51}
$$

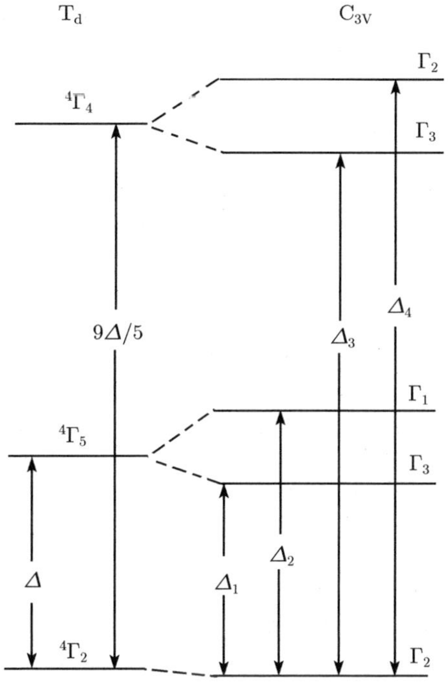

Fig. 1.4. Orbital energy levels of Co^{2+} in the T_d group and C_{3v} group crystal fields.

where

$$\begin{aligned}\Delta_4 &= 3\left[324a^2 - 4a(3b+c) + (3b+c)^2\right]^{1/2},\\ \Delta_3 - \Delta_1 &= \left[(16a-3b-8c)^2 + 320a^2\right]^{1/2}.\end{aligned} \tag{1.52}$$

Actually, the parameters a, b, and c are obtained from the calculation with Eqs. (1.51) and (1.52), by use of the energies of the absorption peaks of the absorption spectrum. For example, for the Co^{2+} ion in CdSe, the splitting energies estimated from the absorption spectrum at 80 K are: $\Delta_1 = 2800\,\text{cm}^{-1}$, $\Delta_3 = 5200\,\text{cm}^{-1}$, $\Delta_4 = 6000\,\text{cm}^{-1}$, $\Delta_2 = 3300\,\text{cm}^{-1}$. Similarly, we can also consider the SOC effect further (see Villeret *et al.*, 1990).

1.7. Experimental Study of the Properties of Magnetic Ions in Semiconductors

The solid-solution $A_{1-x}M_xB$ forms when the II–VI semiconductors AB are doped with magnetic ions M, which have been studied early in the

1940s of the last century. The paramagnetic resonant researches of the ionic crystal doped with magnetic ions were made simultaneously. Because the crystal field of the ionic crystal is rather simple, every ion constituting the crystal can be looked as the point charge, both the theoretical and the experimental researches have made a mighty advance and formed the systematic theory (Abragam and Bleaney, 1970). The semiconductor is different from the ionic crystal in two aspects: (1) The magnetic ions have large differences in the chemical properties from the III–V semiconductors, so their solubility in the III–V semiconductors is relatively small, in general $x < 0.1$, less than 10%. But the magnetic ions can be incorporated in sizable amounts in the II–VI host without affecting substantially the crystallographic quality of the resulting material (e.g. up to nearly 80% of Mn can be accommodated in CdTe). Later in 1988, people called this kind of semiconductors doped with magnetic ions as DMS. (2) The semiconductors are covalent crystals, each semiconductor atom cannot be looked as point charge; hence, strictly speaking, the crystal field theory is not valid for semiconductors. But from the point of symmetry, the potential field formulas given by the crystal field theory still hold, only the coefficients in the formulas are empirical and should be determined by the experiments.

Earlier the most widely studied DMS is the Mn^{2+} ions embedded in the CdTe. The CdTe crystal has the zincblende structure and the T_d symmetry. The basis state of the Mn^{2+} ion is the $^6S_{5/2}$ state, its 6-fold degenerate state of $S = 5/2$ splits in the T_d crystal field into one 4-fold degenerate Γ_8 state and one 2-fold degenerate Γ_7 state. It is found from the experiments that the splitting energy between these two levels is very small, only $0.0084\,\mathrm{cm}^{-1}$. So later on, the Mn^{2+} ions in the semiconductor are seen as free ions.

Next we introduce the experiments of the Fe^{2+} ions in CdTe (Slack *et al.*, 1969). The $Cd_{1-x}Fe_xTe$ samples were grown from the melt using a restricted molten-zone technique. The starting materials were high purity Cd, Te, and Fe, because Fe will substrate Cd, and the Te proportion is larger than Cd. The solubility of Fe^{2+} ions in the CdTe is limited, so the total Fe concentration given by the standard chemical analysis does not equal the actual concentration of Fe, including also the precipitated Fe. After the Fe-doped CdTe samples were annealed at 800°C, the equilibrium Fe^{2+} concentration is $3.5 \times 10^{18}\mathrm{cm}^{-3}$ to $4 \times 10^{19}\mathrm{cm}^{-3}$.

The basis state of the free Fe^{2+} ion is ^5D (Table 1.2), which splits into ^5E and $^5T_2$ states in the CdTe T_d crystal field. The energy of the ^5E state is lower, and the splitting energy of the ^5E and $^5T_2$ levels is $\sim 2500\,\mathrm{cm}^{-1}$. When the SOC is included, the ^5E level splits into Γ_1, Γ_4,

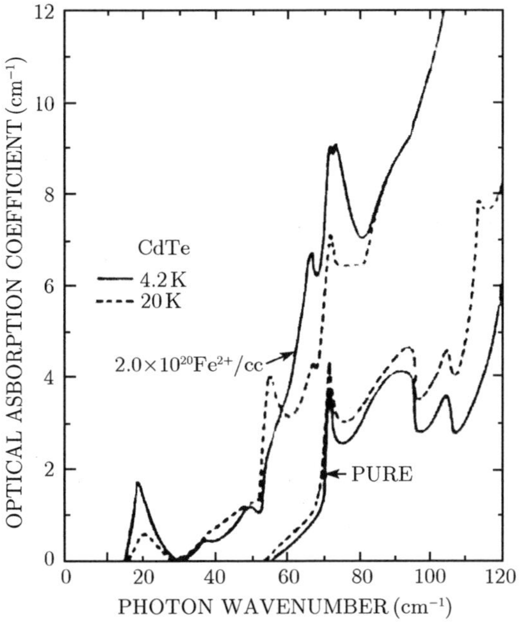

Fig. 1.5. Far-infrared absorption coefficients versus photon wavenumber at 4.2 and 20 K for both pure and Fe-doped CdTe.

Γ_3, Γ_5, and Γ_2 levels, the level spacing is $\sim 20\,\text{cm}^{-1}$ (Fig. 1.2). By using the far-infrared spectroscopy ($0 < \nu < 120\,\text{cm}^{-1}$) and the near-infrared spectroscopy ($1000 < \nu < 13{,}000\,\text{cm}^{-1}$) we can measure the transitions between five levels of the ^{5}E state, and those between ^{5}E and $^5T_2$ levels, respectively, and determine the positions of levels.

Figure 1.5 is the far-infrared absorption spectrum at 4.2 and 20 K for both pure CdTe and CdTe : Fe^{2+} crystals (Slack *et al.*, 1969). The absorption peaks of the pure CdTe can be assigned to phonons at critical points in the Brillouin zone, for example: -2TA(L) $= 71.3\,\text{cm}^{-1}$, -2TA(X) $= 92\,\text{cm}^{-1}$, $-$LA(L) $= 104.5\,\text{cm}^{-1}$, $-$LO + TA $= 114\,\text{cm}^{-1}$, while the phonon energies given by the theoretical calculation are: TA(L) $= 34\,\text{cm}^{-1}$, TA(X) $= 41.5\,\text{cm}^{-1}$, LA(L) $= 102\,\text{cm}^{-1}$, LO $= 150\,\text{cm}^{-1}$, TO $= 140\,\text{cm}^{-1}$. Comparing two results, we found that the agreement between the experimental and theoretical results is good.

The far-infrared absorption peak positions of the CdTe : Fe^{2+} are $\nu = 18.6$, 54.8, 66.7, and $73.2\,\text{cm}^{-1}$, which are assigned to transitions between levels of the Fe^{2+} state. From Fig. 1.2, the $\nu = 18.6\,\text{cm}^{-1}$ peak is caused by the Γ_1–Γ_4 transition, the 66.7 and 73.2 cm^{-1} peaks are the Γ_1–T_5 transitions, because the Γ_5 level is a doublet level. The 54.8 cm^{-1} peak is the transition of Γ_4 to Γ_5 up level. The calculation gives that the transition of the Γ_1–Γ_3 levels is forbidden, so the position of the Γ_3 level cannot be determined by the experiment. It can also be seen from Fig. 1.2 that the agreement between the experimental and calculated results is not good. We can modulate the parameters λ and Dq so that the level spacing between Γ_1–Γ_4 levels is in agreement with the experimental value, but the positions of other levels are not right. This just indicates that the crystal field theory based on the point charge model is not strict.

The near-infrared absorption spectrum of the $Cd_{1-x}Mg_xTe{:}Fe^{2+}$ crystal at 2 and 14 K is shown in Fig. 1.6 (Slack *et al.*, 1969). In Fig. 1.6, the absorption band in the range of $2000 < \nu < 3000\,cm^{-1}$ is assigned to the ^5E to $^5T_2$ transition, and the band $\nu \sim 11,500\,cm^{-1}$ to the ^5E–$^3T_1$ transition. The weak band at $\nu \sim 14{,}000\,cm^{-1}$ to the ^5E–$^3T_2$ transition, i.e. to a state derived from the ^3P, ^3H free-ion excited levels.

Splitting of the five levels of the ^5E state of Fe^{2+} in CdTe in a $\langle 100 \rangle$ applied magnetic field is shown in Fig. 1.7 (Slack *et al.*, 1969). The Γ_4 level is triply degenerate in the zero magnetic field, while the Γ_1 level is a singlet. The Γ_4 level splits into three component levels in an applied magnetic field. The transitions from Γ_1 level to three splitting levels of Γ_4 were observed in experiment, the original $\nu = 18.6\,cm^{-1}$ absorption peak in the zero magnetic field splits into ν_h, ν_c, and ν_l three peaks, where the ν_c is the central peak, and the ν_h and ν_l peaks have higher and lower

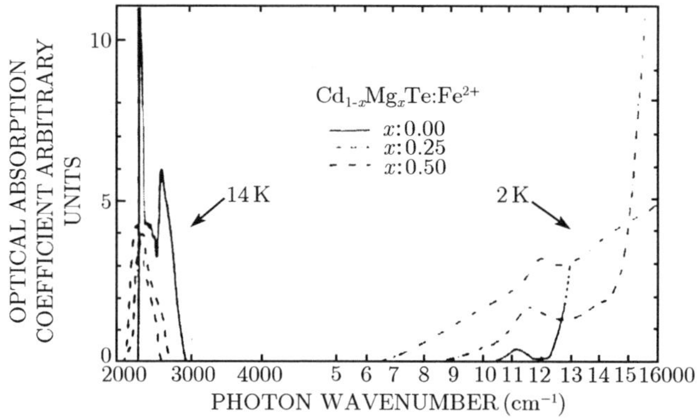

Fig. 1.6. Near-infrared absorption spectrum of the $Cd_{1-x}Mg_xTe{:}Fe^{2+}$ crystal at 2 and 14 K.

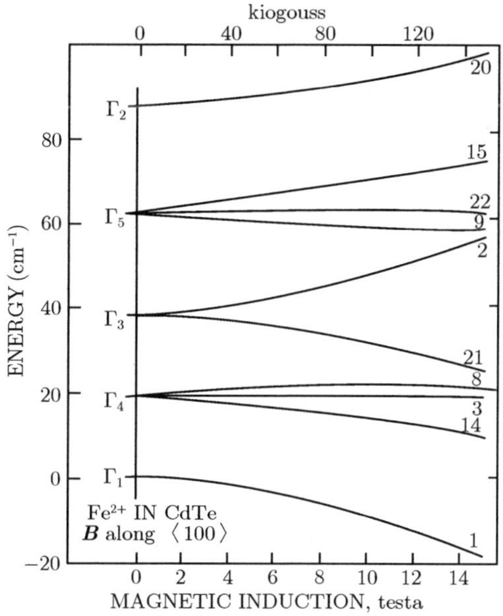

Fig. 1.7. Splitting of the five levels of the ^{5}E state of Fe^{2+} in CdTe in a $\langle 100\rangle$ applied magnetic field.

energies than the ν_c peak, respectively. The energy difference between ν_h and ν_l peaks is linearly proportional to the magnetic field strength, which results in the effective g-factor of the Γ_4 level $g(\Gamma_4) = 1.02 \pm 0.03$, while the theoretical value is $g(\Gamma_4) = 0.940$. The theoretical g-factor of the Γ_5 level is $g(\Gamma_5) = 1.305$, while experimentally two peaks of the Γ_1–Γ_5 transitions at the zero magnetic field are rather broad; the splitting caused by the magnetic field has not been observed, so no experimental determination of $g(\Gamma_5)$ was made.

Due to the splitting of the energy levels of the magnetic ion in the magnetic field, we can measure its magnetic susceptibility, which is defined as

$$\chi = \lim_{H \to 0} \frac{\partial M}{\partial H}, \tag{1.53}$$

where H is the applied magnetic field and M is the magnetization. The magnetization equals

$$M(H) = N_0 k_B T \frac{\partial \ln Z}{\partial H}, \tag{1.54}$$

$$Z = \sum_i \exp\left(-\frac{E_i}{k_B T}\right), \tag{1.55}$$

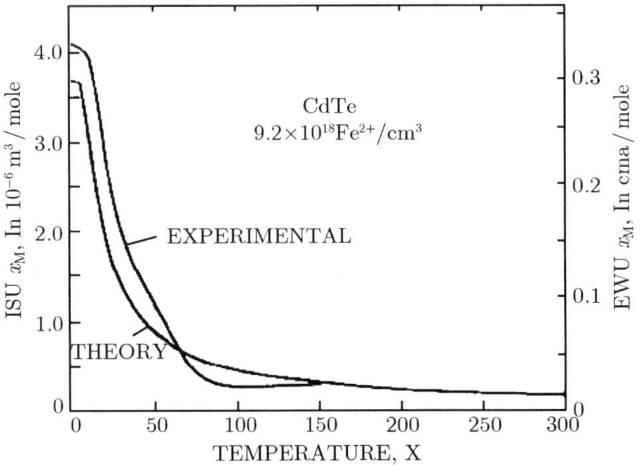

Fig. 1.8. Experimental and theoretical magnetic susceptibilities of Fe^{2+} in CdTe versus temperature.

where Z is the partition function and N_0 is the ion number in the unit volume. From Eqs. (1.53) and (1.54), we obtain

$$\chi = N_0 k_B T \lim_{H \to 0} \frac{\partial^2 \ln Z}{\partial H^2}. \tag{1.56}$$

The experimental and theoretical magnetic susceptibilities of Fe^{2+} in CdTe versus temperature are shown in Fig. 1.8 (Slack *et al.*, 1969), where the magnetic ion density $N_0 = 9.2. \times 10^{18}\,\text{cm}^{-3}$ In Fig. 1.8, the temperature can divide into three regions: (1) $T < 5\,\text{K}$, χ is dominated by the Γ_1 level; (2) $5\,\text{K} < T < 40\,\text{K}$, χ is dominated by the Γ_1 and Γ_4 levels; (3) $T > 40\,\text{K}$, the computed χ versus T is sensitive to the Γ_3, Γ_5, and Γ_2 levels. The agreement between theory and experiment is not as good for $T > 40\,\text{K}$ as at lower temperatures.

References

Abragam A, Bleaney B. (1970) *Electron Paramagnetic Resonance of Transition Ions.* Clarendon, Oxford.

Edmonds AR. (1957) *Angular Momentum in Quantum Mechanics.* Princeton University Press.

Giriat W, Furdyna JK. (1988) in *Diluted Magnetic Semiconductors in Semiconductors and Semimetals*, Vol. 25, p. 1, Furdyna JK and Kossnt J (eds). Academic Press, Boston.

Hatchings MT. (1964), in *Solid State Phys*, Vol. **16**: 227, Seitz F and Turnbull D (eds.), Academic, New York.
Heine V. (1970), *Group Theory in Quantum Mechanics*, Pergamon Press, London-Oxford.
Koster GF, Dimmock JO, Wheefer RG, Statz H. (1963) *Properties of the Thirty-Two Point Groups*, MIT Press, Cambridge, MA.
Slack GA, Roberts S, Vallin JT. (1969) *Phys Rev* **187**: 511.
Stevens KWH. (1952) *Proc Phys Soc* **65**: 209.
Villeret M, Rodriguez S, Kartheuser E. (1990) *Physica B* **162**: 89.

Chapter 2

PROPERTIES OF DMSs

The semiconductors doped with magnetic ions form a new kind of ternary alloys having magnetic properties. Because the concentration of the doped magnetic ions is generally not high, and the magnetic properties are not strong, they are called DMS, or semimagnetic semiconductors. The crystalline structures, concentration ranges, and energy gaps of some DMSs are listed in Table 2.1 (Liu and Zhang, 1994), where ZB represents the zincblende structure and WZ the wurtzite structure.

After the 1960s, people started the growth of DMSs and study of their physical properties. From Table 2.1, we see that for each DMS, there is a definite variable range of the magnetic ion concentration x, beyond this range there will appear mixed phases. For example, $Cd_{1-x}Mn_xSe$, $0 < x \leq 0.5$, its phase diagram is shown in Fig. 2.1 (Cook, 1968). In the range of $0 < x < 0.5$, $Cd_{1-x}Mn_xSe$ forms uniform wurtzite structure. At $x = 1$, it forms MnSe rock salt structure, and in the range of $0.5 < x < 0.95$, it forms two phases co-existence of wurtzite and rock salt structures.

The most important progress of the DMS is the successful growth of DMS thin films, heterstructures, and superlattices, especially in 1984 the $Cd_{1-x}Mn_xTe$ thin film and the $Cd_{1-x}Mn_xTe/Cd_{1-y}Mn_yTe$ superlattice were grown by the molecular beam epitaxy (MBE) method. From then, this field developed very fast, including the growth of $Zn_{1-x}Mn_xSe$, $Hg_{1-x}Mn_xTe$, etc., thin films and superlattices by the MBE method.

Up to now, most of DMSs are doped with Mn^{2+} ions; this kind of DMS was studied in more detail. But there are also some DMSs doped with Fe^{2+} or Co^{2+} ions (Table 2.1). There is an important difference between the DMSs doped with Fe^{2+} and Co^{2+} ions: the permanent magnetic moment of the Fe^{2+} ions is zero at low temperatures, and in the dilute magnetic limit ($x < 0.01$), the $A^{II}_{1-x}Fe_xB^{VI}$ alloy shows Van Vleck paramagnetism.

Table 2.1. Crystalline structures, concentration ranges, and energy gaps of some DMSs.

Material	Crystalline structure	Concentration range	Energy gap
$Zn_{1-x}Mn_xS$	ZB	$0 < x \leq 0.10$	Wide
	WZ	$0.10 < x \leq 0.45$	Wide
$Zn_{1-x}Mn_xSe$	ZB	$0 < x \leq 0.30$	Wide
	WZ	$0.30 < x \leq 0.57$	Wide
$Zn_{1-x}Mn_xTe$	ZB	$0 < x \leq 0.86$	Wide
$Cd_{1-x}Mn_xS$	WZ	$0 < x \leq 0.45$	Wide
$Cd_{1-x}Mn_xSe$	WZ	$0 < x \leq 0.50$	Wide
$Cd_{1-x}Mn_xTe$	ZB	$0 < x \leq 0.77$	Wide
$Hg_{1-x}Mn_xS$	ZB	$0 < x \leq 0.37$	Narrow
$Hg_{1-x}Mn_xSe$	ZB	$0 < x \leq 0.38$	Narrow
$Hg_{1-x}Mn_xTe$	ZB	$0 < x \leq 0.50$	Narrow
$Zn_{1-x}Co_xS$	ZB	$0 < x \leq 0.12$	Wide
$Zn_{1-x}Co_xSe$	ZB	$0 < x \leq 0.06$	Wide
$Cd_{1-x}Co_xSe$	ZB	$0 < x \leq 0.22$	Wide
$Zn_{1-x}Fe_xSe$	ZB	$0 < x \leq 0.30$	Wide
$Cd_{1-x}Fe_xSe$	WZ	$0 < x \leq 0.15$	Wide
$Hg_{1-x}Fe_xSe$	ZB	$0 < x \leq 0.15$	Narrow
$Hg_{1-x}Fe_xTe$	ZB	$0 < x \leq 0.12$	Narrow
$(Zn_{1-x}Mn_x)_3As_2$	—	$0 < x \leq 0.10$	Narrow
$(Cd_{1-x}Mn_x)_3As_2$	—	$0 < x \leq 0.20$	Narrow
$Pb_{1-x}Mn_xTe$	—	$0 < x \leq 0.10$	Narrow

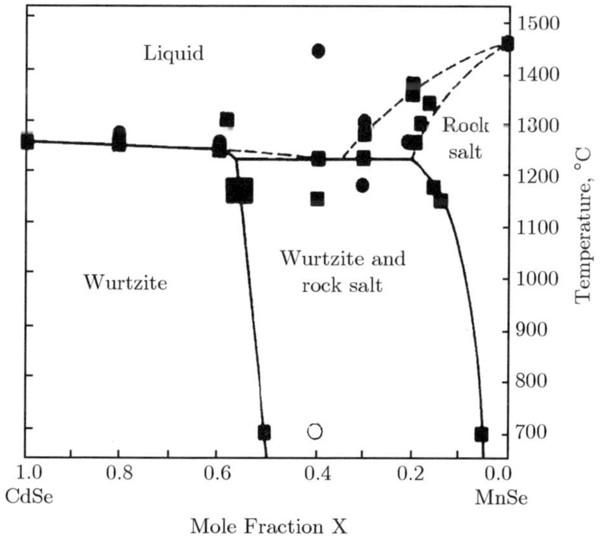

Fig. 2.1. Phase diagram of CdSe–MnSe.

2.1. Effective-mass Theory of Semiconductors in the Magnetic Field

Effective-mass Hamiltonian of an arbitrary band in the applied magnetic field can be written as (Roth *et al.*, 1959),

$$H_{ij} = g\mu_{\rm B}(\boldsymbol{s}\cdot\boldsymbol{B})_{ij} + \sum_{\alpha\beta} D_{ij}^{\alpha\beta} P_\alpha P_\beta, \tag{2.1}$$

where

$$\begin{aligned} D_{ij}^{\alpha\beta} &= \frac{1}{2m_0}\delta_{ij}\delta_{\alpha\beta} + \frac{1}{2m_0^2}\sum_n{}' \frac{\pi_{in}^\alpha \pi_{nj}^\beta}{E_0 - E_n} \quad (\alpha,\beta = x,y,z) \\ \pi_{in}^\alpha &= \langle i|p_\alpha|n\rangle \\ \boldsymbol{P} &= \boldsymbol{p} + \frac{e\boldsymbol{A}}{c}. \end{aligned} \tag{2.2}$$

Because the operators P_α and P_β are not commutable, the Hamiltonian H is divided into symmetric and antisymmetric two parts,

$$H_{ij} = H_{ij}^{(S)} + H_{ij}^{(A)}, \tag{2.3}$$

where the symmetric part

$$\begin{aligned} H_{ij}^{(S)} &= \sum_{\alpha\beta} \bar{D}_{ij}^{\alpha\beta}\{P_\alpha, P_\beta\} \\ \bar{D}_{ij}^{\alpha\beta} &= \frac{1}{2}(D_{ij}^{\alpha\beta} + D_{ij}^{\beta\alpha}) \\ \{P_\alpha \cdot P_\beta\} &= \frac{1}{2}(P_\alpha P_\beta + P_\beta P_\alpha). \end{aligned} \tag{2.4}$$

By use of the commutation relation

$$\boldsymbol{P}\times\boldsymbol{P} = \left(\frac{e\hbar}{ic}\right)\boldsymbol{B}, \tag{2.5}$$

we obtain the antisymmetric part,

$$H_{ij}^{(\rm A)} = \frac{1}{m_0^2}\frac{e\hbar}{2ic}\boldsymbol{B}\cdot\sum_n{}' \frac{\boldsymbol{\pi}_{in}\times\boldsymbol{\pi}_{nj}}{E_0 - E_n} + g\mu_{\rm B}(\boldsymbol{s}\times\boldsymbol{B})_{ij}. \tag{2.6}$$

For the CB minimum near $\boldsymbol{k} = 0$, the effective-mass Hamiltonian including the spin is

$$H = \frac{1}{2m^*}\left(\boldsymbol{p} + \frac{e}{c}\boldsymbol{A}\right)^2 + g^*\mu_{\rm B}\boldsymbol{s}\cdot\boldsymbol{B}, \tag{2.7}$$

where m^* and g^* are the effective mass and the effective g-factor of the electron, respectively.

$$\frac{1}{2m^*} = \frac{1}{2m_0} + \frac{1}{m_0^2} \sum_n \frac{|\pi_{sn}^x|^2}{E_0 - E_n}, \tag{2.8}$$

$$g^* = g + \frac{2}{m_0 i} \sum_n \frac{(\boldsymbol{\pi}_{sn} \times \boldsymbol{\pi}_{ns}) \cdot \boldsymbol{n}}{E_0 - E_n}, \tag{2.9}$$

$\boldsymbol{n}$ is the unit vector in the magnetic field direction. The second term in Eq. (2.9) equals zero in the case of no SOI, which is proportional to the electron spin $\boldsymbol{\sigma}$ due to the property of time inversion: $\boldsymbol{\sigma}$ changes the sign in the operation of the time inversion operator $-i\sigma_y c$.

2.2. DMSs of a Wide Band Gap

2.2.1. *Magnetic energy levels of wide bandgap semiconductors*

Generally, II–VI semiconductors are of wide gap. An important property of DMSs in the magnetic field is that the magnetic ions can cause giant splitting of the CB and the valence band of the semiconductor, resulting in a series of special physical phenomena. For this, we first consider the energy level splitting of semiconductors without magnetic ions in the magnetic field. For the wide gap semiconductors, the difference between Landau levels produced by the magnetic field is smaller than the energy gap E_g, so we can calculate the CB and the valence band separately.

The effective-mass Hamiltonian of the CB is Eq. (2.7). If magnetic field is assumed in the z-direction, then the generalized moment operator $\boldsymbol{P}$ satisfies the commutation relation (see Eq. (2.5), taking $\hbar = 1$),

$$[P_x, P_y] = \frac{e}{ic} B. \tag{2.10}$$

Introducing the creating and destruction operators,

$$a^+ = \sqrt{\frac{c}{2eB}}(P_x + iP_y), \quad a = \sqrt{\frac{c}{2eB}}(P_x - iP_y). \tag{2.11}$$

They satisfy the commutation relation,

$$[a, a^+] = 1. \tag{2.12}$$

Let the wave function of the harmonic oscillator be u_n, and the operators a^+ and a have the following properties,

$$a^+ a u_n = n u_n, \quad a u_n = \sqrt{n} u_{n-1}, \quad a^+ u_n = \sqrt{n+1} u_{n+1}. \tag{2.13}$$

By using Eq. (2.7) and the operators a^+ and a, we obtain the effective-mass equation of the electron,

$$\left[\frac{heB}{m^* c}\left(a^+ a + \frac{1}{2}\right) + \frac{p_z^2}{2m^*} + g^* \mu_{\mathrm{B}} B s_z\right] \phi(\boldsymbol{r}) = E\phi(\boldsymbol{r}). \tag{2.14}$$

The solution of Eq. (2.14) is

$$\phi(\boldsymbol{r}) = u_n(x, y) e^{ikz} \chi_s, \tag{2.15}$$

where $u_n(x, y)$ is the wave function of the harmonic oscillator in the x, y plane and χ_s is the spin wave function. Inserting Eq. (2.15) into Eq. (2.14), we obtain the eigen-energies,

$$E_{n,k,\sigma} = \left(n + \frac{1}{2}\right)\frac{\hbar eB}{m^* c} + \frac{\hbar^2 k^2}{2m^*} + g^* \mu_{\mathrm{B}} B\sigma, \quad n = 0, 1, 2, \ldots, \sigma = \pm 1/2. \tag{2.16}$$

When $k = 0$, Eq. (2.16) gives a series of Landau levels, σ is the spin component of the electron; in the magnetic field, the originally degenerate spin up and down levels split. The effective g-factor of semiconductor g^* depends on the energy gap. Table 2.2 lists the energy gaps and g^*-factors of some typical semiconductors. Generally, when the gap is larger, $g^* \approx 2.0$, near that of the free electron. When the energy gap decreases, g^* decreases gradually and becomes negative. For the semiconductor with a very small gap, InSb, $E_{\mathrm{g}} = 0.18\,\mathrm{eV}$, $g^* = -51.31$. This is caused by the

Table 2.2. Energy gaps (RT) and g^*-factors of semiconductor CB electrons.

	ZnS (zincblende)	ZnS (wurtzite)	ZnSe (zincblende)	CdS (wurtzite)	CdSe (wurtzite)
II–VI compounds					
E_g (eV)	3.56	3.58	2.67	2.50	1.75
g^*	1.88	2.3	1.20	1.78	0.60
III–V compounds	GaAs	GaSb	InP	InAs	InSb
E_{g} (eV)	1.43	0.70	1.35	0.354	0.18
g^*	−0.44	−9.1	1.48	−14.7	−51.31
IV	C	Si	Ge		
E_{g} (eV)	5.48	1.11	0.664		
g^*	2.00	1.999	−1.8		

large interaction between the CB and the valence band, which will be discussed in Sec. 2.3.

The effective-mass Hamiltonian of the valence band in the zero magnetic field is a 6×6 matrix, whose basis functions are the X, Y, Z orbital functions of the valence band top multiplying the spin $s_z = \pm 1/2$ wave functions. If the SOC is not taken into account, the Hamiltonian with the basis functions X, Y, Z can be written as (Luttinger, 1956),

$$H_{\rm h} = \frac{1}{2m_0} \begin{vmatrix} Lp_x^2 + M(p_y^2 + p_z^2) & Np_xp_y & Np_xp_z \\ Np_xp_y & Lp_y^2 + M(p_x^2 + p_z^2) & Np_yp_z \\ Np_xp_z & Np_yp_z & Lp_z^2 + M(p_y^2 + p_x^2) \end{vmatrix}, \tag{2.17}$$

where L, M, N are the effective-mass parameters, which are determined by the $\boldsymbol{k} \cdot \boldsymbol{p}$ perturbation between the valence band and other energy bands at the Γ point (Eq. (2.2)). In the magnetic field, the effective-mass Hamiltonian can be written as the symmetric and antisymmetric two parts, as Eq. (2.3) shows. The symmetric part $H^{(\rm S)}$ is given by Eq. (2.11), in which the $P_\alpha P_\beta$ term is replaced by the $\{P_\alpha P_\beta\}$. The antisymmetric part can be written as,

$$H^{(\rm A)} = \frac{e}{2c}\xi \boldsymbol{I} \cdot \boldsymbol{B}, \tag{2.18}$$

where $\boldsymbol{I}$ is the spin matrix of spin equal to 1,

$$I_x = \begin{vmatrix} 0 & 0 & 0 \\ 0 & 0 & -i \\ 0 & i & 0 \end{vmatrix}, \quad I_y = \begin{vmatrix} 0 & 0 & i \\ 0 & 0 & 0 \\ -i & 0 & 0 \end{vmatrix}, \quad I_z = \begin{vmatrix} 0 & -i & 0 \\ i & 0 & 0 \\ 0 & 0 & 0 \end{vmatrix}. \tag{2.19}$$

Its physical meaning is that the non-diagonal part is equivalent to a Bohr magnetic moment equal to $e/c\xi$, whose energy splits in the magnetic field. Besides, the splitting of the valence band electron magnetic moment in the magnetic field should be added,

$$H_{\rm g} = g_{\rm h}\mu_{\rm B}Bs_z, \tag{2.20}$$

where $g_{\rm h}$ is the g-factor of the valence band electron (hole).

The Hamiltonian (2.17) has not taken into account the hole spin. If the basis functions are taken as $X\uparrow, Y\uparrow, Z\uparrow$ and $X\downarrow, Y\downarrow, Z\downarrow$, the valence band Hamiltonian becomes 6×6-dimensional.

$$H_{\rm h2} = \begin{vmatrix} H_{\rm h} & 0 \\ 0 & H_{\rm h} \end{vmatrix}. \tag{2.21}$$

There is no interaction between hole states with positive and negative spins, the non-diagonal part equals zero. If the SOC effect is taken into account, the SOC Hamiltonian should be added,

$$H_{\text{so}} = \frac{\hbar}{4m_0^2c^2}(\nabla V \times \boldsymbol{p}) \cdot \boldsymbol{\sigma}$$

$$= \begin{vmatrix} 0 & -i\lambda & 0 & 0 & 0 & \lambda \\ i\lambda & 0 & 0 & 0 & 0 & -i\lambda \\ 0 & 0 & 0 & -\lambda & i\lambda & 0 \\ 0 & 0 & -\lambda & 0 & i\lambda & 0 \\ 0 & 0 & -i\lambda & -i\lambda & 0 & 0 \\ \lambda & i\lambda & 0 & 0 & 0 & 0 \end{vmatrix}, \tag{2.22}$$

where $\boldsymbol{\sigma}$ is the Pauli matrix and

$$\lambda = -i\left(\frac{\hbar}{4m_0^2c^2}\right)\langle X|(\nabla V \times \boldsymbol{p})_y|Z\rangle. \tag{2.23}$$

When the SOC interaction is considered (see Eq. (2.22)), there is coupling between the hole states with positive and negative spins. The result is that two groups of energy levels are formed: one is the energy levels of total angular moment $J = 3/2$, which have higher energies $\Delta E_{\text{so}} = \lambda$, are 4-fold degenerate at the Γ point, and are called heavy and light hole bands; other is the energy levels of $J = 1/2$, which have lower energies $\Delta E_{\text{so}} = -2\lambda$, 2-fold degenerate at the Γ point, and called spin–orbit splitting band.

Figure 2.2 is the energy bands of GaAs in the zero magnetic field, where the valence band top consists of the Γ_8 and Γ_7 levels: the Γ_8 are the heavy and light hole levels, 4-fold degenerate at the Γ point; the Γ_7 are the spin–orbit splitting levels, 2-fold degenerate at the Γ point. The energy difference between the Γ_8 and Γ_7 levels is called the spin–orbit splitting energy $\Delta = 3\lambda = 0.34\,\text{eV}$. When $k \neq 0$, the Γ_8 levels split into two levels: one has a smaller curvature and is called the heavy hole band; other has a larger curvature and is called the light hole band.

The calculation of the magnetic levels of valence band in the magnetic field is similar to the calculation of the electron Landau levels. We transform the operators P_x, P_y into a^+ and a (see Eq. (2.11)), expand the effective-mass envelope functions with the harmonic oscillator functions u_n, and use the properties of the operators a^+, a operated on the u_n (Eq. (2.6)), and we can obtain the magnetic levels and corresponding wave functions.

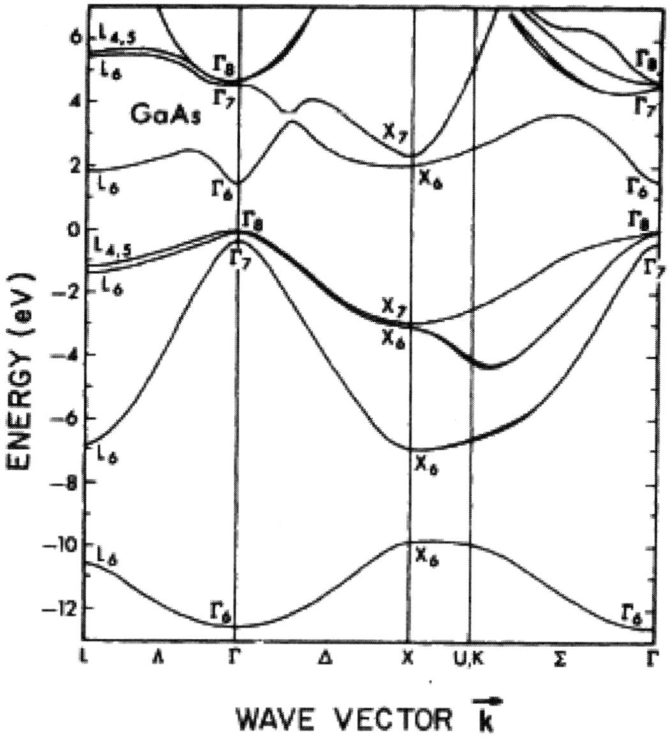

Fig. 2.2. Energy bands of GaAs.

If the spacing of the hole Landau levels is much smaller than the spin–orbit splitting energy Δ, we can only use the Hamiltonian of the heavy and light holes to calculate the magnetic levels.

$$H_{\text{hl}} = \frac{1}{2m_0}\begin{vmatrix} P_1 & Q & R & 0 \\ Q^* & P_2 & 0 & R \\ R^* & 0 & P_2 & -Q \\ 0 & R^* & -Q^* & P_1 \end{vmatrix}. \tag{2.24}$$

The basis functions are the following:

$$\left|\frac{3}{2},\frac{3}{2}\right\rangle = \frac{1}{\sqrt{2}}(X+iY)\uparrow,$$

$$\left|\frac{3}{2},\frac{1}{2}\right\rangle = \frac{1}{\sqrt{6}}(X+iY)\downarrow - \sqrt{\frac{2}{3}}Z\uparrow,$$

$$\left|\frac{3}{2},\frac{1}{2}\right\rangle = -\frac{1}{\sqrt{6}}(X - iY)\uparrow - \sqrt{\frac{2}{3}}Z\downarrow,$$
$$\left|\frac{3}{2},\frac{3}{2}\right\rangle = \frac{1}{\sqrt{2}}(X - iY)\downarrow . \tag{2.25}$$

The terms in the Hamiltonian (2.24) are

$$\begin{aligned} P_1 &= (\gamma_1 + \gamma_2)(p_x^2 + p_y^2) + (\gamma_1 - 2\gamma_2)p_z^2, \\ P_2 &= (\gamma_1 - \gamma_2)(p_x^2 + p_y^2) + (\gamma_1 + 2\gamma_2)p_z^2, \\ Q &= -2\sqrt{3}\gamma_3 p_z(p_x - ip_y), \\ R &= \sqrt{3}[\gamma_2(p_x^2 - p_y^2) - 2i\gamma_3 p_x p_y], \end{aligned} \tag{2.26}$$

where γ_1, γ_2, and γ_3 are called Luttinger effective-mass parameters (Luttinger, 1956), which are related to the effective-mass parameters L, M, and N as follows,

$$\begin{aligned} \gamma_1 &= \frac{L + 2M}{3}, \\ \gamma_2 &= \frac{L - M}{6}, \\ \gamma_3 &= \frac{N}{6}. \end{aligned} \tag{2.27}$$

When a magnetic field is applied in the z-direction, the operators p_x and p_y should be changed into generalized operators P_x and P_y. Luttinger (1956) proved that the Hamiltonian (2.24) can also be divided into symmetric and antisymmetric parts. The symmetric part is just Eq. (2.24), where the operators $p_\alpha p_\beta$ are replaced by $\{P_\alpha P_\beta\}$, and the antisymmetric part is

$$H_{\mathrm{a}} = \frac{e}{2c}\kappa \boldsymbol{J} \cdot \boldsymbol{B}, \tag{2.28}$$

where κ is the antisymmetric effective-mass parameter determined generally by the experiment. $\boldsymbol{J}$ is the spin matrix of $J = 3/2$.

$$J_x = \frac{1}{2}\begin{vmatrix} 0 & \sqrt{3}i & 0 & 0 \\ -\sqrt{3}i & 0 & 2i & 0 \\ 0 & -2i & 0 & \sqrt{3}i \\ 0 & 0 & -\sqrt{3}i & 0 \end{vmatrix},$$

$$J_y = \frac{1}{2}\begin{vmatrix} 0 & \sqrt{3} & 0 & 0 \\ \sqrt{3} & 0 & 2 & 0 \\ 0 & 2 & 0 & \sqrt{3} \\ 0 & 0 & \sqrt{3} & 0 \end{vmatrix},$$

$$J_z = \frac{1}{2}\begin{vmatrix} 3 & 0 & 0 & 0 \\ 0 & 1 & 0 & 0 \\ 0 & 0 & -1 & 0 \\ 0 & 0 & 0 & -3 \end{vmatrix}. \tag{2.29}$$

The electron and hole Landau levels calculated by this method are shown in Fig. 2.3 (Lax *et al.*, 1960). From Fig. 2.3, we see that the electron Landau levels are denoted by the principal quantum number n and the spin quantum number $s_z = \pm 1/2$. From Table 2.2, the electron g-factor of Ge $g_e = -1.8$, so that the $s_z = -1/2$ level is higher than the $s_z = 1/2$ level. The hole Landau levels are divided into four groups, roughly corresponding to $m_J = \pm 3/2$ and $m_J = \pm 1/2$ levels, produced by the heavy and light hole bands, respectively. Because the energy spacing of the Landau levels $\sim eB/m^*c$, which is inversely proportional to the effective mass, the spacing of the heavy hole Landau levels is smaller and that of the light hole Landau levels is larger. Because the non-diagonal terms of the Hamiltonian (2.24) do not equal zero, there is coupling between the heavy and light hole states. Figure 2.3 is only a simplified

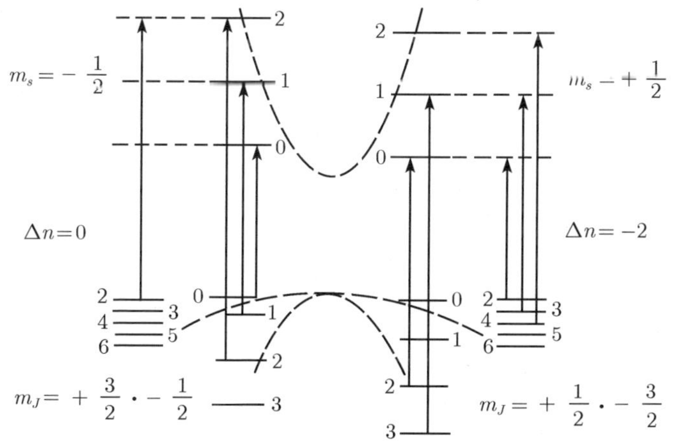

Fig. 2.3. Electron and hole Landau levels and allowed transitions of Ge in the magnetic field.

denotation and classification, actually each state of Landau level is composed of mixing of different basis functions (see Eq. (2.25)), but has a main component.

Experimentally we can measure the transition energies between electron and hole Landau levels with the cyclotron resonance and magnetic optical spectrum. The cyclotron resonance measures the transitions between the CB Landau levels or the valence band Landau levels, and the magnetic optical spectrum measures the transitions from the valence Landau levels to the CB Landau levels. Theoretically we can calculate the selection rules of the electron transition for different magnetic field directions, polarization direction of light. The arrows in Fig. 2.3 indicate the possible transitions when the electric field direction of light $E//B//z$.

2.2.2. *Magnetic interaction in DMSs*

For definition, we assume that the magnetic ion in DMSs is Mn^{2+} ion. From Chap. 1, the basis state of the Mn^{2+} ion is the $S_{5/2}$ state, i.e. its total orbital angular momentum $L = 0$, and does not split in the crystal field. Thus, the Mn^{2+} ion is equivalent to a free ion, whose behavior can be described by the paramagnetism theory. For other magnetic ions, for example, Fe^{2+}, Co^{2+} ions, the total angular momentum of the their basis state $L \neq 0$, which splits in the crystal field. From Sec. 1.7, the basis state of the Fe^{2+} ion is ^5D, which splits into ^5E and $^5T_2$ states in the T_d symmetry crystal field of CdTe. The splitting energy of ^5E and $^5T_2$ states $\sim 250\,\text{cm}^{-1} = 0.31\,\text{eV}$. After the SOC effect is taken into account, the ^5E level splits further into $\Gamma_1, \Gamma_4, \Gamma_3, \Gamma_5$, and Γ_2 levels, the energy spacing $\sim 20\,\text{cm}^{-1}$ (Fig. 1.2). The basis state is the singlet Γ_1; in the low temperature, the permanent magnetic moment of the Fe^{2+} ion is zero, so that in the dilute magnetic limit ($x < 0.01$), $A^{II}_{1-x}Fe_xB^{VI}$ alloy shows Van Vleck paramagnetism. We will discuss this point later, now we only discuss the DMSs doped with Mn^{2+} ions.

For the DMSs, because the concentration of the magnetic ion is low ($x < 0.1$), especially for III–V compounds, the orbit of the d electron is much local in the space, and the direct interaction between magnetic ions is much small. From the 1950s, Zener (1951) started to propose that the interaction between the local magnetic moments of the magnetic ion d electrons is through the indirect interaction by the conduction electrons in the semiconductor, s or p electrons. This indirect interaction can be described by a simple Heisenberg-like formula (Furyna, 1982).

$$H_{\rm sd} = \sum_n J(\boldsymbol{r} - \boldsymbol{R}_n)\boldsymbol{s} \cdot \boldsymbol{S}_n, \tag{2.30}$$

where $\boldsymbol{s}$ is the spin operator of an s electron at the $\boldsymbol{r}$ point, $\boldsymbol{S}_n$ is the total spin of the magnetic ion 3d shell at the $\boldsymbol{R}_n$ point. $J(\boldsymbol{r} - \boldsymbol{R}_n)$ is the interaction constant, which concentrates mainly near $\boldsymbol{R}_n$ and approaches zero quickly when $\boldsymbol{r}$ deviates from the $\boldsymbol{R}_n$ point due to the localization of the 3d electron. Because the wave function of the conduction electron is spread over the space, which interacts with a lot of Mn ions, we can use the thermal average of the spin operator of the Mn ion $\langle S\rangle$ to replace $\boldsymbol{S}_n$. In the zero magnetic field, $\langle S\rangle = 0$. In the magnetic field along the z-direction, $\langle S\rangle$ has only one component $\langle S_z\rangle$, which is not equal to zero.

Therefore, by use of the virtual crystal approximation and the molecular average field approximation, the exchange Hamiltonian (2.31) becomes

$$H_{\rm sd} = x\sigma_z\langle S_z\rangle \sum_{\boldsymbol{R}} J(\boldsymbol{r} - \boldsymbol{R}), \tag{2.31}$$

where $\boldsymbol{R}$ represents the coordinate of the positive ions. Thus, the obvious advantage of Eq. (2.31) is having the lattice periodicity, so that Bloch functions can be used as basis functions. Let

$$J(\boldsymbol{r}) = \sum_{\boldsymbol{R}} J(\boldsymbol{r} - \boldsymbol{R}). \tag{2.32}$$

$J(\boldsymbol{r})$ has the lattice periodicity, localizes near the positive ion lattice, and is isotropic. The exchange integrations not equal to zero of $J(\boldsymbol{r})$ between the valence electron wave functions are

$$\begin{aligned} \alpha &= \langle S|J(\boldsymbol{r})|S\rangle, \\ \beta &= \langle X|J(\boldsymbol{r})|X\rangle = \langle Y|J(\boldsymbol{r})|Y\rangle = \langle Z|J(\boldsymbol{r})|Z\rangle, \end{aligned} \tag{2.33}$$

where $|S\rangle$ and $|X\rangle$, $|Y\rangle$, $|Z\rangle$ are Bloch functions of the CB bottom and the valence band top, respectively.

For the basis state of the Mn^{2+} ion $3d^5$ electrons $S_{5/2}$, according to the quantum theory of paramagnetism,

$$\langle S_z\rangle = -SB_S\left[\frac{gS\mu_{\rm B}B}{k_{\rm B}T}\right], \quad S = \frac{5}{2}, \tag{2.34}$$

where g is the g-factor of the free electron, $g = 2$. $B_S(y)$ is the Brillouin function with the argument S,

$$B_S(y) = \frac{2S+1}{2S}\coth\left[\frac{(2S+1)y}{2S}\right] - \frac{1}{2S}\coth\left(\frac{y}{2S}\right). \tag{2.35}$$

In the semiconductors instead of Eq. (2.34), $\langle S_z \rangle$ is generally taken as (Dai *et al.*, 1994),

$$\langle S_z \rangle = -SB_S \left[\frac{gS\mu_B B}{k_B(T+T_0)} \right], \tag{2.36}$$

where T_0 is an experience parameter, representing the antiferromagnetic interaction between Mn^{2+} ions. For $Zn_{1-x}Mn_xSe$, $T_0 = 1.4\,K$ (Dai *et al.*, 1994).

From Eqs. (2.31)–(2.33), we can obtain the splitting energies of the CB and the valence band in the magnetic field due to the s–d and p–d interactions between the magnetic ions and the conduction electrons, respectively. For the CB,

$$\Delta E_c = \pm \frac{1}{2} \bar{x} N_0 \alpha \langle S_z \rangle, \tag{2.37}$$

where N_0 is the density of the ions in semiconductors, and generally $N_0\alpha$ and $N_0\beta$ are called exchange integrations of the CB and the valence band, respectively, whose values are determined by the experiment. For example, for $Zn_{1-x}Mn_xSe$, $N_0\alpha = -0.27\,eV$, $N_0\beta = 0.90\,eV$. $\bar{x}$ is the effective concentration of the Mn ions, which is smaller than the actual concentration x (Lee *et al.*, 1996). If we take $\bar{x} = 0.04, T_0 = 1.4\,K$, $T = 1.5\,K$, $B = 1\,T$, then we obtain

$$\frac{gS\mu_B B}{k_B(T+T_0)} = 1.1582, \quad B_{5/2}(1.1582) = 0.48.$$

Then from Eq. (2.37), we obtain $\Delta E_c = \pm 6.48\,meV$. Thus, the originally degenerate CB of positive and negative spins split, and the splitting energy is about 13 meV, the positive spin energy band is higher than the negative spin band. The above result is calculated for $Zn_{1-x}Mn_xSe$, but represents the general characteristic of DMSs.

Next we consider the splitting of the valence band, temporarily we do not take into account the SOC effect. The basis functions of the valence band top are $X\uparrow, Y\uparrow, Z\uparrow, X\downarrow, Y\downarrow$, and $Z\downarrow$, and the energy-splitting matrix of the basis functions due to the p–d interaction is

$$\Delta E_v = \begin{vmatrix} +A & 0 & 0 & 0 & 0 & 0 \\ 0 & +A & 0 & 0 & 0 & 0 \\ 0 & 0 & +A & 0 & 0 & 0 \\ 0 & 0 & 0 & -A & 0 & 0 \\ 0 & 0 & 0 & 0 & -A & 0 \\ 0 & 0 & 0 & 0 & 0 & -A \end{vmatrix}, \tag{2.38}$$

where

$$A = \frac{1}{2}\bar{x}N_0\beta\langle S_z\rangle. \tag{2.39}$$

Because $\langle S_z\rangle < 0$ (see Eq. (2.36)), $N_0\beta > 0$; thus $A < 0$. From Eq. (2.38), we see that the 6-fold degenerate valence band splits into two groups: one's spin is up, the energy equals A, and other's spin is down, the energy equals $-A$. Thus, the energies of the group of spin down are higher than those of spin-up group, and the splitting energy of the two groups is $|2A|$.

After the SOC effect is taken into account, the valence band splits into two groups of energy bands, one is the heavy and light hole band, whose total angular momentum is $J = 3/2$. Another is the spin–orbit splitting band, $J = 1/2$, and which is lower than the heavy and light hole band by $\Delta_{\rm so}$. The basis functions of the heavy and light hole bands are given in Eq. (2.25), and the basis functions of the spin–orbit splitting band is given in the following,

$$\begin{aligned} \left|\frac{1}{2},\frac{1}{2}\right\rangle &= \frac{1}{\sqrt{3}}(X+iY)\downarrow + \frac{1}{\sqrt{3}}Z\uparrow, \\ \left|\frac{1}{2},-\frac{1}{2}\right\rangle &= -\frac{1}{\sqrt{3}}(X-iY)\uparrow + \frac{1}{\sqrt{3}}Z\downarrow. \end{aligned} \tag{2.40}$$

Taking six functions in Eqs. (2.25) and (2.40) as the basis functions, we calculate the energy-splitting matrix caused by the p–d interaction (Eq. (2.38)) as follows:

$$\Delta E_{\rm v} = \begin{vmatrix} A & 0 & 0 & 0 & 0 & 0 \\ 0 & \frac{A}{3} & 0 & 0 & -\frac{2\sqrt{2}}{3}A & 0 \\ 0 & 0 & -\frac{A}{3} & 0 & 0 & \frac{2\sqrt{2}}{3}A \\ 0 & 0 & 0 & -A & 0 & 0 \\ 0 & -\frac{2\sqrt{2}}{3}A & 0 & 0 & -\frac{A}{3} & 0 \\ 0 & 0 & \frac{2\sqrt{2}}{3}A & 0 & 0 & \frac{A}{3} \end{vmatrix}. \tag{2.41}$$

From Eq. (2.41), we see that when the SOC effect is taken into account, the splitting of the valence band is not so simple as the CB (as Eq. (2.37)). There are couplings between bands, the non-diagonal matrix elements do not equal zero. If the spin–orbit splitting energy is far larger than the s–d splitting energy $\Delta_{\rm so} \gg A$, then the spin–orbit splitting band can be neglected, and only the heavy and light hole bands be considered. Then the non-diagonal terms in Eq. (2.41) can be neglected, and the splitting energy of the heavy and light hole bands can simply be written as

$$\Delta E_{\rm v} = \left(\frac{2}{3}A\right) M_J = \frac{1}{3}\bar{x}N_0\beta M_J\langle S_z\rangle, \tag{2.42}$$

where M_J is the component of the hole state angular momentum in the magnetic field. Because $A < 0$, the energy of the $M_J = -3/2$ state is the highest, the next levels are $M_J = -1/2, 1/2$ and $3/2$ states, respectively. In the spin–orbit splitting band, the energy of the $M_J = 1/2$ state is higher than that of the $M_J = -1/2$ state.

Let $N_0\beta = 0.90\,\text{eV}$, $\bar{x} = 0.04$, $B = 1\,\text{T}$, the calculation obtains

$$\Delta E_{\rm v}\left(M_J = \frac{3}{2}\right) = -21.6\,\text{meV}$$

$$\Delta E_{\rm v}\left(M_J = \frac{1}{2}\right) = -7.2\,\text{meV}.$$

The total splitting energy of the four valence bands is 43.2 meV. The splitting energies of the CB and the valence band are far larger than the general Zeeman splitting energy of electrons in the magnetic field. Let $g = 2$, $B = 1\,\text{T}$, the Zeeman splitting energy

$$\Delta E_{\rm Z} = g\mu_{\rm B}B = 0.116\,\text{meV}.$$

Therefore, we call the energy splitting caused by the exchange interaction between the magnetic ion d electrons and the conduction electrons in the DMSs the giant Zeeman splitting. The giant Zeeman splitting energy depends on the Brillouin function $B_S(y)$ (see Eq. (2.35)). When the magnetic field is very small, or the temperature is very high, we expand the $B_S(y)$ in y and obtain

$$B_S(y) \approx \frac{1}{3}\frac{S+1}{S}y. \tag{2.43}$$

Thus, the splitting energy is proportional to the magnetic field. $B_{5/2}(y)$ as a function of y is shown in Fig. 2.4. From Fig. 2.4, we see that when $y < 2, B_{5/2}(y)$ increases with y linearly, when $y \geq 2, B_{5/2}(y)$ approaches the saturate value 1. That means when the magnetic field is very strong, all ions tend to one direction, $\langle S_z\rangle = -S$.

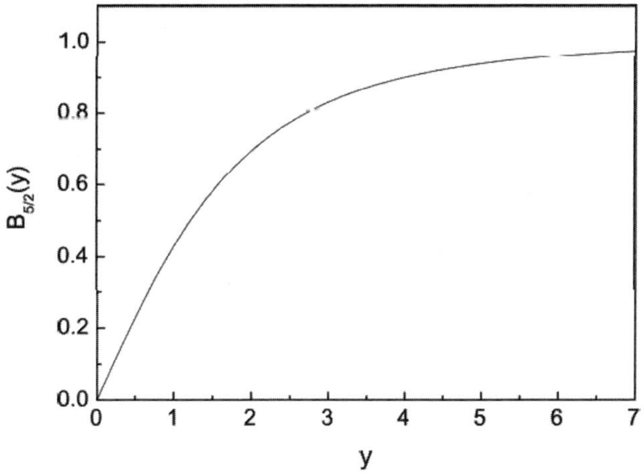

Fig. 2.4. $B_{5/2}(y)$ as a function of y.

2.2.3. *DMSs of the wurtzite structure*

The above discussions are all for the DMSs of the zincblende structure, but many DMSs are the II–VI compounds of the wurtzite structure, for example, CdSe, CdS, ZnS, etc. These materials are hexagonal symmetric, the growth $c(z)$-axis is not equivalent to the perpendicular x and y axes. Thus, the three states at the valence band top are not degenerate; there is a crystal field splitting energy between the Z state and the X, Y states.

$$-\Delta_c = \langle Z|H_{cr}|Z\rangle.$$

Generally, $\Delta_c > 0$; thus, the Z state is lower than the X and Y states by Δ_c. Let $1/\sqrt{2}(X + iY) \uparrow, 1/\sqrt{2}(X - iY) \uparrow, Z \uparrow, 1/\sqrt{2}(X + iY) \downarrow$, $1/\sqrt{2}(X - iY) \downarrow, Z \downarrow$ be the basis functions, at $\boldsymbol{k} = 0$, the SOC Hamiltonian can be written as

$$H_{so} = \begin{vmatrix} \lambda & 0 & 0 & 0 & 0 & 0 \\ 0 & -\lambda & 0 & 0 & 0 & \sqrt{2}\lambda \\ 0 & 0 & -\Delta_c & -\sqrt{2}\lambda & 0 & 0 \\ 0 & 0 & -\sqrt{2}\lambda & -\lambda & 0 & 0 \\ 0 & 0 & 0 & 0 & \lambda & 0 \\ 0 & \sqrt{2}\lambda & 0 & 0 & 0 & -\Delta_c \end{vmatrix}, \qquad (2.44)$$

where $\lambda = \Delta_{so}/3$, defined by Eq. (2.23). From Eq. (2.44), we see that the $m_J = \pm 3/2$ states $1/\sqrt{2}(X + iY) \uparrow$ and $1/\sqrt{2}(X - iY) \downarrow$ do not couple with other states, whose energies equal λ. There are couplings between $Z \uparrow$ state and $1/\sqrt{2}(X + iY) \downarrow$ state, $Z \downarrow$ state and $1/\sqrt{2}(X + iY) \downarrow$ state, and the corresponding secular equation is (2-fold degenerate),

$$\begin{vmatrix} -\Delta_c - E & -\sqrt{2}\lambda \\ -\sqrt{2}\lambda & -\lambda - E \end{vmatrix} = 0. \tag{2.45}$$

Solving Eq. (2.45), we obtain the energies,

$$E = \frac{1}{2}[-(\lambda + \Delta_c) \pm \sqrt{\Delta_c^2 - 2\Delta_c\lambda + 9\lambda^2}]. \tag{2.46}$$

Next we discuss two cases: (1) $\lambda \gg \Delta_c$, i.e. the crystal field splitting energy is very small,

$$E \approx \frac{1}{2}\left[-(\lambda + \Delta_c) \pm 3\lambda\left(1 - \frac{\Delta_c}{9\lambda}\right)\right] = \begin{cases} \lambda - \dfrac{2}{3}\Delta_c \\ -2\lambda - \dfrac{1}{3}\Delta_c. \end{cases} \tag{2.47}$$

This case is similar to the zincblende structure case: the valence band splits into the heavy and light hole band and the spin–orbit splitting band, the splitting energy is $3\lambda = \Delta_{so}$. The heavy hole band does not change with energy λ, the light hole band shifts down by $2\Delta_c/3$, and the spin–orbit splitting band shifts down $\Delta_c/3$. In the zero-order approximation, their wave functions are given by Eqs. (2.25) and (2.40).

(2) $\Delta_c \gg \lambda$. From Eq. (2.46),

$$E \approx \frac{1}{2}\left[-(\lambda + \Delta_c) \pm \Delta_c\left(1 - \frac{\lambda}{\Delta_c}\right)\right] = \begin{cases} -\lambda \\ -\Delta_c. \end{cases} \tag{2.48}$$

There are two levels (2-fold degenerate) under the heavy hole level $E = \lambda$, the spacing between them and the heavy hole level is 2λ and $\lambda + \Delta_c$, and the zero-order wave functions are $(X + iY) \downarrow /\sqrt{2}, (X - iY) \uparrow /\sqrt{2}$ and $Z \uparrow, Z \downarrow$, respectively.

In the first case, p–d energy splitting caused by the magnetic ions is still Eq. (2.41), because the basis functions have not changed. Experimentally, the splitting of the four bands of $J = 3/2$ in the magnetic field can be observed, but at $B = 0$ there is a small energy splitting $2\Delta_c/3$. In the second case, let the zero-order wave functions $(X + iY)\uparrow/\sqrt{2}, (X - iY)\uparrow/\sqrt{2}, (X + iY)\downarrow/\sqrt{2}, (X - iY)\downarrow/\sqrt{2}$,

$Z\uparrow, Z\downarrow$ be the basis functions, the p–d energy-splitting matrix is

$$H_{\mathrm{h}} = \begin{vmatrix} \frac{\Delta_{\mathrm{so}}}{3}+A & 0 & 0 & 0 & 0 & 0 \\ 0 & -\frac{\Delta_{\mathrm{so}}}{3}+A & 0 & 0 & 0 & 0 \\ 0 & 0 & -\frac{\Delta_{\mathrm{so}}}{3}-A & 0 & 0 & 0 \\ 0 & 0 & 0 & \frac{\Delta_{\mathrm{so}}}{3}-A & 0 & 0 \\ 0 & 0 & 0 & 0 & -\Delta_{\mathrm{c}}+A & 0 \\ 0 & 0 & 0 & 0 & 0 & -\Delta_{\mathrm{c}}-A \end{vmatrix}. \tag{2.49}$$

Thus, each group of levels splits into two levels of energies $\pm A$.

In literatures according to m_J these three groups of levels are represented as one Γ_9 and two Γ_7 levels, as shown in Fig. 2.5. In the spectra, they are also called A, B, and C levels, named by the exciton peaks associated with these levels.

From the above discussion, we see that the meaning of the Γ_9 level is more clear, but the energies and wave functions of two Γ_7 levels depend on the relative magnitude of Δ_{c} and Δ_{so}. When $\Delta_{\mathrm{c}} \approx \Delta_{\mathrm{so}}$, the case is more complicated, the energies and the wave functions can only be obtained by solving the secular equation. Table 2.3 lists the Δ_{c} and Δ_{so} of some wurtzite semiconductors.

From Table 2.3 we see that for CdSe, $\Delta_{\mathrm{so}} \gg \Delta_{\mathrm{c}}$, for ZnS and CdS, $\Delta_{\mathrm{c}} \approx \Delta_{\mathrm{so}}$; and for ZnO, $\Delta_{\mathrm{c}} \gg \Delta_{\mathrm{so}}, \Delta_{\mathrm{so}} < 0 (\lambda < 0)$, so the order of two Γ_9 and Γ_7 levels in Fig. 2.5 should be reversed.

2.2.4. *Experimental observations*

Magnetoreflection and magnetotransmission spectra have been used to study the giant Zeeman splitting effect in DMSs. For DMSs of zincblende

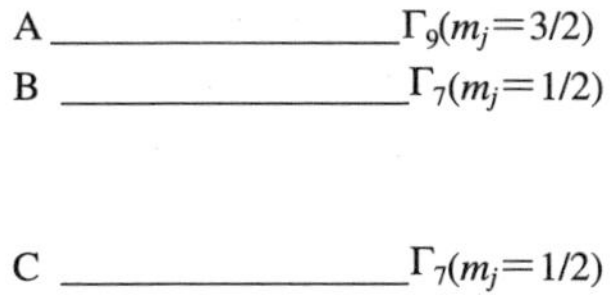

Fig. 2.5. Levels of valence band top of wurtzite semiconductors ($\Delta_{\mathrm{c}} \gg \lambda$).

Table 2.3. Δ_c and Δ_{so} of some wurtzite semiconductors (unit: meV).

	ZnO	ZnS	CdS	CdSe
Δ_c	42	58	28	39
Δ_{so}	−5	86	68	418

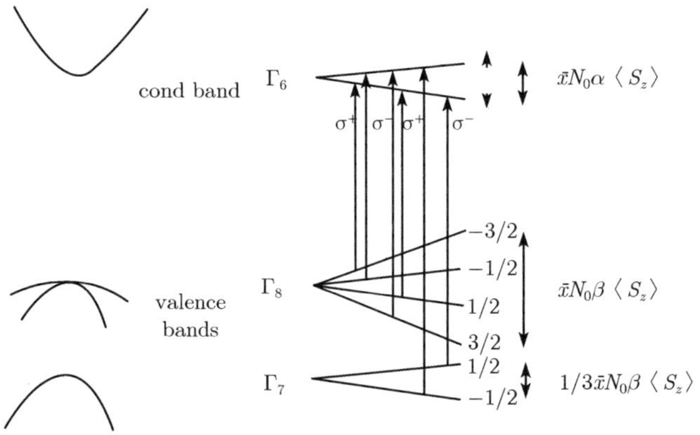

Fig. 2.6. Schematic diagram of energy bands, giant Zeeman splitting, and allowed transitions of zincblende DMS in the magnetic field.

structure in the magnetic field, their energy bands, giant Zeeman splitting, and allowed transitions are shown in Fig. 2.6 (Gaj, 1988).

In Fig. 2.6, the arrows denote the allowed transitions, σ^+, σ^- represent the right and left polarized lights transmitting in the magnetic field direction (Faraday configuration), respectively. π represents the light polarized in the magnetic field direction (Voigt configuration), and transmitting in the direction perpendicular to the magnetic field. The transition selection rules are: for $\sigma^+, \Delta m_J = +1$; for $\sigma^-, \Delta m_J = -1$; for $\pi, \Delta m_J = 0$.

From the basis functions of valence bands (Eqs. (2.25) and (2.40)), we can calculate the relative strengths of all transitions. The dipole transition operators of the σ^+, σ^- and π transitions are $(p_x \pm ip_y)/\sqrt{2}$ and p_z, respectively. The calculated relative strengths of transitions from left to right in Fig. 2.6 are listed as follows, where the unit is $P^2 = |\langle X|p_x|S\rangle|^2$.

$$\sigma^+: 1, \frac{1}{3}; \quad \sigma^-: \frac{1}{3}, 1; \quad \sigma^+: \frac{2}{3}; \quad \sigma^-: \frac{2}{3};$$
$$\pi: \frac{2}{3}, \frac{2}{3}, \frac{1}{3}, \frac{1}{3}. \tag{2.50}$$

According to the effective-mass theory, the wave functions of the semiconductor energy levels in the magnetic field should be the multiplications of the effective-mass envelope functions and the Bloch functions of the CB bottom or valence band top, $|S\rangle \uparrow, |S\rangle \downarrow$ and Eqs. (2.25) and (2.40). The optical transition matrix element equals

$$Q_{nn'} = \int \langle S|\phi_n^*(\boldsymbol{r})p_\alpha\phi_{n'}(\boldsymbol{r})|P\rangle \mathrm{d}\boldsymbol{r}, \tag{2.51}$$

where $|S\rangle$ and $|P\rangle$ are the Bloch functions of the CB and valence band, $\phi_n(\boldsymbol{r})$ and $\phi_{n'}(\boldsymbol{r})$ are their envelope functions, respectively. In the magnetic field, $\phi_n(\boldsymbol{r})$ and $\phi_{n'}(\boldsymbol{r})$ are the harmonic oscillation functions in the (x, y) plane as shown in Eq. (2.15). Because $\phi_n(\boldsymbol{r})$ and $\phi_{n'}(\boldsymbol{r})$ are smoothly varying functions over a unit cell, Eq. (2.51) can be simplified as follows:

$$Q_{nn'} = \int \phi_n^*(\boldsymbol{r})\phi_{n'}(\boldsymbol{r})\mathrm{d}\boldsymbol{r} \cdot \langle S|p_\alpha|P\rangle. \tag{2.52}$$

$\langle S|p_\alpha|P\rangle$ is the transition matrix element of Bloch functions discussed in Eq. (2.50), while $\int \phi_n^*(\boldsymbol{r})\phi_{n'}(\boldsymbol{r})\mathrm{d}\boldsymbol{r}$ is the overlap integration of the envelope functions. Because $\phi_n(\boldsymbol{r})$ is the harmonic oscillation function, the transition selection rules are obtained.

$$\Delta n = 0, \quad \Delta k_z = 0, \tag{2.53}$$

where n is the number of Landau levels and k_z is the wave vector of the wave function in the magnetic field direction. Generally, the transition occurs between the first Landau level ($n = 0$); thus, the transition energy equals nearly the transition energy between band edges.

If the spin–orbit splitting energy Δ_{so} is far larger than the giant Zeeman splitting energy A, then the coupling terms of the heavy and light hole bands and the spin–orbit splitting bands in the energy-splitting matrix (2.41) can be neglected. If the effect of the magnetic field on the band electron state is neglected, then the transition energies from the heavy and light hole states to the electron states can be written as follows (E_0 is the transition energy in the zero magnetic field):

$$\begin{aligned}
E_{1\pm} &= E_0 \mp \frac{1}{2}(\alpha - \beta)\bar{x}N_0\langle S_z\rangle, && \text{(relative intensity: 3)},\\
E_{2\pm} &= E_0 \pm \frac{1}{2}\left(\alpha + \frac{\beta}{3}\right)\bar{x}N_0\langle S_z\rangle, && \text{(relative intensity: 1)},\\
E_{3\pm} &= E_0 \pm \frac{1}{2}\left(\alpha - \frac{\beta}{3}\right)\bar{x}N_0\langle S_z\rangle, && \text{(relative intensity: 2)},
\end{aligned} \tag{2.54}$$

where the up and down signs in the first and second formulas correspond to the σ^+ and σ^- polarization transitions (Fig. 2.6), respectively, and the third formula corresponds to the π polarization transitions. The transition energies from the spin–orbit splitting bands to the CBs are as follows (E_1 is the transition energy in the zero magnetic field):

$$\begin{aligned} E_{4\pm} &= E_1 \pm \frac{1}{2}\left(\alpha - \frac{\beta}{3}\right)\bar{x}N_0\langle S_z\rangle, \quad \text{(relative intensity: 2)}, \\ E_{5\pm} &= E_1 \pm \frac{1}{2}\left(\alpha + \frac{\beta}{3}\right)\bar{x}N_0\langle S_z\rangle, \quad \text{(relative intensity: 2)}, \end{aligned} \tag{2.55}$$

where the first formula corresponds to the σ^+ and σ^- polarization transitions, and the second formula to the π polarization. All these transitions are shown in Fig. 2.6. Because different authors took different signs for $\langle S_z\rangle, \alpha$, and β, the signs of equations in different references are confused. Here we take $H_{ex} = xN_0J\sigma_z\langle S_z\rangle, \alpha < 0, \beta > 0$ and $\langle S_z\rangle < 0$ (Dai *et al.*, 1994), while early authors (for example, Gaj *et al.*, 1979; Gaj, 1988) took $H_{ex} = -xN_0J\sigma_z\langle S_z\rangle, \alpha > 0, \beta < 0$, and $\langle S_z\rangle < 0$.

Let $A = |xN_0J\alpha\langle S_z\rangle/6|, B = |xN_0J\beta\langle S_z\rangle/6|$, then from Eqs. (2.54) and (2.55) Gaj obtained the energies and relative intensities of free excitons in the DMSs of zincblende structure, and listed in Table 2.4 (Gaj, 1988).

The transition energies shown in Fig. 2.6 are generally measured with the optical transmission or reflection spectra. In spite of its advantage, for example, audio-visual, easy to identify and explain, the transmission technique is used less frequently, because it requires much more effect to make the sample very thin. The absorption coefficient in the excitonic region of

Table 2.4. Energies and relative strengths of free excitons in the DMS of zincblende structure.

Valence band	CB	Energy relative to the zero field	Polarization	Relative intensity
$\|3/2, -3/2\rangle$	$-1/2$	−3A−3B	σ^+	3
$\|3/2, -1/2\rangle$	$1/2$	3A–B	σ^+	1
$\|3/2, 1/2\rangle$	$-1/2$	−3A+B	σ^-	1
$\|3/2, 3/2\rangle$	$1/2$	3A+3B	σ^-	3
$\|1/2, -1/2\rangle$	$1/2$	3A+B	σ^+	2
$\|1/2, 1/2\rangle$	$-1/2$	−3A–B	σ^-	2
$\|3/2, 1/2\rangle$	$1/2$	3A+B	π	2
$\|3/2, -1/2\rangle$	$-1/2$	−3A–B	π	2
$\|1/2, 1/2\rangle$	$1/2$	3A–B	π	1
$\|1/2, -1/2\rangle$	$-1/2$	−3A+B	π	1

a direct gap semiconductor is of the order of 10^5–$10^6\,\text{cm}^{-1}$, samples with a thickness below 1 μm are necessary. Thus, the reflectivity measurements are commonly used in DMS to determine Zeeman splittings of the free exciton ground state. The intensity of the reflected light depends on the refractive index of semiconductor n, i.e. the real part of the dielectric function $\varepsilon_r(\omega)$. The imagine part of the dielectric function is the absorption coefficient $\alpha(\omega)$. The $\varepsilon_r(\omega)$ can be calculated from $\alpha(\omega)$ through the K–K relation. It is proved that the peaks in the reflection spectrum correspond to the peaks in the absorption spectrum, i.e. the transition energies.

Figure 2.7 is the exciton reflection spectra of $Cd_{0.98}Mn_{0.02}Te$ at 7.1 K and 1.1 T (Gaj *et al.*, 1978). From Fig. 2.7, we see that for three cases of polarizations, the peak positions of the reflection spectra are different. The transition energies are measured roughly from Fig. 2.7 as follows:

$$\begin{aligned}
\sigma^+: \quad & E_{1+} = E_0 - 3A - 3B = 1.612\,\text{eV},\\
& E_{2+} = E_0 + 3A - B = 1.624\,\text{eV},\\
\sigma^-: \quad & E_{1-} = E_0 + 3A + 3B = 1.637\,\text{eV},\\
& E_{2-} = E_0 - 3A + B = 1.625\,\text{eV},\\
\pi: \quad & E_{3+} = E_0 + 3A + B = 1.630\,\text{eV},\\
& E_{3-} = E_0 - 3A - B = 1.619\,\text{eV}.
\end{aligned}$$

From the above results, we obtain $E_0 = 1.625\,\text{eV}$, $A = 0.001\,\text{eV}$, and $B = 0.003\,\text{eV}$. Thus, $|N_0\alpha/N_0\beta| \approx 1/3$, the absolute values of $N_0\alpha$ and $N_0\beta$ are calculated from $\langle S_z\rangle$ of Eq. (2.36), knowing the magnetic field

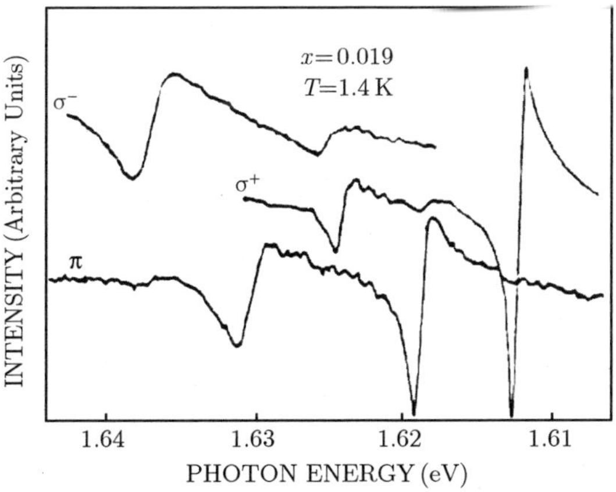

Fig. 2.7. Exciton reflection spectra of $Cd_{0.98}Mn_{0.02}Te$ at 7.1 K and 1.1 T.

Table 2.5. Experimentally measured exchange constants $N_0\alpha$ and $N_0\beta$ of various large gap DMSs.

DMS	Crystal structure	$N_0\alpha$(eV)	$N_0\beta$(eV)
$Cd_{1-x}Mn_xTe$	ZB	−0.22	0.88
$Zn_{1-x}Mn_xTe$	ZB	−0.19	1.09
$Zn_{1-x}Mn_xSe$	ZB	−0.29	1.40
$Cd_{1-x}Mn_xSe$	WZ	−0.23	1.27
$Cd_{1-x}Mn_xS$	WZ		−5.6

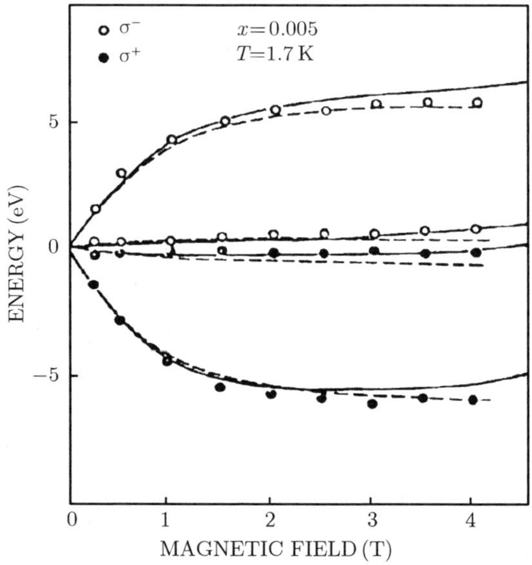

Fig. 2.8. $E_{1\pm}$ and $E_{2\pm}$ four transition energies of $Cd_{1-x}Mn_xTe$ at $x = 0.005$, $T = 1.7\,K$ versus magnetic field. Dots are experimental results and curves are theoretical results.

strength B, temperature T and T_0, and $\bar{x}$ (different from x). In this case, it is obtained $N_0\alpha = -0.22\,eV$, $N_0\beta = 0.88\,eV$ (Arciszewska and Nawrocki, 1982). Table 2.5 lists the experimentally measured exchange constants $N_0\alpha$ and $N_0\beta$ of various large gap DMSs (Gaj, 1988).

Figure 2.8 shows the $E_{1\pm}$ and $E_{2\pm}4$ transition energies (see Eq. (2.54)) of $Cd_{1-x}Mn_xTe$ at $x = 0.005$, $T = 1.7\,K$ versus magnetic field (Gaj *et al.*, 1979). The middle two curves correspond to the small intensity transitions $E_{2\pm}$. From Fig. 2.8, we see that the transition energies saturate when the magnetic field is large, similar to the form of the Brillouin function (Fig. 2.4).

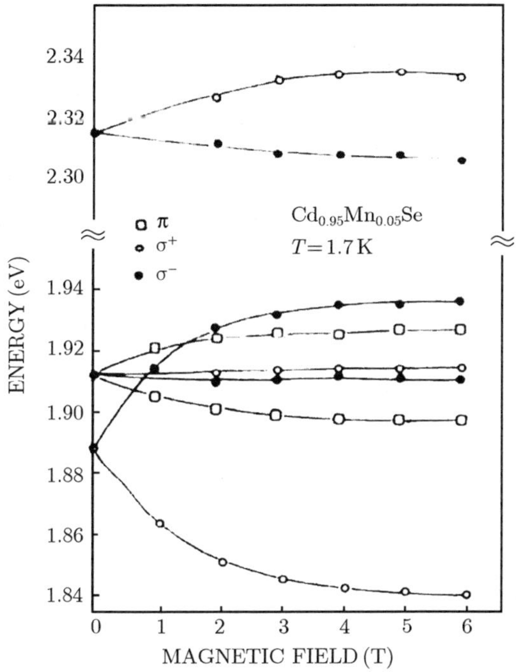

Fig. 2.9. Transition energies of wurtzite $Cd_{0.95}Mn_{0.05}Se$ at 1.7 K versus magnetic field in the Faraday configuration. Magnetic field is parallel to the crystal hexagonal axis.

Figure 2.9 shows the transition energies of wurtzite $Cd_{0.95}Mn_{0.05}Se$ at 1.7 K versus magnetic field in the Faraday configuration. Magnetic field is parallel to the crystal hexagonal axis (Arciszewska and Nawrocki, 1982). From Table 2.3, we obtain the spin–orbit splitting energy of CdSe $\Delta_{so} = 418\,\text{meV}$, and the crystal field splitting energy $\Delta_c = 39\,\text{meV}$, $\Delta_{so} \gg \Delta_c$. The bands of valence band top are shown in Fig. 2.5. The wave functions of bands are basically same as those of the zincblende semiconductors, for the Γ_9 level, wave functions are $|3/2, \pm 3/2\rangle$, for the Γ_7 level, wave functions are $|3/2, \pm 1/2\rangle$, and for the lowest Γ_7 level, wave functions are $|1/2, \pm 1/2\rangle$; thus, Table 2.4 is still suitable.

From Fig. 2.9, we see that the high energy transitions on the top are the transitions from the spin–orbit splitting band Γ_7 to the CB, the low energy transitions on the bottom are the transitions from the Γ_9 band to the CB, and the middle energy transitions are the transitions from the light hole band Γ_7 to the CB. The resulting splitting pattern and selection rules are

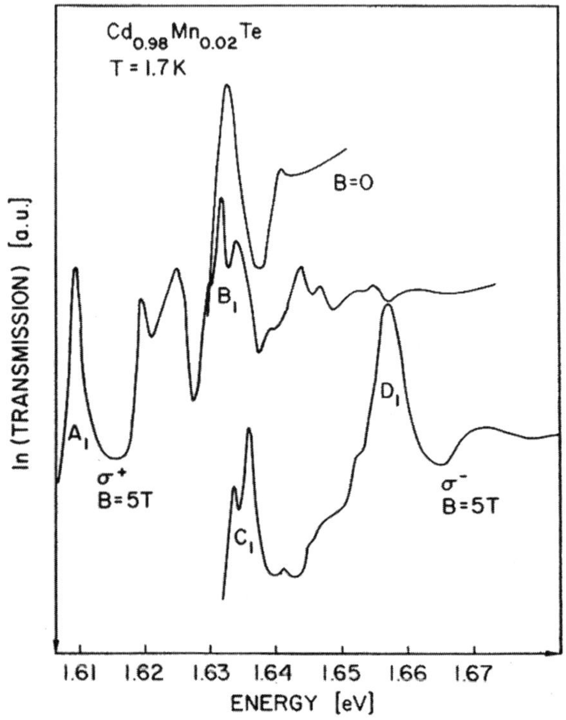

Fig. 2.10. Transmission spectra of $Cd_{0.98}Mn_{0.02}Te$ at 1.7 K, $B = 0$ and $B = 5$ T.

then analogous to those in cubic crystals, except for a displacement of the two strong σ components relative to the remaining ones due to the crystal axial anisotropy (compared with Fig. 2.8).

The reflection spectra can only give the transition energies of band ground states (Fig. 2.7), while the transmission spectra can give more details. Figure 2.10 is the transmission spectra of $Cd_{0.98}Mn_{0.02}Te$ at 1.7 K, $B = 0$ and $B = 5$ T (Gaj, 1988). From Fig. 2.10, we see that there are four transitions of polarizations σ^+ and σ^- (Table 2.4), their principal peaks are represented by A_1 (hole $m_J = -3/2$ to electron $s = -1/2$), B_1 ($m_J = -1/2$ to $s = 1/2$), C_1 ($m_J = 1/2$ to $s = -1/2$), and D_1 ($m_J = 3/2$ to $s = 1/2$), respectively. Besides the principal peaks, there are rich structures. Figure 2.11 is the absorption energies of $Cd_{0.95}Mn_{0.05}Te$ at $T = 1.64$ K versus magnetic field (Gaj, 1988). From Fig. 2.11, we see that above the principal components of the exchange split exciton ground states $A_1, \ldots, D_1$, a number of weaker lines are visible that can be denoted

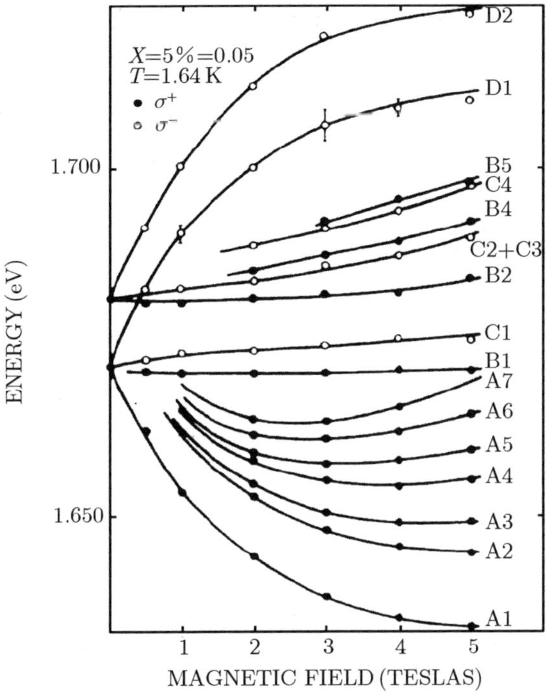

Fig. 2.11. Absorption energies of $Cd_{0.95}Mn_{0.05}Te$ at $T = 1.64\,K$ versus magnetic field.

$A_2, \ldots, A_6$, $B_2, \ldots, B_4$, etc. We have thus four sets of absorption lines A_i through D_i. As said in Sec. 2.2.1, in the magnetic field except the giant Zeeman splitting caused by the magnetic ions, each band can form a series of Landau levels, the spacing between Landau levels ΔE_{Landau} is smaller than the giant Zeeman splitting energy ΔE_{Zeeman}. The exciton state can be formed for each Landau level, so that many absorption peaks can be observed in the transmission spectra (Fig. 2.11). Each exciton state consists of a series of discrete energy levels, whose spacing $\Delta E_{\text{exciton}}$ is once more smaller than ΔE_{Landau}; thus,

$$\Delta E_{\text{Zeeman}} > \Delta E_{\text{Landau}} > \Delta E_{\text{exciton}}.$$

Due to the common effect of all these factors, many absorption peaks were observed in the transmission spectra. The authors assigned the A1, A2, A3 peaks in Fig. 2.11 to the three discrete exciton states of $n = 0$ Landau levels, and the A4, A5 peaks to the exciton states of $n = 2$ Landau level by the theoretical calculation (Twardowski and Ginter, 1982). As for the B, C, D series due to the overlap of all peaks it is difficult to assign them.

2.3. Narrow Bandgap DMSs

2.3.1. *Magnetic energy levels of narrow bandgap semiconductors*

In the above section, we discussed wide gap semiconductors. Because the energy gap between the valence band and the CB is large, and the interaction between them is weak, we can treat the CB and the valence band separately, and write their individual Hamiltonians.

The narrow gap semiconductors have narrow, even negative energy gaps, for example, InAs, InSb, HgTe, whose energy gaps are 0.418, 0.2355, and −0.3025 eV, respectively. Negative energy gap represents that the CB bottom (Γ_6 energy level) is located below the valence band top (Γ_8 energy level). In this case, the interaction between the CB and the valence band is large, resulting in large nonparabolic CB and valence bands (when k is larger, $E(k)$ is no longer $\sim k^2$); thus, the CB and the valence band should be considered together.

Assume that the main contribution to the electron effective-mass (Eq. (2.8)) and g-factor (Eq. (2.9)) comes from the valence band and that from other bands can be neglected. Insert the six valence band wave functions Eqs. (2.25) and (2.40) into Eqs. (2.8) and (2.9), we obtain

$$\frac{1}{2m^*} = \frac{1}{2m_0} + \frac{1}{m_0^2}\frac{3E_g + 2\Delta}{3E_g(E_g + \Delta)}|\langle S|p_x|X\rangle|^2, \tag{2.56}$$

$$g^* = g - \frac{4}{m_0}\frac{\Delta}{3E_g(E_g + \Delta)}|\langle S|p_x|X\rangle|^2. \tag{2.57}$$

From Eqs. (2.56) and (2.57), we see that because E_g is relatively small, the effective mass and the g-factor are much different from those of free electrons. For example, for the electron of InSb $m^* = 0.0136\, m_0$, $g^* = -51.31$.

In general case, we need to consider the CB and the valence band together and include the contributions to the effective-mass parameters from other bands. Take the Bloch functions of the CB bottom $|S\uparrow\rangle$, $|S\downarrow\rangle$ and the valence band top $|3/2, 3/2\rangle, |3/2, 1/2\rangle, |3/2, -1/2\rangle, |3/2, -3/2\rangle$, and $|1/2, 1/2\rangle$, $|1/2, -1/2\rangle$ as the basis functions, we write the eight bands effective-mass Hamiltonian (Efros and Rosen, 1998).

$$
H = \begin{vmatrix}
E_{\mathrm{g}} + \dfrac{\alpha p^2}{2m_0} & 0 & \dfrac{i}{\sqrt{2}} V p_+ & \sqrt{\dfrac{2}{3}} V p_z & \dfrac{i}{\sqrt{6}} V p_- & 0 & \dfrac{i}{\sqrt{3}} V p_z & \dfrac{1}{\sqrt{3}} V p_- \\
0 & E_{\mathrm{g}} + \dfrac{\alpha p^2}{2m_0} & 0 & -\dfrac{1}{\sqrt{6}} V p_+ & i\sqrt{\dfrac{2}{3}} V p_z & -\dfrac{1}{\sqrt{2}} V p_- & \dfrac{i}{\sqrt{3}} V p_+ & -\dfrac{1}{\sqrt{3}} V p_z \\
-\dfrac{i}{\sqrt{2}} V p_- & 0 & -(P+Q) & -L & -M & 0 & -\dfrac{i}{\sqrt{2}} L & i\sqrt{2} M \\
\sqrt{\dfrac{2}{3}} V p_z & -\dfrac{1}{\sqrt{6}} V p_- & -L^* & -(P-Q) & 0 & -M & i\sqrt{2} Q & -i\sqrt{\dfrac{3}{2}} L \\
-\dfrac{i}{\sqrt{6}} V p_+ & -i\sqrt{\dfrac{3}{2}} V p_z & -M^* & 0 & -(P-Q) & L & i\sqrt{\dfrac{3}{2}} L^* & i\sqrt{2} Q \\
0 & -\dfrac{1}{\sqrt{2}} V p_+ & 0 & -M^* & L^* & -(P+Q) & i\sqrt{2} M^* & \dfrac{i}{\sqrt{2}} L^* \\
-\dfrac{i}{\sqrt{3}} V p_z & -\dfrac{i}{\sqrt{3}} V p_- & \dfrac{i}{\sqrt{2}} L^* & -i\sqrt{2} Q & -i\sqrt{\dfrac{3}{2}} L & -i\sqrt{2} M & -\Delta - P & 0 \\
\dfrac{1}{\sqrt{3}} V p_+ & -\dfrac{1}{\sqrt{3}} V p_- & -i\sqrt{2} M^* & i\sqrt{\dfrac{3}{2}} L^* & -i\sqrt{2} Q & -\dfrac{i}{\sqrt{2}} L & 0 & -\Delta - P
\end{vmatrix}, \tag{2.58}
$$

where

$$\begin{aligned} V &= -\frac{i}{m_0}\langle S|p_z|Z\rangle, \quad p_\pm = p_x \pm ip_y, \\ P &= \frac{\gamma_1}{2m_0}p^2, \quad Q = \frac{\gamma}{2m_0}(p_x^2 + p_y^2 - 2p_z^2), \\ L &= -\frac{i\sqrt{3}\gamma}{m_0}p_z p_-, \quad M = \frac{\sqrt{3}\gamma}{2m_0}p_-^2. \end{aligned} \tag{2.59}$$

The electron momentum part in Eq. (2.58) is $\alpha p^2/2m_0$, where α is the electron effective-mass parameter. It equals the electron effective-mass m_e^* minus the contribution of the valence band, because the contribution of the valence band has been included in the non-diagonal terms of Eq. (2.58). Thus, from Eq. (2.56),

$$\begin{aligned} \alpha &= \frac{m_0}{m_\mathrm{e}^*} - \frac{E_p}{3}\left(\frac{2}{E_\mathrm{g}} + \frac{1}{E_\mathrm{g}+\Delta}\right) \\ E_p &= \frac{2}{m_0}|\langle S|P_x|X\rangle|^2 = 2m_0V^2. \end{aligned} \tag{2.60}$$

Similarly, the hole effective-mass parameters γ_1, γ should equal the Luttinger parameters γ_1^L, γ^L minus the contribution of the CB. Consider the heavy hole state, assume that the basis function is $|3/2, 3/2\rangle$, and the envelope function is e^{ikz}, we obtain its diagonal energy,

$$E_{3/2}(k) = -(P+Q) = -\frac{1}{2m_0}(\gamma_1 - 2\gamma)k_Z^2.$$

The interaction matrix element with the CB is $iVp_+/\sqrt{2}$ (see Eq. (2.58)), because $p_+e^{ikz} = 0$; thus,

$$\gamma_1 - 2\gamma = \gamma_1^L - 2\gamma^L. \tag{2.61}$$

Next we consider the light hole state, its basis function is $|3/2, 1/2\rangle$. Except the diagonal term, its energy also includes the second-order perturbation energy caused by the non-diagonal term,

$$E_{1/2}(k) = -\frac{1}{2m_0}(\gamma_1 + 2\gamma)k_z^2 - \frac{2V^2k_z^2}{3E_\mathrm{g}} = -\frac{1}{2m_0}(\gamma_1^\mathrm{L} + 2\gamma^\mathrm{L})k_z^2.$$

Thus,

$$\gamma_1 + 2\gamma + \frac{2E_p}{3E_\mathrm{g}} = \gamma_1^\mathrm{L} + 2\gamma^\mathrm{L}. \tag{2.62}$$

From Eqs. (2.61) and (2.62), we obtain

$$\gamma_1 = \gamma_1^{\mathrm{L}} - \frac{E_p}{3E_{\mathrm{g}}}, \quad \gamma = \gamma^{\mathrm{L}} - \frac{E_p}{6E_{\mathrm{g}}}. \tag{2.63}$$

Because we assumed that the CB and the valence bands are isotropic, we need only to calculate the energy band dispersion relation $E(k)$ in a special direction, for example, in the z-direction. From Eq. (2.58), we can obtain the secular equation of the energy bands. For the heavy hole band,

$$E_{\mathrm{hh}}(k) = -(\gamma_1 - 2\gamma)E_K, \quad E_K = \frac{k_z^2}{2m_0}. \tag{2.64}$$

There is coupling between the light hole band, spin–orbit splitting band, and the CB: the secular equation is

$$\begin{vmatrix} E_{\mathrm{g}} + \alpha E_K - E & \sqrt{\frac{2}{3}}Vk & \frac{i}{\sqrt{3}}Vk \\ \sqrt{\frac{2}{3}}Vk & -(\gamma_1 - 2\gamma)E_K - E & i\sqrt{2}(-2\gamma E_K) \\ -\frac{i}{\sqrt{3}}Vk & i\sqrt{2}(2\gamma E_K) & -\Delta - \gamma_1 E_K - E \end{vmatrix} = 0. \tag{2.65}$$

Expanding the determinant (2.65), we obtain

$$\begin{aligned} &[E - E_{\mathrm{g}} - \alpha E_K]\{[E + \gamma_1 E_K + \Delta][E + (\gamma_1 + 2\gamma)E_K] - 8(\gamma E_K)^2\} \\ &\quad - E_p E_K \left(E + \frac{2\Delta}{3}\right) - E_p(\gamma_1 - 2\gamma)E_K^2 = 0. \end{aligned} \tag{2.66}$$

If the contribution from remote bands to the valence band energy dispersion is neglected, then $\gamma_1 = 0, \gamma = 0$, Eq. (2.66) becomes

$$(E - E_{\mathrm{g}} - \alpha E_K)(E + \Delta)E - E_p E_K \left(E + \frac{2}{3}\Delta\right) = 0. \tag{2.67}$$

Equation (2.67) is a third-power algebra equation of E, has three roots, corresponding to the CB, light hole band, and the spin–orbit splitting band, respectively. The three bands calculated from Eq. (2.66) or (2.67) have large nonparabolicity, i.e. deviates from $\sim k^2$.

Just because the CB, light hole band, and the spin–orbit splitting band have large nonparabolicity, the Landau energy levels calculated by the numerical method are no longer equal spacing.

2.3.2. *Magnetic optical spectra of $Hg_{1-x}Mn_xTe$*

HgTe is a narrow gap semiconductor and has a negative energy gap. The schematic diagrams of the band structures of three narrow gap semiconductors InSb(a), HgTe(b), and HgSe(c) are shown in Fig. 2.12 (Bastard, 1979), where for InSb the CB bottom Γ_6 lies above the valence band top Γ_8, the energy gap ε_0 is positive; for HgTe and HgSe, the Γ_6 level lies below the Γ_8 level, the gap is negative. The energy gap and the effective-mass parameters of the negative gap semiconductor are all negative, for example, HgTe, $E_{\mathrm{g}} = -0.3025\,\mathrm{eV}$, $m_{\mathrm{e}}^* = -0.3\,\mathrm{m}_0$, $\gamma_1^L = -12.8, \gamma^L = -8.4$ so that the curvature of the CB is toward down and that of the valence band is toward up, as shown in Fig. 2.12(b).

Because the band structure of HgTe is different from that of normal semiconductors, its magnetic levels have special characteristics. Figure 2.13 shows the magnetic levels and three kinds of possible transitions σ^+, σ^- and π of HgTe at $k_H = 0$ (k_H is the wave vector along the magnetic field direction) (Guldner *et al.*, 1973), where the Landau levels are denoted by $0^+, 1^+, 2^+$ and $0^-, 1^-, 2^-, +$ and $-$ represent the spin directions. From Fig. 2.13, we see that the Landau levels of the CB and the valence band (light hole) upset in the energy direction.

From Eqs. (2.56) and (2.57), because $E_{\mathrm{g}} < 0, m_{\mathrm{e}}^* < 0$, $g^* > 0$, and $g^* \approx 30$. In Fig. 2.13, the energies of the electron Landau levels with spin +

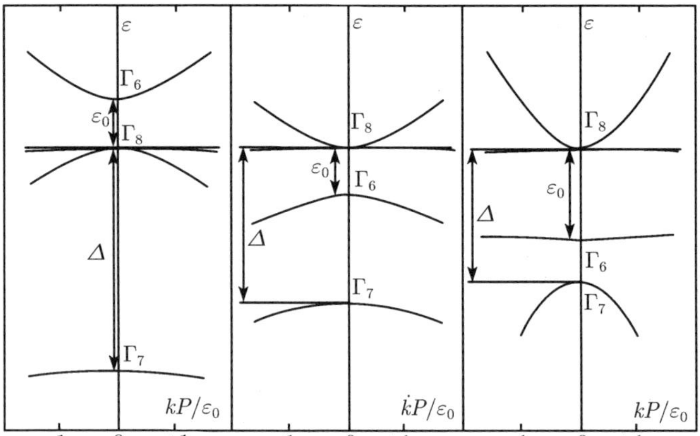

Fig. 2.12. Schematic diagrams of the band structures of three narrow gap semiconductors InSb(a), HgTe(b), and HgSe(c).

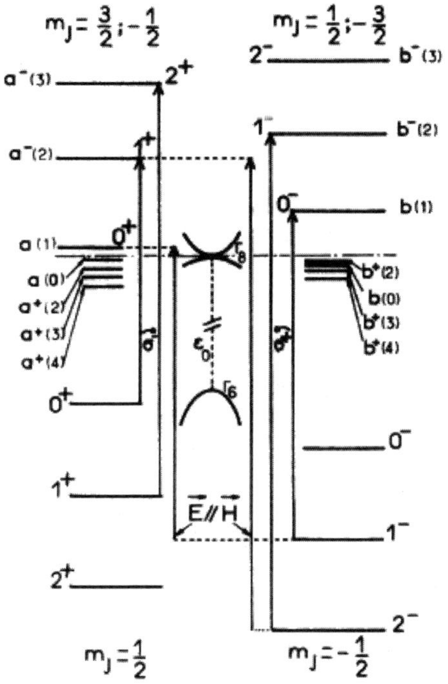

Fig. 2.13. Magnetic levels and three kinds of possible transitions σ^+, σ^- and π of HgTe at $k_H = 0$.

are higher than the energies of those with spin −. The hole Landau levels divide into two sets, $m_J = 3/2, -1/2$ and $m_J = 1/2, -3/2$, represented by $a^{\pm}(n)$ and $b^{\pm}(n)$, respectively, where $a(0)$ and $b(0)$ are ground states, other states are excited states, corresponding to the heavy hole (+) and light hole (−) states, respectively. The optical transition selection rules can be calculated from the wave functions of energy levels, as denoted in Fig. 2.13.

$$\begin{aligned} \sigma^+ &: \quad (n)^+ \to a^-(n), \quad (n)^- \to b^-(n), \\ \sigma^- &: \quad (n-2)^+ \to a^-(n), \quad (n-2)^- \to b^-(n), \\ \pi &: \quad (n)^- \to a^-(n), \quad (n-2)^+ \to b^-(n). \end{aligned} \tag{2.68}$$

Figure 2.14 is variations of electron effective masses m^* and g^* factors of $Hg_{1-x}Cd_xTe$ (×) and $Hg_{1-x}Mn_xTe$ (o, $T = 4.2$ K; •, $T = 2$ K) with the energy gap ε_0 (Bastard, 1979). When the doping concentration x increases, the $|\varepsilon_0|$ decreases; thus, the $|\varepsilon_0|$ reflects the variation of x. In the upper figure of Fig. 2.14, the electron effective masses m^* decrease as the $|\varepsilon_0|$

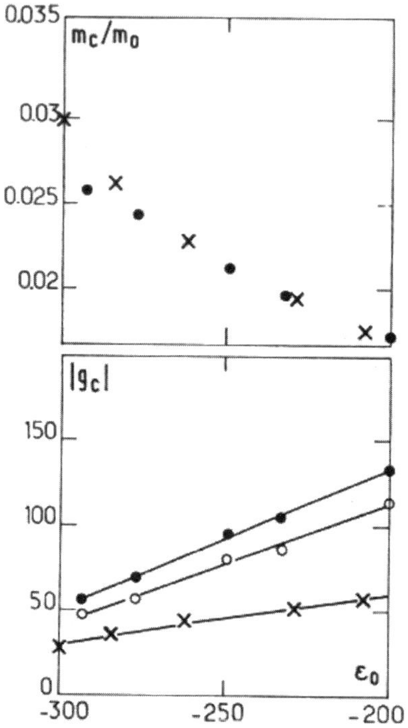

Fig. 2.14. Variations of electron effective masses m^* and g^*-factors of $Hg_{1-x}Cd_xTe$ ($\times$) and $Hg_{1-x}Mn_xTe$ (o, $T = 4.2\,K$; $\bullet$, $T = 2\,K$) with the energy gap ε_0.

decreases and are independent of doping Cd or Mn. In the below figure of Fig. 2.14, $|g^*|$ increase as $|\varepsilon_0|$ decreases, and the $|g^*|$-factor of $Hg_{1-x}Mn_xTe$ is apparently larger than that of $Hg_{1-x}Cd_xTe$, which is caused by the giant Zeeman splitting produced by the Mn ions. From Eq. (2.57), when $|E_g|$ decreases, $|g^*|$ increases. For electron, the giant Zeeman splitting energy is positive, added to the Zeeman splitting energy, resulting in an increase of $|g^*|$. Furthermore, the $|g^*|$ increases faster when the temperature is lower as expected from Eq. (2.36), the average magnetic moment $\langle S_z \rangle$ of paramagnetic ions.

Figure 2.15 is the fan chart of Γ_6–Γ_8 magneto-absorption energies of $Hg_{1-x}Mn_xTe$ ($x = 0.008$) at $T = 2\,K$ versus magnetic field (Bastard, 1979). In Fig. 2.15, the symbol 0^+1^+ denotes the transition from electron 0^+ level to hole $1^+[a^+(2)]$ level (Fig. 2.13); others analogize, and they all satisfy the transition selection rules (2.68). From Fig. 2.15, we see that the transition energies vary with the magnetic field strength basically according to the linear relation, but the opposite bending of σ^+ and σ^- lines

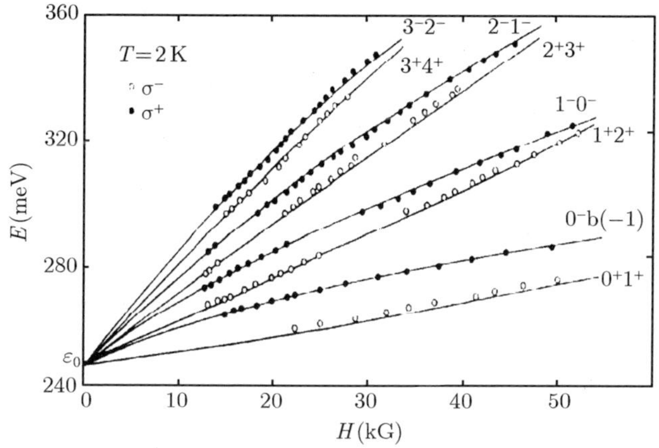

Fig. 2.15. Fan chart of Γ_6–Γ_8 magneto-absorption energies of $Hg_{1-x}Mn_xTe$ ($x = 0.008$) at $T = 2$ K versus magnetic field. Solid lines: theory; symbols: experiments.

reflects the behavior of $\langle S_z \rangle$. From the comparison between theory and experiments, one gets for $Hg_{1-x}Mn_xTe$, the magnetic exchange interaction energies $N_0\alpha = -0.75$ eV, $N_0\beta = 1.5$ eV.

2.4. Microstructures of DMSs

2.4.1. *DMS superlattices (Faraday configuration)*

Suppose a superlattice, where the potential well material is DMS, the barrier material is general semiconductor, and the magnetic field is perpendicular to the interface, i.e. parallel to the growth direction (Faraday configuration). The electron cyclotron movement caused by the magnetic field is in the nonconfined x–y plane of the superlattice, so that the results of the magnetic energy levels in bulk material discussed in Sec. 2.3 are still suitable in this case, only taking $k_z = 0$. Because there is a confined potential in the z-direction $V(z)$, the energy bands of the electron and hole in the z-direction split into discrete energy levels. For electrons, the movement equations in the z and x, y directions is independent, so that each discrete level has a series of Landau levels. For hole, because the non-diagonal terms of the hole effective-mass Hamiltonian (Eq. (2.17) or (2.24)) consist of $p_x p_z$ and $p_y p_z$ terms, the movements in the two directions are not independent. Strictly, the magnetic energy levels should be calculated by the numerical method. But for the ground state or low excited states, in the zero-order approximation one can treat the two movements as independent. The ground states of the heavy and light hole states in the superlattice have a series of Landau levels. In the experiments the σ^+ and σ^- polarized transition peaks can also be observed with the optical reflection or absorption spectra. The Zeeman splitting energies are the same as those in the bulk material, only the peak energies blue shift relative to those in the bulk material; the shifting energy equals the summation of the confined state energies of the electron and the hole in the quantum well. Thus, the meaning using DMS as the quantum well material is less.

The goal of early proposal of the DMS superlattice is to use the giant Zeeman splitting of electron and hole states with up and down spins in the magnetic field to modulate the band offset. Figure 2.16 is the schematic

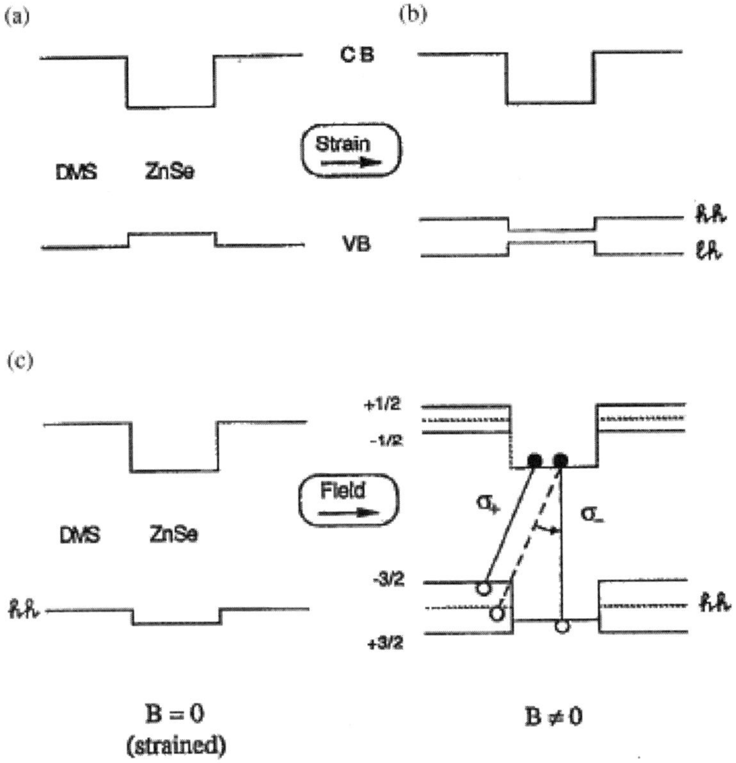

Fig. 2.16. Schematic diagram of energy bands of ZnSe/$Zn_{1-x}Mn_xSe$ superlattice grown on the GaAs substrate. (a) Band alignment at no strain and $B = 0$. (b) Strain splits heavy and light hole levels. (c) Heavy hole level defines VB edge. (d) When a magnetic field is applied, large splitting of DMS levels dominates the VB alignment.

diagram of energy bands of $ZnSe/Zn_{1-x}Mn_xSe$ superlattice grown on the GaAs substrate (Jonker *et al.*, 1991). From Fig. 2.16, we see that at zero magnetic field the superlattice is type II (Fig. 2.16(c)). The application of a magnetic field splits the band edges in both the barriers and the wells. Since the CB offset is large and the CB spin splitting relatively small, the electrons remain confined to the ZnSe well (Fig. 2.16(d)). The reverse is true in the valence band, where the offset is small and the exchange interaction and sequent spin splittings are four to six times larger than that in the CB. As a magnetic field is applied, the $m_J = -3/2$ hole level shifts up, so that the spin-down holes remain confined to the DMS barrier. However, the $m_J = +3/2$ level shifts down and overcomes the zero field offset, so that the spin-up holes become localized in the ZnSe wells. This behavior is called the spin-dependent I/II type transform of the superlattice caused by the magnetic field. From Fig. 2.16, the necessary condition for this transform is that the valence band offset ΔE_v is far smaller than the giant Zeeman splitting of the valence band.

Figure 2.17 is the magnetic field dependence of the heavy hole (hh) and light hole (lh) transitions for a $Zn_{0.91}Mn_{0.09}Se/ZnSe/Zn_{0.91}Mn_{0.09}Se/$ GaAs quantum well sample (Jonker *et al.*, 1991). The most striking feature of Fig. 2.17 is the pronounced asymmetry in the heavy hole spin splitting (hh), in contrast to the completely symmetric behavior exhibited by a uniform material (Fig. 2.8). The σ_+ component exhibits a strong field dependence similar to the corresponding transition in a bulk sample, the σ_- component exhibits relatively little field dependence, negligible compared with the giant Zeeman splitting of DMS materials. Furthermore, the σ_+ component is systematically weaker than the σ_- component, and the heavy hole intensity ratio σ_+/σ_- decreases as the well width increases. This can be explained from Fig. 2.16(d), the σ_+ exciton is the transition from the $m_J = -3/2$ hole state in $Zn_{1-x}Mn_xSe$ layer to the electron state in the ZnSe layer, because the energy of the $m_J = -3/2$ state increases with the magnetic field due to the giant Zeeman splitting, the exciton energy decreases. The σ_- exciton is the transition from the hole state in the ZnSe layer to the electron state in the same layer, and the transition energy changes a little equal to the Zeeman splitting. The observation that σ_+/σ_- is less than unity simply reflects the indirect versus direct nature of the transitions, as the electron and hole wave function overlap is much smaller for the spin-down holes localized in the barriers than for the spin-up holes localized in the wells. Increasing the well width further reduces this overlap for the spin-down holes, further reducing the intensity ratio σ_+/σ_-.

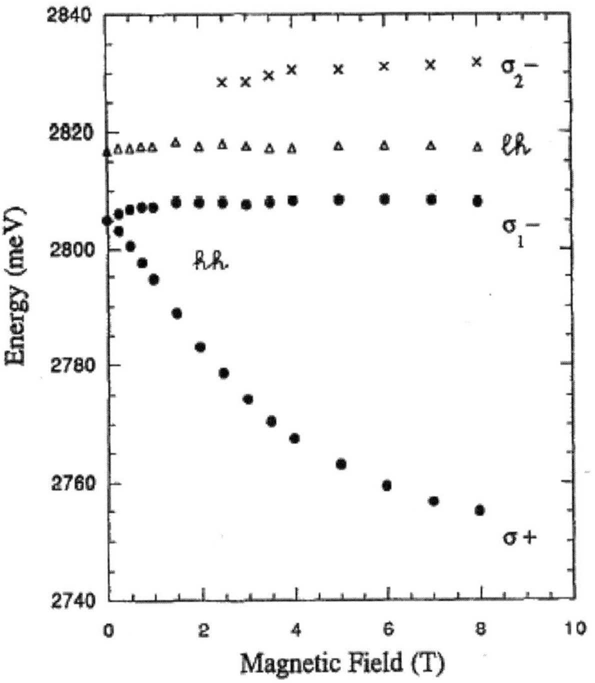

Fig. 2.17. Magnetic field dependence of the heavy hole (hh) and light hole (lh) transitions for a $Zn_{0.91}Mn_{0.09}Se/ZnSe/Zn_{0.91}Mn_{0.09}Se/GaAs$ quantum well sample.

This magnetic field-induced I/II types transform of the DMS superlattice is not general. Figure 2.18 is the exchange-induced splittings of the 1S and 2S excitons in $CdTe/Cd_{1-x}Mn_xTe$ ($x = 0.075$) multi-quantum wells in the Faraday configuration at 2 and 20 K (Jackson *et al.*, 1994). In Fig. 2.18, the asymmetry of the energy splittings of σ^+ and σ^- transitions is not apparent, their intensities differ a little. The splitting energies are much smaller than the giant Zeeman splitting energies, at $T = 2$ K, $B = 8$ T, the heavy hole 1S splitting energy is less than 10 meV. While in $Cd_{0.925}Mn_{0.075}Te$ the splitting energy of σ^+ and σ^- transitions should be 90 meV; thus, the excitons are produced in the CdTe well layer, not in the $Cd_{0.925}Mn_{0.075}Te$ barrier layer. The energy splitting is caused by the change of the band offset due to the giant Zeeman splitting; the barrier height of the $s = +1/2$ electron and $m_J = +3/2$ heavy hole states increases and that of the $s = -1/2$ electron and $m_J = -3/2$ heavy hole states decreases (Fig. 2.16(d)). Because of the difference of the confined potential, the transition energy of the σ^- exciton ($s = +1/2$ to $m_J = +3/2$)

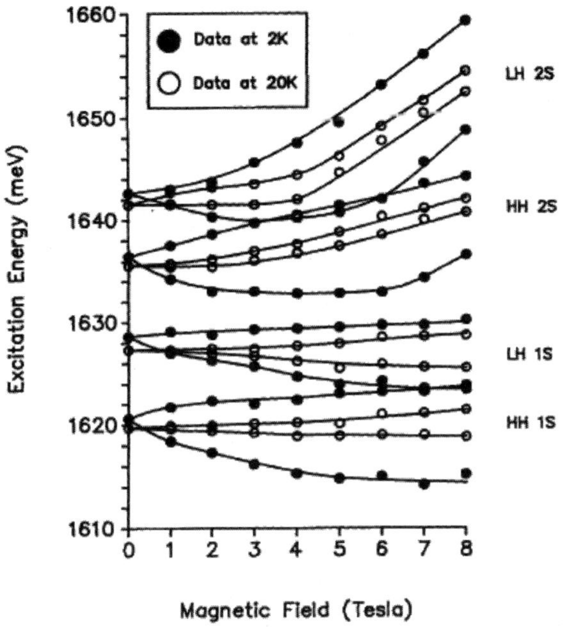

Fig. 2.18. Exchange-induced splittings of the 1S and 2S excitons in CdTe/Cd_{1-x} Mn_xTe ($x = 0.075$) multi-quantum wells in the Faraday configuration at 2 and 20 K.

increases and that of the σ^+ exciton ($s = -1/2$ to $m_J = -3/2$) decreases, resulting in splitting.

For the CdTe/$Cd_{0.925}Mn_{0.075}$Te superlattice, the total band offset $\Delta E_c + \Delta E_v = 120$ meV. At $T = 2$ K, $T_0 = 3.1$ K, $B = 8$ T, the $m_J = \pm 3/2$ heavy hole bands shift up and down 40 meV (taking $N_0\alpha = -0.22$ eV, $N_0\beta = 0.88$ eV). From Fig. 2.16(d), if $\Delta E_v < 40$ meV, the I/II type transform will occur; if $\Delta E_v > 40$ meV, then the superlattice will keep as the I type. From the above experimental results, it is found that the CdTe/$Cd_{0.925}Mn_{0.075}$Te superlattice is obviously different from the ZnSe/$Zn_{0.91}Mn_{0.09}$Se superlattice, and concluded that $\Delta E_v > 40$ meV, the I/II type transform has not occurred.

Define the partition ratio $P_v = \Delta E_v/(\Delta E_c + \Delta E_v)$, then we obtain $P_v > 0.35$. To check this result, one calculated the binding energies and the transition energies as functions of magnetic field of the $(e, h)^\pm$ excitons for the case of $P_v = 0.2$ (Jackson *et al.*, 1994), shown in Fig. 2.19. From Fig. 2.19, we see that at $B = 2$ T there is an abrupt change on the σ^+ curve, the binding energy falls rapidly and the transition energy also falls

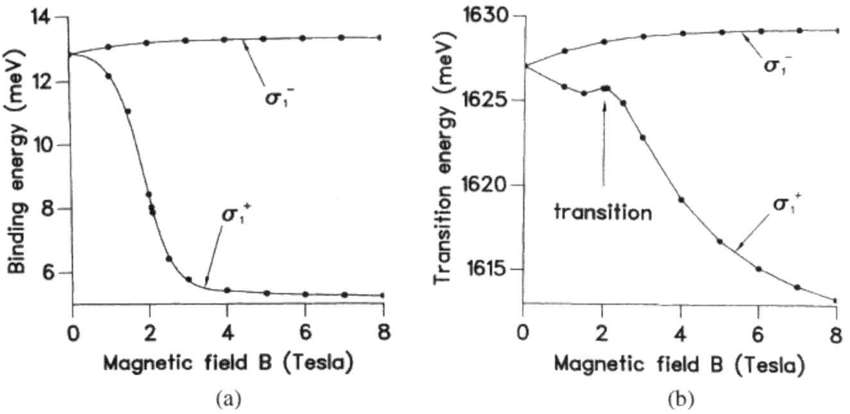

Fig. 2.19. (a) Binding energies. (b) Transition energies as functions of magnetic field of the $(e, h)^{\pm}$ excitons for $P_v = 0.2$.

rapidly. That means after $B > 2\,\mathrm{T}$, the superlattice transforms from I type to II type, where the electron and the hole are confined in different layers. This has not been observed in the experiments; thus, it is proved that $P_v > 0.35$.

2.4.2. *DMS superlattice (Voigt configuration)*

As mentioned in Sec. 2.4.2, the case of the Faraday configuration (magnetic field perpendicular to the interface) is rather simple; in the zero-order approximation, the cyclotron movement in the x–y plane caused by the magnetic field can be separated from the movement in the z-direction, i.e. the quantum confinement effect can be treated independently. The case of the Voigt configuration (magnetic field parallel to the interface) is more complicated, because the cyclotron movement and the confinement of the quantum well are all in the same direction, the z-direction. In this direction, there are either the confined potential of quantum well or the parabolic potential caused by the magnetic field, so that the theoretical analysis is more difficult.

Next we introduce the theory of DMS superlattice in the Voigt configuration (Wu *et al.*, 2002). For definition, we consider the ZnSe/$Zn_{1-x}Mn_xSe$ superlattice, whose energy bands are shown in Fig. 2.16. The hole effective-mass Hamiltonian is Eq. (2.24), where it is assumed that the superlattice growth direction is in the y-direction and the magnetic direction is in the z-direction. The gauge of the magnetic field is taken as $\boldsymbol{A} = (-By, 0, 0)$, and the electron Hamiltonian can be written as

$$H_e = \frac{1}{2m_e^*}[\beta^2(y-y_0)^2 + p_y^2 + p_z^2] + V_e(y), \tag{2.69}$$

where $\beta = eB/c$, the cyclotron center $y_0 = \hbar k_x/\beta$, the electron intrinsic Zeeman splitting is neglected because it is too small. From Eq. (2.69), we see that in the y-direction, except the periodic potential of the superlattice $V_e(y)$, there is a parabolic potential caused by the magnetic field. This so simple problem cannot be solved analytically, only be solved numerically.

The hole effective-mass Hamiltonian can be written as (Wu *et al.*, 2002),

$$H_h = \frac{1}{2m_0}\begin{vmatrix} P_1 + 3\hbar eB\kappa & Q & R & \\ Q^* & P_2 + \hbar eB\kappa & 0 & R \\ R^* & 0 & P_2 - \hbar eB\kappa & -Q \\ 0 & R^* & -Q^* & P_1 - 3\hbar eB\kappa \end{vmatrix} + H_s + V_h(y), \tag{2.70}$$

where

$$\begin{aligned} P_1 &= (\gamma_1+\gamma_2)[\beta^2(y-y_0)^2 + p_y^2] + (\gamma_1 - 2\gamma_2)p_z^2, \\ P_2 &= (\gamma_1-\gamma_2)[\beta^2(y-y_0)^2 + p_y^2] + (\gamma_1 + 2\gamma_2)p_z^2, \\ Q &= -i2\sqrt{3}\gamma_3[\beta(y-y_0)p_z - ip_yp_z], \\ R &= \sqrt{3}\gamma_2[\beta^2(y-y_0)^2 - p_y^2] - i2\gamma_3\left[\frac{\beta(y-y_0)p_y + i\hbar eB}{2}\right]. \end{aligned} \tag{2.71}$$

$V_{e,h}(y)$ is the periodic potential of superlattice for electron and hole, and the band offsets are determined by the strain and the giant Zeeman splitting of $Zn_{1-x}Mn_xSe$

$$\begin{aligned} \Delta V_{exch}^e &= -\bar{x}N_0\alpha\sigma_Z\langle S_Z\rangle, \\ \Delta V_{exch}^h &= -\frac{1}{3}\bar{x}N_0\beta\sigma_Z\langle S_Z\rangle. \end{aligned} \tag{2.72}$$

It is assumed that the ZnSe layers have no strain, and the $Zn_{1-x}Mn_xSe$ layers are subjected to a double axial compressive strain. Thus, in the DMS layers, the strain components $\varepsilon_{xx} = \varepsilon_{zz} = (a_1 - a_2)/a_2, \varepsilon_{yy} = -2\varepsilon_{zz}C_{12}/(C_{11} + C_{12})$, where a_1 and a_2 are the lattice constants of ZnSe and $Zn_{1-x}Mn_xSe$, C_{11} and C_{12} are elastic compliance constant. The strain

Hamiltonian is written as (Wu *et al.*, 2002),

$$H_s = -\frac{b_v}{2}(\varepsilon_{zz} - \varepsilon_{yy}) \begin{vmatrix} 1 & 0 & -\sqrt{3} & 0 \\ 0 & -1 & 0 & -\sqrt{3} \\ -\sqrt{3} & 0 & -1 & 0 \\ 0 & -\sqrt{3} & 0 & 1 \end{vmatrix}, \tag{2.73}$$

where b_v is the shear deformation potential constant of the valence band. The matrix (2.73) is not diagonal, not like the case of perpendicular magnetic field. In this case, for each m_J component there is no definite band offset ΔV_s. Only in the ZnSe layers, $H_s = 0$.

Because the electron and hole Hamiltonians all include parabolic potential terms $\beta^2(y - y_0)^2$, the electron envelope function in the y-direction is expanded by a series of sine functions defined in a hard-wall well with its width D much larger than the cyclotron–orbit radius,

$$\phi(\boldsymbol{r}) = e^{i(k_x x + k_Z z)} \cdot \sqrt{\frac{2}{D}} \sum_n C_n \sin\frac{n\pi y}{D}. \tag{2.74}$$

Using this expansion, we can calculate the needed energies and wave functions of ground and low excited states. The hole envelope function has four components, each component is expanded as Eq. (2.74).

Let $k_z = 0$, then the non-diagonal term in the hole Hamiltonian (2.70) $Q = 0$, and the Hamiltonian decouples into two independent 2×2 matrices, which can be solved separately. In this way, we obtain two sets of energy levels, denoted by the $m_J = (3/2, -1/2)$ and $m_J = (1/2, -3/2)$ hole states, respectively. Their envelope functions are

$$\Phi_h = \begin{pmatrix} a\phi_{3/2}\psi_{3/2} \\ b\phi_{-1/2}\psi_{-1/2} \end{pmatrix}, \quad \Phi_h = \begin{pmatrix} c\phi_{1/2}\psi_{1/2} \\ d\phi_{-3/2}\psi_{-3/2} \end{pmatrix}, \tag{2.75}$$

respectively, where $\psi_{3/2}, \psi_{1/2}, \psi_{-1/2}$, and $\psi_{-3/2}$ are the Bloch functions of the valence band top (Eq. (2.25)), $\phi_{3/2}, \phi_{1/2}, \phi_{-1/2}$, and $\phi_{-3/2}$ are the corresponding envelope functions Eq. (2.74). a, b, c, and d are the normalization constants, $a^2 + b^2 = 1, c^2 + d^2 = 1$. The squared optical transition element is

$$G = \frac{2}{m_0}|\boldsymbol{\varepsilon} \cdot \langle \Phi_e|\boldsymbol{p}|\Phi_h\rangle|^2, \tag{2.76}$$

where $\boldsymbol{\varepsilon}$ is the unit vector of the electric field direction of light wave. The $\sigma^{\pm}$ polarizations correspond to $\boldsymbol{\varepsilon} \cdot \boldsymbol{p} = (p_x \pm ip_y)/\sqrt{2}$, the π polarization corresponds to $\boldsymbol{\varepsilon} \cdot \boldsymbol{p} = p_z$. From Eqs. (2.25) and (2.75), one obtains the light transition intensities of various polarizations,

$$
\begin{aligned}
G(\sigma^+) &= \frac{2P^2}{m_0} d^2 |\langle \phi_e^{-1/2} | \phi_h^{-3/2} \rangle|^2, \\
G(\sigma^+) &= \frac{2P^2}{m_0} \frac{b^2}{3} |\langle \phi_e^{1/2} | \phi_h^{-1/2} \rangle|^2, \\
G(\sigma^-) &= \frac{2P^2}{m_0} \frac{c^2}{3} |\langle \phi_e^{-1/2} | \phi_h^{1/2} \rangle|^2, \\
G(\sigma^-) &= \frac{2P^2}{m_0} a^2 |\langle \phi_e^{1/2} | \phi_h^{3/2} \rangle|^2, \\
G(\pi) &= \frac{2P^2}{m_0} \frac{2c^2}{3} |\langle \phi_e^{1/2} | \phi_h^{1/2} \rangle|^2, \\
G(\pi) &= \frac{2P^2}{m_0} \frac{2b^2}{3} |\langle \phi_e^{-1/2} | \phi_h^{-1/2} \rangle|^2,
\end{aligned} \tag{2.77}
$$

where $P = \langle S | p_x | X \rangle$.

Figure 2.20 shows the different spin components of CB and valence-band edges in the DMS layer as functions of the magnetic field. Dashed lines indicate the ZnSe layer band edge (Wu *et al.*, 2002). The parameters in calculation are $x = 0.04, T = 1.5\,\text{K}$, $T_0 = 1.4\,\text{K}$. The strain-free band offset between the ZnSe and $Zn_{1-x}Mn_xSe$ layers is 30 meV, and the valence band partition ratio $P_v = 0.2$. The strain-free lattice constant is $a_1 = 5.666\,\text{Å}$ for ZnSe and $a_2 = (5.666 + 0.268x)\,\text{Å}$ for $Zn_{1-x}Mn_xSe$ $(x = 0.1)$. In zero

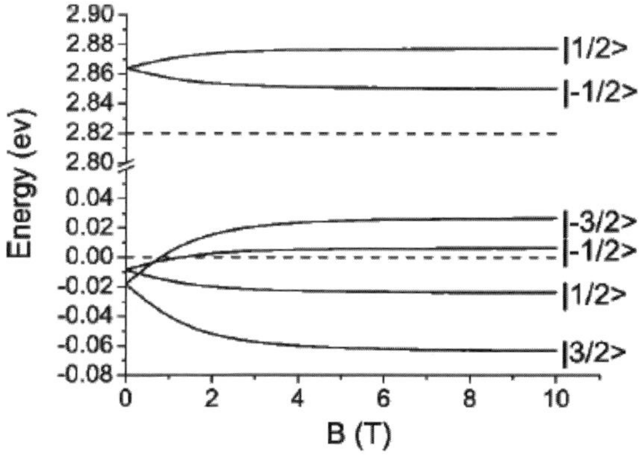

Fig. 2.20. Different spin components of CB and valence-band edges in the DMS layer as functions of the magnetic field. Dashed lines indicate the ZnSe layer band edge.

field, the hydrostatic strain pushes the CB offset from 24 to 44 meV, which is much larger than the saturation value (~13.5 meV) of the giant Zeeman shift for spin down electrons so that the DMS layer remains a potential barrier under any magnetic field strength. However, for the spin $-3/2$ and $-1/2$ components of the valence band, the DMS layer will transform from the potential barrier to potential well with increasing magnetic field larger than 1 T. The corresponding exciton state transforms from I type to II type. At B =0, the splitting between the $m_J = \pm 1/2$ and $m_J = \pm 3/2$ bands is caused by the strain.

Figure 2.21 shows the electron and hole magnetic energy bands versus y_0 at (a) $B = 0.5\,\mathrm{T}$ and (b) $B = 10\,\mathrm{T}$ (Wu *et al.*, 2002), where $y_0 = 0$ corresponds to the center of the DMS layer, and $y_0 = 0.5\,\mathrm{L}$ to the center of the ZnSe layer. In Fig. 2.21, the energy origin is taken at the bottoms of quantum wells (ZnSe layer), denoted by the dashed lines in Fig. 2.20. The negative hole energy is taken as positive, i.e. the coordinate axis is reversed. In the case of $B = 0.5\,\mathrm{T}$, the band alignment is of type I and the DMS layer is a potential barrier for both electrons and holes. The first several Landau subbands are almost dispersionless since the magnetic length l_B is much larger than the period of the DMS superlattice L. In the case of B =10T, however, the Landau subbands show an obvious dispersion. The energy of the hole ground state reaches a minimum when the cyclotron center y_0 locates at the origin of the y-axis (the center of the DMS layer). It is because when the magnetic field increases larger than 0.7 T, the DMS layer becomes a potential well for holes with components $J_z = -3/2$ and $-1/2$ (Fig. 2.20), so that a type-II band alignment is achieved, while the ZnSe layer is always the potential well for electrons. In Fig. 2.21, the solid lines denote the spin-down states for electron and the HH states for hole. The dashed lines denote the spin-up states for electron and the LH states for hole. The HH states are composed of the second set of wave functions in Eq. (2.75), and the main components of low excited states are $\phi_{\mathrm{h}}^{-3/2} (d \gg c)$. The LH states are composed of the first set of wave functions in Eq. (2.75), and the main components of low excited states are $\phi_{\mathrm{h}}^{-1/2} (b \gg a)$. The energies of the HH states are always lower than those of the LH states. The large energy splitting between the HH ground state and the LH ground state is caused by the giant Zeeman splitting in the DMS layer.

Figure 2.22 shows the electron and hole energy levels at $y_0 = 0$ versus magnetic field (Wu *et al.*, 2002). The variation of the electron energies with the magnetic field is similar to the general fan chart; the electron energy levels increase with increasing magnetic field, but the lower-energy levels

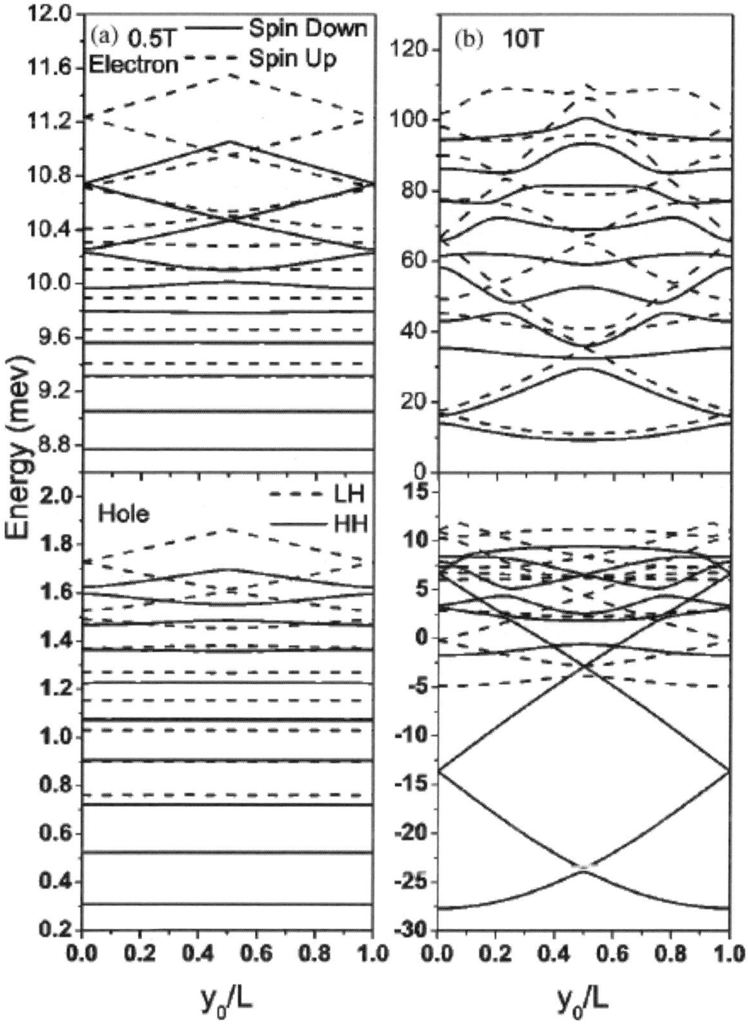

Fig. 2.21. Electron and hole magnetic energy bands versus y_0 at (a) $B = 0.5\,\mathrm{T}$ and (b) $B = 10\,\mathrm{T}$.

of the HH hole state decrease rapidly at small magnetic field. When the magnetic field increases the ground state energy saturates, and the higher excited states exhibit minima and strong anticrossing behavior. The level of the HH ($m_J = -3/2$) state decreases much more than that of the LH ($m_J = -1/2$) state, in agreement with Eq. (2.42) of the giant Zeeman splitting.

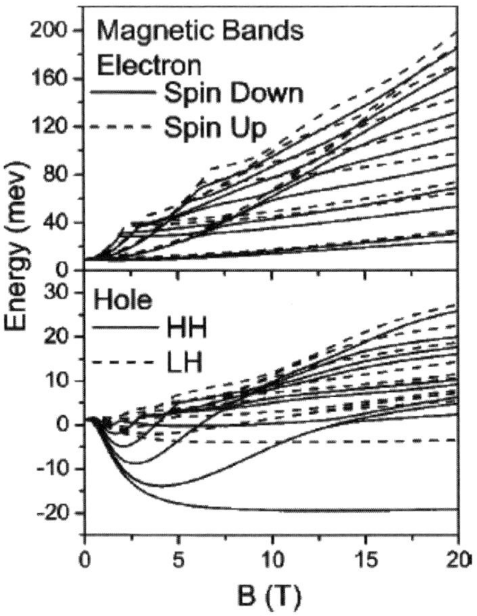

Fig. 2.22. Electron and hole energy levels at $y_0 = 0$ versus magnetic field.

Figure 2.23 shows the wave function distributions of the electron and hole ground states at different magnetic fields (Wu *et al.*, 2002). In the small magnetic field $B = 0.2\,\mathrm{T}$ the electron and HH, LH hole states are all located in the potential well layers over a lot of periods, which means that the cyclotron radius $L_B = \hbar/eB$ is much larger than the period L. When $B = 0.7\,\mathrm{T}$, the superlattice transforms from I type to II type, the maxima of the electron and hole wave functions are located in the well and barrier layers separately. The overlap of the electron and hole wave functions decreases. When the magnetic field is much large $B = 15\,\mathrm{T}$, the wave functions become more and more localized, but the overlap of the electron and hole wave functions increases.

2.4.3. *DMS quantum dots*

The first example of a DMS confined in all three dimensions (DMS QD) was reported in 1991 (Wang *et al.*, 1991). $Zn_{0.93}Mn_{0.07}S$ clusters of $\sim$25 Å diameter are successfully synthesized inside a glass matrix and fully characterized by chemical analysis, X-ray diffraction (XRD), extended X-ray absorption fine structure (EXAFS), and photoluminescence spectroscopy.

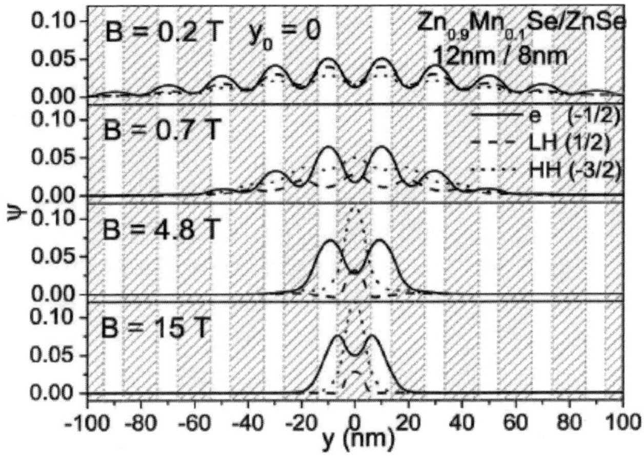

Fig. 2.23. Wave function distributions of the electron and hole ground states at different magnetic field. Shadowed area indicates the DMS layer.

The XRD experiment shows that it has a zincblende structure. The EXAFS experiment shows that the Mn–S is formed with a bond length of 2.45 Å and a coordination number of 4.6. This Mn–S bond length is slightly longer than the Zn–S bond length, 2.35 Å. The EXAFS results therefore support that Zn^{2+} is randomly substituted by Mn^{2+} in the ~25 Å $Zn_{0.93}Mn_{0.07}S$ clusters and the Mn^{2+} ions are located inside the clusters. The exciton peak of ZnS at wave length 346 mm (3.58 eV) and the luminescence peak at 594 nm (2.09 eV) due to the $^4T_1$–$^6A_1$ transition of Mn^{2+} ions are observed, and the exciton peak blue shifts 0.23 eV relative to the bulk material, caused by the quantum confinement effect.

Then it was found (Bhargava *et al.*, 1994) that the $Zn_{1-x}Mn_xS$ QDs with diameters of 27–33 Å yield an external photoluminescent quantum efficiency 18% and a luminescent decay at least five orders of magnitude faster than the corresponding Mn^{2+} transition in the bulk crystals, reaching 3.7 ns. This indicates that due to the quantum confinement effect the electron and hole wave functions are localized, which provide an effective path of energy transfer through the sp–d interaction, resulting in the high luminescent efficiency at RT.

The high-quality, highly fluorescent $Zn_{1-x}Mn_xSe$ QDs have been reported (Norris *et al.*, 2001). The luminescence, magnetic circular dichroism (MCD), and electron paramagnetic resonance (EPR) were measured to confirm that the Mn^{2+} ions are embedded inside the nanocrystal. Optical measurements show that the efficient emission from the Mn^{2+} ions is obtained, with a quantum yield of 22% at 295 K and 75% below 50 K. The EPR experiment measures the hyperfine interaction with the ^{55}Mn nuclear spin ($I = 5/2$). Since the hyperfine splitting is strongly dependent on the local environment, EPR spectra can be used to determine the location of the Mn. From Fig. 2.24(a), we see that EPR experiments exhibit a six-line spectrum with a hyperfine splitting of $60.4 \times 10^{-4}\,\text{cm}^{-1}$, consistent with the value ($61.7 \times 10^{-4}\,\text{cm}^{-1}$) for Mn substituted at Zn sites in the cubic ZnSe lattice. Thus, it is concluded that most of the Mn ions substrate for the Zn in the QDs.

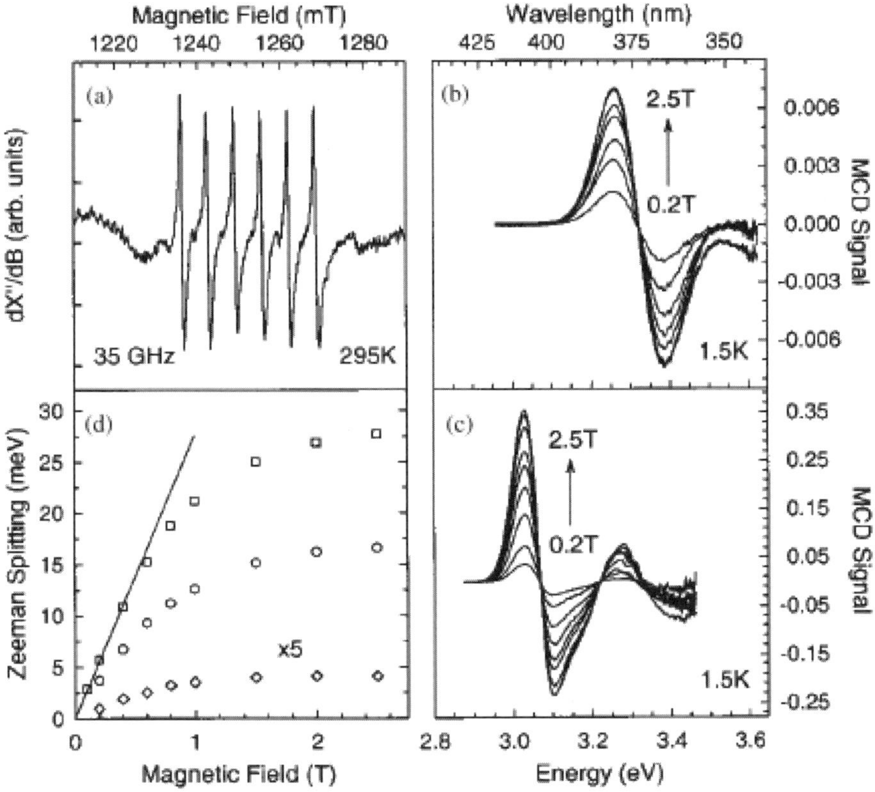

Fig. 2.24. (a) EPR spectrum of the $D = 29.0\,\text{Å}$, $C_1 = 1.0\%$ sample at 35 GHz and 295 K. (b) and (c) MCD spectra of $D = 29.0\,\text{Å}$, $C_1 = 1.0\%$ and $D = 53.0\,\text{Å}$, $C_1 = 2.5\%$ samples versus magnetic field from 0.2 to 2.5 T at 1.5 K. (d) Zeeman splitting energies versus magnetic field.

The experimental results of the MCD have been reported first successfully (Norris *et al.*, 2001). The MCD experiment measures the ratio $[I^+(E)-I^-(E)]/[I^+(E)+I^-(E)]$ when an excited light along the magnetic field direction irradiates on the QD, $I^+(E)$ and $I^-(E)$ are the energy-dependent transmitted intensities of right and left polarized beams, respectively. Figures 2.24 (b) and 2.24(c) are the MCD spectra of $D = 29.0$ Å, $C_1 = 1.0\%$ and $D = 53.0$ Å, $C_1 = 2.5\%$ samples versus magnetic field from 0.2 to 2.5 T at 1.5 K. From the peak positions, we can obtain the energies of absorption maxima E^+ and E^-, then the relation of the Zeeman splitting energy and the magnetic field. The results are shown in Fig. 2.24(d); the three curves from up to down correspond to $D = 53$ Å, $C_1 = 2.5\%$; $D = 42$ Å, $C_1 = 2.5\%$; $D = 29$ Å, $C_1 = 1.0\%$; respectively. From Fig. 2.24(d), we see that the variations of the Zeeman splitting energies with the magnetic field are similar to those of bulk materials (for example, Fig. 2.8); in the small magnetic field, the splitting energy is proportional to the magnetic field, when the magnetic field is large, it saturates. From the slope in the figure, we obtain $g_{\text{eff}} = 475$, while for the undoped QD $g_{\text{eff}} = -1.1$.

Next we consider the giant Zeeman splitting effect in QDs caused by the sp–d interaction (Zhu and Xia, 2006). The calculation of the magnetic energy levels in QDs is described in Sec. 10.1; for the DMS QDs, we should take into account the exchange interaction terms of electron Eq. (2.37) and of hole Eqs. (2.38) and (2.39).

We calculated the electron structures and the giant Zeeman splittings of $Zn_{1-x}Mn_xSe$ quantum spheres, taking the parameters $N_0\alpha = -0.27$ eV, $N_0\beta = 0.9$ eV, $T_0 = 1.4$ K, $T = 1.5$ K. Figures 2.25 and 2.26 show the electron and hole energy levels of undoped ZnSe and $Zn_{1-x}Mn_xSe$ quantum spheres ($\bar{x} = 0.018$) versus radius at $B = 2$ T, respectively (Zhu and Xia, 2006). Comparing Figs. 2.25 and 2.26, we found that in the same conditions the level splitting of the $Zn_{1-x}Mn_xSe$ sphere is much larger than that of the ZnSe sphere. It is noticed that the energy unit is $\varepsilon_0 = (\hbar/R)^2/2m_0$ in the figures, so that when the radius R is larger, the energy value in units of ε_0 is larger. Figure 2.27 shows hole S state energy levels in Fig. 2.26 in units of meV versus radius. From Fig. 2.27, we see that as the radius increases the hole levels approach the valence band top ($E = 0$). The splitting energies versus radius are shown in the inset of Fig. 2.27, which do not change much with the radius, only decreases slightly, and are not equal. This is different from the giant Zeeman spltting in the bulk material, where the splitting energies of the valence band are equal. From Eq. (2.42), we

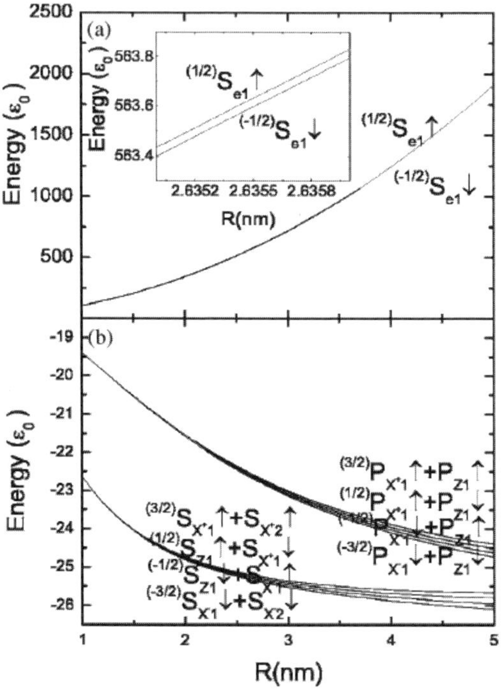

Fig. 2.25. Electron and hole energy levels of undoped ZnSe quantum spheres versus radius at $B = 2\,\mathrm{T}$. Inset is the enlarged energies of electron states of ${}^{1/2}\mathrm{S}_{\mathrm{e}1} \uparrow$ and ${}^{-1/2}\mathrm{S}_{\mathrm{e}1} \downarrow$ ranging from $R = 2.635$ to $2.636\,\mathrm{nm}$.

calculated the splitting energy of the valence band in $\mathrm{Zn}_{1-x}\mathrm{Mn}_x\mathrm{Se}$, and obtained $\Delta E_{\mathrm{v}} = 10.1\,\mathrm{meV}$.

Figure 2.28 shows the theoretical calculated energy difference of transitions from electron $s_z = +1/2$ to hole $m_J = -1/2$ and from $s_z = -1/2$ to $m_J = -3/2$ states versus magnetic field B for three $\mathrm{Zn}_{1-x}\mathrm{Mn}_x\mathrm{Se}$ quantum spheres (Zhu and Xia, 2006), where the dots are experimental results (Norris *et al.*, 2001). In general, the effective Mn ion concentration $\bar{x}$ is different from the real concentration x, since besides the Mn ions substituting the cation ions, they may occupy other positions, or form the cluster, which has no contribution to the average magnetic moment $\langle S_z \rangle$. $\bar{x}$ is generally smaller than x, their difference depends on the condition of the material growth. In Fig. 2.28, the three curves correspond to $\bar{x} = 0.032, 0.019$, and 0.004, while the Mn concentrations in the growth process are $x = 0.025$, 0.025, and

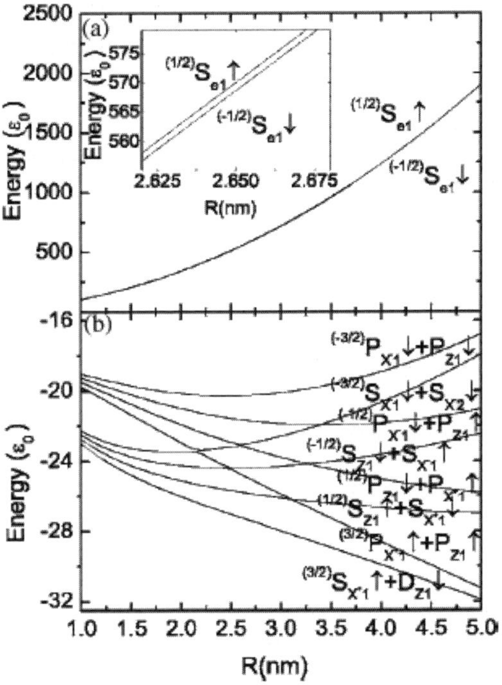

Fig. 2.26. Electron and hole energy levels of $Zn_{1-x}Mn_xSe$ quantum spheres ($\bar{x} = 0.018$) versus radius at $B = 2\,T$.

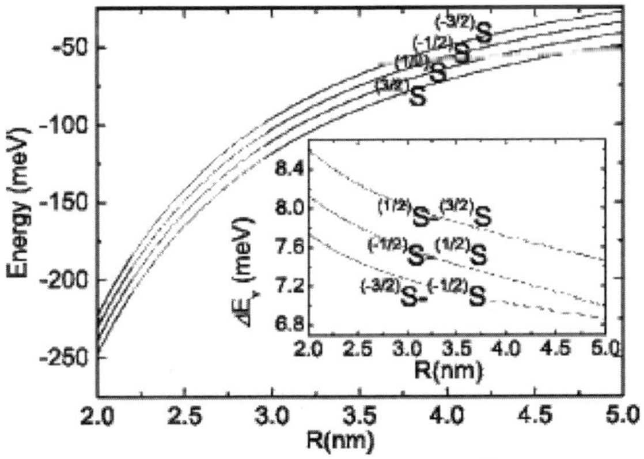

Fig. 2.27. Hole S state energy levels in Fig. 2.26 in units of meV versus radius. Inset shows the hole splitting energies of the four levels versus radius.

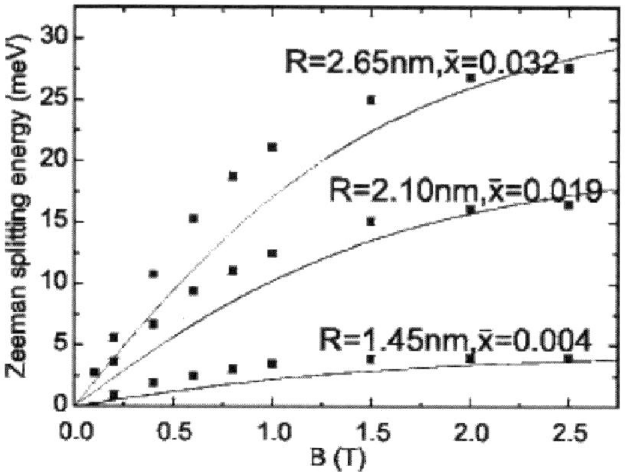

Fig. 2.28. Theoretical calculated energy difference of transitions from electron $s_z = +1/2$ to hole $m_J = -1/2$ and from $s_z = -1/2$ to $m_J = -3/2$ states versus magnetic field B for three $Zn_{1-x}Mn_xSe$ quantum spheres.

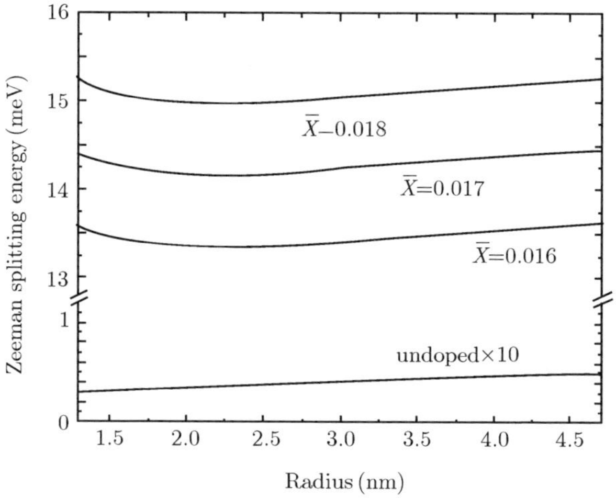

Fig. 2.29. Energy differences of two transitions $(s_z, m_J) = (1/2, -1/2)$ to $(-1/2, -3/2)$ versus radius at $B = 2\,\mathrm{T}$ for different $\bar{x}$.

0.01, respectively. According to the theoretical calculations, the giant Zeeman splitting energies do not change basically with the radius, determined mainly by $\bar{x}$.

Figure 2.29 shows the energy differences of two transitions $(s_z, m_J) = (1/2, -1/2)$ to $(-1/2, -3/2)$ versus radius at $B = 2\,\mathrm{T}$ for different $\bar{x}$ (Zhu and Xia, 2006). From Fig. 2.29, we see that the giant Zeeman splitting energies are determined mainly by $\bar{x}$ and have less relation with the radius. The lowest curve is for undoped QD ($\times 10$), so that in the case of $R = 2.1\,\text{nm}$, $\bar{x} = 0.018$, the giant Zeeman splitting energy is 14.9785 meV, about 45 times of 0.0337 meV in the undoped case.

2.4.4. *MP effect*

It is proposed that due to the strong localization of electron and hole wave functions in the QDs, there is a short-range touching interaction between the electron, hole, and the magnetic ions (Bhattacharjee and Guillaume, 1997),

$$H_{\mathrm{e,h}} = \frac{\alpha_{\mathrm{e,h}}}{2}\frac{V_0}{4}(\boldsymbol{\sigma}\cdot\boldsymbol{S})\delta(\boldsymbol{r}_{\mathrm{e,h}} - \boldsymbol{R}), \tag{2.78}$$

where $\boldsymbol{S}$ is the spin of the Mn ion at $\boldsymbol{R}(S = 5/2)$, $\boldsymbol{\sigma}$ is the Pauli matrix of the electron spin, $\alpha_{\mathrm{e,h}}$ is the electron (or hole) exchange integration, and V_0 is the unit cell volume. This interaction can induce an effective magnetic field parallel to the magnetic moment of the Mn ion, and the magnetic field depends on the position of the Mn ion in the QD $\boldsymbol{R}$. For the electron, hole–ion interaction,

$$B_{\mathrm{MP}}(r) = \frac{1}{g_{\mathrm{Mn}}\mu_{\mathrm{B}}}\left[\alpha s_{\mathrm{z}}|\phi_{\mathrm{e}}(r)|^2 + \frac{\beta}{3}J_{\mathrm{z}}|\psi_{\mathrm{h}}(r)|^2\right], \tag{2.79}$$

where $\phi_{\mathrm{e}}(r)$ and $\psi_{\mathrm{h}}(r)$ are electron and hole radial wave functions, respectively. Through the short-range touching interaction, one kind of composite particle of electron, hole, and Mn ions with same spin orientation can be formed, which is called magnetic polaron (MP). Bhattacharjee and Guillaume (1997) gave a set of electron and hole coupling radial equations of electron and hole including the Zeeman energy caused by $B_{\mathrm{MP}}(r)$ (Eq. (2.79)). They solved the set of equations for the MP ground state exactly through an iterative numerical procedure. The resulting self-consistent solution $\{\phi_{\mathrm{e}}(r), \psi_{\mathrm{h}}(r)\}$ shows an orbital contraction due to e–h Coulomb attraction and MP formation. The polaron binding energy and the total spin are calculated as functions of temperature for different QD radii. In order to compute the polaron energy in the presence of an external magnetic field (B), they replaced B_{MP} by $B + B_{\mathrm{MP}}$ in the equations. The true polaron binding energy in B is then obtained after subtracting off the Zeeman shift of the e–h excitation energy: $E_{\mathrm{Z}} = 1/2(\alpha - \rho\beta)M(B)/(g_{\mathrm{Mn}}\mu_{\mathrm{B}})$ from E_{p} given by

$$E_p = \int_0^R 4\pi r^2 \mathrm{d}r B_{\mathrm{MP}}(r) M(B_{\mathrm{MP}}(r)). \tag{2.80}$$

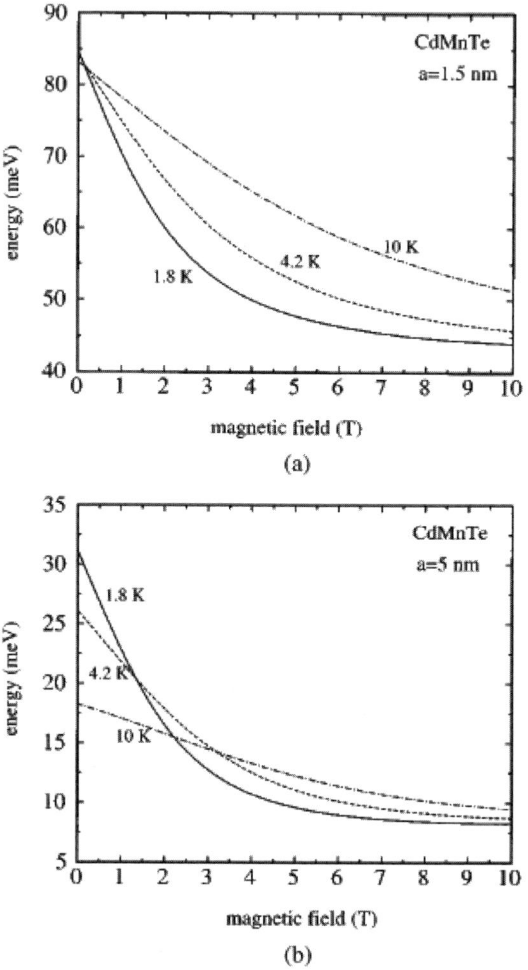

Fig. 2.30. Polaron binding energy versus magnetic field in a $Cd_{0.89}Mn_{0.11}Te$ QD of radius 1.5 nm (a) and 5 nm (b) at different temperatures: $T = 1.8, 4.2$, and 10 K.

Figure 2.30 shows the polaron binding energy versus magnetic field in a $Cd_{0.89}Mn_{0.11}Te$ QD of radius 1.5 nm (a) and 5 nm (b) at different temperatures: $T = 1.8, 4.2$, and 10 K. From Fig. 2.30, we see that the polaron energies decrease with increasing B, and the decrease is faster at lower temperatures. In small size QD (Fig. 2.30(a)), there remains a substantial binding energy even at 10 T, because the polaronic effective field in the core region is much larger than B and can orient Mn spins

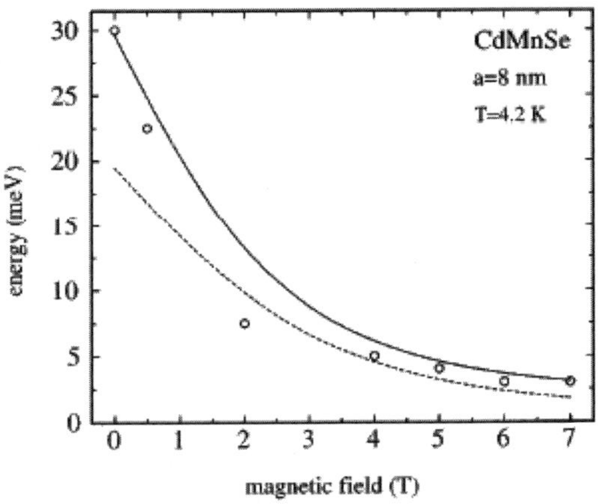

Fig. 2.31. Polaron binding energy versus magnetic field in a $Cd_{0.9}Mn_{0.1}Te$ QD of radius 8 nm at $T = 4.2\,K$.

further. On the other hand, in large dots (Fig. 2.30(b)), the binding energy tends to decrease faster toward a small value.

Figure 2.31 shows the polaron binding energy versus magnetic field in a $Cd_{0.9}Mn_{0.1}Te$ QD of radius 8 nm at $T = 4.2\,K$ (Bhattacharjee and Guillaume, 1997). The solid line is the result of the exact theoretical calculation, the dashed line is the result of a simplified model, and the circles represent the experimental results of photoluminescence Stokes shift (Yanata *et al.*, 1993). The exact solution yields a significantly better agreement with experiment.

Quasi-zero-dimensional magnetic polarons in single $Cd_{0.93}Mn_{0.07}Te/Cd_{0.6}Mg_{0.4}Te$ QDs have been studied by photoluminescence spectroscopy (Maksimov *et al.*, 2000). The structures consists of three monolayers of $Cd_{0.93}Mn_{0.07}Te$ embedded in nonmagnetic $Cd_{0.6}Mg_{0.4}Te$ barriers grown by MBE. To achieve the high spatial resolution required for measuring single QDs, Al masks with apertures down to 150 nm in diameter have been prepared by electron beam lithography and lift-off technique. The obtained DMS single QD has the shape of a disc. Figure 2.32 shows the magnetic field dependence of the $Cd_{0.93}Mn_{0.07}Te$ single QD photoluminescence (PL) peak energy for σ^+ and σ^- polarizations, respectively (Maksimov *et al.*, 2000). The solid and dashed lines are the theoretical results. From Fig. 2.32, we see that the variation of the PL peak energies near $B = 0$ does not like the case of general giant Zeeman splitting (Figs. 2.8 and 2.11); it seems that there is an intrinsic magnetic field B_{MP} as shown in Eq. (2.79). In

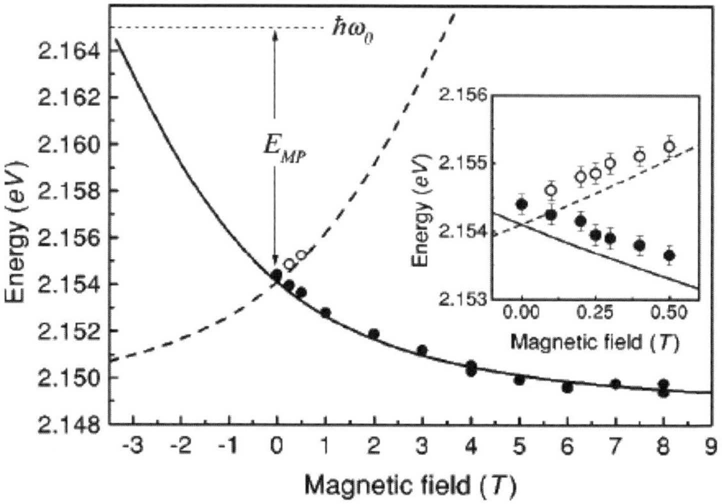

Fig. 2.32. Magnetic field dependence of the $Cd_{0.93}Mn_{0.07}Te$ SQD PL peak energy for σ^+ (solid points) and σ^- (open points) polarizations. Solid and dashed lines are theoretical results, and inset is the magnified representation in the range of low magnetic field.

order to explain the magnetic field-dependent PL spectra of DMS single QDs, the formation of MP due to an alignment of magnetic ion spins in the strong exchange field of the band carriers has to be taken into account. From Fig. 2.32, it is estimated that $B_{\mathrm{MP}} \approx 3.5\,\mathrm{T}$.

Chang and Xia (2002) considered a DMS disc-like QD embedded in the nonmagnetic semiconductor material. The MP Hamiltonian can be written as

$$\begin{aligned} H_{\mathrm{MP}} = {} & \frac{1}{2m_{\mathrm{e}}^*}\left[\boldsymbol{p}_{\mathrm{e}} + \frac{e\boldsymbol{A}(\boldsymbol{r}_{\mathrm{e}})}{c}\right]^2 + V_{\mathrm{e}}(\rho, z) + g_{\mathrm{e}}^*\mu_{\mathrm{B}}B \\ & + \frac{1}{2m_{\mathrm{h}}^*}\left[\boldsymbol{p}_{\mathrm{h}} - \frac{e\boldsymbol{A}(\boldsymbol{r}_{\mathrm{h}})}{c}\right]^2 + V_{\mathrm{h}}(\rho, z) + g_{\mathrm{h}}^*\mu_{\mathrm{B}}B + V_{\mathrm{exch}} - \frac{e^2}{\varepsilon|\boldsymbol{r}_{\mathrm{e}} - \boldsymbol{r}_{\mathrm{h}}|}, \end{aligned} \tag{2.81}$$

where it is assumed that the valence band is a simple parabolic band with the effective mass m_{h}^*. The exchange potential,

$$V_{\mathrm{exch}} = V_{\mathrm{exch}}^{\mathrm{e}} + V_{\mathrm{exch}}^{\mathrm{h}} = \bar{x}N_0\alpha s_z\langle S_z\rangle + \frac{1}{3}\bar{x}N_0\beta m_J\langle S_z\rangle. \tag{2.82}$$

The magnetic field in the QD $B = B_{\mathrm{MP}} + B_{\mathrm{ext}}$, B_{ext} is the applied magnetic field, and B_{MP} is given in Eq. (2.80).

For the σ^+ and σ^- polarized magnetic polarons, the transition energies

$$E = E_g(T) + E_{MP}^{\pm}, \tag{2.83}$$

where E_g is the energy gap of the corresponding bulk material,

$$E_g(T) = E_g(0) - \frac{aT^2}{b+T}, \tag{2.84}$$

which is temperature-dependent.

Figure 2.33 shows the MP transition energy versus magnetic field in the $Cd_{0.93}Mn_{0.07}Te/Cd_{0.6}Mg_{0.4}Te$ DMS QD for both σ^+ and σ^- transitions at $T = 2\,K$, $\bar{x} = 0.045$ (Chang and Xia, 2002). From Fig. 2.33, we see that the theoretical calculation is in good agreement with the experimental results (Maksimov *et al.*, 2000). Here the QD radius $R = 4\,nm$ is used as a fitting parameter due to the lack of more detailed description of the experimental samples in (Maksimov *et al.*, 2000). With increasing magnetic field, the energy of the $\sigma^+(\sigma^-)$ MP decreases (increases) rapidly for small magnetic field and saturates for strong magnetic field. Notice that the energy dependence for the MP changes significantly with changing QD radius (see the inset of Fig. 2.33). This is because the total magnetic field B

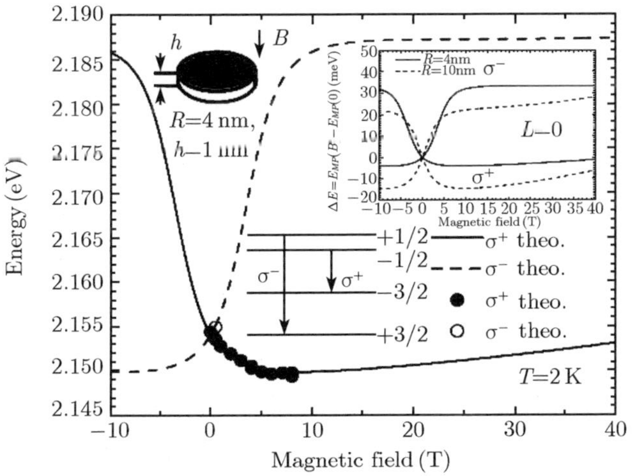

Fig. 2.33. MP transition energy versus magnetic field in the $Cd_{0.93}Mn_{0.07}Te/Cd_{0.6}Mg_{0.4}Te$ DMS QD for both σ^+ and σ^- transitions at $T = 2\,K$, $\bar{x} = 0.045$. The radius of the QD $R = 4\,nm$ and the height $h = 1\,nm$. The inset shows the MP transition energy versus magnetic field for different QD radii.

is determined by the external magnetic field B_{ext} and the carrier-induced magnetic field B_{MP} (see Eq. (2.80)). The carrier-induced magnetic field decreases with increasing QD radius through the electron and hole wave functions.

It is found experimentally that when $B_{\text{ext}} = 0$, the σ^+ MP transition energy increases first with increasing temperature, reaches a maximum, then decreases when the temperature is higher than a certain value (Maksimov *et al.*, 2000). The calculated MP transition energy as a function of the temperature is shown in Fig. 2.34 by the solid line. This behavior can be understood by the temperature dependence of the exchange interaction determined by the average spin of the magnetic ion $\langle S_z \rangle$. The experimental results may also arise from the shrinkage of the band gap with increasing temperature. When the temperature dependence of the energy gap Eq. (2.84) is taken into account, the theoretical results (dot-dashed line in Fig. 2.34) is in agreement with the experimental results. From the inset in Fig. 2.34, we see that the behavior of the σ^- transition is qualitatively different from that of the σ^+ transition especially at low temperatures. This phenomenon arises from the fact that both the sp–d exchange interaction for the σ^- transition and the band gap shrinkage decrease with increasing temperature.

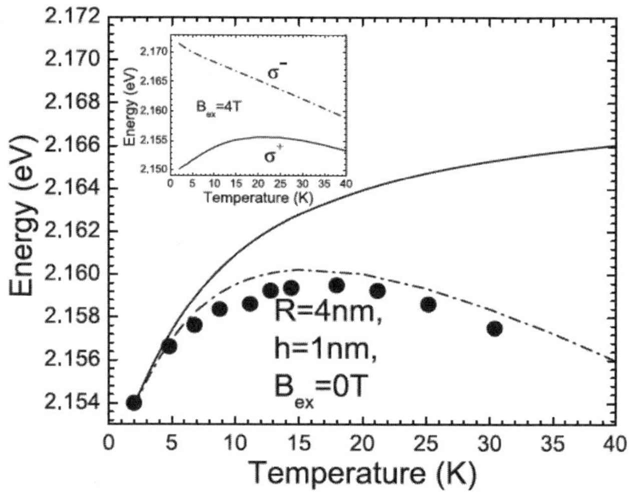

Fig. 2.34. σ^+ MP transition energy in the DMS QD at $B_{\text{ext}} = 0$ versus temperature. Solid points are experimental results. The inset shows the MP transition energies at $B_{\text{ext}} = 4\,\text{T}$ versus temperature.

2.4.5. *DMS quantum wires*

Experimentally various semiconductor wires, such as ZnO nanowires (Huang *et al.*, 2001) and ZnO nanobelts (Pan *et al.*, 2001), have been grown successfully. The calculation of the magnetic energy levels of quantum wires will be described in Sec. 10.2.

Figure 2.35 shows the electron energy levels of $Zn_{1-x}Mn_xO$ ($\bar{x} = 0.0018$) quantum wire of $R = 3\,\text{nm}$ at $B = 2\,\text{T}$ versus the wave vector along the z-direction k (Zhu and Xia, 2007). From Fig. 2.35, we see that the originally degenerate ground state and low excited states split in the magnetic field, where the (n, L) denotes the principal and angular momentum quantum numbers, respectively. The ground state splits into $(1,0)\downarrow$ and $(1,0)\uparrow$ states, and the energy of the $(1,0)\downarrow$ state is the lowest. The originally 4-fold degenerate first excited state splits into four states with $L = \pm 1$, $s_z = \pm 1$ under the common interactions of Landau splitting and Zeeman splitting.

Because the cyclotron radii of the low excited states are smaller than the radius of the quantum wire, the energy of the state with the angular momentum L and the spin component s_z can be approximately written as,

$$E(L, s_z) = (3 + |L| + L)\hbar\omega_0 + \bar{x}N_0\alpha s_z\langle S_z\rangle. \tag{2.85}$$

From Eq. (2.85), we see that the difference of the levels is $2\hbar\omega_0$, where $\omega_0 = eB/2m^*c$, and that of the $s_z = \pm 1/2$ levels is $\bar{x}N_0\alpha\langle S_z\rangle$. The former

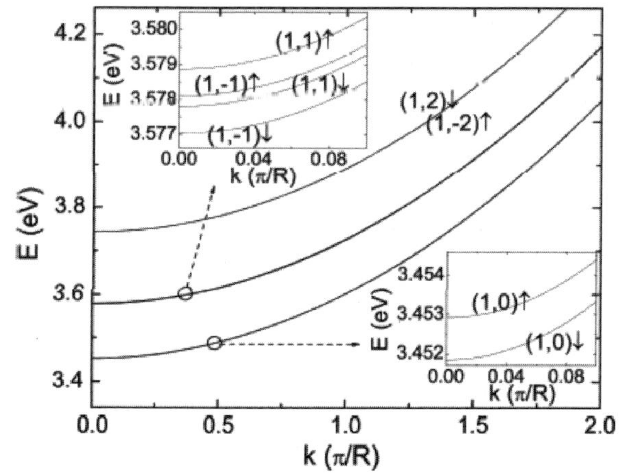

Fig. 2.35. Electron energy levels of $Zn_{1-x}Mn_xO$ ($\bar{x} = 0.0018$) quantum wire of $R = 3\,\text{nm}$ at $B = 2\,\text{T}$ versus the wave vector along the z-direction k. The inset shows the magnified levels.

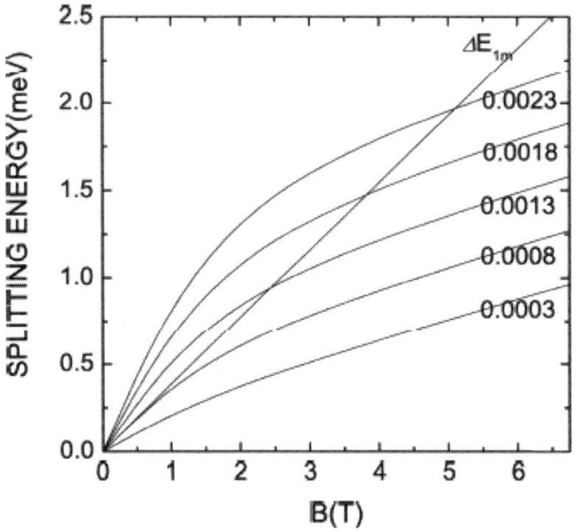

Fig. 2.36. Variation of the Landau energy level splitting $L = \pm 1$ and the Zeeman splitting $s_z = \pm 1$ for several effective Mn^{2+} concentrations with the magnetic field.

is proportional to the magnetic field B, and the latter is determined by $\langle S_z \rangle$ (Eq. (2.36)), which saturates in the large magnetic field, as shown in Fig. 2.36 (Zhu and Xia, 2007). The Zeeman level splitting ΔE_Z depends on the effective concentration of magnetic ions $\bar{x}$. From Fig. 2.36, we see that when $\bar{x} < 0.0003$, ΔE_Z is always smaller than the Landau level splitting ΔE_L. Thus the $(1,1)\downarrow$ state is always higher than the $(1,-1)\uparrow$ states. In the case of Fig. 2.35, $\bar{x} = 0.0018, B = 2\,\mathrm{T}$, this time $\Delta E_Z > \Delta E_L$, the $(1,1)\downarrow$ state is lower than the $(1,-1)\uparrow$ state.

Figure 2.37 shows the hole levels of $R = 3\,\mathrm{nm}$, $\bar{x} = 0.0018$ quantum wire at $B = 2\,\mathrm{T}$ versus the wave vector k, (a) $J = \pm 1/2$, (b) $J = \pm 3/2$ (Zhu and Xia, 2007). Due to the magnetic exchange interaction, the levels of $\pm J$ split, in general the negative J levels (dot lines), are higher the positive J levels (solid lines) due to the $V_{\mathrm{exch,h}}$ in Eq. (2.82) with $N_0\beta = 2.7\,\mathrm{eV}$. But some of hole energy levels are in opposite order due to the quantum confinement effect and the coupling effect between hole states. The hole ground state is $(1,-1)X^+\downarrow + (1,1)X^-\downarrow$, and the electron ground state is $(1,0)S\downarrow$ (Fig. 2.35), where X^+ and X^- represent the Bloch functions of the valence band top. Because the overlap integrations of the hole state envelope functions $(1,-1), (1,1)$ and the electron state envelope function $(1,0)$ are zero, the transition is indirect. The calculation found that when the radius of the quantum wire $R > 16.6\,\mathrm{nm}$, the hole ground state becomes $(1,0)X^+\downarrow$, the direct transition will occur. Thus, there exists a critical

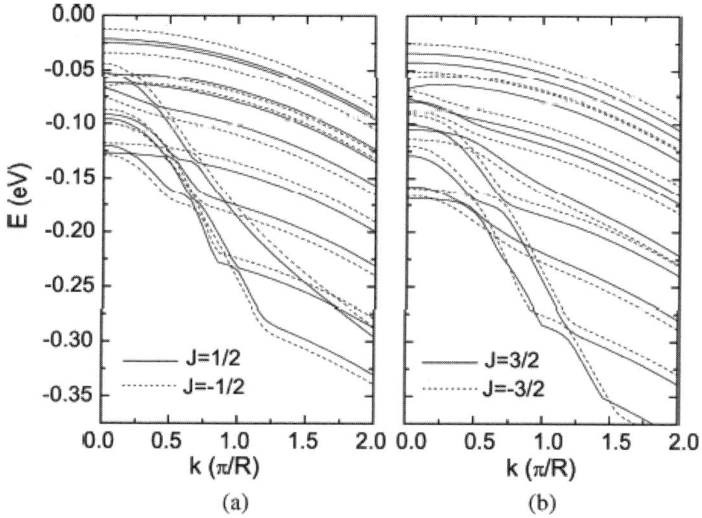

Fig. 2.37. Hole levels of $R = 3\,\text{nm}$, $\bar{x} = 0.0018$ quantum wire at $B = 2\,\text{T}$ versus the wave vector k, (a) $J = \pm 1/2$, (b) $J = \pm 3/2$.

radius, and when the radius of the quantum wire is smaller than the critical radius, there will be a dark exciton effect, as observed in many QDs.

The splitting energies of hole ground states $(1,-1)X^+ \downarrow +(1,1)X^- \downarrow$ and $(1,1)X^- \uparrow +(1,-1)X^+ \uparrow$ versus radius of quantum wire R, (a) undoped, (b) doped Mn ions $\bar{x} = 0.0018$, are shown in Fig. 2.38. From Fig. 2.38, we see that the splitting energies increase with increasing radius, and they will approach the value in the bulk material, but the increments are small for both undoped and doped cases ($< 1\,\text{meV}$). It indicates that the splitting energies of hole states are basically independent of the radius of quantum wires in the external field, i.e. the quantum confinement effect. In contrast, this splitting is greatly enhanced by the magnetic exchange interaction in the Mn-doped quantum wires. The splitting energy of quantum wire with $R = 11.4\,\text{nm}$ and $\bar{x} = 0.0018$ is $9.13\,\text{meV}$, 127 times that in the undoped case ($0.072\,\text{meV}$), as depicted in Fig. 2.38.

The transitions from the electron states $(1,0)S \uparrow$ and $(1,0)S \downarrow$ to the hole states $(1,-1)X^+ \downarrow +(1,1)X^- \downarrow, (1,1)X^- \uparrow +(1,-1)X^+ \uparrow$, and $(1,0)Z \uparrow, (1,0)Z \downarrow$ will emit light of polarizations σ^+, σ^- and π, respectively. Figure 2.39 shows the transition energies of $\sigma^{\pm}$ and π transitions versus magnetic field for (a) $R = 3\,\text{nm}$, $\bar{x} = 0.0018$; (b) $R = 3\,\text{nm}$, $\bar{x} = 0.0056$; (c) $R = 1.5\,\text{nm}$, $\bar{x} = 0.0018$ (Zhu and Xia, 2007). From Fig. 2.39, we see that the splitting energies as functions of magnetic field are all similar

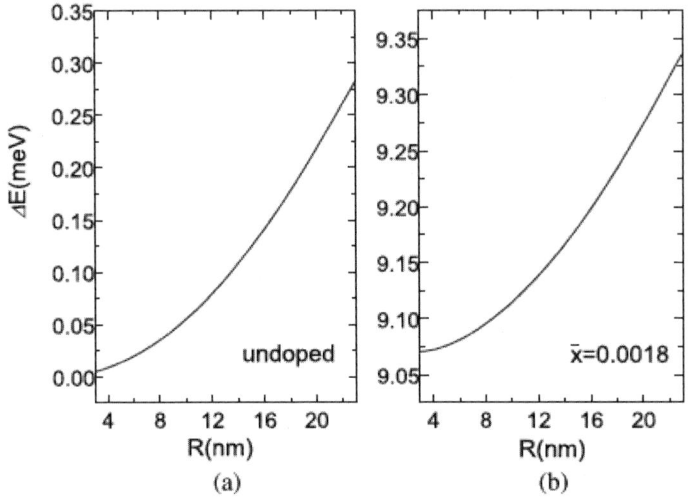

Fig. 2.38. Splitting energies of hole ground states $(1,-1)X^+\downarrow+(1,1)X^-\downarrow$ and $(1,1)X^-\uparrow+(1,-1)X^+\uparrow$ versus radius of quantum wire R, (a) undoped and (b) doped Mn ions $\bar{x}=0.0018$.

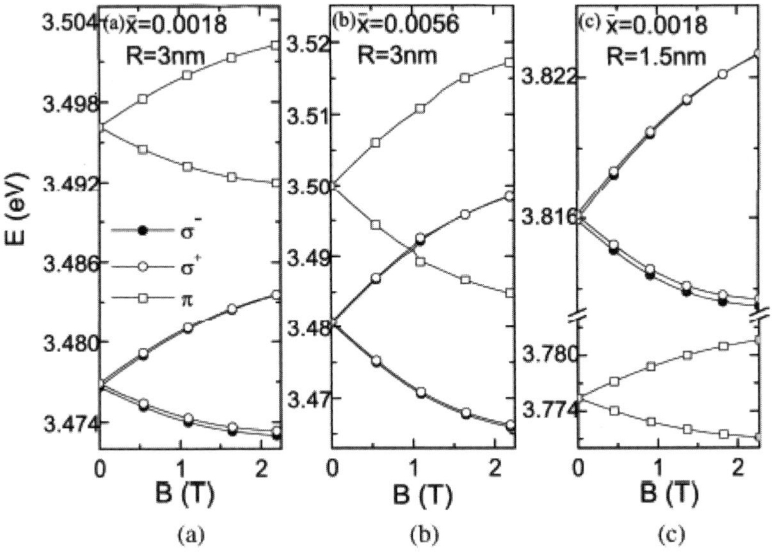

Fig. 2.39. Transition energies of $\sigma^\pm$ and π transitions versus magnetic field for (a) $R=3\,\text{nm}$, $\bar{x}=0.0018$; (b) $R=3\,\text{nm}$, $\bar{x}=0.0056$; (c) $R=1.5\,\text{nm}$, $\bar{x}=0.0018$.

to the $\langle S_z \rangle$, and the saturate splitting energy is much larger at higher Mn-doped concentration case, as depicted in Fig. 2.39(b). When the radius of the quantum wire is smaller than 2.3 nm, the hole ground state becomes $(1,0)Z\uparrow$ and $(1,0)Z\downarrow$, so that the energies of the π polarized transitions are lower than those of the $\sigma^{\pm}$ polarized transitions, as depicted in Fig. 2.39(c).

2.5. Transport Properties of DMSs

The study of the transport phenomena of semiconductor in a strong magnetic field (in particular, the Shubnikov–de Haas effect, SdH) is a useful tool for investigating the energy band structure. Because the Landau quantized energy is proportional inversely to the effective mass, this method is suitable especially to narrow gap semiconductors, mainly the n-type semiconductors.

The magnetic field positions of the maxima (or minima) observed in the SdH effect are mainly determined by the crossing of the Fermi level with consecutive Landau levels. Their temperature dependence is, in usual materials, only very slight. But in $Hg_{1-x}Mn_xTe$, a very strong temperature shift of the maxima of the SdH effect has been observed (Jaczynski *et al.*, 1978). The shift is particularly pronounced in the case of spin split peaks, which cannot be accounted by the small temperature variation of the effective mass.

In order to understand the temperature shift of the SdH peaks, Fig. 2.40 shows the Landau subband energies in $Hg_{0.975}Mn_{0.025}Te$ at $B = 4$ T versus the wave vector along the magnetic field k_{100}, (a) $T = 1.4$ K, (b) $T = 30$ K (Kossut, 1988). It is noticed that when $x = 0.025$, the energy gap is still negative, similar to the magnetic energy levels of HgTe (Fig. 2.13). The uppermost heavy hole subband (indicated by an arrow in Fig. 2.40) is shifted upwards and overlap with the conduction subbands. The band gap in DMS opens only at elevated temperature, as shown in Fig. 2.40(b). The "CB" belongs to the Γ_8 level, where $a(n)$ and $b(n)$ correspond to the spin up and down states. The Zeeman splitting energy (Eq. (2.86)) increases greatly the energy splitting of $a(n)$ and $b(n)$ levels.

$$\Delta E = \frac{1}{2}\bar{x}N_0\beta\langle S_z\rangle. \tag{2.86}$$

In Fig. 2.40, the splitting of the $a(0)$ and $b(0)$ levels are denoted by the thick arrows. From Fig. 2.40, we see that when the temperature arises from

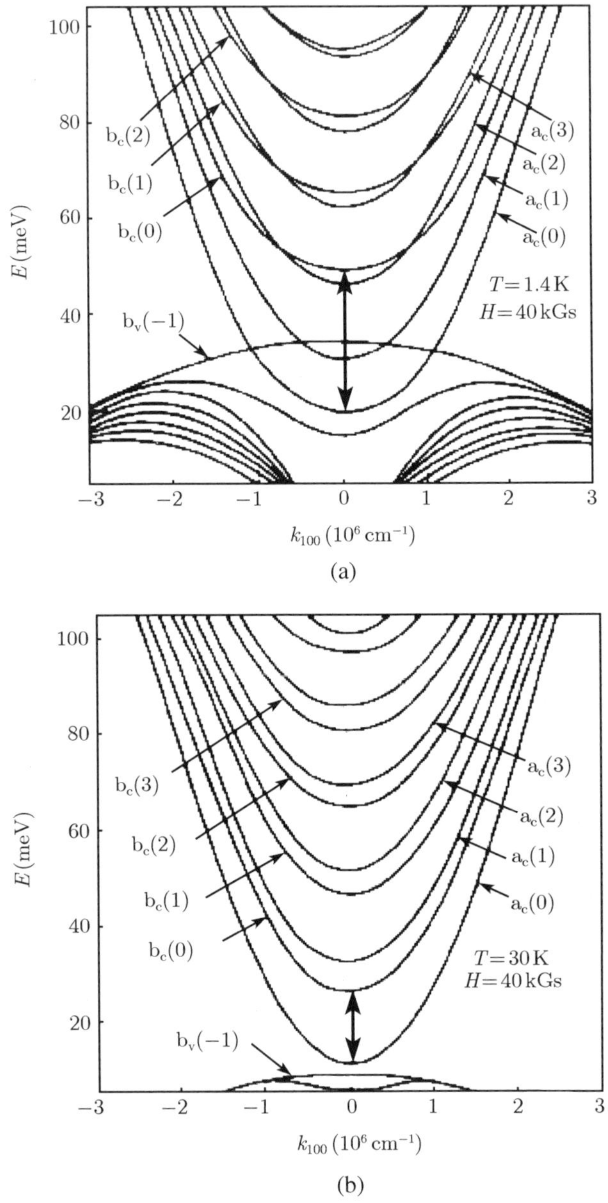

Fig. 2.40. The Landau subband energies in $Hg_{0.975}Mn_{0.025}Te$ at $B = 4\,T$ versus the wave vector along the magnetic field k_{100}, (a) $T = 1.4\,K$, (b) $T = 30\,K$. Thick arrows indicate spin splitting of the lowest CB Landau level. The valence band Landau level $b_v(-1)$ is indicated by a thin arrow.

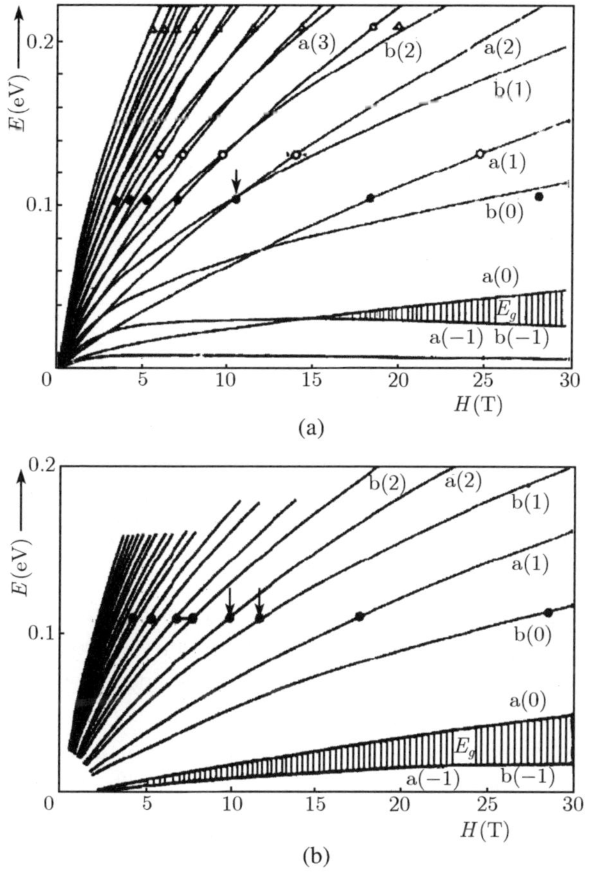

Fig. 2.41. Energy levels of $Hg_{0.98}Mn_{0.02}Te$ versus magnetic field, at (a) $T = 4\,K$, (b) $T = 36\,K$. The black, white, and triangle dots represent the positions of SdH peaks for the electron densities of 4.6×10^{17}, 7.5×10^{17}, and $2.4 \times 10^{18}\ cm^{-3}$, respectively.

1.4 to 30 K, the splitting energy decreases rapidly, due to decrease of the Brillouin function $B_{5/2}$ of $\langle S_z \rangle$ (Eqs. (2.34) and (2.35)). That means that the elevation of temperature results in the decrease in the average magnetic momentum of magnetic ions.

Figure 2.41 shows the energy levels of $Hg_{0.98}Mn_{0.02}Te$ versus magnetic field, at (a) $T = 4\,K$, (b) $T = 36\,K$ (Byszewski *et al.*, 1979). When $T = 4\,K$, the $a(2)$ and $b(1)$ levels intersect as shown by the short vertical arrow in Fig. 2.41(a); thus, only one SdH peak was observed. When $T = 36\,K$, these two levels separate as shown by the short vertical arrows in Fig. 2.41(b);

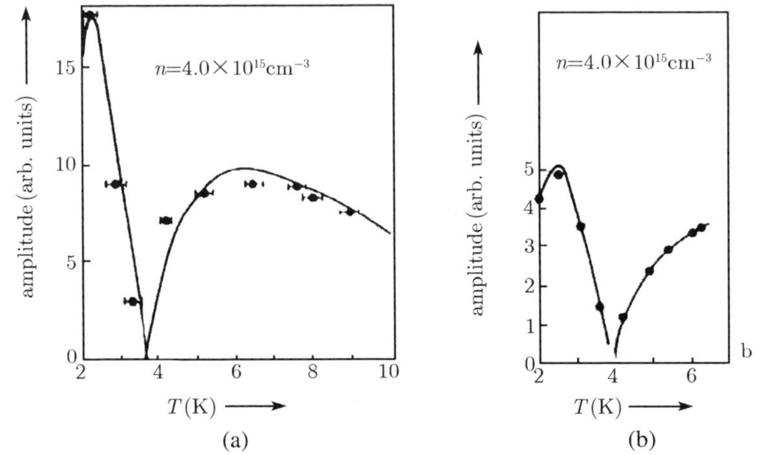

Fig. 2.42. Normalized amplitude of SdH oscillations versus temperature for $Hg_{1-x}Mn_xTe$, $n = 4.0 \times 10^{15}\,cm^{-3}$, (a) $x = 0.02$, (b) $x = 0.03$. Dots are experimental results, solid line is the theoretical result.

thus, two SdH peaks were observed. The magnetic field at which the intersection occurs depends sensitively on temperature (through the $\langle S_z \rangle$ temperature dependence); this fact explains the sensitive temperature dependence of the SdH peaks.

Figure 2.42 shows the normalized amplitude of SdH oscillations versus temperature for $Hg_{1-x}Mn_xTe$, $n = 4.0 \times 10^{15}\,cm^{-3}$, (a) $x = 0.02$, (b) $x = 0.03$ (Jaczynski *et al.*, 1978). Usually, in nonmagnetic semiconductors, the amplitude of a given magnetoresistance (MR) maximum decreases monotonically when the temperature increases. But in $Hg_{1-x}Mn_xTe$, the amplitude behaves in a drastically nonmonotonic manner (Fig. 2.42). It drops quickly with increasing temperature, passes a region of very small values, and becomes appreciable once again at even higher temperature. Then, after passing through a maximum, it starts to decrease again. Only the last portion of this behavior, in the highest temperature range, could be consistently understood in terms of the theory usually applicable to nonmagnetic semiconductors. According to the theory of the SdH effect, the amplitude of the SdH oscillation can be expressed in the form of a harmonic series (Kossut, 1988),

$$\frac{\Delta\rho}{\rho_0} \sim \sum_{r=1}^{\infty} \frac{\sqrt{B}X_r \exp[-arT_D/B]}{\sqrt{r}\sinh X_r} \cos(\pi\nu r) \cos\left[2\pi r\left(\frac{P_{\rm SH}}{B} - \gamma\right) \pm \frac{\pi}{4}\right], \tag{2.87}$$

where $a = 2\pi^2 m^* k_B/e\hbar, T_D$ is a temperature describing the broadening of the Landau levels, $X_r = arT/B$, and ν is the spin splitting factor defined as

$$\nu = \frac{1}{2} g^* \frac{m^*}{m_0}. \tag{2.88}$$

The oscillation character of the MR is given by the last cosine factoring Eq. (2.87) with the fundamental period P_{SH}. For the spherical Fermi surface, this can be related to the electron concentration n by

$$P_{SH} = \frac{1}{2} \frac{hc}{e} (3\pi^2 n)^{2/3}. \tag{2.89}$$

For the DMS, the $\cos(\pi\nu r)$ term in Eq. (2.87) can influence the temperature behavior of the amplitude quite significantly. When the argument of the cosine equals $i\pi/2$, where i is an odd integer, the amplitude may vanish. For the first harmonic in Eq. (2.87), i.e. for $r = 1$, the condition for the zeros is

$$g^* = \frac{m_0}{m^*} i, \quad i = \pm 1, \pm 3, \ldots. \tag{2.90}$$

It turns out from the theoretical calculation that it is possible to meet condition (2.90) in the temperature range where the SdH effect is studied, thus the temperature behavior of the amplitude as shown in Fig. 2.42 is appeared.

2.6. Fe^{2+} Ion-doped DMSs — Van Vleck Paramagnetism

Except the DMS doped with Mn ions, there is DMS doped with Fe ions. At the beginning, people looked the Fe ions as same as the Mn ions, and thought that the Fe ion also has the Curie–Weiss paramagnetism, and its thermo-average magnetic momentum is given by the Brillouin function (Eqs. (2.34) and (2.35)). But the results of the magneto-optical experiments of $Hg_{1-x}Fe_xTe$ show obvious difference from those of $Hg_{1-x}Mn_xTe$. Figure 2.43 shows the energies of the transmission minima versus magnetic field (Guldner *et al.*, 1980). σ^+ polarization for $Hg_{1-x}Fe_xTe$, $x \sim 0.016$, $T = 4.2\,K$. From Fig. 2.43, we see that the energies of transmission minima are proportional to the magnetic field and do not saturate at large magnetic field. When the temperature decreases from 4.4 to 2 K, the curves do not change. These two points are contrast to the magnetic field and temperature dependences predicted by the Brillouin function. The authors still used the original theory to explain the phenomena, assuming that $T_0 \gg 4.4\,K$ in Eq. (2.35).

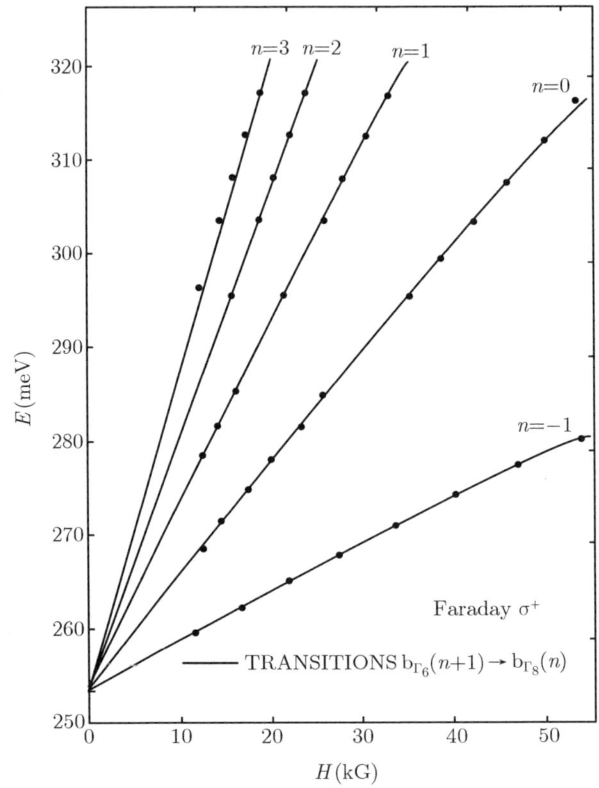

Fig. 2.43. Energies of the transmission minima versus magnetic field. σ^+ polarization for $Hg_{1-x}Fe_xTe$, $x \sim 0.016$, $T = 4.2\,K$. Solid lines: theoretical fit. Symbols: experiments.

Until 2 years later, in an international conference of narrow gap semiconductors, this problem was clarified (Serre *et al.*, 1982). The energy levels of the Fe^{2+} ion in the zincblende crystal field and SOI versus magnetic field is shown in Fig. 1.7. When the magnetic field is small, there is only Zeeman splitting term $\mu_B(L_z+g_{Fe}S_z)B$, proportional to B. When the magnetic field is large ($B \sim 10\,T$), the splitting energy includes the B^2 term. It has been introduced in Chap. 1 that at $B = 0$, the lowest five levels of the Fe^{2+} ion in the cubic crystal field are $\Gamma_1, \Gamma_4, \Gamma_3, \Gamma_5$, and Γ_2, and their degeneracies are $1, 3, 2, 3$, and 1, respectively. From the theoretical calculation, it turns out that the spacing of these five levels are equal, equal to $6\lambda^2/\Delta$, where λ is the SOC parameter, $\Delta = |10Dq|$ is the splitting energy of ^5E and $^5T_2$ states caused by the crystal field. For HgTe crystal, $\lambda = 12\,meV$, $\Delta = 0.3\,eV$. From Fig. 1.7, we see that the ground state is a singlet, i.e. there is no permanent moment at $B = 0$. Consequently, there is no linear Zeeman shift. By using the perturbation theory, we obtain the ground state energy varying with B^2,

$$E_0(B) = E_0 - \frac{4}{3}\mu_B^2 \frac{\Delta}{\lambda^2} B^2. \tag{2.91}$$

Thus, the magnetic moment of Fe ion is field-induced and varies linearly with B. The magnetic susceptibility is therefore independent of the magnetic field and of temperature, as long as $k_B T \ll 6\lambda^2/\Delta$. The Fe^{2+} magnetism is of Van Vleck type, while not Curie–Weiss type.

2.7. Giant Faraday Rotation and KR

DMSs display a unique magneto-optical property, for instance, MCD, Giant Faraday rotation and KR. Giant Faraday rotation, and KR have a wide potential application in the future semiconductor devices. Time-resolved Faraday rotation and KR (TRFR and TRKR) are powerful tools to study spin dynamics in semiconductors.

2.7.1. *Magneto-optical property of magnetic semiconductors*

Giant Faraday rotation or KR comes from the giant Zeeman spin splitting in CB and valence band. Figure 2.44 shows that the transition energies of the right and left circular light are no longer degenerate because of the spin splittings of electron and hole states. Magnetic semiconductors have a large oscillator strength of interband transition and giant Zeeman splitting,

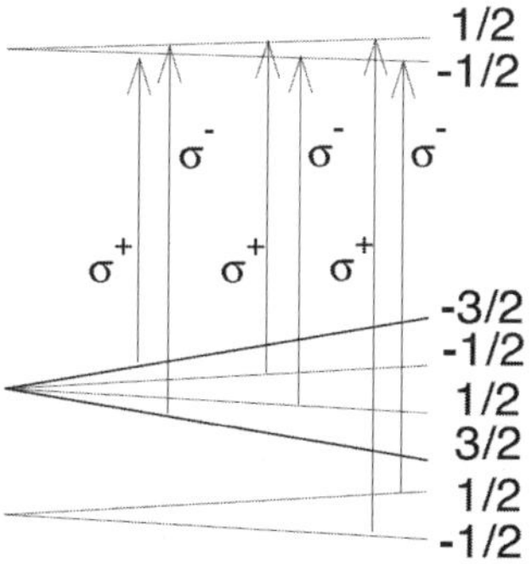

Fig. 2.44. Schematic of the band splitting.

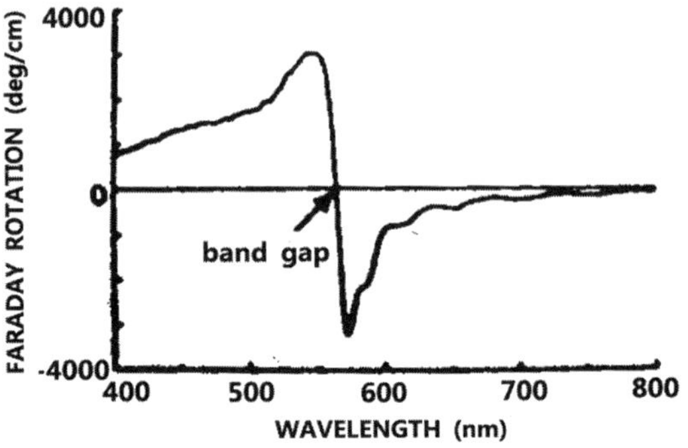

Fig. 2.45. Faraday rotation spectrum of CdMnTe.

and consequently show large Faraday rotation and KR (Gaj *et al.*, 1978; Turner *et al.*, 1983; Koyanagi *et al.*, 1987; Dillon *et al.*, 1990; Fumagalli and Munekata, 1996; Sinova *et al.*, 2003). For example, Zeeman splitting in $Ga_{1-x}Mn_xAs$ can approach 100 meV, and the oscillator strength of the interband transition is of the same order of magnitude as that in GaAs.

The Faraday rotation of $Cd_{1-x}Mn_xTe$ thin film is shown in Fig. 2.45. From this figure, one can see that the Faraday rotation approaches the maximum in the vicinity of the band gap, which can be up to 3×10^3 deg/cm, in the same order of magnitude as that of $Y_3Fe_5O_{12}$, slightly smaller than that of the transition atom-doped material: YIG($\sim 10^4$ deg/cm) (Shinagawa, 1999). One can also tune the bandgap of $Cd_{1-x}Mn_xTe$ by changing the Mn concentration, and consequently change the Faraday rotation at a specific wavelength.

Figure 2.46 shows the Faraday rotation of GaMnAs in infrared regime for different hole densities (Sinova *et al.*, 2003). From the figure, one can find that the DMS display giant Faraday rotation, which can be beyond $2.4. \times 10^5$ deg/cm, much larger than that in the visible regime. Different from the Faraday rotation in the visible regime, usually the Faraday rotation of magnetic semiconductors in the infrared regime is not determined by the interband transition, but the intersubband transition. For p-type GaMnAs, the Faraday rotation in the infrared regime is generated by the transitions among the spin-split-off band, light-hole band and heavy-hole band; therefore, it depends sensitively on the hole density. This giant Faraday rotation arises from the giant Zeeman splitting and the resulting big difference of the Fermi energies. For DMS GaMnAs material, Zeeman spin splitting is about 100 meV, and the splitting at the top of the valenceband

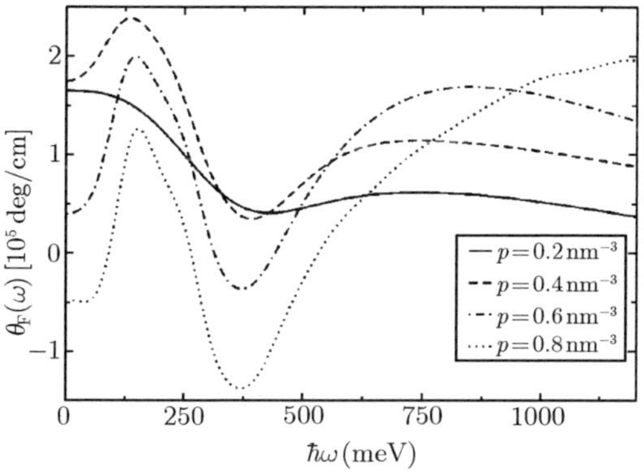

Fig. 2.46. Faraday rotation in $Ga_{1-x}Mn_xAs$ for different hole densities.

is also approximately 100 meV, comparable with the photon energy in the infrared regime, consequently leads to a giant Faraday rotation.

Szczytko *et al.* (2001) found that the doping destroys the spatial translation symmetry in heavily doped $Ga_{1-x}Mn_xAs$ and relaxes the restriction of the momentum conservation in the optical transition. Usually, the interband absorption coefficient is

$$\alpha(\omega) = \int \frac{d\boldsymbol{k}_c}{(2\pi)^3}\frac{d\boldsymbol{k}_v}{(2\pi)^3}\boldsymbol{M}^2_{c,v}(\boldsymbol{k}_c,\boldsymbol{k}_v) \times \delta[\hbar\omega - E_g + E_v(\boldsymbol{k}_v) - E_c(\boldsymbol{k}_c)][f_v(E_v(\boldsymbol{k}_v)) - f_c(E_c(\boldsymbol{k}_c))], \quad (2.92)$$

where E_g is the band gap, $E_v(\boldsymbol{k}_v)$, $E_c(\boldsymbol{k}_c)$ are the energy dispersions of the CB and valence band, respectively, $f_v(E_v(\boldsymbol{k}_v)), f_c(E_c(\boldsymbol{k}_c))$ are the occupation numbers of electron and hole, respectively. The photon energy $\hbar\omega = E_g - E_v(\boldsymbol{k}_v) + E_c(\boldsymbol{k}_c)$. The momentum conserves in the optical transition for small doping case

$$\boldsymbol{M}^2_{c,v}(\boldsymbol{k}_c,\boldsymbol{k}_v) = A^2(2\pi)^3\delta(\boldsymbol{k}_c - \boldsymbol{k}_v), \quad (2.93)$$

where

$$A^2 = \frac{2\pi e^2\hbar E_p}{6cm_0\phi\varepsilon_0\sqrt{\varepsilon_1}E_\omega},$$

here $E_\omega = \hbar\omega; E_p = 2m_0P^2\hbar^2$, P is the interband Kane matrix element, ε_1 is the dielectric constant, m_0 the mass of free electron, c is the speed of light, and e is the electron charge. For heavily doped material, electron momentum is not a good quantum number, and δ function is to broaden a Gaussian function. Meanwhile

$$\boldsymbol{M}^2_{\mathrm{c,v}}(\boldsymbol{k}_\mathrm{c}, \boldsymbol{k}_\mathrm{v}) = A^2 \left(\frac{2\pi}{\sigma_k^2}\right)^{3/2} \exp\left[-\frac{(\boldsymbol{k}_\mathrm{c} - \boldsymbol{k}_\mathrm{v})^2}{2\sigma_k^2}\right]. \quad (2.94)$$

Low-dimensional semiconductor structures show high light-emitting efficiency and mobility due to quantum confinement effect (Xia and Zhu, 1995). Low-dimensional DMS structures also have attracted much attention, especially in CdTe/CdMnTe (Bicknell *et al.*, 1987; Williams *et al.*, 1989); ZnMnSe/ZnSe (Bylsma *et al.*, 1985), InMnAs/InAlAs/GaAlSb quantum well (Munekata *et al.*, 1993, Fumagalli and Munekata, 1996). Due to the quantum confinement effect, the sp–d exchange interaction between Mn ions and electrons varies with changing of the size of quantum structures; Faraday rotation and KR spectrum also changes. Figure 2.47 is the schematic of 2D InMnAs hole gas, The samples were grown on the GaSb buffer layer with the thickness ranging from 400 to 600 nm, which relaxes the strain caused by the lattice mismatch between AlSb and GaAs. InAlAs layers with different thicknesses ranging from 0 to 24 nm were grown to tune the barrier height of the heterostructure between AlSb and InMnAs, to adjust the p–d exchange interaction between Mn ions and 2D hole gas. The KR spectrum is measured as shown in Fig. 2.48, where d_{sp} is the thickness of the InAlAs layer, at $B = 3\,\mathrm{T}$ and $T = 5.5\,\mathrm{K}$. When $d_{\mathrm{sp}} = 3.5\,\mathrm{nm}$ (the peak α_1) and $d_{\mathrm{sp}} = 7\,\mathrm{nm}$ (the peak α_2), the Kerr angles approach

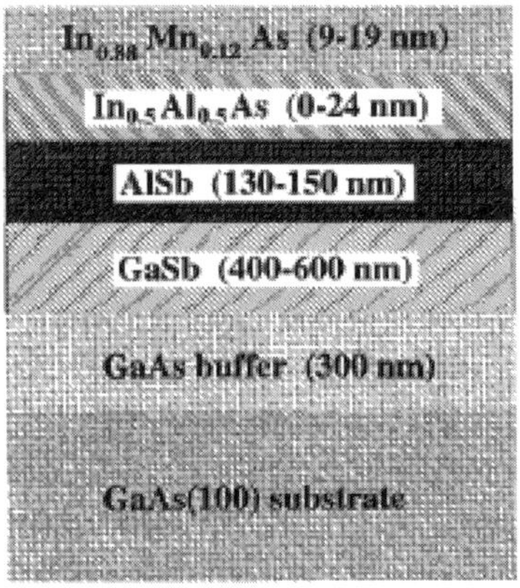

Fig. 2.47. Schematic of the InMnAs/InAlAs/GaAlSb heterostructure.

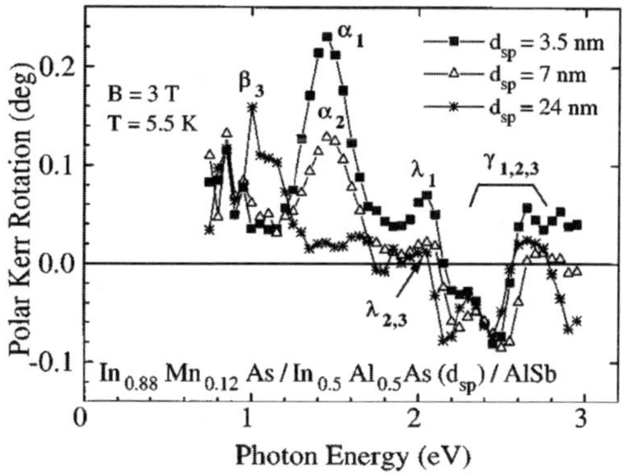

Fig. 2.48. KR in the InMnAs/InAlAs/GaAlSb heterostructure.

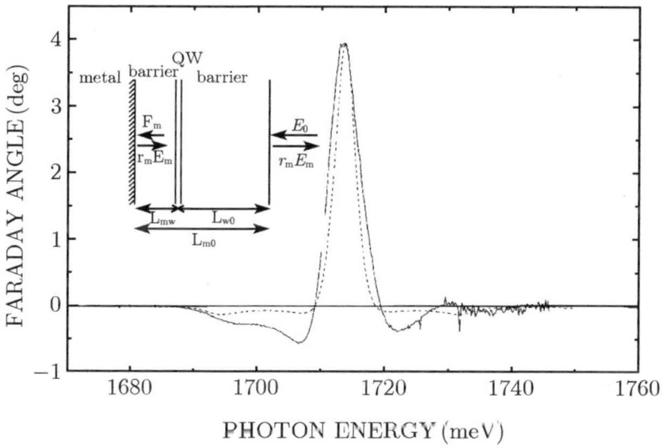

Fig. 2.49. FR spectrum of $Cd_{0.94}Mn_{0.06}Te$ single quantum well.

the maximum in the vicinity of 1.45 eV. But the peak disappear when d_{sp} = 24 nm. This feature indicates that the p–d exchange interaction exists between the Mn ion and the holes in the InMnAs layer and the 2D hole gases formed in the AlSb layer, which plays a dominant role.

Microcavity or photonic crystal can be used to enhance Faraday rotation and KR (Inoue *et al.*, 1998; Gourdona *et al.*, 2002). In microcavity, Faraday rotation and KR can be enhanced one to three orders of magnitude due to localization and interference effect of the light. The inset in Fig. 2.49 is the schematic of a microcavity, which consists of a 10 nm thick $Cd_{0.94}Mn_{0.06}Te$ single quantum well with metal mirrors.

The theoretical result agrees well with the experiment, shown in Fig. 2.49. For a single quantum well, Faraday rotation can reach $4°(4 \times 10^6$ deg/cm). This giant Faraday rotation has widely potential applications in magneto-optical device, e.g. light isolator.

2.7.2. *TRFR and TRKR in magnetic semiconductors*

TRFR or TRKR spectrum is a powerful tool to detect spin or magnetic moment dynamics (Awschalom and Samarth, 2002). It can also be used to measure the exchange interaction between magnetic ions and electrons in magnetic semiconductors. Experiments demonstrated that the frequency of electron spin processing in DMS can approache THz, which is much higher than that in conventional semiconductors (~GHz) due to the strong sp–d exchange interaction (Myers *et al.*, 2003). Experimental setup is shown schematically in Fig. 2.50(a). Changing Cd concentration gradually in ZnSe/ZnCdSe quantum wells, one can obtain a quantum well with a parabolic confining potential. In the center of the ZnCdSe quantum well,

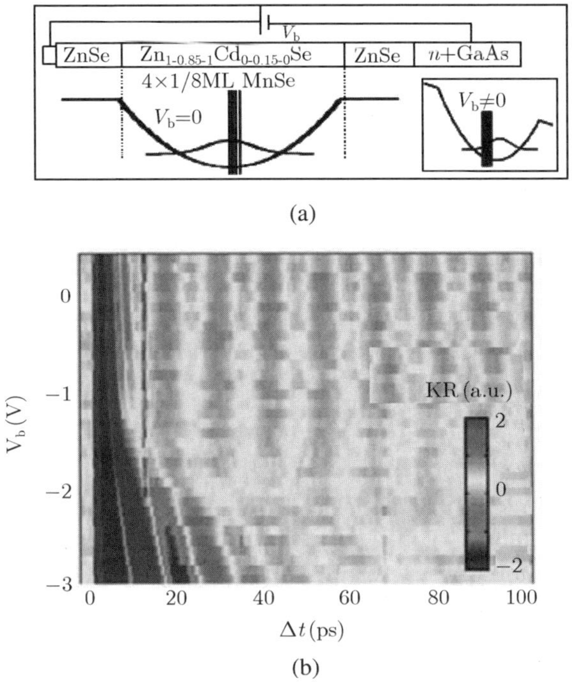

Fig. 2.50. (a) Schematic of MnSe layers inserted parabolic potential well. (b) KR of sample as a function of V_b and Δt at 5 K and $B = 3$ T.

there are a few MnSe layers. One can tune the sp–d exchange interaction between Mn ions and electrons by applying a bias cross the quantum well, i.e. changing the barrier height and symmetry. Figure 2.50(b) is the TRKR spectrum. The minimum of the confining potential shifts from the center as the external electric fields increase, therefore the overlap between the electron wavefunction and MnSe layers decreases, consequently leading to the weakening of the sp–d exchange interaction, lowering of the electron spin processing, and increasing of the SRT.

Brunetti *et al.* (2006) studied the spin processing and relaxation of exciton polariton in CdMnTe quantum wells embedded in a microcavity. The exciton polariton is a new kind of collective excitation arising from the coupling between the exciton and electromagnetic field. Because the photon is massless, the mass of the exciton polariton is less than the effective mass of the electron; consequently, it has higher frequency of spin processing, which can be up to THz. Figure 2.51 shows the experimental results. The left panel (a) is the KR spectrum for different magnetic fields. The right panel (b) is the Fourier transformation of the KR spectrum.

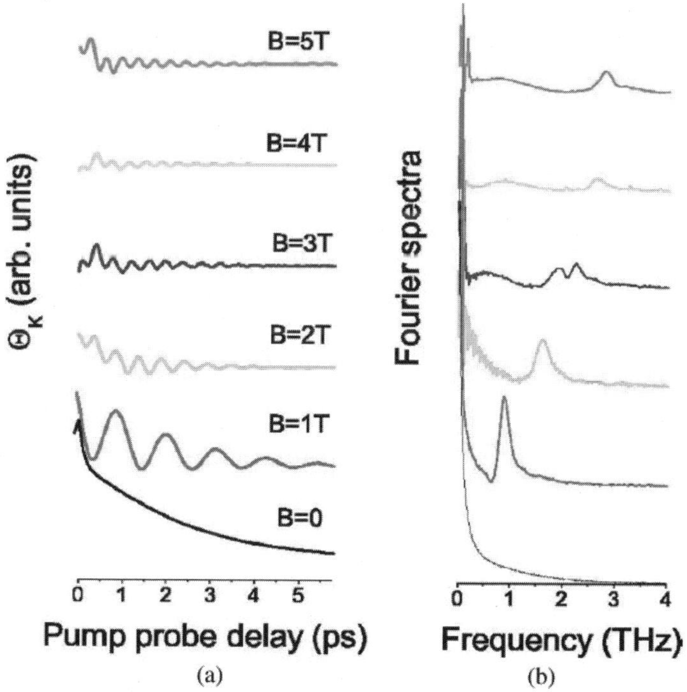

Fig. 2.51. (a) KR of exciton polariton in CdMnTe quantum wells embedded in a microcavity. (b) The Fourier transformation.

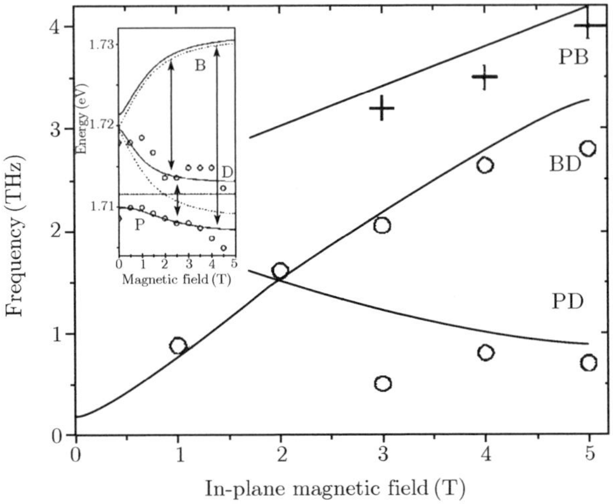

Fig. 2.52. The frequency of exciton polariton versus magnetic fields. The inset is the dispersion of polariton.

Figure 2.52 describes how the processing frequency of exciton polariton changes with increasing magnetic fields. The solid line is the numerical calculation, and the dashed line is experimental measurement. The inset of Fig. 2.52 shows the energy of exciton polariton as a function of magnetic fields, in which the solid (dashed) line denotes the theoretical results with and without the coupling between exciton and electromagnetic field and the circles are the experiment results. The energies of the photon-like and dark exciton branches decrease and those of the bright exciton branch increases as the external magnetic fields increase. The coupling between exciton and electromagnetic field is varied, consequently changing the processing frequency of exciton polariton.

2.8. Light-Induced Magnetization

Comparing with ferromagnetic metal, one of the important properties of magnetic semiconductors is the spin orientation of magnetic ions manipulated by light. Generally, the magnetization of magnetic ions can be controlled by an external magnetic field, another way is the light-induced magnetization (Krenn *et al.*, 1985; Awschalom *et al.*, 1987; Krenn *et al.*, 1989; Beschoten *et al.*, 1999), which is to control the magnetization of magnetic ions utilizing the spin states of optically generated carriers. CP light generates spin-polarized carriers in magnetic semiconductors, and the spin-polarized carriers align the magnetic ions via the sp–d exchange interaction. Polarized magnetic ions affect the spin polarization and the Zeeman splitting of electron and hole states. Actually, these two subsystems, electron and magnetic ions, influence each other; the interplay leads to the oscillation of the magnetization of magnetic ions, which can be detected by the time-resolved FR spectrum. From the TRFR spectrum, the important parameters, e.g. the strength of the sp–d exchange interaction, SRT of optically generated carriers and magnetic ions, the oscillator strength of the interband transition, and the density of optically generated carriers.

The spin-polarized carriers generated by a CP light from the heavy hole (hh) and light hole (lh) bands are (Mackowski *et al.*, 2004)

$$\begin{aligned}(G_+ - G_-)_{\mathrm{hh}} &= \frac{P_i(\mathrm{hh})|M_{\mathrm{hh}}|^2 D_{\mathrm{hh}}}{|M_{\mathrm{hh}}|^2 D_{\mathrm{hh}} + |M_{\mathrm{lh}}|^2 D_{\mathrm{lh}}} G, \\ (G_+ - G_-)_{\mathrm{lh}} &= \frac{P_i(\mathrm{lh})|M_{\mathrm{lh}}|^2 D_{\mathrm{hh}}}{|M_{\mathrm{hh}}|^2 D_{\mathrm{hh}} + |M_{\mathrm{lh}}|^2 D_{\mathrm{lh}}} G,\end{aligned} \tag{2.95}$$

where $P_i(\mathrm{hh})(P_i(\mathrm{lh}))$, $|M_{\mathrm{hh}}|(|M_{\mathrm{lh}}|)$, $D_{\mathrm{hh}}(D_{\mathrm{lh}})$ are the initial spin polarization, the interband transition element, and the joint density of states, respectively, and G is the total carrier generation rate. Therefore one can obtain the rate equation of electron

$$\frac{\partial(n_+ - n_-)}{\partial t} = G_+ - G_- - \frac{n_+ - n_-}{\tau} - \frac{n_+ - n_-}{T_{\mathrm{eph}}} - \frac{n_+ - n_-}{T_{\mathrm{eMn}}}, \tag{2.96}$$

where n_+ and n_- are spin-up and spin-down electron density and τ, T_{eph}, and T_{eMn} are the electron recombination time, electron–phonon and electron–magnetic ion relaxation times, respectively. The stationary solution of the above equation is

$$n_+ - n_- = \frac{G_+ - G_-}{1/\tau + 1/T_{\mathrm{eph}} + 1/T_{\mathrm{eMn}}}. \tag{2.97}$$

In the case of $T_{\mathrm{eMn}} \ll \tau, T_{\mathrm{eph}}$, we obtain

$$n_+ - n_- \cong (G_+ - G_-)T_{\mathrm{eMn}}. \tag{2.98}$$

The effective magnetic field of spin-polarized carriers on magnetic ions is

$$B^* = -\frac{n_+ - n_-}{2g\mu_{\mathrm{B}}}\alpha, \tag{2.99}$$

where $\alpha = \langle S|J|S\rangle$ is the s–d exchange interaction strength. While for the magnetic ions, one can get the rate equation and stationary solution

$$\begin{aligned}\frac{\partial(N_+ - N_-)}{\partial t} &= \frac{n_+ - n_-}{T_{\mathrm{eMn}}} - \frac{n_+ - n_-}{T_{\mathrm{Mnph}}}, \\ N_+ - N_- &\cong (n_+ - n_-)T_{\mathrm{Mnph}}/T_{\mathrm{eMn}},\end{aligned} \tag{2.100}$$

where T_{Mnph} is the magnetic ion–phonon relaxation time.

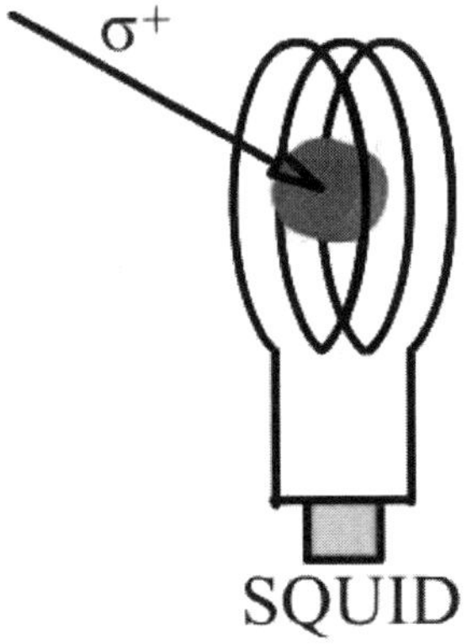

Fig. 2.53. Schematic of experimental setup.

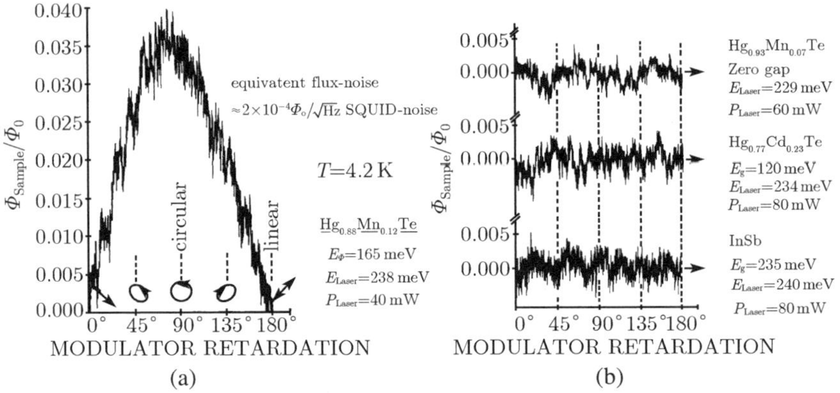

Fig. 2.54. Experimental results for different materials.

Experimental setup is shown schematically in Fig. 2.53. First one can generate the carriers optically using a laser with periodically changed polarization plane, then measure the magnetic moment of the sample using the superconducting quantum interference device (SQUID) technique. The magnetic flux ϕ_{sample} changes with variation of the pumping light polarization (Fig. 2.54) (Krenn *et al.*, 1985). The result in Fig. 2.54(a) is measured in $Hg_{0.88}Mn_{0.12}Te$ sample, while Fig. 2.54(b) shows the magnetic flux for $Hg_{0.93}Mn_{0.07}Te$, $Hg_{0.77}Cd_{0.23}Te$, InSb from up to down. From Fig. 2.54(a) one can see that the magnetic flux in $Hg_{0.88}Mn_{0.12}Te$ sample increases when the polarization of the pumping light changes from linear polarization to circular polarization and approaches the maximum when the pumping light is CP. Thus, one can confirm that this effect is caused by the light-induced magnetization, not by the thermal effect induced by the pumping light. For nonmagnetic material $Hg_{0.77}Cd_{0.23}Te$ and InSb, the light-induced magnetization cannot be observed because of the weak interaction between carriers and nonmagnetic ions. For zero-gap $Hg_{0.93}Mn_{0.07}Te$, the light-induced magnetization also cannot be detected since the optically generated carriers have a large momentum and a very short spin lifetime when the photon energy is much larger than the band gap.

Another important issue in the light-induced magnetization is the intensity of the pumping light on magnetization. The light-induced magnetization can be enhanced by increasing the intensity of pumping light, since the spin-polarized carrier density increases with increasing the intensity of the pumping light. On the other hand, the increase in the pumping power increases the sample temperature, leading to the saturation of the light-induced magnetization, shown schematically in Fig. 2.55 (Krenn *et al.*, 1989).

Due to the quantum confinement effect in low-dimensional semiconductor structures, the sp–d exchange interaction becomes stronger compared with that in bulk materials, and consequently the light-induced magnetization is also enhanced (Boukari *et al.*, 2002; Mackowski *et al.*, 2004).

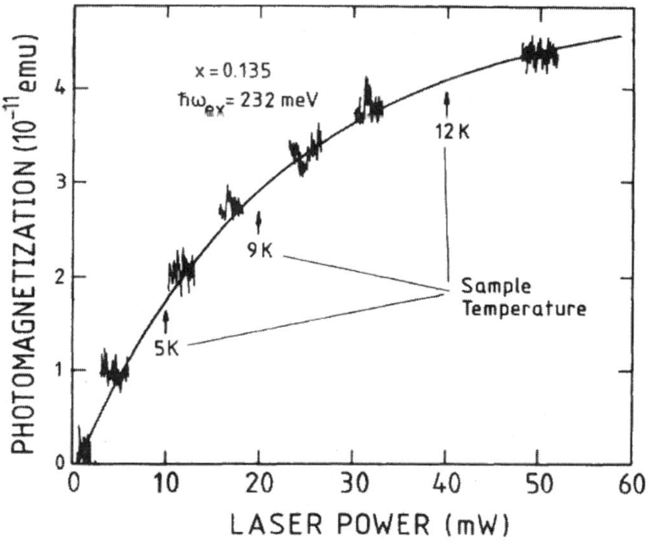

Fig. 2.55. Magnetization versus the intensity of pumping light.

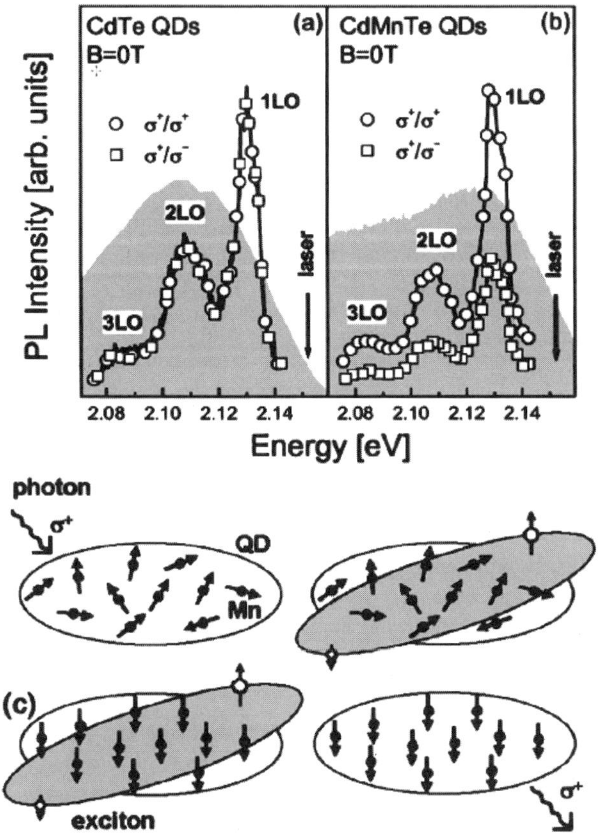

Fig. 2.56. Resonantly excited Photoluminescence (PL) of (a) CdTe QDs and (b) CdMnTe QDs measured at $B = 0\,\mathrm{T}$ and $T = 4.2\,\mathrm{K}$. The shaded areas represent the nonresonantly excited PL and the arrows mark the laser excitation energy. (c) The schematic of the formation of the optically-induced exciton magnetic-polarons in CdMnTe QDs.

Figures 2.56(a) and 2.56(b) are the PL spectra of CdTe and CdMnTe QDs, respectively. The underlying physics can be understood from Fig. 2.56(c); the spin-polarized excitons in QDs are generated by a right CP pumping light and polarize the magnetic ions in QDs via the sp–d exchange interaction, and the back action of spin-polarized Mn ions on the excitons makes the emitted photon become polarized. Comparing Figs. 2.56(a) and 2.56(b), one can see that the PL spectrum is nonpolarized for nonmagnetic semiconductor QDs. But for CdMnTe QDs, the PL spectrum displays a strong polarization and demonstrate experimentally the light-induced magnetization, while the light-induced magnetization is never observed in the bulk CdMnTe material. This phenomenon clearly proves that the quantum confinement effect can enhance the light-induced magnetization greatly.

References

Arciszewska M, Nawrocki M. (1982) In *Physics of Semiconducting Compounds (11th Conf.)*, p. 225, *Pol Acad Sci*, Warsaw, Poland.

Awschalom DD, Warnock J, von Molnár S. (1987) *Phys Rev Lett* **58**: 812.

Awschalom DD, Samarth N. (2002) In *Semiconductor Spintronics and Quantum Computation*, edited by Awschalom DD, Loss D, Samarth N Springer, Berlin.

Bastard G. (1979) *Inst Phys Conf Ser* **43**: 231.

Beschoten B, Crowell PA, Malajovich I, Awschalom DD, Matsukura F, Shen A, Ohno H. (1999) *Phys Rev Lett* **83**: 3073.

Bhargava RN, Gallagher D, Hong X, Nurmikko A. (1994) *Phys Rev Lett* **72**: 416.

Bhattacharjee AK, Guillaume CB. (1997) *Phys Rev B* **55**: 10613.

Bicknell RN, Giles NC, Schetzina JF. (1987) *Appl Phys Lett* **50**: 691.

Boukari H, Kossacki P, Bertolini M, Ferrand D, Cibert J, Tatarenko S, Wasiela A, Gaj JA, Dietl T. (2002) *Phys Rev Lett* **88**: 207204.

Brunetti A, Vladimirova M, Scalbert D, André R, Solnyshkov D, Malpuech G, Shelykh IA, Kavokin AV. (2006) *Phys Rev B* **73**: 205337.

Bylsma RB, Becker WM, Bonsett TC, Kolodziejski LA, Gunshor RL, Yamanishi M, Datta S. (1985) *Appl Phys Lett* **47**: 1039.

Byszewski P, Szlenk K, Kossut J, Galazka RR. (1979) *Phys Stat Sol* (*b*) **95**: 359.

Chang K, Xia JB. (2002) *J Phys C* **14**: 13661.

Cook WR. (1968) *Ceramic J Am Soc* **51**: 518.

Dai N, Ram-Mohan LR, Luo H, Yang GL, Zhang FC, Dobrowolska M, Furdyna JK. (1994) *Phys Rev B* **50**: 18153.

Dillon JF, Furdyna Jr JK, Debska U, Mycielski A. (1990) *J Appl Phys* **67**: 4917.

Edmonds AR. (1957) *Angular Momentum in Quantum Mechanics*, p. 75, Princeton.

Efros AI L, Rosen M. (1998) *Phys Rev B* **58**: 7120.

Fumagalli P, Munekata H. (1996) *Phys Rev B* **53**: 15045.

Furdyna JK. (1982) *J Appl Phys* **53**: 7636.

Gaj JA. (1988) In *Semiconductors and Semimetals*, p. 275 Vol. 25, Furdyna JK, Kossu J (eds). Acadmic Press, Boston.

Gaj JA, Galazka RR, Nawrocki M. (1978) *Solid State Commun* **25**: 193.

Gaj JA *et al.* (1979) in *Physics of Semiconductor* (14th Int. Conf. Edinburg, 1978), 1113, Wilson BLH (ed.) The Institute of Physics, Bristol.

Gourdona C, Lazarda G, Jeudya V, Testelina C, Ivchenkob EL, Karczewskic G. (2002) *Solid State Commun* **123**: 299.

Guldner Y, Riganx C, Grynberg M, Mycielski A. (1973) *Phys Rev B* **8**: 3875.
Guldner Y, Rigaux C, Menant M, Mullin DP, Furdyna JK. (1980) *Solid State Commun* **33**: 133.
Huang MH, Wu Y, Feick H, Tran N, Weber E, Yang P. (2001) *Adv Mater* **13**: 113.
Inoue M, Arai K, Fujii T, Abe M. (1998) *J Appl Phys* **83**: 6768.
Jackson SR, Nicholls JE, Hagston WE, Harrison P, Stirner T, Hogg JHC. (1994) *Phys Rev B* **50**: 5392.
Jaczynski M, Kossut J, Galazka RR. (1978) *Phys Stat Sol (b)* **88**: 73.
Jonker BT, Liu X, Chou WC, Petrou A. (1991) *J Appl Phys* **69**: 6097.
Kossut J. (1988) In *Semiconductors and Semimetals*, p. 183, Vol. 25, Furdyna JK, Kossut J (eds.) Academic Press.
Koyanagi T, Matsubara K, Takaoka H, Takagi T. (1987) *J Appl Phys* **61**: 3020.
Krenn H, Zawadzki W, Bauer G. (1985) *Phys Rev Lett* **55**: 1510.
Krenn H, Kaltenegger K, Dietl T, Spalek J, Bauer G. (1989) *Phys Rev B* **39**: 10918.
Lax B, Zwerdling S. (1960) in *Progress in Semiconductors*, p. 222, Gibson AF, Kröger FA, Burgess RE (eds). Heywood & Company, London.
Lee S, Dobrowolska M, Furdyna JK, Luo H, Ram-Mohan LR. (1996) *Phys Rev B* **54**: 16939.
Liu YH, Zhang LS. (1994) *Adv Phys* **14**: 82.
Luttinger JM. (1956) *Phys Rev* **102**: 1030.
Mackowski S, Gurung T, Nguyen TA, Jackson HE, Smith LM, Karczewski G, Kossut J. (2004) *Appl Phys Lett* **84**: 3337.
Maksimov AA, Bacher G, McDonald A, Kulakovskii VD, Forchel A, Becker CR, Landwehr G, Molenkamp LW. (2000) *Phys Rev B* **62**: R7767.
Munekata H, Zaslavsky A, Fumagalli P, Gambino RJ. (1993) *Appl Phys Lett* **63**: 2929.
Myers RC, Ku KC, Li X, Samarth N, Awschalom DD. (2003) *Phys Rev B* **72**: 041302.
Norris DJ, Yao N, Charnok FT, Kennedy TA. (2001) *Nano Lett* **1**: 3.
Pan ZW, Dai ZR, Wang ZL. (2001) *Science* **291**: 1947.
Roth LM, Lax B, Zwerdling S. (1959) *Phys Rev* **114**: 90.
Serre H, Bastard G, Rigaux C, Mycielski J. (1982) in *Physics of Narrow Gap Semiconductors*, Lecture Notes in Physics, Vol. 152, p. 321. Gornic E, Heirich H, Palmetshofer L (eds.) Springer-Verlag, Berlin.
Shinagawa K. (1999) In *Magneto-Optics*, Chap. 5, Sugano S, Kojima N (eds). Spinger-Verlag, Berlin.
Sinova J, Jungwirth T, Kucera J, MacDonald AH. (2003) *Phys Rev B* **67**: 235203.
Szczytko J, Bardyszewski W, Twardowski A. (2001) *Phys Rev B* **64**: 075306.
Turner AE, Gunshor RL, Datta S. (1983) *Appl Opt* **22**: 3152.
Twardowski A, Ginter J. (1982) *Phys Stat Sol (b)* **110**: 47.
Wang Y, Herron N, Moller K, Bein T. (1991) *Solid State Commun* **77**: 33.
Williams GM, Cullis AG, Whitehouse CR, Ashenford DE, Lunn B. (1989) *Appl Phys Lett* **55**: 1303.

Wu HB, Chang K, Xia JB. (2002) *Phys Rev B* **65**: 195204.
Xia JB, Zhu BF. (1995) In *Semiconductor Superlattice Physics*, pp. 1–16, Shanghai Scientific & Technical Publishers, Shanghai.
Yanata K, Suzuki K, Oka Y. (1993) *J Appl Phys* **73**: 4595.
Zener C. (1951) *Phys Rev* **81**: 440.
Zhu YH, Xia JB. (2007) *Phys Rev B* **75**: 205113.
Zhu YH, Zhang XW, Xia JB. (2006) *Phys Rev B* **73**: 165326.

Chapter 3

FERROMAGNETIC SEMICONDUCTORS

Ferromagnetism in magnetic semiconductors has been studied widely several tens of years ago, for example, $CdCr_2Se_4$, EuO, and sulfide compounds of lead ($Pb_{1-x-y}Sn_yMn_xTe$), etc. Recently, ferromagnetism has been discovered in Mn-doped III–V compounds, such as $In_{1-x}Mn_xAs$ (Ohno *et al.*, 1992), and $Ga_{1-x}Mn_xAs$ (Ohno *et al.*, 1996). These magnetic compounds attract utmost interests as they combine the two large branches in condensed matter physics: semiconductor and magnetism. They are also attractive from the technological point of view, since they can add a new degree of freedom associated with the spin of carriers to the electronic as well as optical semiconductor devices. The experiment found that the Curie temperature in $Ga_{1-x}Mn_xAs$, $x \sim 0.05$ reaches as high as 110 K (Matsukura *et al.*, 1998).

The average field theory proposed by Dietl *et al.* (2001) predicted that a large kind of magnetic-ion-doped III–V or II–VI compounds have ferromagnetism and have a rather high Curie temperature. Later, the first-principle calculations also predicted that there exists ferromagnetic ground state in many different materials. These predictions excite the upsurge of crystal growth of FMSs, including: magnetic II–VI, III–V nitride, phosphide, arsenide compounds, and magnetic IV elements. In some cases, for example, (Zn, Mn)Te and (Cd, Mn)Te, the theoretical predictions have been confirmed by experiments (Haury, 1997; Ferrand, 2001; Boukari, 2002). Recently, as high as 940 K Curie temperature in (Ga, Mn)N has been reported (Sakaki *et al.*, 2002), and the substitution of the Ga atoms by the Mn ions was confirmed by many addition measurements, in order to exclude the virtual ferromagnetism caused by the external factors such as Mn ion cluster.

In order to synthesize ferromagnetic III–V compounds, the difficulty of the low solubility of magnetic ions should be first overcome. Solubility of magnetic ions in II–VI compounds is no problem, because the Mn^{2+}

ions and the II negative ions are equivalent, the magnetic ions can be easily introduced into the host materials to very high concentrations. By using nonequilibrium growth conditions of low-temperature MBE (LT-MBE), a III–V-based DMS, (In, Mn)As, was successfully prepared and shown to exhibit hole-induced ferromagnetic ordering at low temperatures (Ohno *et al.*, 1992). The key is that the InAs substrate temperature descends from 500°C to 250°C, this way can avoid the formation of thermodynamic stable compound MnAs and synthesize uniform alloys. The temperature 250°C is high enough to provide thermal energy needed for single crystal epitaxy growth.

The Mn ions in alloy provide not only the magnetic moment, but also the holes. The Mn ions act as acceptors to make InMnAs p-type. It is found later that the holes become mediators of ferromagnetic interaction between magnetic ions, resulting in the ferromagnetism.

3.1. FMS $Ga_{1-x}Mn_xAs$

$Ga_{1-x}Mn_xAs$ is a typical FMS, whose Curie temperature is rather high and whose magnetic and transport properties have been studied widely and deeply.

The first problem is that if the Mn ions are incorporated into the alloy uniformly and substrate the Ga atoms. In order to determine the lattice constants of grown (Ga, Mn)As layers, θ–2θ scans of XRD were performed at RT using Cu $K\alpha$ radiation (Ohno *et al.*, 1996). Clear diffraction peaks from (Ga, Mn)As were observed for all the samples with different Mn compositions. The lattice constant a is calculated from the peak position of (Ga, Mn)As (004) diffraction pattern assuming that (Ga, Mn)As is fully strained and that (Ga, Mn)As has the same elastic constant as GaAs. The lattice constants a versus Mn concentration x in $Ga_{1-x}Mn_xAs$ and $In_{1-x}Mn_xAs$ films are shown in Fig. 3.1 (Ohno, 1998). As can be seen from Fig. 3.1, a increases linearly with x, showing that it follows Vegard's law. The extrapolated lattice constant for GaAs (0.566 nm) and MnAs (0.598 nm) are in good agreement with the known GaAs lattice constant (0.56533 nm) and with the MnAs lattice constant extrapolated from the (In, Mn)As work (0.601 nm), suggesting that all the Mn ions were incorporated in the zincblende alloy.

The magnetization of the film is measured by superconducting quantum interference device (SQUID) magnetometer, which shows the presence of a ferromagnetic order in (Ga, Mn)As films. Figure 3.2 shows the magnetic

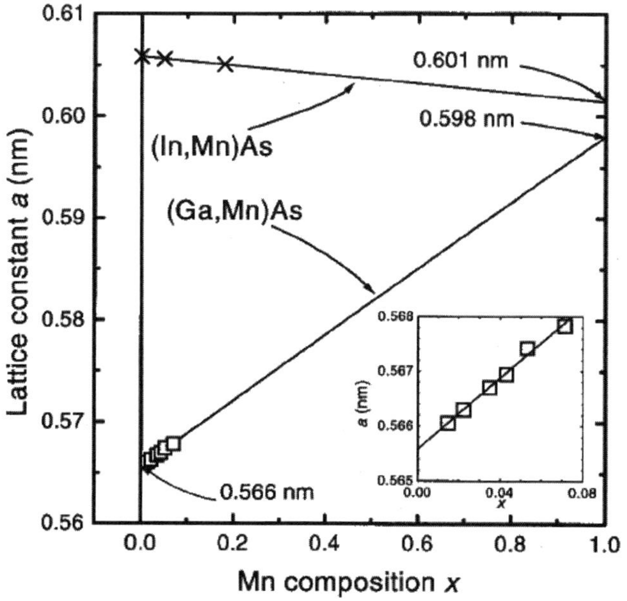

Fig. 3.1. Lattice constants a versus Mn concentration x in $Ga_{1-x}Mn_xAs$ and $In_{1-x}Mn_xAs$ films.

field dependence of magnetization M at 5 K for a $Ga_{1-x}Mn_xAs$ ($x = 0.035$) (Ohno *et al.*, 1996). If the (Ga, Mn)As film is grown on the GaAs substrate, it has a compressive strain in the surface plane, so the direction in the plane is the easy magnetization direction. In contrast, if the (In, Mn)As film is grown on the InAs substrate, it has a tensile strain in the surface plane, so the direction perpendicular to the plane is the easy magnetization direction. A clear square hysteresis loop observed in Fig. 3.2 indicates a well-orderd ferromagnetic structure present in the film. The inset in Fig. 3.2 shows the temperature dependence of the remanent magnetization M_r of the same sample. There is a clear transition to the ordered state below 60 K; thus, the Curie temperature is 60 K. Later, people found that the Curie temperature cannot be determined only by this method. If the Mn atoms form the MnAs cluster in GaAs, the magnetic hysteresis loop can also be measured by SQUID. Since the paper of Ohno *et al.* (1996) was published, there were many papers declaring they discovered FMSs with high Curie temperatures, but the results were caused by the MnAs clusters.

The concentration of hole introduced by the Mn ions in $Ga_{1-x}Mn_xAs$ is high, reaches 10^{18}–10^{20} cm^{-3}. Figure 3.3 shows the magnetic field

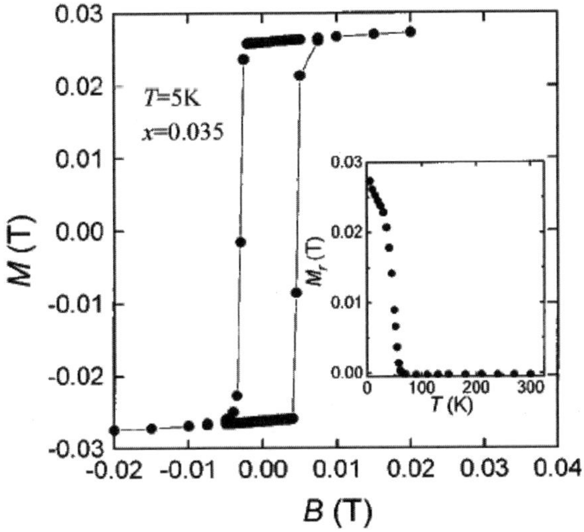

Fig. 3.2. Magnetic field dependence of magnetization M at 5 K for a $Ga_{1-x}Mn_xAs$ ($x = 0.035$). The field is applied parallel to the sample surface. Inset shows the temperature dependence of the remanent magnetization M_r of the sample.

dependence of Hall resistivity ρ_{Hall} and resistivity ρ of $Ga_{1-x}Mn_xAs$ ($x = 0.053$) with temperature as a parameter (Matsukura *et al.*, 1998). From Fig. 3.3(a), we see that when the temperature is low, ρ_{Hall} increases rapidly with increasing magnetic field and saturates at low field. Negative MR is observed in the whole temperature range and is most pronounced at intermediate temperatures (30–110 K). ρ_{Hall} of the magnetic materials can be expressed as,

$$\rho_{Hall} = R_0 B + R_s M, \tag{3.1}$$

where R_0 is the ordinary Hall coefficient, R_s the anomalous Hall coefficient, and M the magnetization of the sample. From the comparison between the results of magnetization and magnetotransport measurements (Ohno *et al.*, 1996), $R_s = c\rho$, and $c = 1.2$–$1.9\,T^{-1}$. When the magnetic field is small, the first term on the right-hand side of Eq. (3.1) is negligible relative to the second term; thus,

$$\rho_{Hall} = c\rho M.$$

From the ρ_{Hall} and ρ obtained from the magnetotransport measurements, we can determine the spontaneous magnetization M_s. The temperature dependence of M_s is shown in the inset of Fig. 3.3(a). From the figure, we

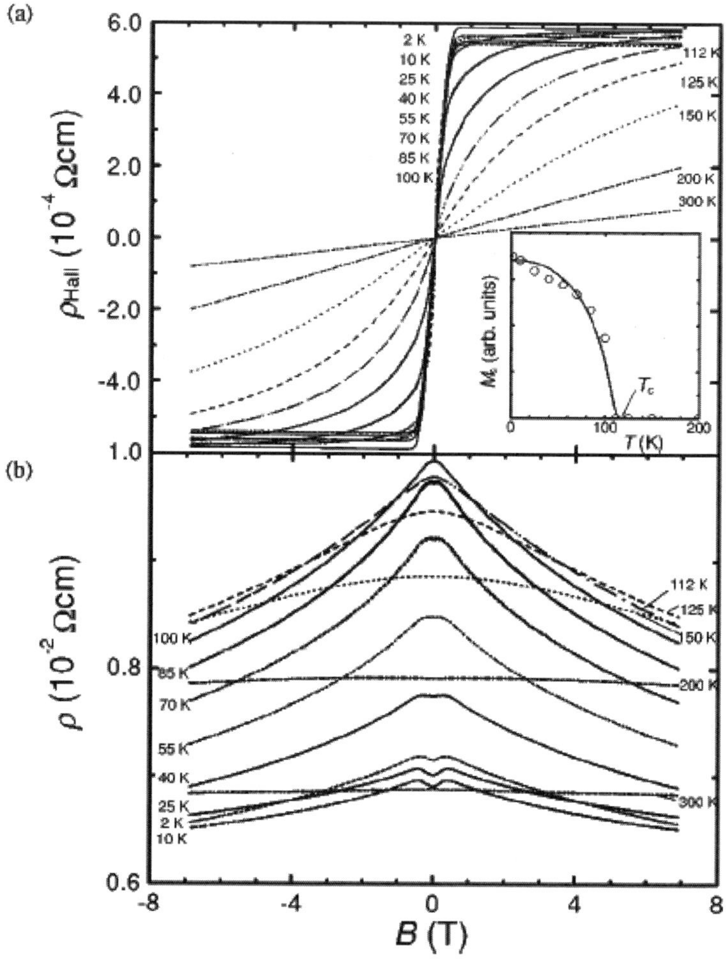

Fig. 3.3. Magnetic field dependence of Hall resistivity $\rho_{\rm Hall}$ and resistivity ρ of $Ga_{1-x}Mn_xAs$ ($x = 0.053$) with temperature as a parameter. The inset shows the temperature dependence of the spontaneous magnetization M_s determined from magneto-transport measurement, solid line is calculation by a mean-field theory.

see that when $T = 110\,\mathrm{K}$, $M_s = 0$; thus, we obtain the Curie temperature $T_C = 110\,\mathrm{K}$.

In the strong magnetic field ($B = 5{-}7\,\mathrm{T}$) and at low temperature ($T < 10\,\mathrm{K}$), $\rho_{\rm Hall}$ varies linearly with B. The slope of the $\rho_{\rm Hall}{-}B$ curve is used to determine R_0. The conduction type is p-type, and the hole concentration p of the sample shown in Fig. 3.3 is $1.5 \times 10^{20}\,\mathrm{cm}^{-3}$.

The results in Fig. 3.3 were measured when the magnetic field is perpendicular to the sample plane, i.e. the magnetic field is not in the easy magnetization direction. Samarth (2004) measured the magnetic dependence of the Hall resistivity $R_{\rm Hall}$ in the $Ga_{1-x}Mn_xAs$ layer with tensile strain, in this case, the perpendicular magnetic field is in the easy magnetization direction. At $T = 120\,K$, they observed the magnetic hysteresis of $R_{\rm Hall}$ and determined the Curie temperature of the sample $T_C = 135\,K$.

Figure 3.3(b) shows the magnetic field dependence of resistivity ρ of $Ga_{1-x}Mn_xAs$ ($x = 0.053$) with temperature as a parameter. Figure 3.4 shows the temperature dependence of ρ at zero field for $x = 0.015 - 0.071$ (Matsukura *et al.*, 1998). From Fig. 3.4, we see that the resistivities of mediate x samples increase with increasing temperature, reach a maximum near the T_C, and decrease with T continuously increasing. For low and high x samples, ρ is high at low temperature and decrease monotonously with T increasing. The former shows metal characteristic and the latter shows insulator characteristics. The inset in Fig. 3.4 shows ρ of $x = 0.053$ sample versus T at different magnetic fields. The peaks shift to higher temperature when the magnetic field increases. At $B = 7\,T$, ρ decreases about 20% relative to the zero-field case.

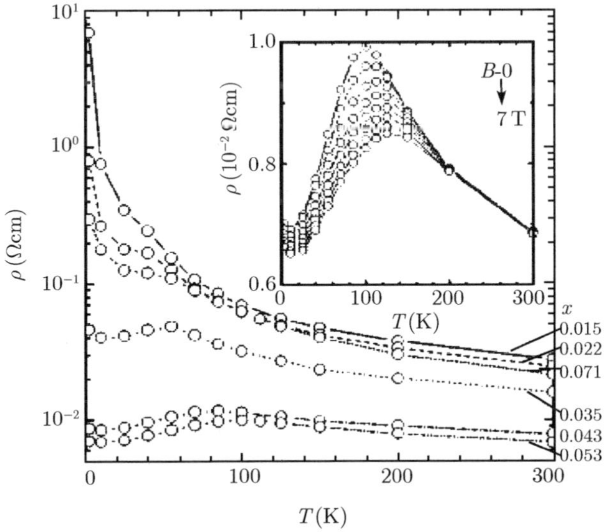

Fig. 3.4. Temperature dependence of ρ at zero field for $x = 0.015$–0.071. The inset shows ρ of $x = 0.053$ sample versus T at different magnetic fields.

The fact that ρ has maximum near T_C for mediate x samples indicates that the scattering of carriers by magnetic spin fluctuation via exchange interaction is the main scattering mechanism. The observed negative resistance in Fig. 3.3(b) (a metallic sample) can be understood as the reduction of scattering by aligning spins by B. Carriers are scattered by the exchange interaction of the following form,

$$H = -J_{\mathrm{pd}} \sum_i \delta(\boldsymbol{r} - \boldsymbol{R}_i) \boldsymbol{S}_i \cdot \boldsymbol{s}, \tag{3.2}$$

where $\boldsymbol{S}_i$ is the spin of Mn at $\boldsymbol{R}_i$ and $\boldsymbol{s}$ is the spin of carrier at $\boldsymbol{r}$. From Eq. (3.2), the resistivity by the spin-disorder scattering is obtained (Matsukura *et al.*, 1998),

$$\rho_s = 2\pi^2 \frac{k_{\mathrm{F}}}{pe^2} \frac{m^2 J_{\mathrm{pd}}}{h^3} n_s [S(S+1) - \langle S \rangle^2], \tag{3.3}$$

where k_{F} is Fermi wave vector, m the hole effective mass, n_s the density of Mn, $S = 5/2$ is Mn spin, $\langle S \rangle$ the thermal average of $\boldsymbol{S}$.

From Eq. (3.3), we see that when the magnetic field increases, $\langle S \rangle$ increases and ρ_{s} decrease. Equation (3.3) can fit well the experimental ρ–B curves (Fig. 3.2(b)) and ρ–T curves (Fig. 3.4). $\langle S \rangle$ was calculated from $M_{\mathrm{Hall}} = n_{\mathrm{s}} g \mu_{\mathrm{B}} \langle S \rangle$ obtained from the magnetotransport, k_{F} was calculated from p assuming a spherical Fermi surface, and the hole effective mass m was taken as $0.5\,m_0$. The exchange interaction parameter $J_{\mathrm{pd}} = 150 \pm 40\,\mathrm{eV\,Å^3}$ was obtained from the fitting.

Equation (3.3) accounts mainly for the variation of resistivity below the Curie temperature T_{C}. When the temperature is higher than T_{C}, the interaction between magnetic ions is mainly paramagnetic interaction, and the resistivity is obtained as (Ohno, 2002),

$$\rho_1 = 2\pi^2 \frac{k_{\mathrm{F}}}{pe^2} \frac{m^2 \beta^2}{h^3} \frac{k_{\mathrm{B}} T}{g^2 \mu_{\mathrm{B}}^2} (2\chi_{\perp}(T, B) + \chi_{||}(T, B)). \tag{3.4}$$

That means at a high temperature, the resistivity is proportional to the transverse and longitudinal magnetic susceptibilities $\chi_{\perp} = \partial M / \partial B$, $\chi_{||} = M/B$. According to the Weiss theory, $1/\chi = C(T - T_{\mathrm{C}})$, when the temperature increases, χ, i.e. ρ decreases. From the fit of Eq. (3.4) to the high-temperature parts of T and B dependences of ρ, $|N_0 \beta| = 1.5 \pm 0.2\,\mathrm{eV}$ is obtained, where N_0 is the density of cation sites.

The Curie temperature obtained from the magnetotransport measurement that the temperature at which the spontaneous magnetization equals zero is not precise. Next we introduce a method more precisely to determine the Curie temperature: Arrot plot method (Arrot, 1957). According to the

Weiss–Brillouin ferromagnetism phenomenal theory, the magnetization of a ferromagnetic material,

$$M(T, B) = \frac{g\mu_B J}{a^3} \mathrm{B}_J \left(\frac{g\mu_B J(B + \lambda M)}{k_B T} \right), \tag{3.5}$$

where a is the average distance of neighbor spins. It differs from the ordinary magnetization formula of the paramagnetic material Eq. (2.34), in the argument of the Brillouin function the magnetic field B is replaced by $B + \lambda M$, where λ is a parameter. Because the ferromagnetic material has strong spontaneous magnetization, λ is of 10^4–10^5 orders of magnitude. The Curie temperature can be calculated from Eq. (3.5), at $B = 0$; near the Curie temperature, the spontaneous magnetization is very small, thus the $B_J(x)$ can be expanded in x. Take the first term, we obtain

$$M = \frac{g\mu_B J}{a^3} \frac{J+1}{3J} \cdot \frac{g\mu_B J\lambda M}{k_B T_C}.$$

The Curie temperature,

$$T_C = \frac{J(J+1)}{3k_B a^3} (g\mu_B)^2 \lambda. \tag{3.6}$$

If $T_C = 100\,\mathrm{K}$, $a = 100\,\mathrm{nm}$, then form Eq. (3.6), we obtain $\lambda = 4.03 \times 10^4$.

Let $M_0 = g\mu_B J/a^3$, then from Eq. (3.5),

$$\frac{g\mu_B J(B + \lambda M)}{k_B T} = B_J^{-1} \left(\frac{M}{M_0} \right). \tag{3.7}$$

Near the Curie temperature, $M/M_0 \ll 1$, the right-hand side of Eq. (3.7) can be expanded in M/M_0. Because $y = B_J(x)$ is an odd function of x, $x = B_J^{-1}(y)$ is also an odd function of y, and Eq. (3.7) can be written as

$$\frac{\mu B}{k_B T} + \lambda \frac{\mu M}{k_B T} = \alpha \frac{M}{M_0} + \beta \left(\frac{M}{M_0} \right)^3 + \gamma \left(\frac{M}{M_0} \right)^5 + \cdots, \tag{3.8}$$

where $\mu = g\mu_B J$. When $T = T_C$, we have

$$\lambda \frac{\mu M}{k_B T_C} = \alpha \frac{M}{M_0}. \tag{3.9}$$

That is Eq. (3.6); thus,

$$\frac{\mu B}{k_B T_C} = \beta \left(\frac{M}{M_0} \right)^3 + \gamma \left(\frac{M}{M_0} \right)^5 + \cdots. \tag{3.10}$$

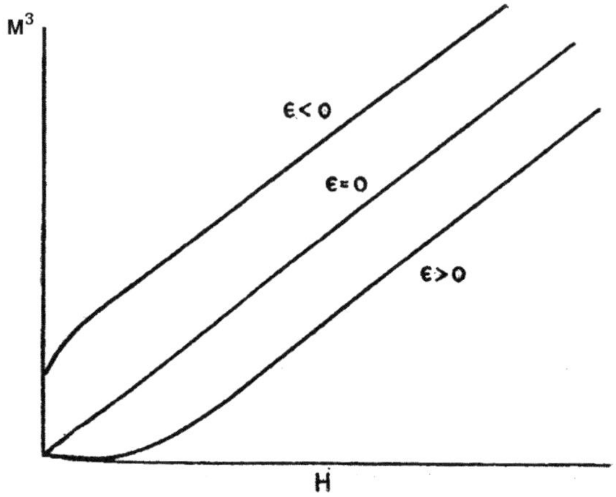

Fig. 3.5. Arrot plot — M^3 as functions of B for temperatures just below T_C ($\varepsilon < 0$), at T_C ($\varepsilon = 0$), and above T_C ($\varepsilon > 0$).

One plots M^3 as functions of B, if $T = T_C$, this will be a line crossing the coordinate origin, as shown in Fig. 3.5 ($\varepsilon = 0$) (Arrot, 1957). If $T \neq T_C$, then Eq. (3.8) can be written as

$$\frac{\mu B}{k_B T} = \varepsilon\alpha\frac{M}{M_0} + \beta\left(\frac{M}{M_0}\right)^3 + \gamma\left(\frac{M}{M_0}\right)^5 + \cdots, \tag{3.11}$$

where $\varepsilon = (1 - T_C/T)$. Figure 3.5 shows also the M^3–B curves for the case of $\varepsilon > 0$ and $\varepsilon < 0$; this figure is called the Arrot plot. The Curie temperature can be determined precisely by using the Arrot plot. Now one generally takes B/M as a variable and plots M^2–B/M; the result is the same.

Figure 3.6 is the Arrot plot of $Ga_{1-x}Mn_xAs$ ($x = 0.053$) sample (Ohno, 1998). The Hall resistance R_{Hall} and sheet resistance R_{sheet} are obtained from the magnetotransport measurement, and R_{Hall}/R_{sheet} is proportional to the magnetization M_0 (see Eq. (3.1)). From Fig. 3.6, we see that when $T < 110\,K$, the M^2 curve is similar to the $\varepsilon < 0$ case, and the extrapolated straight lines of the saturation magnetization intersect with the ordinate. When $T > 110\,K$, the curves correspond to the $\varepsilon > 0$ case, and the extrapolated straight lines of the saturation magnetization intersect with the abscissa. The Curie temperature is then determined $T_C = 110\,K$.

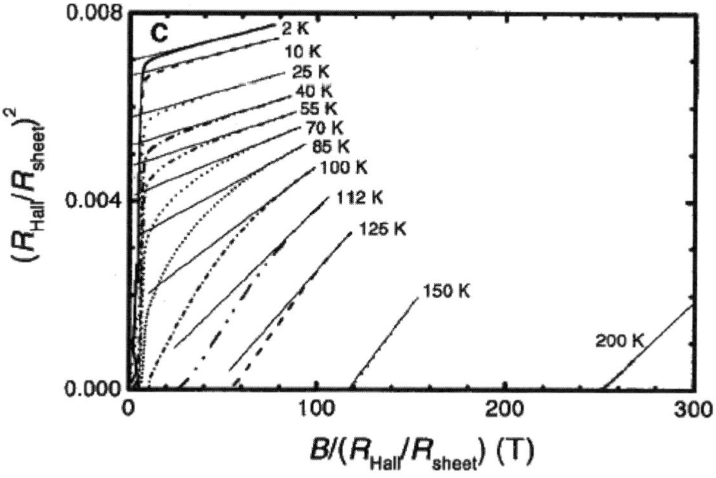

Fig. 3.6. Arrot plot of $Ga_{1-x}Mn_xAs$ ($x = 0.053$) sample.

3.2. Other FMSs

$In_{1-x}Mn_xAs$ was one of the FMSs discovered most early, though its Curie temperature is not high $T_C = 7.5$ K. Figure 3.7(a) shows the temperature dependence of the remanent Hall resistivity divided by the resistivity $\rho_{Hall}(0)/\rho(0)$, and Fig. 3.7(b) shows the temperature dependence of remanent magnetization M_0 measured by a SQUID magnetometer (Ohno *et al.*, 1992). Comparison of Figs. 3.7(a) and 3.7(b) shows that $\rho_{Hall}(0)/\rho(0)$ is indeed proportional to M_0, from which the constant c in $R_s = c\rho$ (Eq. (3.1)) is determined as 6.3. ρ_{Hall}/ρ equals zero at $T = 7.5$ K, but the M_0 measured by SQUID does not equal zero at 7.5 K and approaches zero slowly when the temperature increases. This indicates that the determination of the Curie temperature by SQUID is not precise.

$In_{1-x}Mn_xSb$ is another narrow gap FMS. Figure 3.8(a) shows the temperature dependence of zero-field resistivity for $In_{1-x}Mn_xSb$/InSb/CdTe ferromagnetic layers with various Mn concentrations, Fig. 3.8(b) shows Hall resistivity versus magnetic field at 1.4 K in four samples (Wojtowicz *et al.*, 2003). From Fig. 3.8(a), we see that while the resistivity of the nonmagnetic sample decreases monotonically with decreasing temperature ($In_{1-y}Be_ySb$ film, denoted by open circles) due to increasing hole mobility, the ρ versus T dependence for all magnetic layers is nonmonotonic. The resistivities of the three $In_{1-x}Mn_xSb$ layers grown with Mn cell temperature T_{Mn} of 700°C,

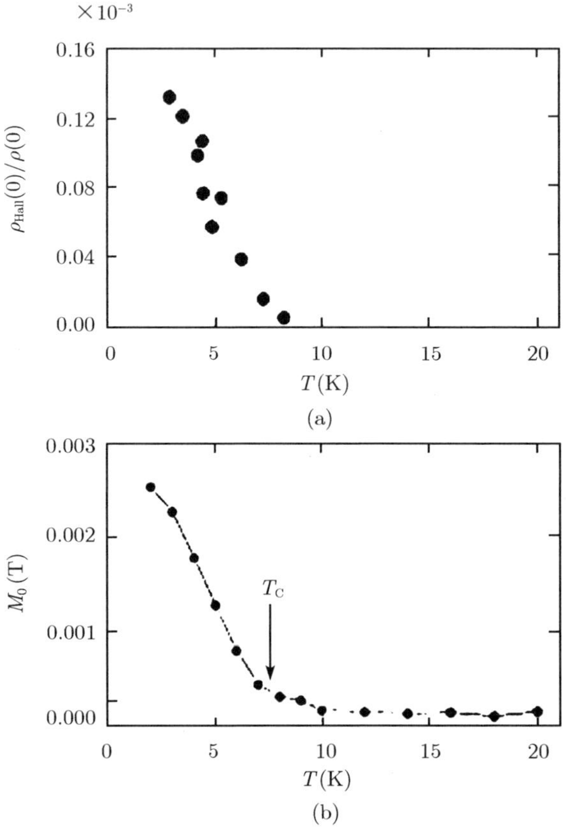

Fig. 3.7. (a) Temperature dependence of the remanent Hall resistivity of $In_{1-x}Mn_xAs$ divided by the resistivity $\rho_{\text{Hall}}(0)/\rho(0)$. (b) Temperature dependence of remanent magnetization of the film.

710°C, and 720°C show clear maxima at 5.4, 7.5, and 9.2 K, respectively. These temperatures serve as a good estimation of T_C. From Fig. 3.8(b), we see that at $T = 1.4\,\text{K}$ the ρ_{Hall} of all samples show magnetic hysteresis loops and the amplitude of the anomalous Hall resistivities and the width of the hyteresis loops increase with increasing Mn concentration. Because the $In_{1-x}Mn_nSb$ layers are grown on the CdTe/GaAs substrate and have small tensile strain, the perpendicular direction is the easy magnetization direction.

Figure 3.9 shows field dependence of magnetization measured by SQUID in $In_{0.972}Mn_{0.028}Sb$ (Wojtowicz *et al.*, 2003). From Fig. 3.9, we see that when $T < 10\,\text{K}$, clear hysteresis loops are observed. The hysteresis

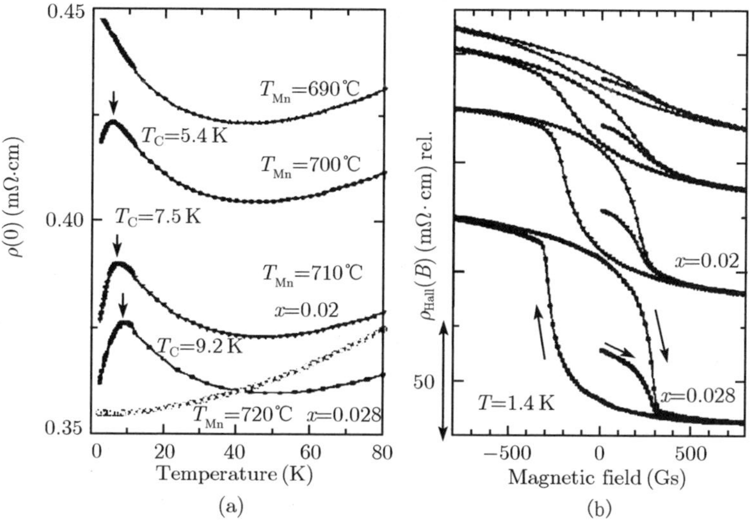

Fig. 3.8. (a) Temperature dependence of zero-field resistivity for $In_{1-x}Mn_xSb/InSb/CdTe$ ferromagnetic layers with various Mn concentrations determined by the Mn effusion temperature T_{Mn}. (b) Hall resistivity vs magnetic field at 1.4 K in four samples.

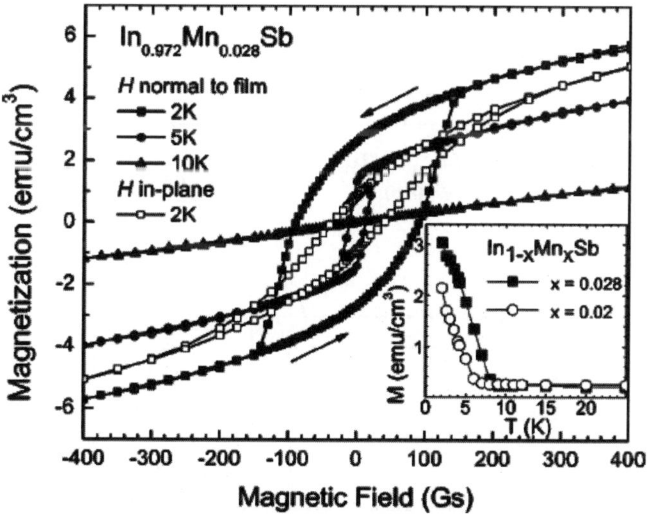

Fig. 3.9. Field dependence of magnetization measured by SQUID in $In_{0.972}Mn_{0.028}Sb$ at various temperatures with the field applied either perpendicular or parallel to the layer plane. The inset shows the temperature dependence of magnetization for $In_{1-s}Mn_xSb$ with $x = 0.02$ and 0.028 at $B = 10$ Gs.

loops for the perpendicular magnetic field are more obvious than that for the parallel field, indicating that the easy axis of magnetization is perpendicular to the layer plane, because the magnetic layer has a small tensile strain. The inset shows the temperature dependence of magnetization for $In_{1-s}Mn_xSb$ with $x = 0.02$ and 0.028 in the perpendicular magnetic field $B = 10\,Gs$. The Curie temperatures determined from these plots are $T_C = 7$ and 8.5 K for two samples. When the temperature approaches T_C, the M decreases rapidly, but does not equal zero. The remanent M is estimated to be produced by the MnSb clusters in samples.

The Mn^{2+} ions in III–V compounds introduce not only magnetic moments, but also holes and through the mediate role of the hole produce ferromagnetism. Though the Mn^{2+} ions are doped in II–VI compounds more easily than in III–V compounds, the charges carried by Mn^{2+} ions are the same as that carried by the substituted bivalent ions; thus, the Mn^{2+} ions cannot provide hole. Therefore, the Mn^{2+}-doped II–VI semiconductors are generally paramagnetic, not ferromagnetic, as discussed in Chap. 2.

Figure 3.10 shows the magnetic field dependence of magnetization in $Zn_{1-x}Cr_xTe$ ($x = 0.20$) at various T measured by SQUID, the inset is the Arrot plot (Saito *et al.*, 2003). From Fig. 3.10, we see that until $T = 300\,K$, the magnetization curves still have hysteresis loops. From the Arrot, we obtain the Curie temperature $T_C = 300 \pm 10\,K$.

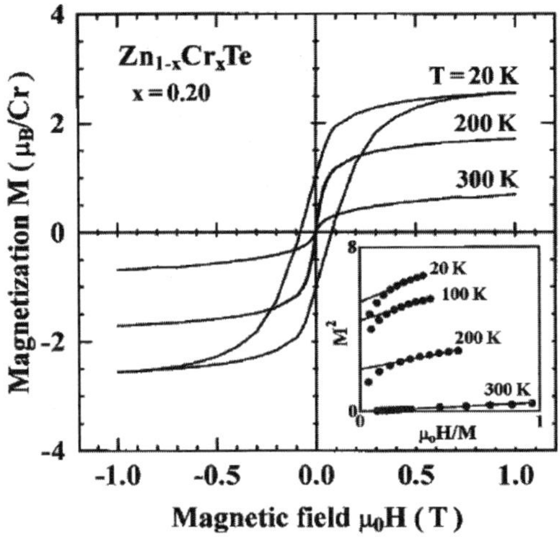

Fig. 3.10. Magnetic field dependence of magnetization in $Zn_{1-x}Cr_xTe$ ($x = 0.20$) at various T. The inset shows the Arrot plot.

To confirm that the observed ferromagnetism originated from the $Zn_{1-x}Cr_xTe$ DMS, not the CrTe clusters, the authors analyzed the magnetic circular dichroism (MCD) spectral shapes (see Sec. 2.4.3 and Fig. 2.24). The intensity of MCD spectral line varies linearly with the giant Zeeman splitting energy, i.e. proportional to the magnetization. The intensity of MCD spectral line depends sensitively on the critical points (CPs) of the energy band; near the CP, the intensity of optical transitions increases a lot. Thus, by use of the analysis of the MCD spectral line shape, one can distinguish the contributions of the DMS and the magnetic cluster. The authors' MCD studies clearly showed that $Ga_{1-x}Mn_xAs$ and $In_{1-x}Mn_xAs$ are real ferromagnetic DMSs. Conversely, The DMSs of ZnO:Co, ZnO:Ni, and GaN:Mn are, in reality, paramagnetic. The observed ferromagnetism comes from some unidentified materials.

MCD spectra are measured in a transmission Faraday configuration, and the GaAs substrates of the samples were removed by chemical etching. Figure 3.11 shows the MCD spectra of various samples at $T = 20\,\text{K}$, $B = 1\,\text{T}$ (Saito *et al.*, 2003), where (a) RT CrTe film, (b) 100 nm thick ZnTe film, (c) 80 nm thick $Zn_{1-x}Cr_xTe$ ($x = 0.20$) film on ZnTe buffer with several monolayers, and (d) 400 nm thick $Zn_{1-x}Cr_xTe$ ($x = 0.20$) film on 200 nm thick ZnTe buffer. The MCD spectrum of RT ferromagnetic

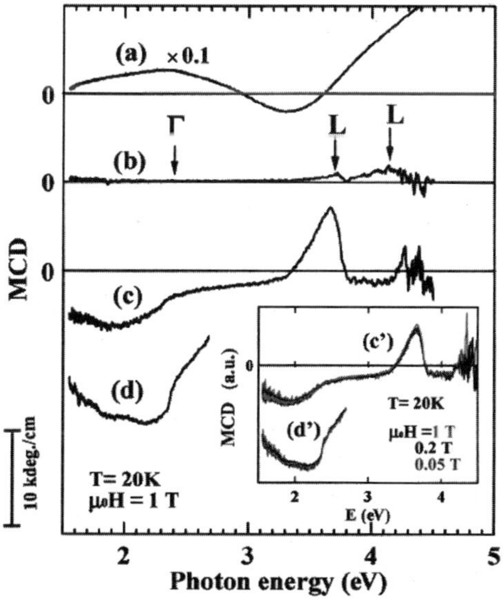

Fig. 3.11. MCD spectra of various samples at $T = 20\,\text{K}$, $B = 1\,\text{T}$.

CrTe film (Fig. 3.11(a)) shows a broad spectrum reflecting its metallic nature, which is completely different from that of ZnTe and $Zn_{1-x}Cr_xTe$. A 100 nm thick ZnTe film (Fig. 3.11(b)) shows weak MCD signals around the Γ (2.4 eV) and L (3.7 and 4.2 eV) CPs due to the diamagnetic Zeeman effect. An 80 nm thick $Zn_{1-x}Cr_xTe$ ($x = 0.20$) film on a thin ZnTe buffer (Fig. 3.11(c)) shows pronounced MCD structures at photon energies corresponding to the L-CPs of ZnTe, indicating that $Zn_{1-x}Cr_xTe$ shares a common band structure with ZnTe. But the $Zn_{1-x}Cr_xTe$ film shows a positive MCD peak at 3.7 eV and a negative signal at 4.2 eV, different from those of ZnTe. This is the general characteristic of the Zeeman splitting induced by the sp–d exchange interaction, and the $Zn_{1-x}Cr_xTe$ film was therefore confirmed to be a DMS.

The inset in Fig. 3.11(c′) shows the normalized MCD spectra of $Zn_{1-x}Cr_xTe$ ($x = 0.20$) film in various magnetic fields. If the film consists of deposits of magnetic atoms, the shape of the MCD will change with the magnetic field, because $Zn_{1-x}Cr_xTe$ and deposits have different magnetization processes and optical spectrum shapes. From Fig. 3.11(c′), we see that the MCD spectra in various magnetic fields are basically the same, indicating that they come from a single material $Zn_{1-x}Cr_xTe$.

The magnetic field dependence of MCD intensity at Γ point, $E = 2.2$ eV coincides with the magnetization curves measured using a SQUID in Fig. 3.10. The Arrot plot can also be obtained by the MCD intensities, as shown in Fig. 3.12 (Saito *et al.*, 2003). The Curie temperature obtained from the MCD Arrot is same as that from the magnetization Arrot plot $T_C = 300 \pm 10$ K. The resistivity of the $Zn_{1-x}Cr_xTe$ film shows semiconducting behavior, that is, a monotonous increase with decreasing T. The resistivity of the film at RT and 100 K are 58 and 280 Ωcm, respectively, which is several orders higher than that of $Ga_{1-x}Mn_xAs$.

Another DMS with higher Curie temperature is $Ga_{1-x}Mn_xP$ (Overberg *et al.*, 2001), which is doped with C to increase the hole concentration. XRD, reflection high-energy electron diffraction (RHEED), and transmission electron microscopy (TEM), etc., were performed to check the crystal quality of $Ga_{1-x}Mn_xP$, the results are similar to those of GaP and Gap:C and reveal that no second phase precipitates or superparamagnetic Mn clusters are found. Magnetization versus magnetic field and versus temperature were measured by SQUID magnetometry, these data indicate that the observed ferromagnetism persists to approximately 250 K. Ion implantation of Mn ions into C-doped GaP has been reported to induce ferromagnetism behavior above RT for optimized Mn concentration near 3 at%. Figure 3.13

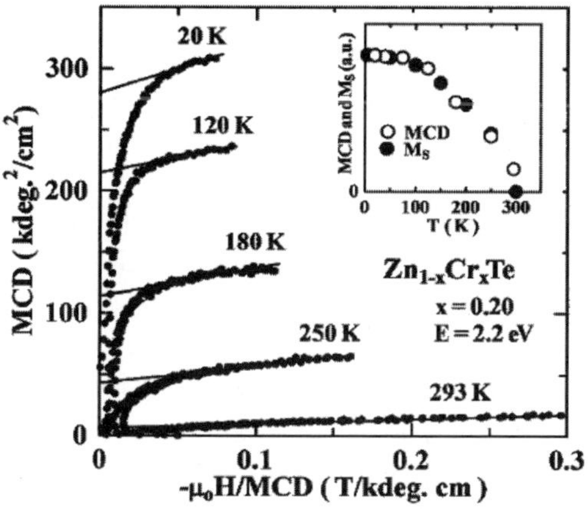

Fig. 3.12. Arrot plot of MCD intensity of $Zn_{1-x}Cr_xTe$ ($x = 0.20$) film at $E = 2.2\,\text{eV}$ and various T. The inset shows the temperature dependence of MCD intensity (open circles) extrapolated to $\mu_0 H = 0\,\text{T}$ obtained by the Arrot plot, together with that of M_s (solid circles).

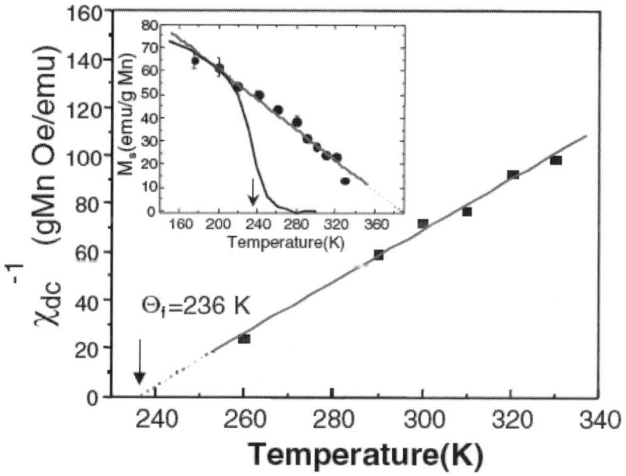

Fig. 3.13. Temperature dependence of inverse susceptibility $\chi^{-1}(T)$ and saturated magnetization (inset) for $Ga_{0.93}Mn_{0.06}P$ (3 at%).

shows the temperature dependence of inverse susceptibility $\chi^{-1}(T)$ and saturated magnetization (inset) for $Ga_{0.93}Mn_{0.06}P$ (3 at%) (Theodoropoulou *et al.*, 2002). The magnetic susceptibility $\chi = \mathrm{d}M_{\mathrm{ZFC}}/\mathrm{d}H$, where M_{ZFC} is the magnetization measured at small magnetic field (500 Oe), which is proportional to the magnetic field. The Curie temperature is identified $T_C = 236\,K$ (vertical arrow in Fig. 3.13) by a linear extrapolation of $\chi^{-1}(T)$ to zero. The hysteresis and accompanying saturated moment extend with diminishing magnitude to substantially higher temperature: 330 and 390 K for a linear extrapolation of $M_s(T)$ to zero (inset in Fig. 3.13).

The higher Curie temperature is due to ion implantation, resulting in Mn clusters of uniform size in samples. Taking into account the Mn clusters and using the spin wave stiffness D of spin wave excitations (magnons), the authors estimated the upper bound of T_C as 385 K.

Up to now, the reported DMS with highest Curie temperature is $Ga_{1-x}Mn_xN$ (Thaler *et al.* 2002). No second phases are found for Mn levels below ~10% for growth temperature of ~750°C. The (Ga, Mn)N retains n-type conductivity under these conditions. Figure 3.14 shows (a) B–H measurement at $T = 300\,K$ for $Ga_{1-x}Mn_xN$ ($x = 0.09$) (closed circles) and the sapphire substrate (open circles), (b) temperature dependence

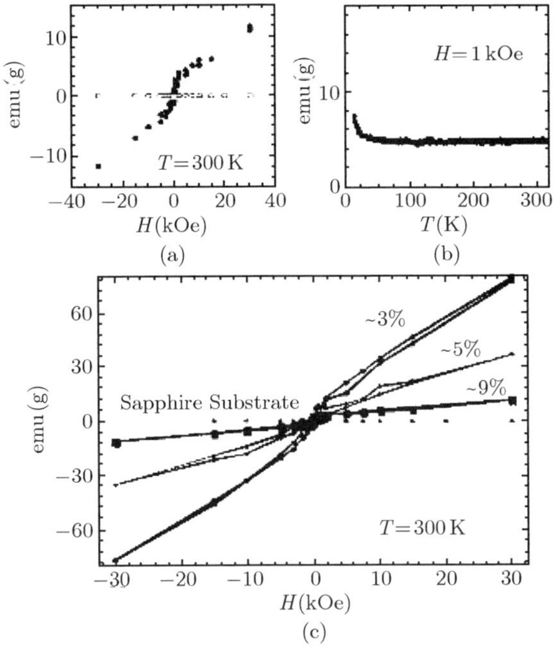

Fig. 3.14. (a) B–H measurement at RT for $Ga_{1-x}Mn_xN$ with 9% Mn (closed circles) and the sapphire substrate (open circles). (b) Temperature dependence of the magnetization for the same sample at $B = 1\,kOe$. (c) B–H measurement at RT for GaMnN grown at 700°C with various Mn concentrations (Thaler *et al.*, 2002).

of the magnetization for the same sample at $B = 1\,\text{kOe}$, and (c) B–H measurement at $T = 300\,\text{K}$ for GaMnN grown at 700°C with various Mn concentrations. As seen in Fig. 3.14(c), B–H measurements indicate weak but discernable remanent magnetization and hysteresis indicative of ferromagnetic ordering for all of the Mn concentrations (~3–9%) studied. The 3% sample shows the highest moment per Mn ion. Higher Mn concentrations lead to enhanced antiferromagnetic coupling, resulting in a lower moment for Mn.

3.3. Fermi-level Engineering

Dietl *et al.* (2001) proposed a mean-field theory about the Curie temperature of FMSs, which suggests that the ferromagnetic coupling between magnetic ions is mediated by holes, so that when the hole concentration is higher, the theoretically predicted Curie temperature is higher. This theory is valid for some FMSs, for example, Mn-doped III–V compounds, but is not valid for $Zn_{1-x}Cr_xTe$. As mentioned in the last section, the Curie temperature of $Zn_{1-x}Cr_xTe$ ($x = 0.20$) is 300 K (Saito *et al.*, 2003), but it has high resistance. The ferromagnetic origin cannot be the carrier-induced mechanism, thus it is not clear. For this, Ozaki *et al.* (2005) investigated the electric and magnetic properties on p-type $Zn_{1-x}Cr_xTe$ doped with N as an acceptor and how the hole concentration influences the Curie temperature.

First the XRD and high-resolution TEM (HRTEM) measurements confirmed that the (Zn, Cr)Te:N layers are almost the same as the (Zn, Cr)Te layers, and any precipitates of other phases were not observed. The concentration of the substitutional nitrogen atoms $[N_{Te}]$ remained within the same order of $10^{20}\,\text{cm}^{-3}$ for a series of samples ($x = 0$–0.09). The RT Hall measurement found that the hole concentration decreases with increasing x, this is because Cr produces an impurity energy level in the middle of the band gap of ZnTe, which can trap the hole provided by N acceptor, i.e. play a compensation role. Figure 3.15 shows the temperature dependence of the resistivity ρ of p-$Zn_{1-x}Cr_xTe$ with various x (Ozaki *et al.*, 2005). With the increase of x, the ρ increases by several order of magnitudes, indicating that the hole density p decreases by three orders of magnitudes. The temperature variation of ρ exhibits the transition from metallic to insulating (MIT) behavior, and the transition takes place between $x = 0.01$ ($p = 1.1 \times 10^{20}\,\text{cm}^{-3}$) and $x = 0.014$ ($p = 1.6 \times 10^{19}\,\text{cm}^{-3}$), in agreement with the critical hole density for MIT $p_c \approx 10^{19}\,\text{cm}^{-3}$ according to Mott's theory. Therefore, when the Cr composition x increases at a fixed N

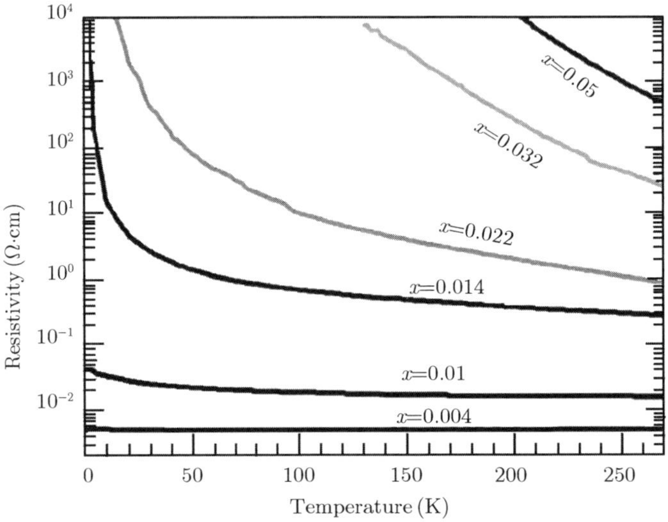

Fig. 3.15. Temperature dependence of the resistivity of p-$Zn_{1-x}Cr_xTe$ with various x.

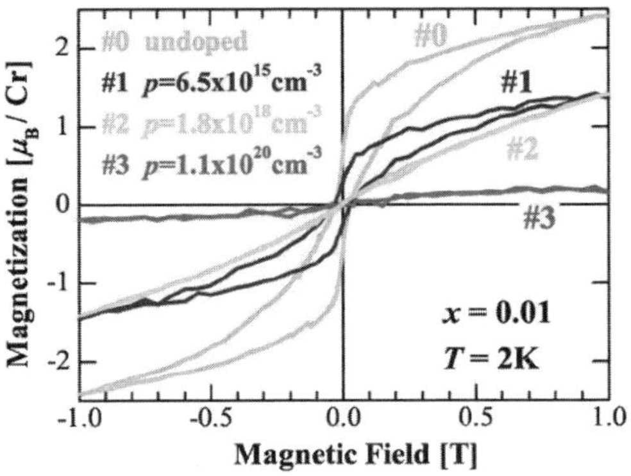

Fig. 3.16. Magnetic field dependence of magnetization for p-$Zn_{099}Cr_{0.01}Te$ with various hole density at $T = 2\,K$.

concentration, the MIT is expected to take place at a certain point where the density of the Cr deep level exceeds that of the N acceptor level.

Figure 3.16 shows the magnetic field dependence of magnetization for p-$Zn_{099}Cr_{0.01}Te$ with various hole concentrations at $T = 2\,K$ (Ozaki *et al.*, 2005). A clear hysterisis loop is observed in the undoped sample no. 0, but this hysteresis becomes less pronounced and disappears with increasing hole concentration p in the order of nos. 1–3. The suppression of the ferromagnetic hysteretic behaviors due to the N doping is also observed at higher Cr compositions. The Arrot analysis shows that the ferromagnetic transition does not take place down to the lowest temperature of 2 K in all the p-doped samples ($x = 0.02$–0.09). This is in sharp contrast to the undoped samples in which the Curie temperature T_C increases almost linearly with x from $T_C = 5$ K ($x = 0.01$) to 125 K ($x = 0.09$) (Ozaki *et al.*, 2004).

The carrier-induced ferromagnetism mechanism can obviously not explain the experimental results in (Zn, Cr)Te. Sato and Katayama-Yoshida (2002) proposed the double exchange interaction as an origin of the ferromagnetism in Cr-doped II–VI compounds based on the first-principle calculation. According to their calculation, Cr 3d electrons create non-bonding state e and antibonding state t_a in the band gap of ZnTe. Since the Fermi level is located in the middle of the t_a state, the ferromagnetic alignment of Cr spins is stabilized due to the band broadening through the electron hopping (double exchange mechanism). When the acceptor concentration increases, the Fermi level deviates from the center of the t_a state and the density of states at the Fermi level decreases, which results in the decline of the ferromagnetic interaction based on the double exchange mechanism. This modulation method of material ferromagnetism is called "Fermi-level engineering."

Another example of the Fermi-level engineering is GaMnN (Reed *et al.*, 2005), in which the dependence of the ferromagnetism on the Fermi level for n- or p-GaMnN is studied. Three distinct types of GaMnN films were grown: undoped, Si doped, and Mg doped. The Si concentration [Si] ranges from 1×10^{18} to $2 \times 10^{20}\,\text{cm}^{-3}$, and the Mg concentration [Mg] ranges from 5×10^{19} to $2 \times 10^{20}\,\text{cm}^{-3}$. No variation in Mn concentration was observed in GaMnN films due to Si or Mg doping. Figure 3.17 shows the magnetic field dependence of magnetization for GaMnN:Si samples with fixed Mn density ($10^{20}\,\text{cm}^{-3}$) and various Si concentrations at $T = 300$ K (Reed *et al.*, 2005). The data indicate that high Si concentrations in the GaMnN:Si films almost eliminate the FM response of the films, whereas a much stronger FM response is observed for low and moderate Si concentrations. In the undoped case, the Fermi level is just in the impurity band of Mn, there is a partial occupancy of electrons and holes in this Mn impurity band, and ferromagnetic behavior occurs in sample a as shown in Fig. 3.17. In the Si heavy-doped case ($n = 2 \times 10^{19}\,\text{cm}^{-3}$), the E_F is close to the CB, the deep Mn band is expected to be completely filled with electrons, leaving no available holes to mediate the magnetic exchange interaction. This results

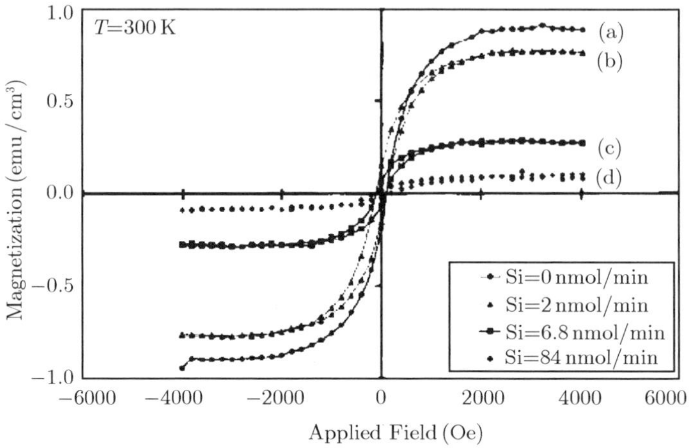

Fig. 3.17. Magnetic field dependence of magnetization for GaMnN:Si samples with fixed Mn density (10^{20} cm^{-3}) and various Si concentrations at $T = 300$ K.

in nonmagnetic behavior as shown in sample d. Reducing the silane flow by an order of magnitude results in two effects: (1) the electron concentration in the CB decreases and the film resistivity increases; (2) E_F moves closer to the Mn energy band which becomes partially filled since [Si]$\ll$[Mn], and ferromagnetic behavior results as shown in samples b and c.

The analysis of GaMnN:Mg is similar, and the Mg acceptor level is ~1.2 eV below the Mn band; electrons in the Mn impurity band are absorbed by the normally empty Mg acceptor states, thus not favorite the Mn–Mn exchange interaction. The introduction of Mg acceptor states at a concentration approaching the Mn concentration ($\sim 2 \times 10^{20}$ cm^{-3}) should nearly eliminate the film's FM response. On the other band, the FM properties of nonoptimal GaMnN films can be enchanced in some samples by the addition of Mg. When the grown GaMnN films have a weakened FM response, the addition of Mg acceptor levels compensates a portion of the electrons filling the Mn impurity band, leaving a partially filled band which is required for ferromagnetism. Figure 3.18 shows the magnetic field dependence of magnetization for (h) undoped GaMnN sample and (g) Mg-doped sample [Mg] $= 5 \times 10^{19}$ cm^{-3} (Reed *et al.*, 2005). In the nonoptimal GaMnN sample, only 2.45% of Mn ions contribute to ferromagnetism. The addition of Mg increases the contributing Mn concentration to 10.42%, a more 4-fold increase.

Therefore, the concept of Fermi-level engineering can be used to explain the experimental results and reveal a means of controlling the

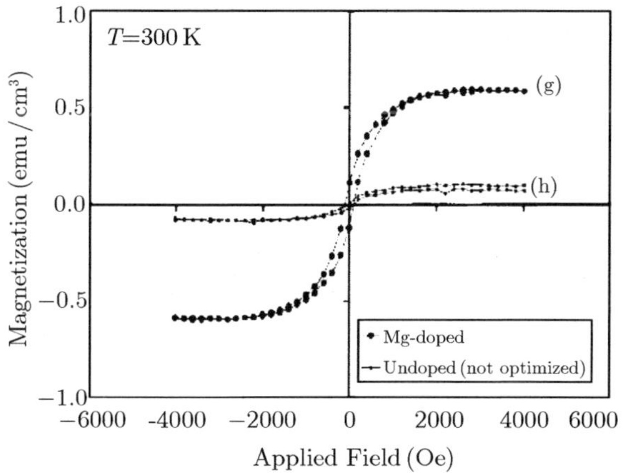

Fig. 3.18. Magnetic field dependence of magnetization for (h) undoped GaMnN sample and (g) Mg-doped sample [Mg] = 5×10^{19} cm^{-3}.

magnetic properties of GaMnN and other films. Ferromagnetism can only be observed if the E_F is located within or near the Mn impurity band, so that carriers are available in this band to mediate the FM properties.

3.4. Influence of Clusters on Ferromagnetism

The alloy phase diagram has a gap of solubility, in which the uniform alloy cannot be grown, and the spinodal decomposition takes place, i.e. the alloy divides into high- and low-density regions of one composition (for example, Mn). As a result, the correlated nanocrystals are formed in the main phase. For example, in the epitaxy-grown (Ga, In)N, the In-rich QDs is found embedded in the In-low matrix. The DMSs have the similar case.

Several methods for introducing MnAs nanoclusters into GaAs, such as Mn ion implantation followed by subsequent thermal annealing, or annealing of the low-temperature MBE-grown (Ga, Mn)As epilayer have been reported (Moreno *et al.*, 2002). The GaAs:MnAs granular material, consisting of Mn(Ga)As nanoclusters embedded in GaAs has a 2-fold advantage of being ferromagnetic/superparamagnetic (depending on cluster size) at RT, while keeping excellent compatibity with III–V heterostructures. Besides, the GaAs:MnAs granular material exhibits a giant magneto-optical effect. Its magneto-optical intensity is greater than that of dilute (Ga, Mn)As at RT. The Verdet constant at 2000 Oe has been shown to be 16 times

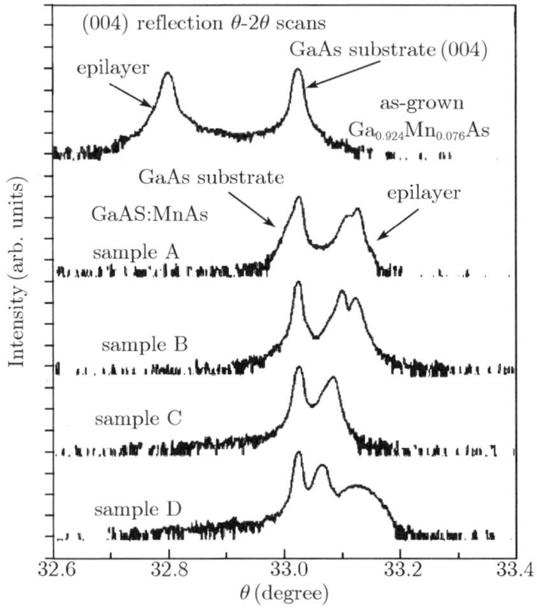

Fig. 3.19. XRD θ–2θ scans for the (004) reflection from an as-grown $Ga_{0.924}Mn_{0.076}As$ film and four GaAs:MnAs granular films annealed under various conditions.

larger than that of (Cd, Mn, Hg)Te, which was commercially developed for an optical isolator. GaAs:MnAs is thus very suitable for application to magneto-optical devices, monolithically integrated with semiconductor laser diodes.

Figure 3.19 shows XRD spectra of the as-grown $Ga_{0.924}Mn_{0.076}As$ film and GaAs:MnAs granular films (Yokoyama *et al.*, 2005). Some peaks at 33.05–33.20° in the spectra of GaAs:MnAs (samples A–D) include the information of the GaAs:MnAs granular layers containing MnAs clusters. Whereas the epilayer peak of the as-grown $Ga_{0.924}Mn_{0.076}As$ is on the lower angle side of the GaAs substrate peak, the epilayer peaks of GaAs:MnAs are on the higher angle side when the annealing temperature is above 500°C. The annealing temperature of samples A and B is 500°C, the MnAs cluster has the zincblende structure; the annealing temperature of samples C and D is 600°C, the MnAs cluster has the wurtzite structure (NiAs type). The XRD peaks of the two types of samples in Fig. 3.19 are also different.

Figure 3.20 shows the (a) magnetization curves of GaAs:MnAs sample A (ZB) and sample C (NiAs-WZ) at 300 K and (b) temperature dependence

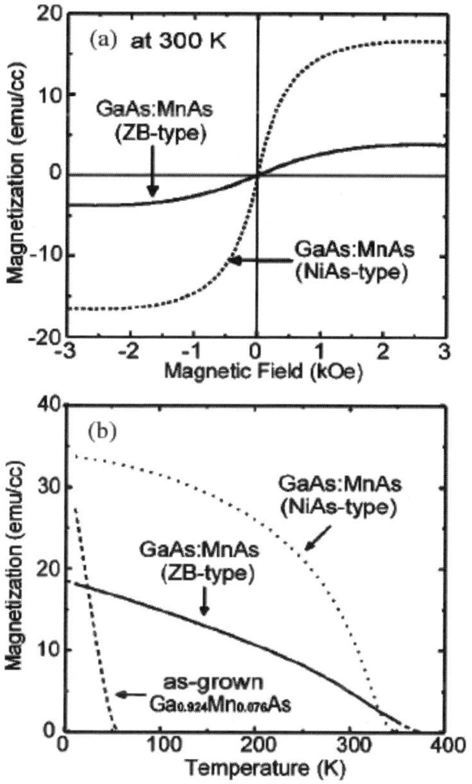

Fig. 3.20. (a) Magnetization curves of GaAs:MnAs sample A (ZB) and sample C (NiAs-WZ) at 300 K. (b) Temperature dependence of magnetization at 2000 Oe for two samples and as-grown $Ga_{1-x}Mn_xAs$.

of magnetization at 2000 Oe for two samples and as-grown $Ga_{1-x}Mn_xAs$ (Yokoyama *et al.*, 2005). The magnetization measurements reveal that samples A, B, C, and D are superparamagnetic or ferromagnetic with very small coercivity, and that the saturated magnetizations of samples A and B are much smaller than those of samples C and D. Assuming that the magnetization comes only from the MnAs cluster, the magnetization per Mn atom is estimated to be 0.3, 0.1, 1.2, and 1.0μ_B for samples A, B, C, and D, respectively. The Curie temperature T_C of the ZB-type MnAs clusters is estimated to be 360 K, which is remarkably higher that of the NiAs-type hexagonal MnAs clusters 313 K. The magnetic-optical intensities (Kerr ellipticity) of NiAs-type hexagonal MnAs clusters are larger than those of the ZB-type MnAs clusters.

3.5. QDs of FMSs

The above-studied materials are all FMS bulk materials. If the QDs and quantum wires of the FMS are synthesized, how does the quantum confinement effect influence the ferromagnetism? The InMnAs QDs has been grown by the MBE method, which has a Curie temperature much higher than that of the bulk material (Jeon *et al.*, 2002). The results of the AFM images show that the densities of the $In_{1-x}Mn_xAs$ QDs grown with Mn/In flux ratios of 0, 0.2, 0.3, and 0.4 are 2.36×10^{10}, 4.0×10^9, 3.5×10^9, and $1.62 \times 10^9\,cm^{-2}$, respectively. The $In_{1-x}Mn_xAs$ QD growth mechanism is a combination of Mn atom and Stranski–Krastanov mode growths and is different from that for InAs QDs due to the addition of Mn. The formation of the self-assembled $In_{1-x}Mn_xAs$ QDs originates from the surfactant effect of the Mn atoms acting as nuclei. Therefore, the Mn atoms are contained only in the $In_{1-x}Mn_xAs$ QDs. Even though the Mn content of the $In_{1-x}Mn_xAs$ QDs is 16%, exceeding that in bulk material, the QDs maintain single crystalline phases.

The x content of Mn in $In_{1-x}Mn_xAs$ QDs is demonstrated by the energy-dispersive X-ray spectroscopy (EDX). Figure 3.21 shows the EDX spectrum of $In_{0.84}Mn_{0.16}As$ QD (Jeon *et al.*, 2002). The EDX spectrum demonstrates that the stoichiometry of the QD is $In_{1-x}Mn_xAs$, and the ratio of In and Mn compositions is 84:16. Thus, the measured sample is $In_{0.84}Mn_{0.16}As$.

The 2D and 3D AFM images of $In_{0.79}Mn_{0.21}As$ QDs show that the dots are very dense and homogeneous and the average diameter and height of the

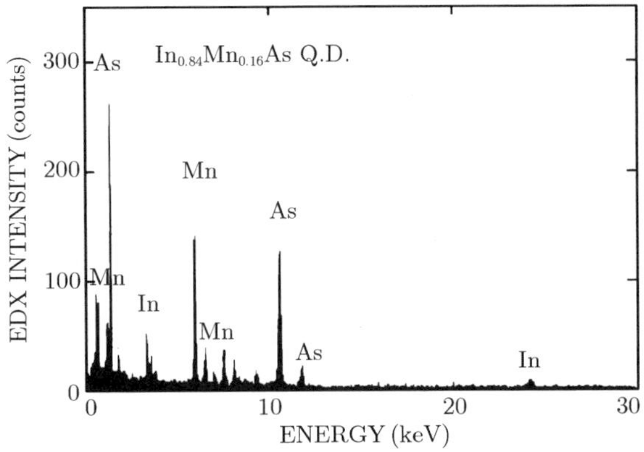

Fig. 3.21. EDX spectrum of $In_{0.84}Mn_{0.16}As$ QD.

dots are about 31.5 and 2.2 nm, respectively (Chen *et al.*, 2005). The average diameter and height of the dots synthesized by another group are 35–37 and 13–15 nm, respectively (Holub *et al.*, 2004). The Mn plays the role of surfactant, alters the surface kinetics, and enhances the adatom migration rates to result in larger dots and a smaller density in the presence of Mn.

Because the total number of Mn atoms in the InMnAs QD is less, it is difficult to measure their magnetic properties. Therefore, InAs:Mn QD multilayers were grown by LT-MBE on semi-insulating (100)-oriented GaAs substrates, between InAs:Mn layers are 50 nm of undoped GaAs. Figure 3.22 shows the (a) magnetization curves for a 10-layer InAs:Mn QD

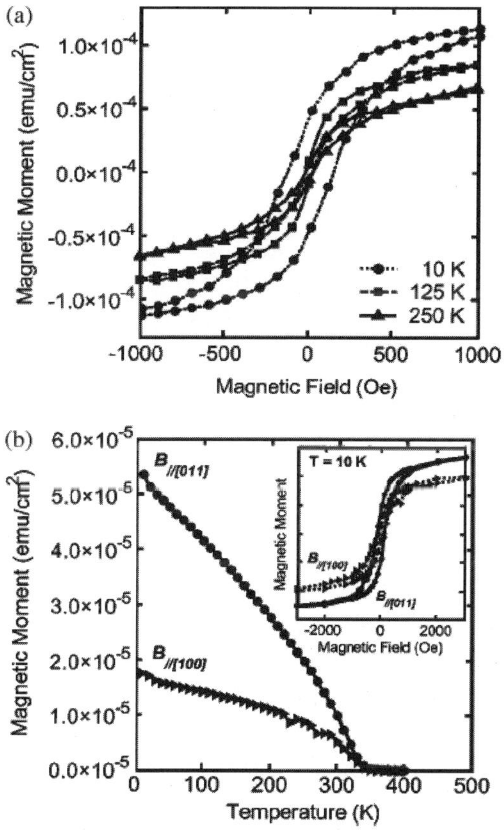

Fig. 3.22. (a) Magnetization curves for a 10-layer InAs:Mn QD sample at $T = 10$, 125, and 250 K. (b) Magnetic moment per unit area versus temperature for the same sample at $B = 50$ Oe. The inset shows the hysteresis loops for the same sample with in-plane and out-of-plane magnetization at $T = 10$ K.

sample at $T = 10$, 125, and 250 K, and (b) magnetic moment per unit area versus temperature for the same sample at $B = 50$ Oe (Holub *et al.*, 2004). Hysteresis is observed when the fields are applied either parallel or perpendicular to the sample surface, as shown in the inset of Fig. 3.22(b), but is less pronounced for the case of a perpendicularly applied field. This result suggests that magnetic anisotropy favors in-plane magnetization, as expected for compressively strained InAs:Mn QDs grown atop a LT-GaAs buffer layer. The temperature at which the magnetic moment vanishes under a small magnetic field is taken as the Curie temperature T_C, which, as seen in Fig. 3.22(b), is as high as 350 K in a sample grown using a Mn/In flux ratio of 0.15. From other reports, $T_C = 400$ K for $In_{0.84}Mn_{0.16}As$ QDs (Jeon *et al.*, 2002), $T_C = 290$ K for $In_{0.79}Mn_{0.21}As$ QDs (Chen *et al.*, 2005), whereas $T_C = 7.5$ K for InMnAs bulk material (Ohno *et al.*, 1992). The reason that the Curie temperature of the QD is remarkably higher than that of the bulk material may be: (1) the Mn content in QDs ($x \leq 0.21$) is much larger than that in the bulk material ($x \leq 0.10$). (2) Quantum confinement effect strengthens the interaction between the magnetic ions and the carriers. (3) The degree of disorder in QDs is larger than in bulk, due not only to substitutional disorder but also to inhomogeneous Mn incorporation among neighboring QDs. When Mn positional disorder is accounted for in a mean-field theory, the Curie temperature can increase significantly with the degree of disorder, as holes migrate toward regions of higher Mn concentration to lower their total energy, thereby stabilizing a ferromagnetic state at higher temperatures than in a uniform system (Berciu and Bhatt, 2004). (4) Cluster effect. In high T_C samples, the possibility of existence of GaMn or $MnAs_nGa_m$ clusters cannot be excluded. The Curie temperatures of these clusters are: MnAs, $T_C = 318$ K; GaMn, $T_C > 400$ K; $MnAs_nGa_m$, $318 < T_C < T_{C\,GaMn}$. (5) Polaron effect, this will be discussed in Sec. 3.7.

Zheng *et al.* (2007) grew $In_{1-x}Cr_xAs$ QDs by LT-MBE, the Cr/In flux ratio is 0.026–0.18. Figure 3.23 shows the high resolution transmission electron microscopy (HRTEM) images and the energy dispersive spectroscopy (EDS) spectrum of QDs in $In_{1-x}Cr_xAs$ sample B (Zheng *et al.*, 2007). The sample B is grown with a Cr/In flux ratio of 0.18, and 18.2 ML of InAs:Cr is deposited on a 2.0 ML-InAs wetting layer. The average base length and height of the dots are about 130 and 27 nm, respectively, and the dot density is approximately $4 \times 10^9\,\text{cm}^{-2}$. Compared with Mn-doped InAs QDs, the Cr-doped InAs QDs grown under the same condition are much larger, which indicates that the formation process of Cr-doped InAs

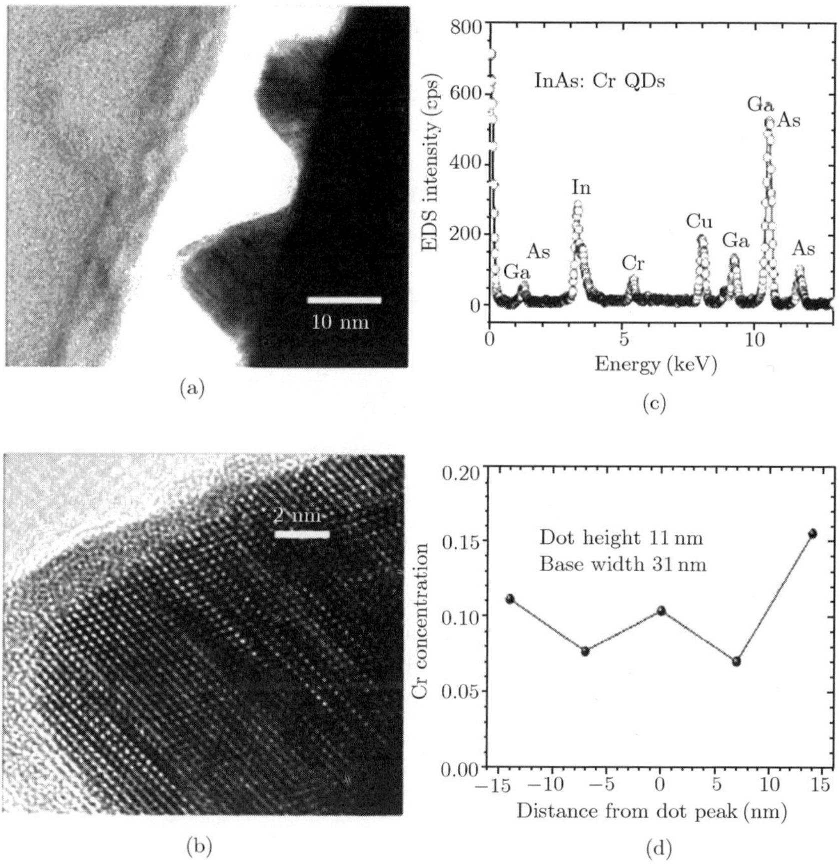

Fig. 3.23. (a) HRTEM image of two QDs in $In_{1-x}Cr_xAs$ sample B. (b) HRTEM image of the larger QD in (a). (c) EDS spectrum of one QD. (d) Cr concentration distribution along the base in a single QD.

QDs is probably different from that of Mn-doped InAs QDs. The HRTEM images in Fig. 3.23 reveal that the structure of all the Cr-doped InAs QDs with Cr/In flux ratio of 0.026–0.18 keeps zincblende, but the crystalline quality degrades with increasing Cr concentration. The sample B has higher Cr concentration, which still maintains zincblende structure, and becomes partly relaxed, and there are a number of obvious defects in it. The sample C with a Cr/In flux ratio of 0.026 has more perfect structure, thus has eight periods of InAs:Cr QDs, every period including 2.0 ML InAs wetting layer, 2.6 ML InAs:Cr, and 25 nm GaAs spacer. The EDS data of sample B in Fig. 2.23(c) reveal the existence of In, As, Cr and

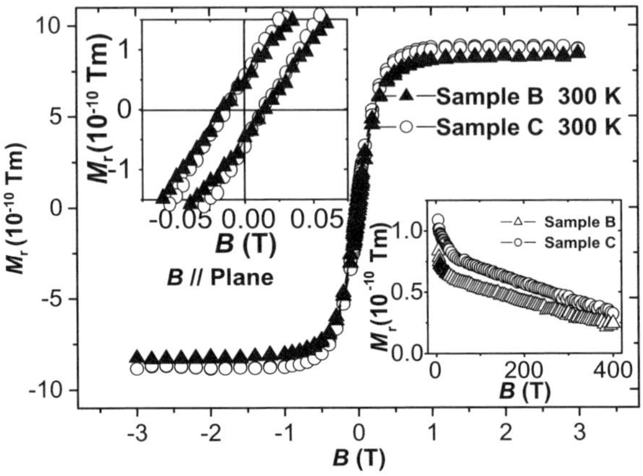

Fig. 3.24. (a) Magnetic field dependence of magnetization of samples B and C at $T = 5$ and 300 K under a magnetic field parallel to the sample plane. The inset shows the enlargement of hysteresis loops around zero magnetic field. (b) Temperature dependence of remanent magnetization of samples B, C, and D (undoped InAs QD) under zero magnetic field.

other elements in the QDs. The composition ratio of Cr/In in sample B is 0.18 got from Fig. 3.23(c), in agreement with the flux ratio 0.18. In order to observe variation of the Cr composition within a single QD, the map-scan mode of EDS analysis along the base of QD as shown in Fig. 3.23(d) reveal that Cr atoms are distributed nearly uniformly in a single QD. The EDS analyses also show that some Cr atoms diffuse into the InAs wetting layer and the GaAs buffer layer.

Figure 3.24 shows the magnetization curves of samples B and C, and the temperature dependence of remanent magnetization of samples B, C, and D. From Fig. 3.24, we see that the Curie temperatures of the samples B and C exceed 400 K.

3.6. Mean-field Theory of FMSs

3.6.1. *Microscopic theory of ferromagnetism*

Heisenberg first attributed the interaction between spins in ferromagnets to the electron exchange effect. The combination of the electron Coulomb interaction and the Pauli exclusion principle produces an effective interaction potential,

$$V = -J(R_{12})\boldsymbol{S}_1 \cdot \boldsymbol{S}_2, \qquad (3.12)$$

where $J(R)$ is the exchange energy depending on the distance R_{12} between spins $\boldsymbol{S}_1$ and $\boldsymbol{S}_2$. If J is positive, then this spin–spin interaction provides a ferromagnetism model.

There are three cases: (1) direct exchange interaction, the basis state wave functions of neighboring atoms overlap, such as CrO_2 and $CrBr_3$; (2) super-exchange interaction; (3) (RKKY) interaction. Some systems have simultaneously local spin and nonlocal spin (cruising electron), such as rare metals and magnetic semiconductors. In these systems, the s–d interaction Hamiltonian can be written as (Yosida, 1957),

$$H_{\text{sd}} = -\frac{1}{N}\sum_{n}\sum_{kk'} J(\boldsymbol{k}-\boldsymbol{k}')\mathrm{e}^{i(\boldsymbol{k}-\boldsymbol{k}')\cdot\boldsymbol{R}_n}\{(a^*_{k'+}a_{k+} - a^*_{k'-}a_{k-})S^z_n + a^*_{k'+}a_{k-}S^-_n + a^*_{k'-}a_{k+}S^+_n\}, \tag{3.13}$$

where $a^*_{k\pm}$, $a_{k\pm}$ are the creation and annihilation operators for the electrons with wave vector $\boldsymbol{k}$ and + or − spin, $\boldsymbol{S}_n$ the spin operator of the magnetic ion at $\boldsymbol{R}_n$, and $J(\boldsymbol{k}-\boldsymbol{k}')$ the exchange integration between a conduction electron and the d electron spin of the Mn ion,

$$J(\boldsymbol{k}-\mathbf{k}') = \int \mathrm{d}\boldsymbol{r}_1 \int \mathrm{d}\boldsymbol{r}_2 \phi^*_d(\boldsymbol{r}_1)\mathrm{e}^{-\mathrm{i}\boldsymbol{k}\cdot\boldsymbol{r}_2}u^*_k(\boldsymbol{r}_2)\frac{1}{r_{12}}\phi_d(\boldsymbol{r}_2)\mathrm{e}^{\mathrm{i}\mathbf{k}'\cdot\boldsymbol{r}_1}u_{k'}(\boldsymbol{r}_1). \tag{3.14}$$

The s–d interaction Hamiltonian (3.13) has the following diagonal element,

$$-\frac{1}{N}J(0)(n_+ - n_-)\sum_n S^z_n, \tag{3.15}$$

where n_+ and n_- represent the total number of electrons of + and − spin. This diagonal energy becomes lower as $n_+ - n_-$ increases and tends to polarize the conduction electrons.

The Fermi energies for electrons with + and − spins are given by

$$E^{\pm}_{\text{F}} = \frac{\hbar^2}{2m}k^{\pm 2}_m \mp \frac{1}{N}J(0)\sum_n S^z_n, \tag{3.16}$$

where $k^{\pm}_m$ are the maximum wave vectors of + and − spins, respectively, and they are related to the density $n_\pm$ by

$$n_\pm = \frac{1}{6\pi^2}(k^{\pm}_m)^3. \tag{3.17}$$

Let the total density of electron $2n = n_+ + n_-$, then

$$n = \frac{1}{6\pi^2} k_m^3, \tag{3.18}$$

k_m is the maximum wave vector for the unpolarized electrons. If we put

$$k_m^{\pm} = k_m \pm \Delta k, \tag{3.19}$$

we get

$$E_{\mathrm{F}}^{\pm} = E_{\mathrm{F}} \pm 2E_{\mathrm{F}} \frac{\Delta k}{k_m} \mp J(0) \sum_n S_n^z, \tag{3.20}$$

where E_{F} represents the Fermi energy of the unpolarized electrons. Similarly,

$$n_{\pm} = n \pm 3n \frac{\Delta k}{k_m}. \tag{3.21}$$

The condition $E_{\mathrm{F}}^{+} = E_{\mathrm{F}}^{-}$ gives

$$\frac{\Delta k}{k_m} = \frac{1}{N} J(0) \cdot \frac{1}{2E_{\mathrm{F}}} \sum_n S_n^z. \tag{3.22}$$

Inserting Eq. (3.22) into Eq. (3.21), we obtain

$$n_{\pm} = n \pm \frac{3n}{2E_{\mathrm{F}}} \cdot \frac{1}{N} J(0) \sum_n S_n^z. \tag{3.23}$$

The above electron polarization expression (3.23) is that given by Fröhlich, Nebarro, and Zener (Yosida, 1957). This polarization arises from the first-order perturbation of the energy, which is proportional to $N^{-1}J(0)/E_{\mathrm{F}}$.

The first-order perturbed wave function of the conduction electron is given by

$$\phi_k = \phi_k^0 + \sum_{k'} \frac{H_{kk'}}{E_k - E_{k'}} \phi_{k'}^0. \tag{3.24}$$

By use of the perturbation Hamiltonian (3.13), and approximating ϕ_k^0 with the plane wave, we obtain

$$\begin{aligned} \phi_{k\pm} = \phi_{k\pm}^0 - \frac{2m}{\hbar^2} \cdot \frac{1}{N} \sum_{\boldsymbol{k}'} \frac{J(\boldsymbol{k} - \boldsymbol{k}')}{k^2 - k'^2} \\ \times \left\{ \sum_n \mathrm{e}^{\mathrm{i}(\boldsymbol{k}-\boldsymbol{k}')\cdot R_n^2} \left[\pm S_n^z \phi_{k'\pm}^0 + S_n^{\pm} \phi_{k'\mp}^0 \right] \right\}. \end{aligned} \tag{3.25}$$

The densities of $\pm$ spins can be obtained by calculating the quantities,

$$\rho_\pm(\boldsymbol{r}) = \sum_n \sum_{k=0}^{k_m^\pm} \phi_{k\pm}^* \phi_{k\pm} = \sum_{k=0}^{k_m^\pm}(1) \mp \frac{2m}{\hbar^2} \cdot \frac{1}{N} \sum_{k=0}^{k_m^\pm} \sum_{k'} \frac{J(\boldsymbol{k}-\boldsymbol{k}')}{k^2 - k'^2}$$
$$\times \sum_n [\mathrm{e}^{\mathrm{i}(\boldsymbol{k}-\boldsymbol{k}')\cdot(\boldsymbol{r}-\boldsymbol{R}_n)} + \mathrm{e}^{-\mathrm{i}(\boldsymbol{k}-\boldsymbol{k}')\cdot(\boldsymbol{r}-\boldsymbol{R}_n)}] S_n^z. \tag{3.26}$$

The first term in Eq. (3.26) is written with the use of Eq. (3.23) as

$$\sum_{k=0}^{k_m^\pm}(1) = n_\pm = n \pm \frac{3n}{2E_F} \cdot \frac{1}{N} J(0) \sum_n S_n^z. \tag{3.27}$$

Let $\boldsymbol{k} - \boldsymbol{k}' = \boldsymbol{q}$, and integrate over $\boldsymbol{k}$, the second term of Eq. (3.26) becomes

$$\pm \frac{2m}{N\hbar^2} \cdot \frac{1}{16\pi^2} \sum_q \left\{ k_m^\pm + \frac{4(k_m^\pm)^2 - q^2}{4q} \ln \left| \frac{2k_m^\pm + q}{2k_m^\pm - q} \right| \right\}$$
$$\times \sum_n [e^{i\boldsymbol{q}\cdot(\boldsymbol{r}-\boldsymbol{R}_n)} + e^{-i\boldsymbol{q}\cdot(\boldsymbol{r}-\boldsymbol{R}_n)}] S_n^z. \tag{3.28}$$

Adding Eqs. (3.27) and (3.28), we get

$$\rho_\pm(\boldsymbol{r}) = n \pm \frac{3n}{2E_\mathrm{F}} \cdot \frac{1}{N} J(0) \sum_n S_n^z \pm \frac{3n}{8E_\mathrm{F}} \cdot \frac{1}{N}$$
$$\times \sum_{\boldsymbol{q}} J(\boldsymbol{q}) f(q) \sum_n [e^{i\boldsymbol{q}\cdot(\boldsymbol{r}-\boldsymbol{R}_n)} + e^{-i\boldsymbol{q}\cdot(\boldsymbol{r}-\boldsymbol{R}_n)}] S_n^z, \tag{3.29}$$

where

$$f(q) = 1 + \frac{4k_m^2 - q^2}{4k_m q} \ln \left| \frac{2k_m + q}{2k_m - q} \right|. \tag{3.30}$$

The difference between $k_m^\pm$ and k_m has been neglected in Eq. (3.30). In Eq. (3.29), the second term is the perturbation coming from the first-order perturbation of the energy and it is uniform over the metal. The third term comes from the first-order perturbation of the wave function and has the same order of magnitude as the second term.

The term $\boldsymbol{q} = 0$, which is omitted from the summation of the third term, can easily be shown to be just equal to the second term because $f(0)$ is equal to 2. Therefore, we finally obtain

$$\rho_\pm(\boldsymbol{r}) = n \pm \frac{3n}{8E_\mathrm{F}} \cdot \frac{1}{N} \sum_{\boldsymbol{q}} J(\boldsymbol{q}) f(q) \sum_n [e^{i\boldsymbol{q}\cdot(\boldsymbol{r}-\boldsymbol{R}_n)} + \mathrm{e}^{-\mathrm{i}\boldsymbol{q}\cdot(\boldsymbol{r}-\boldsymbol{R}_n)}] S_n^z, \tag{3.31}$$

where the summation should be taken over all $\boldsymbol{q}$-values including $\boldsymbol{q} = 0$. This theory indicates that due to the s–d exchange interaction Eq. (3.13) the electrons in the solid are polarized, and the density of + and – spin electrons is given by Eq. (3.31), resulting in ferromagnetism. But this theory has not taken into account the temperature effect, so it cannot be used to calculate the Curie temperature.

If we put $J(\boldsymbol{q}) = \text{const}$, and perform the integration over $\boldsymbol{q}$ in Eq. (3.31),

$$\frac{1}{N}\sum_q f(q)e^{iq\cdot R} = -24\pi\frac{n}{N}F(2k_m R), \quad F(x) = \frac{1}{x^4}(x\cos x - \sin x) \tag{3.32}$$

and obtain

$$\rho_\pm(\boldsymbol{r}) = n \mp \frac{(3n)^2}{E_\mathrm{F}}2\pi J(0)\frac{1}{N}\sum_n F[2k_m|\boldsymbol{r} - \boldsymbol{R}_n|]S_n^z. \tag{3.33}$$

Equation (3.33) is just the expression obtained by Ruderman and Kittel. This expression becomes infinite as $\boldsymbol{r}$ tends to R_n. This is due to the approximation of $J(\boldsymbol{q}) = \text{const}$. In reality, $J(\boldsymbol{q})$ would decrease with increasing $\boldsymbol{q}$, especially when $q > 2k_m$, $J(\boldsymbol{q})$ decreases abruptly. Putting $J(\boldsymbol{q})f(q) = 2J(0)$ for $q < 2k_m$ and $J(\boldsymbol{q})f(q) = 0$ for $q > 2k_m$, we obtain

$$\rho_\pm(\boldsymbol{r}) = n \mp \frac{4(3n)^2}{E_\mathrm{F}}J(0)\frac{1}{N}\sum_n 2k_m|\boldsymbol{r} - \boldsymbol{R}_n|F[2k_m|\boldsymbol{r} - \boldsymbol{R}_n|]S_n^z. \tag{3.34}$$

This ρ has a finite value at $\boldsymbol{r} = \boldsymbol{R}_n$. The polarization given by Eqs. (3.33) and (3.34) oscillates and rapidly vanishes as $|\boldsymbol{r} - \boldsymbol{R}_n|$ tends to infinite. Therefore, the polarization of the conduction electrons is concentrated in the neighborhood of the impurity center and not the uniform polarization given by Fröhlich and Nabarro. Because they considered only the first-order perturbation of energy, Ruderman and Kittel considered simultaneously the first-order perturbation of the wave function and obtained the above expressions.

Starting from the interaction expression (3.34) between conduction electrons and the impurity spin at $\boldsymbol{R}_n$, we obtain the effective interaction between spins at $\boldsymbol{R}_n$ and $\boldsymbol{R}_m$,

$$H_\mathrm{eff} = J_\mathrm{sd}^2 \boldsymbol{S}_n \cdot \boldsymbol{S}_m \Phi(R_{nm}), \tag{3.35}$$

where $\Phi(R_{nm})$ is the route function between local spins,

$$\Phi(R) = \frac{m\mu_B^2}{h^3} \cdot \frac{1}{R^4}(\sin 2k_F R - 2k_F R \cos 2k_F R). \tag{3.36}$$

This interaction is called the RKKY interaction.

3.6.2. *Magnetic interaction in DMSs*

Dietl *et al.* (2001) took into account the valence band structure and calculated the Curie temperatures of various DMSs with the Zener model. The main idea is that the p–d exchange interaction between the local ion spins and holes results in the spin splitting of energy bands and declination of the total energy. When the temperature is lower than a critical temperature, the decrease in the total energy exceeds the increase in the free energy for ion magnetization, and the polarization of local spins, i.e. ferromagnetism, occurs.

Suppose that the magnetic ions have an average magnetic moment M, which produces giant Zeeman splitting of valence band (see Sec. 2.2.2). Let the hole concentration be p, the Fermi-energy level can be determined as a function of p and M, $\varepsilon_F(p, M)$. According to the thermodynamics, the free energy of the hole subsystem equals

$$\begin{aligned} F_c(p, M) &= -k_B T \ln Z \\ &= -k_B T \int d\varepsilon N(\varepsilon) \ln[1 + e^{-\varepsilon - \varepsilon_F(p,M)/k_B T}] + p\varepsilon_F(p, M), \end{aligned} \tag{3.37}$$

where $N(\varepsilon) = \partial p/\partial\varepsilon$ is the density of states. At low temperature, the hole liquid is degenerate, then

$$F_c(p, M) = \int_0^p dp' \varepsilon(M, p'). \tag{3.38}$$

The free energy of the local spin subsystem equals

$$F_s(M) = \int_0^M dM_0 H(M_0). \tag{3.39}$$

For Mn^{2+} ions, M_0 can be expressed as a function of the magnetic field H,

$$M_0(H) = g\mu_B \bar{x} N_0 S B_s \left[\frac{g\mu_B S H}{k_B(T + T_{AF})}\right]. \tag{3.40}$$

In the given T, H, and p, when the system reaches balance, the following relation should be satisfied,

$$\frac{\partial F(M)}{\partial M} = \frac{\partial F_{\rm c}(M)}{\partial M} + H_{\rm loc} = 0, \quad H_{\rm loc} = -\frac{\partial F_{\rm c}(M)}{\partial M}. \tag{3.41}$$

Inserting Eq. (3.41) into Eq. (3.40), we obtain

$$M = g\mu_{\rm B}\bar{x}N_0 S B_s \left[\frac{g\mu_{\rm B}S(-\partial F_{\rm c}(M)/\partial M + H)}{k_{\rm B}(T + T_{\rm AF})}\right], \tag{3.42}$$

where H is the externally applied magnetic field. In the case of FMSs, when $H = 0$, there exists a spontaneous magnetic moment M, so that Eq. (3.42) with $H = 0$ is a nonlinear equation of M.

When the temperature approaches the Curie temperature, it can be proved that

$$F_{\rm c}(M) - F_{\rm c}(0) \approx -\frac{\lambda}{2}M^2. \tag{3.43}$$

Thus,

$$-\frac{\partial F_{\rm c}(M)}{\partial M} = \lambda M, \tag{3.44}$$

where λ is a parameter determined by the numerical calculation, similar to the phenomenal parameter in the Weiss theory, representing the amplitude of the ferromagnetism.

Near the Curie temperature the M, i.e. the argument in the Brillouin function in Eq. (3.42) $\sim$0, we can expand $B_s(x)$ for x, and obtain

$$\begin{aligned} M &= g\mu_{\rm B}\bar{x}N_0 S\frac{S+1}{3S}\lambda\frac{g\mu_{\rm B}MS}{k_{\rm B}(T_{\rm C} + T_{\rm AF})}, \\ T_{\rm C} &= \frac{S(S+1)}{3k_{\rm B}}(g\mu_{\rm B})^2\bar{x}N_0\lambda - T_{\rm AF}. \end{aligned} \tag{3.45}$$

Therefore, the Curie temperature is proportional to λ and the effective concentration of magnetic ions $\bar{x}$. The calculation found that λ is proportional to the hole concentration p.

Figure 3.25 shows the calculated Curie temperature as a function of the hole concentration for $Ga_{0.95}Mn_{0.05}As$ (solid line) (Dietl *et al.*, 2001), where straight dashed lines represent the results obtained assuming a large and a small value of the spin–orbit splitting Δ_0, respectively. The dotted line is calculated neglecting the effect of the SOI on the hole spin susceptibility. The parameters taken in the calculation are: $|\beta N_0| = 1.2\,\text{eV}$, $\bar{x} = x$,

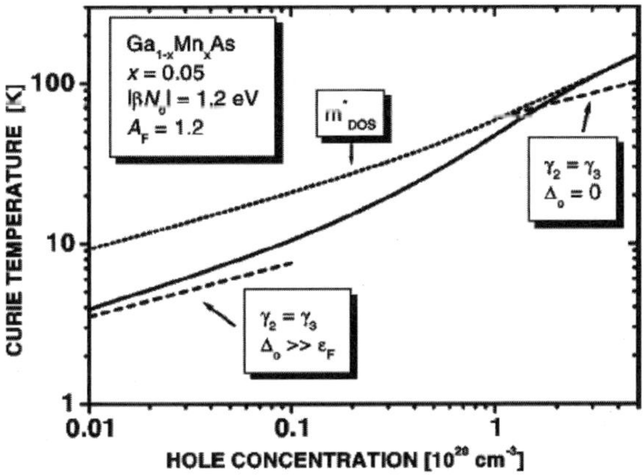

Fig. 3.25. Calculated Curie temperature as a function of the hole concentration for $Ga_{0.95}Mn_{0.05}As$.

$T_{AF} = 0\,K$. From Fig. 3.25, we see that the Curie temperature increases with increasing hole concentration, it is noticed that the two coordinates are all logarithmic. The theory predicts that when $p = 5 \times 10^{20}\,cm^{-3}$, $T_C = 110\,K$, in good agreement with the experimental result.

Figure 3.26 shows the Curie temperatures evaluated for various III–V, II–VI compounds and IV groups containing 5% of Mn^{2+} and hole concentration $3.5 \times 10^{20}\,cm^{-3}$ (Dietl *et al.*, 2000, 2001). At the time the results just published were very exciting, because it predicted that the Curie temperatures of GaN, InN, C, ZnO, etc., all can reach RT (300 K). Latter experimental results were not as ideal as the prediction, except GaN, the DMSs with the Curie temperature at RT are rare; the reason is multiple, indicating that this theory reflects only one aspect of facts.

The above theory is only valid for the Mn^{2+}-doped semiconductors, because the orbital angular momentum of the basis state of Mn^{2+} ion equals zero, the total spin $S = 5/2$, the magnetization is given by Eq. (3.42). Zheng and Xia (2005) extended the mean-field theory of Curie temperature (Dietl *et al.*, 2001) to the other magnetic ions Fe^{2+} and Co^{2+} doping cases. According to Sec. 1.1, the basis state terms of the Fe^{2+} and Co^{2+} ions are $^5D_4$ and $^4F_{9/2}$ (Table 1.2), and their splitting in the semiconductor crystal field is shown in Figs. 1.2 and 1.3, respectively. Thus, the magnetization cannot be given by Eq. (3.42) and is given by

$$M = \bar{x} N_0 k_B T \frac{\partial}{\partial H} \ln Z, \qquad (3.46)$$

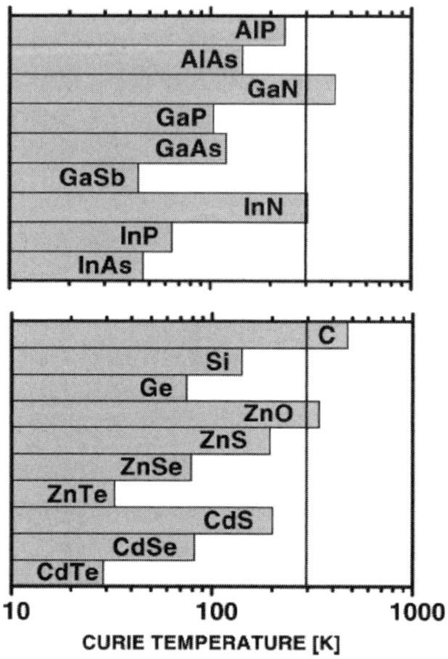

Fig. 3.26. Curie temperatures evaluated for various III–V (top panel) and group IV and II–VI compounds (lower panel) containing 5% of Mn^{2+} and hole concentration $3.5 \times 10^{20}\,\mathrm{cm}^{-3}$.

where Z is the partition function,

$$Z = \sum_i \exp\left(-\frac{E_i}{k_{\mathrm{B}}T}\right). \tag{3.47}$$

Due to the energy level splitting, M does not have an analytical expression, can only be calculated by numerical method.

When $T \to T_{\mathrm{C}}$, $M \to 0$ and $H \to 0$, where H is the local magnetic field induced by holes (see Eq. (3.41)). At this time,

$$M = \chi_0 H, \tag{3.48}$$

where χ_0 is the differential susceptibility for the magnetic ions. From Eq. (3.46), we obtain

$$\chi_0 = \frac{\partial M}{\partial H} = \bar{x} N_0 k_{\mathrm{B}} T \frac{\partial^2}{\partial H^2} \ln Z|_{H \to 0}. \tag{3.49}$$

On the other hand, for the hole subsystem from Eqs. (3.41) and (3.43), we have

$$\chi_0 = \frac{1}{\lambda} = \left.\frac{M^2}{2[F_{\mathrm{c}}(p,0) - F_{\mathrm{c}}(p,M)]}\right|_{M \to 0}. \tag{3.50}$$

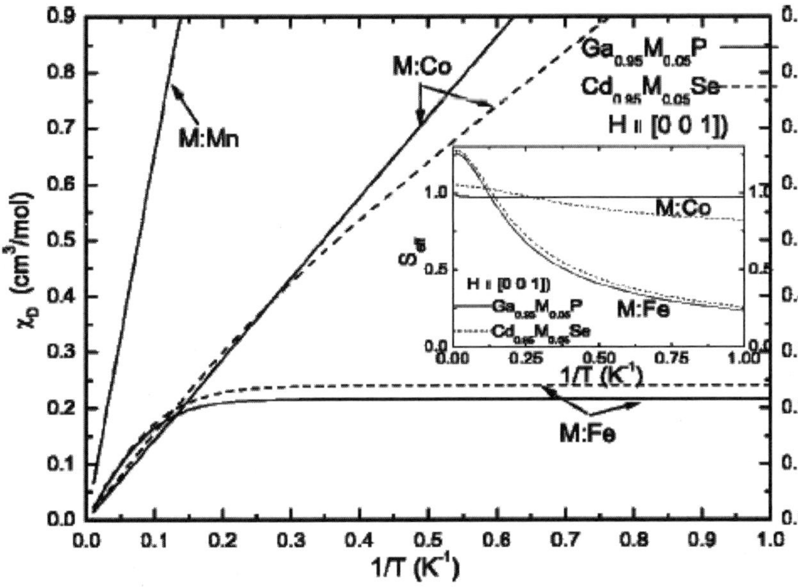

Fig. 3.27. Paramagnetic susceptibilities χ_D as functions of $1/T$ for Mn^{2+}, Fe^{2+}, and Co^{2+} in GaP and CdSe at $H \to 0$. The inset shows the effective spin angular momenta S_{eff} of these DMSs as functions of $1/T$.

Combination of Eqs. (3.49) and (3.50) is just the equation determining the Curie temperature.

It is convenient to introduce an temperature-dependent effective spin angular momentum $S_{\text{eff}}(T)$ such that the paramagnetic susceptibility $\chi_D(T)$ for other magnetic ions takes the same form as that for Mn^{2+} ions,

$$\chi_D = \frac{1}{3k_B T} S_{\text{eff}}(S_{\text{eff}} + 1)\bar{x}N_0(g\mu_B)^2. \tag{3.51}$$

For Mn^{2+} ions, $S_{\text{eff}} = 5/2$, independent of temperature. For Fe^{2+} and Co^{2+} ions, S_{eff} depends on the temperature, the temperature dependence should be determined by numerical calculation of Eq. (3.49). Figure 3.27 shows the paramagnetic susceptibilities χ_D as functions of $1/T$ for Mn^{2+}, Fe^{2+}, and Co^{2+} in GaP and CdSe at $H \to 0$. The inset shows the effective spin angular momenta S_{eff} of these DMSs as functions of $1/T$ (Zheng and Xia, 2005). From Fig. 3.27, we see that the $\chi_D \sim 1/T$ curve for Mn^{2+} ions is a straight line, corresponding to $S_{\text{eff}} = 5/2$. The $\chi_D \sim 1/T$ curve for Co^{2+} ions is basically a straight line, but the corresponding S_{eff} is smaller than that of Mn^{2+}, is about 1 as seen from the inset of Fig. 3.27. The χ_D

saturates as $1/T$ increases, and its S_{eff} decreases with increasing $1/T$ as shown in the inset of Fig. 3.27.

From Eqs. (3.50) and (3.51), the Curie temperature can be obtained,

$$T_{\text{C}} = -\frac{2}{3}S_{\text{eff}}(S_{\text{eff}}+1)\bar{x}N_0(g\mu_{\text{B}})^2 \cdot \left.\frac{F_{\text{c}}(M)-F_{\text{c}}(0)}{k_{\text{B}}M^2}\right|_{M\to 0}. \tag{3.52}$$

Equation (3.52) is a transcendental equation, because S_{eff} is still a function of temperature, except for the Mn^{2+}-doped semiconductors. When M is very small, the last term in Eq. (3.51) can be written as (Dietl *et al.*, 2001),

$$\frac{F_{\text{c}}(M)-F_{\text{c}}(0)}{M^2} = -A_{\text{F}}\rho_s\beta^2 \cdot \frac{1}{2(2g\mu_{\text{B}})^2}, \tag{3.53}$$

where ρ_s is the density of states at the Fermi level, β is the exchange integration of hole, A_{F} is a constant of about 1. Inserting Eq. (3.53) into Eq. (3.52), it can be seen that the Curie temperature of DMSs is mainly decided by three factors: (a) the density of states at the Fermi level ρ_s, which is determined by the hole band structure of the host semiconductor; (b) the exchange energy $\bar{x}N_0\beta$; (c) the paramagnetic susceptibility χ_D or the effective spin angular momentum S_{eff}.

Figure 3.28 shows the calculated Curie temperature as functions of the hole concentration for $Ga_{0.95}M_{0.05}P$ (M = Co, Fe, Mn)

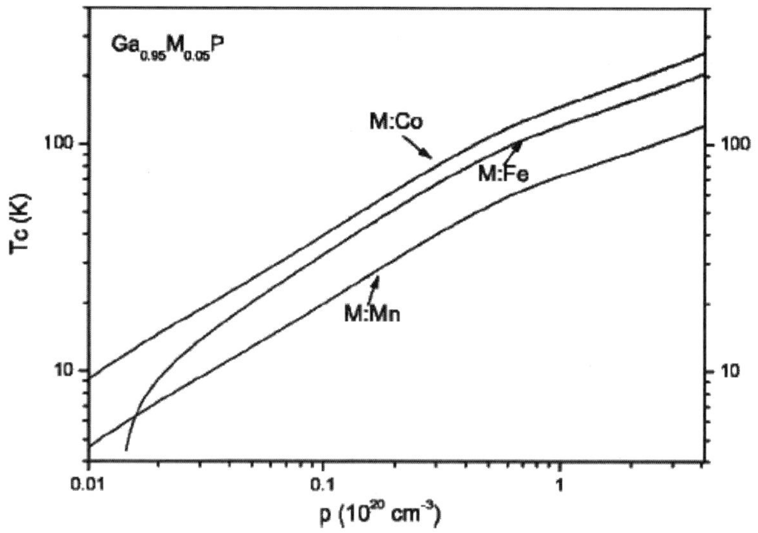

Fig. 3.28. Curie temperature as functions of the hole concentration for $Ga_{0.95}M_{0.05}P$ (M = Co, Fe, Mn).

(Zheng and Xia, 2005). It can be seen that the Curie temperatures decrease in order of Co, Fe, Mn for a wide range of hole concentrations. From Fig. 3.27, we have $S_{\rm eff}$ $(Co^{2+}) < S_{\rm eff}$ $(Fe^{2+}) < S_{\rm eff}$ (Mn^{2+}); on the other hand, $\beta(Co^{2+}) > \beta(Fe^{2+}) > \beta(Mn^{2+})$, because β is inversely proportional to the local spin angular momentum. The β factor dominates for most of the range of hole concentrations (e.g. the Curie temperature of $Ga_{0.95}Co_{0.05}P$ is about twice as high as that of $Ga_{0.95}Mn_{0.05}P$), since the difference between $S_{\rm eff}$ of Co, Fe, and Mn is small. For the case of p not too small, $T_C \propto p^{1/3}$ still holds, and the density of states at the Fermi level ρ_s is roughly proportional to $p^{1/3}$.

Figure 3.29 shows the calculated Curie temperatures as functions of the hole concentration for various semiconductors doped with 5% of Fe^{2+} (Zheng and Xia, 2005). The chemical trend of the Curie temperature can be obtained from the figure: (1) T_C increases with decreasing spin–orbit splitting, which enhances the density of states ρ_s at Fermi surface; (2) T_C increases with decreasing lattice constant of the host semiconductor; (3) T_C increases with increasing p–d exchange energy and increasing hole effective mass. It can be seen from Fig. 3.29 that $Zn_{0.95}Fe_{0.05}S$ has the highest Curie temperature, because of its small lattice constant and large hole effective mass.

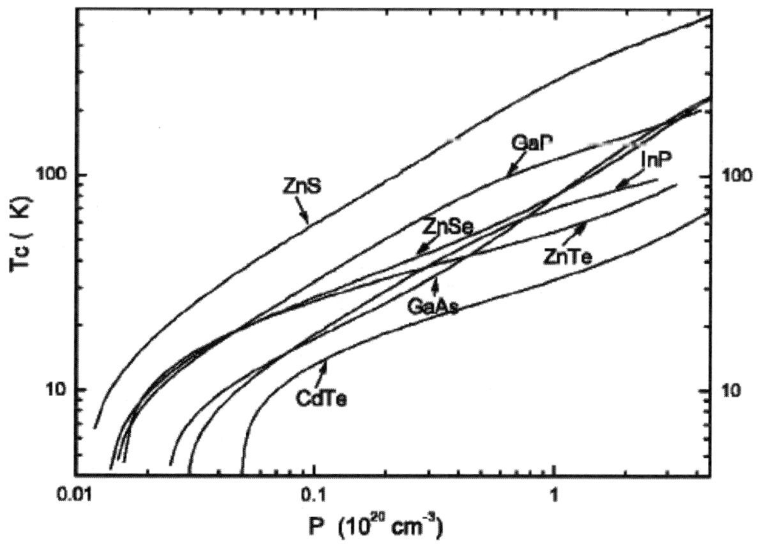

Fig. 3.29. Curie temperatures as functions of the hole concentration for various semiconductors doped with 5% of Fe^{2+}.

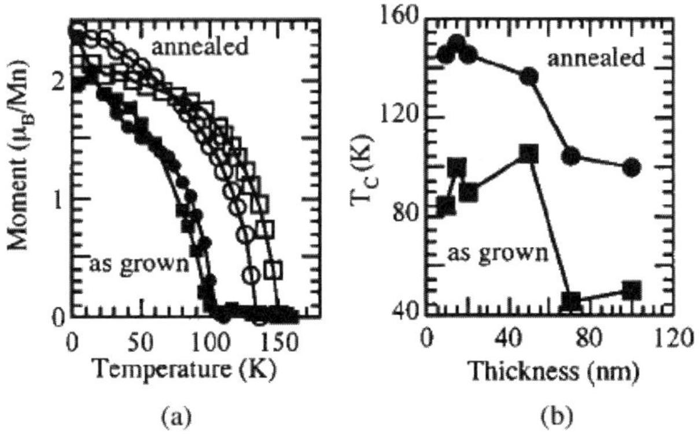

Fig. 3.30. (a) Magnetization versus temperature for $Ga_{1-x}Mn_xAs$ ($x \approx 0.085$) epilayers with thickness $t = 15\,\text{nm}$ (squares) and 50 nm (circular), as grown and after annealing. (b) Curie temperature for the same epilayers with varying thickness.

The experiments of $Ga_{1-x}Mn_xAs$ epilayers confirmed part theoretical predictions, but indicated that the Curie temperatures are related to many other factors (Ku *et al.*, 2003). Figure 3.30 shows (a) magnetization versus temperature for $Ga_{1-x}Mn_xAs$ ($x \approx 0.085$) epilayers with thickness $t = 15\,\text{nm}$ (squares) and 50 nm (circular), as grown and after annealing. (b) Curie temperature for the same epilayers with varying thickness. The data demonstrate that T_C is increased above 110 K after annealing in both samples, reaching a maximum value of ~150 K for the 15 nm-thick sample. Analysis of the Raman spectra shows that annealing increases the hole density of as-grown samples from $p \sim 0.18 \times 10^{21}\,\text{cm}^{-3}$ to $p \sim 1.5 \times 10^{21}\,\text{cm}^{-3}$. An empirical fit shows that $T_C \propto p^{1/3}$ and an extrapolation of this trend indicates that enhancing the T_C of $Ga_{1-x}Mn_xAs$ to 300 K would require an additional order of magnitude increase in p.

Why can annealing increase the Curie temperature? The ion channeling studies gave some insights into the problem. These experiments show that a significant fraction of Mn ions occupies nonsubstitute, interstitial sites (Mn_I) and that annealing decreases the concentration of Mn_I. Since Mn_I acts as a donor and hence compensates holes, optimal annealing increases p and correspondingly enhances T_C. The relatively small increase in the ferromagnetic moment in some samples implies that a large fraction of Mn ions are still not contributing to the ferromagnetic state even after annealing.

Other important factor is the thickness of samples. The highest T_C in both as-grown and annealed samples occurs for $10\,\text{nm} \leq t \leq 50\,\text{nm}$ and that T_C is suppressed for larger t. It seems that the thickness dependence of T_C may be associated with the proximity of a free surface. The theoretical calculation suggests that the Mn defect sites have rather different energies near a free surface. The experiment found that $Ga_{1-x}Mn_xAs$ epilayers capped with a thin GaAs epilayer (~10–50 nm) typically show a decrease in T_C upon annealing, rather than increase, indicating that the nature of the surface of the $Ga_{1-x}Mn_xAs$ directly affects the properties of the entire epilayer.

3.6.3. *FMS quantum wires and quantum slabs*

This section extends the mean-field theory of FMSs from the bulk material to the quantum wires and slabs. The above section studied the hole-mediated ferromagnetism, because of the complexity of the hole bands the Curie temperatures can be calculated only by numerical method. Here we consider the ferromagnetism of quantum slabs and wires mediated by electrons (Zhang *et al.*, 2007). Because the electron CB is a simple parabolic band, the Curie temperatures can be presented by the analytical expressions, and the physical meaning is rather clear. The results show that though in the bulk materials the Curie temperature of the electron-mediated ferromagnetism is low, while in the low-dimensional materials, especially in quantum wires, high Curie temperature can be obtained due to the difference in the density of states between the bulk material and the low-dimensional material.

Because of the s–d exchange interaction of magnetic ions, the electron band of spin + and − in the FMS splits, the splitting energy equals (see Eqs. (2.37) and (3.40)),

$$\Delta E = \bar{x} N_0 \alpha \langle S_z \rangle = \frac{\alpha M}{g\mu_B}. \tag{3.54}$$

The free energy of the electron subsystem in the degenerate case is (see Eq. (3.38)),

$$F_c(M) = \int_0^{E_F} EN(E)\mathrm{d}E, \tag{3.55}$$

where $N(E)$ is the density of states. The Curie temperature is given by Eq. (3.45), where the parameter is calculated by Eq. (3.44).

First we consider the 1D quantum wire case. Suppose that the quantum wire is rather thin, so that the interval between transverse confined energy levels is large, the electrons fill only in the first subband. The density of states of the 1D subband is

$$N(E) = \frac{4}{\pi^2 D^2}\sqrt{\frac{2m^*}{\hbar^2}}\frac{1}{\sqrt{E}}, \tag{3.56}$$

where D is the diameter of quantum wire. From Eq. (3.55), the free energy of nondoped quantum wire is

$$\begin{aligned} F_{\rm c}(0) &= \int_0^{E_{\rm F}} E \cdot \frac{4}{\pi^2 D^2}\sqrt{\frac{2m^*}{\hbar^2}}\frac{1}{\sqrt{E}}{\rm d}E \\ &= \frac{8}{3\pi^2 D^2}\sqrt{\frac{2m^*}{\hbar^2}}E_{\rm F}^{3/2}. \end{aligned} \tag{3.57}$$

In the magnetic ion-doped quantum wire, the subband of spin + and – splits, their subband bottoms shift $-\Delta E/2$ and $\Delta E/2$ relative to the original band bottom, respectively (see Eq. (3.54)), the Fermi energy level becomes $E'_{\rm F}$. The free energies of the subsystem for spin + and – are

$$\begin{aligned} F_{{\rm c}\downarrow}(M) &= \frac{2}{\pi^2 D^2}\sqrt{\frac{2m^*}{\hbar^2}}\int_{\Delta E/2}^{F'_{\rm F}} {\rm d}E\frac{E}{\sqrt{E+\Delta E/2}} \\ &= \frac{4}{3\pi^2 D^2}\sqrt{\frac{2m^*}{\hbar^2}}\left(E'_{\rm F} + \frac{\Delta E}{2}\right)^{1/2}(E'_{\rm F} - \Delta E), \end{aligned} \tag{3.58}$$

$$F_{{\rm c}\uparrow}(M) = \frac{4}{3\pi^2 D^2}\sqrt{\frac{2m^*}{\hbar^2}}\left(E'_{\rm F} - \frac{\Delta E}{2}\right)^{1/2}(E'_{\rm F} + \Delta E). \tag{3.59}$$

Assuming that the electron density is $n_{\rm e}$, we calculate the Fermi energy level as follows,

$$\begin{aligned} n_{\rm e} &= \frac{4}{\pi^2 D^2}\sqrt{\frac{2m^*}{\hbar^2}}\int_0^{E_{\rm F}}\frac{{\rm d}E}{\sqrt{E}} = \frac{8}{\pi^2 D^2}\sqrt{\frac{2m^*}{\hbar^2}}\sqrt{E_{\rm F}}, \\ n_{{\rm e}\downarrow} &= \frac{2}{\pi^2 D^2}\sqrt{\frac{2m^*}{\hbar^2}}\int_0^{E'_{\rm F}+\Delta E/2}\frac{{\rm d}E}{\sqrt{E}} = \frac{4}{\pi^2 D^2}\sqrt{\frac{2m^*}{\hbar^2}}\sqrt{E'_{\rm F}+\frac{\Delta E}{2}}, \\ n_{{\rm e}\uparrow} &= \frac{4}{\pi^2 D^2}\sqrt{\frac{2m^*}{\hbar^2}}\sqrt{E'_{\rm F}-\frac{\Delta E}{2}}. \end{aligned} \tag{3.60}$$

From the conservation of the electron number $n_e \downarrow + n_e \uparrow - n_e = 0$, we obtain

$$E_F' = E_F + \frac{(\Delta E)^2}{16 E_F}. \tag{3.61}$$

From Eqs. (3.57)–(3.59), we obtain

$$\begin{aligned} F_c(M) - F_c(0) &= \frac{4}{3\pi^2 D^2}\sqrt{\frac{2m^*}{\hbar^2}}\left[-\frac{3}{8}\frac{(\Delta E)^2}{E_F^{1/2}}\right] \\ &= -\frac{1}{2\pi^2 D^2}\sqrt{\frac{2m^*}{\hbar^2}}\frac{(\Delta E)^2}{E_F^{1/2}} \\ &= -\frac{1}{\pi^2 D^2}\left(\frac{m^*}{\hbar^2}\right)\left(\frac{\alpha M}{g\mu_B}\right)^2 \frac{1}{k_F}. \end{aligned} \tag{3.62}$$

Inserting Eq. (3.62) into Eq. (3.45) and taking $T_{AF} = 0$, we obtain the Curie temperature,

$$\begin{aligned} T_C &= \frac{2}{3}\bar{x} N_0 S(S+1)\frac{m^*\alpha^2}{k_B \hbar^2 \pi^2 D^2}\cdot\frac{1}{k_F} \\ &= \frac{16}{3}\bar{x} N_0 S(S+1)\frac{m^*\alpha^2}{k_B \hbar^2 \pi^4 D^4}\cdot\frac{1}{n_e}, \end{aligned} \tag{3.63}$$

where $k_F = (1/8)\pi^2 D^2 n_e$ is the Fermi wave vector. It is noted that T_C is inversely proportional to D^4, so T_C will increase dramatically as D decreases. T_C is also inversely proportional to n_e.

When n conduction subbands are filled, the Curie temperature is

$$\begin{aligned} T_C &= \frac{2}{3}\bar{x} N_0 S(S+1)\frac{m^*\alpha^2}{k_B \hbar^2 \pi^2 D^2}\sum_n \frac{1}{k_{Fn}} \\ &= \frac{1}{12}\bar{x} N_0 S(S+1)\frac{\alpha^2}{k_B} N(E_F). \end{aligned} \tag{3.64}$$

We note that T_C is proportional to the density of states at the Fermi level, because the carriers in these states mediate the ferromagnetism.

Next we consider the 2D quantum slab (quantum well or film with high enough barrier) case. The density of states of quantum slab is

$$N(E) = \frac{m^*}{\pi\hbar^2 L}, \tag{3.65}$$

where L is the slab thickness. Because the density of states Eq. (3.65) is a constant, the Fermi levels for the nondoped and doped cases are equal,

$E'_F = E_F$. It is easy to deduce the Curie temperature,

$$\begin{aligned} T_C &= \frac{1}{12}\bar{x}N_0 S(S+1)\frac{m^*\alpha^2}{k_B\hbar^2\pi L} \\ &= \frac{1}{12}\bar{x}N_0 S(S+1)\frac{\alpha^2}{k_B}N(E_F). \end{aligned} \tag{3.66}$$

Equation (3.66) is similar to Eq. (3.64); the Curie temperature is also proportional to the density of states at the Fermi level. We note that in the one filled subband case ($n = 1$), the Curie temperature of the slab is independent of the electron density and inversely proportional to the slab thickness L.

In order to compare with the bulk material, we deduce the Curie temperature of the bulk material,

$$\begin{aligned} T_C &= \frac{1}{12}\bar{x}N_0 S(S+1)\frac{m^*\alpha^2}{k_B\hbar^2\pi^2}k_F = \frac{1}{12}\bar{x}N_0 S(S+1)\frac{\alpha^2}{k_B}N(E_F) \\ &= \frac{\sqrt{3}}{12}\bar{x}N_0 S(S+1)\frac{m^*\alpha^2}{k_B\hbar^2\pi^{4/3}}n_e^{1/3}. \end{aligned} \tag{3.67}$$

From Eqs. (3.64), (3.66), and (3.67), we see that no matter the 1D, 2D, or 3D cases, T_C is always proportional to the density of states at the Fermi level.

Figure 3.31 shows the Curie temperature versus dimensions for three cases (Zhang *et al.*, 2007), where D_0 is the length unit. Comparing the solid lines and dashes lines, we note that the Curie temperature of DMS nanowires is quite larger than that of the corresponding bulk DMSs, due to the density of states of 1D quantum wires. For the wire case, the first decreasing range of T_C with D corresponds to the first filled subband case (Eq. (3.63)), T_C increases dramatically with decreasing D. As D increases, the interval between neighboring transverse confined levels decreases. When the Fermi level intersects one subband bottom, the density of states has a singularity; thus, the T_C curve has a peak. For the slab case (Fig. 3.31(c)), the results are similar, when the Fermi level intersects a subband bottom, the density of states increases a step, and the T_C has also a abrupt increase, but is smaller than that of wires. Except the first filled subband case, the T_C of the slab differs from that of the bulk material slightly.

Figure 3.32 shows the Curie temperature T_C versus electron density n_e for (a) wire, $D = 2D_0$ and, (b) slab, $L = D_0$ (Zhang *et al.*, 2007). Figure 3.32 reflects that the T_C is proportional to the density of states at the Fermi level, for the bulk material $T_C \sim n_e^{3/2}$, as shown in Eq. (3.67). For

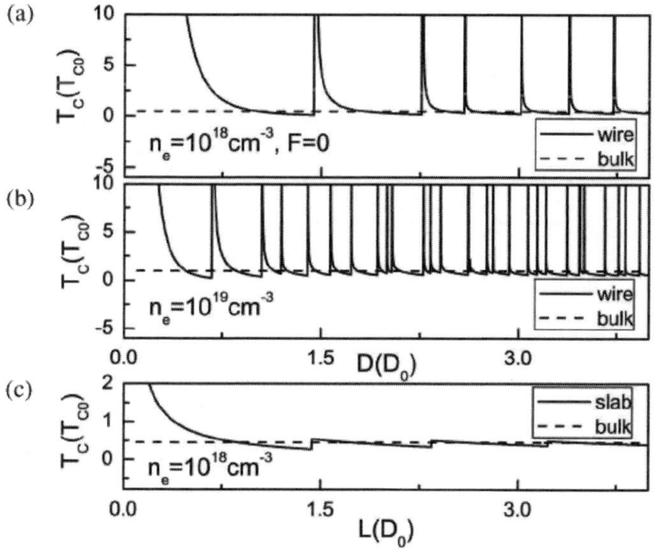

Fig. 3.31. Curie temperature versus dimensions for (a) wire with $n_e = 10^{18}\,\text{cm}^{-3}$, (b) wire with $n_e = 10^{19}\,\text{cm}^{-3}$, and (c) slab with $n_e = 10^{18}\,\text{cm}^{-3}$. The dashed lines are T_C of the corresponding bulk materials.

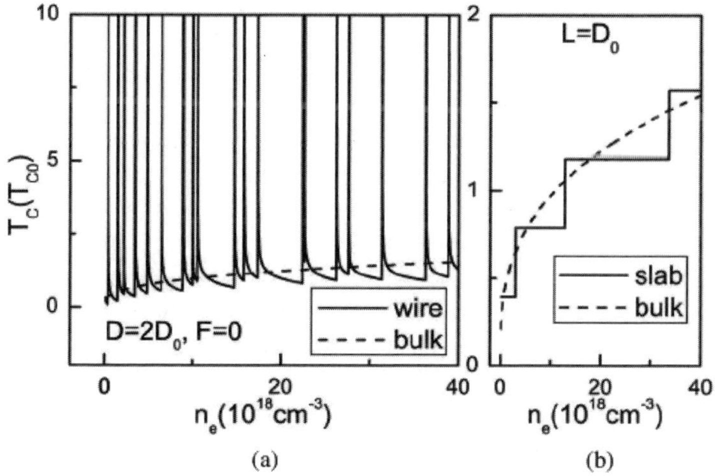

Fig. 3.32. Curie temperature T_C versus electron density n_e for (a) wire, $D = 2D_0$ and (b) slab, $L = D_0$. The dashed lines are T_C of bulk materials.

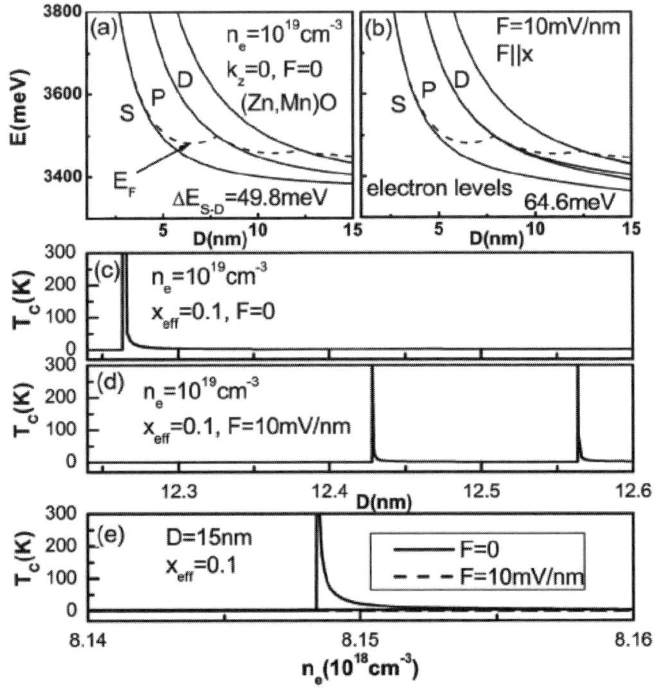

Fig. 3.33. Electron levels of (Zn, Mn)O nanowires with $n_{\rm e} = 10^{19}\,{\rm cm}^{-3}$ at $k_z = 0$ versus diameter D at (a) transverse electric field $F = 0$, (b) $F = 10\,{\rm mV/nm}$; $T_{\rm C}$ versus D at (c) $F = 0$, (d) $F = 10\,{\rm mV/nm}$; (e) $T_{\rm C}$ versus $n_{\rm e}$ at $F = 0$ and $F = 10\,{\rm mV/nm}$.

the wires and slabs, when $n_{\rm e}$ increases, the Fermi level intersects sequently the subband bottom one by one, where the density of states has an abrupt change, and the $T_{\rm C}$ curve shows peaks (wires) or steps (slabs).

An applied electric field perpendicular to the wire will influences the electron states in the wire, the electron filling situation, and the $T_{\rm C}$. Figure 3.33 shows the electron levels versus D, the $T_{\rm C}$ versus D, and the $T_{\rm C}$ versus $n_{\rm e}$ for non- and applied electric field cases (Zhang *et al.*, 2007). It is noticed that the energy difference of S and D levels $\Delta E_{\rm S-D}$ are changed largely by the electric field, especially when the diameter is large. This change of energy differences will affect the Curie temperature dramatically. There are two peaks of $T_{\rm C}$ as a function of D in the $F = 10\,{\rm mV/nm}$ case, whose diameter positions are quite different from that of the peak in the $F = 0$ case. Similarly, the $T_{\rm C}$ peaks appear at different $n_{\rm e}$ for the non- and applied electric field cases. If the ferromagnetic DMS nanowires are used to inject the spin polarized current into the normal semiconductors, we can use a transverse electric field to turn on or turn off the spin current.

3.6.4. *Ferromagnetic semiconductor QDs*

Experimentally, the high Curie temperature $T_C = 350\,\mathrm{K}$ has been discovered in the Mn-doped ZnO QDs (Norberg *et al.*, 2004). The high T_C has also be reported in the heavily doped InMnAs QDs (Chen *et al.*, 2007). Theoretically, the ferromagnetism in the QDs mediated by electrons has been calculated (Fernandez-Rossier and Brey, 2004), and it is found that the T_C depends on the parity of the electron number and the T_C is relatively low.

In the following, we consider the ferromagnetism in the QDs mediated by holes (Zhang *et al.*, 2008). The hole effective-mass Hamiltonian in the QDs is given in Eqs. (10.5) and (10.6) in Sec. 10.1. Assuming that the semiconductor is of wide gap, the interaction between the CB and the valence band can be neglected; thus, the hole Hamiltonian is 6×6-dimensional. For FMSs, the p–d exchange interaction is induced by the spontaneous magnetic moment $\boldsymbol{M}$, but not the applied magnetic field; thus,

$$H_{\mathrm{pd}} = \frac{\beta \boldsymbol{s}_h \cdot \boldsymbol{M}}{(g_{\mathrm{Mn}}\mu_{\mathrm{B}})}. \tag{3.68}$$

At temperature T, the free energy of the hole system is

$$F_{\mathrm{ch}}(M) = \frac{1}{V}\sum_n E_{\mathrm{hn}}(M)\frac{1}{1+\exp[-(E_{\mathrm{hn}}(M)-E_{\mathrm{F}})/k_{\mathrm{B}}T]}. \tag{3.69}$$

The spontaneous magnetic moment of the system is given by the following transcendental equation:

$$M = g_{\mathrm{Mn}}\mu_{\mathrm{B}}\bar{x}N_0 S B_{\mathrm{S}}\left[\frac{S g_{\mathrm{Mn}}\mu_{\mathrm{B}}(-\partial F_{\mathrm{c}}(M)\partial M}{k_{\mathrm{B}}(T+T_{\mathrm{AF}})}\right]. \tag{3.70}$$

If the hole density p and the temperature T is given, M can be solved from Eq. (3.70).

Figures 3.34(a), 3.34(b), 3.34(e), and 3.34(d) show the magnetization $M(T,p)$ versus temperature for $Zn_{1-x}Mn_xO$ QD ($\bar{x} = 0.05$) at $M||z$ and $M||x$, respectively (Zhang *et al.*, 2008). The calculation found that when M is very small, $F_c(M) - F_c(0)$ approaches zero not as $\sim \lambda M^2$ as shown in Eq. (3.43); thus the formula of the Curie temperature Eq. (3.45) is no more suitable. In fact, because the energy levels in QD are discrete, there

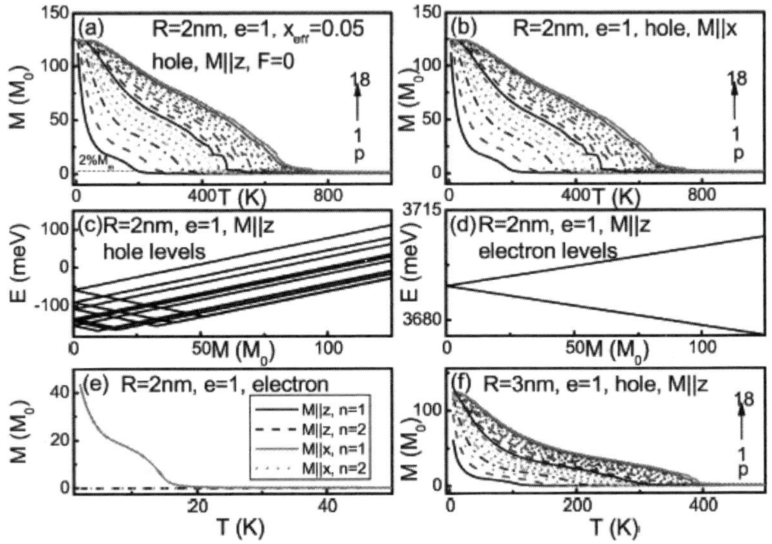

Fig. 3.34. Magnetization $M(T,p)$ versus temperature for $R = 2\,\text{nm}$ $Zn_{1-x}Mn_xO$ QD ($\bar{x} = 0.05$) at (a) $M||z$, (b) $M||x$, and various curves correspond to the hole number $p = 1$–18. (c) Hole levels and (d) electron levels versus M; $M(T,n)$ versus T for (e) electron, $R = 2\,\text{nm}$, (f) hole, $R = 3\,\text{nm}$.

is no subband; therefore, $F_c(M)$ is only a linear function of M. When the temperature increases, the M approaches zero smoothly, but not equal to zero. Thus, we define the Curie temperature of QDs such that at T_C the M reaches a minimum $M(T_C,p) = M_{\min}$. From Fig. 3.34, we see that the T_C and magnetization of $R = 2\,\text{nm}$ QD are much larger than those of $R = 3\,\text{nm}$ QD, for former of $p = 18$, $T_C \approx 700\,\text{K}$, which decreases with increasing R. The Curie temperature of hole-mediated ferromagnetism is much higher than that of the electron-mediated one.

Figures 3.34(c) and 3.34(d) show hole levels and electron levels versus M, respectively (Zhang *et al.*, 2008). When $M > 50\ M_0$, all hole levels increase as M increases, so that the magnetization increases with the hole number at relatively low temperature. When T is high, there are many step decreases of the magnetization due to the level crossing in the small M range. The lowest two electron levels split as M increases and do not cross with upper levels. When there are two electrons in the dot, the magnetization cannot reduce the free energy, and the calculated self-consistent magnetization is nearly zero as shown in Fig. 3.34(e). When there is one electron, the magnetization is large at low temperature; the

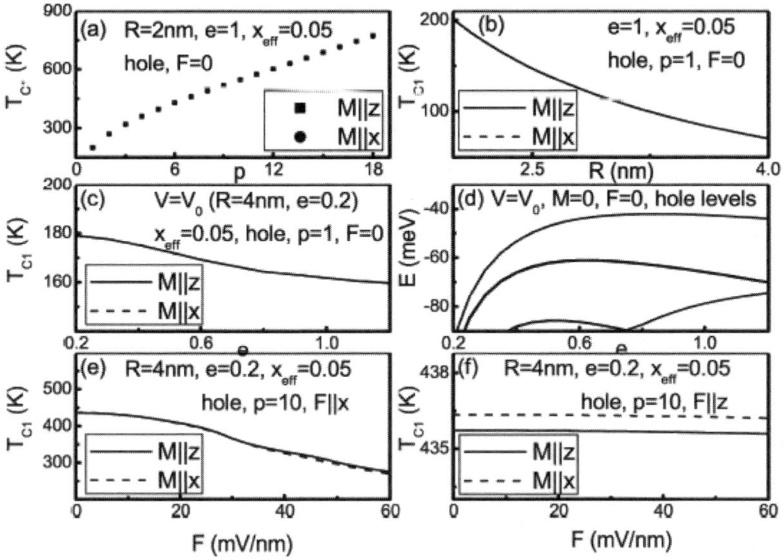

Fig. 3.35. (a) T_C versus hole number p for a $R = 2\,\text{nm}$, $\bar{x} = 0.05$ QD. (b) T_C versus radius R for $p = 1$. (c) T_C versus aspect ratio of ellipsoid e for fixed volume V_0. (d) Hole levels versus e for fixed volume. (f) Applied electric field $F||x$ (g) $F||z$, T_C versus F for ellipsoid of $R = 4\,\text{nm}$, $e = 0.2$, $\bar{x} = 0.05$, $p = 10$.

Curie temperature is about 20 K. So the dependence of electron-mediated ferromagnetism on the parity of the number of electrons is confirmed for the ground-state electron.

Figure 3.35 shows T_C of QDs as functions of various parameters (Zhang *et al.*, 2008). From Fig. 3.35, we see that the T_C increases as the hole number p in the dot increases, similar to the bulk material, but is larger than that of the bulk material. The T_C decreases as the radius increases, because the volume of the dot V increases and the free energy decreases (see Eq. (3.69)). The T_C relates also to the shape of the dot; for the same volume V, T_C decreases with increasing e (longitudinal axis/transverse axis), so that the flat ellipsoid favors the higher Curie temperature, due to the decrease in the energy differences between the hole levels. The transverse electric field $F||x$ has a large effect on the T_C of the flat ellipsoid ($e = 0.2$) as shown in Fig. 3.35(e), the T_C can be turned from much higher than the RT to lower than it. In the $F||z$ case, the change of T_C is relatively small. Because the longitudinal length is much smaller than the transverse length for a flat ellipsoid ($e = 0.2$), the effect of the transverse electric field on the hole levels is much larger than that of the longitudinal electric field.

3.7. First-principle Calculation of FMSs

There have been two theories about FMSs. The initial theory is to use model Hamiltonian, in which certain specific magnetic interactions between ions are postulated. The most prominent supposition for the interaction type between spin-polarized 3d impurity ions in semiconductors is assumed to be similar to the interaction of nuclear spins mediated by the conduction electrons in metals. This type of interaction is called RKKY interaction, which predicted ferromagnetism in various 3d impurity-doped compound semiconductors, as discussed in the last section. Another theory is the first-principle calculation of the band structures. Instead of prejudging the preferred type of magnetic interactions between ions, it articulates fundamental electron–ion and electron–electron interactions and lets the magnetic interactions emerge as solutions to the fundamental many-electron Hamiltonian. The availability of density-functional approximations, accurate pseudopotentials, and the ubiquitous computer packages freed the theorists from having to guess at the output how a particular impurity would affect a particular host crystal.

Sato and Katayama-Yoshida (2002) investigated systematically the stability of the ferromagnetic state in II–VI and III–V compounds based DMSs and predicted the candidates for high-T_C ferromagnets based on first-principle calculations. Here we introduce Zunger group's works (Mahadevan and Zunger, 2004). They have carried out first-principle electronic structure calculations using density functional theory, within the momentum space total energy pseudopotential method, using ultrasoft pseudopotentials as implemented in the VASP code. They studied transition-metal (TM) impurities V, Cr, Mn, Fe, Co, and Ni in 64-atom supercells of zincblende GaSb, GaAs, GaP, and GaN. The d partial density of states as well as the local moment at the TM were calculated within a sphere of radius 1.2 Å about the atom and have been broadened with a Gaussian of 0.2 eV full width at half-maximum (FWHM). The GGA exchange functional is used throughout the work.

Figure 3.36 shows the TM projected density of states for substitute V–Ni impurities in GaAs for various symmetries (Mahadevan and Zunger, 2004). The GGA band gap of pure GaAs is found to be 0.3 eV, and all the V–Ni impurities introduce levels into the band gap. The two sets of t_+ and t_- levels for each impurity are suggestive of bonding/antibonding combinations arising from hybridization between the TM ion and the neighboring dangling bonds. Bonding states with a large wave function amplitude on

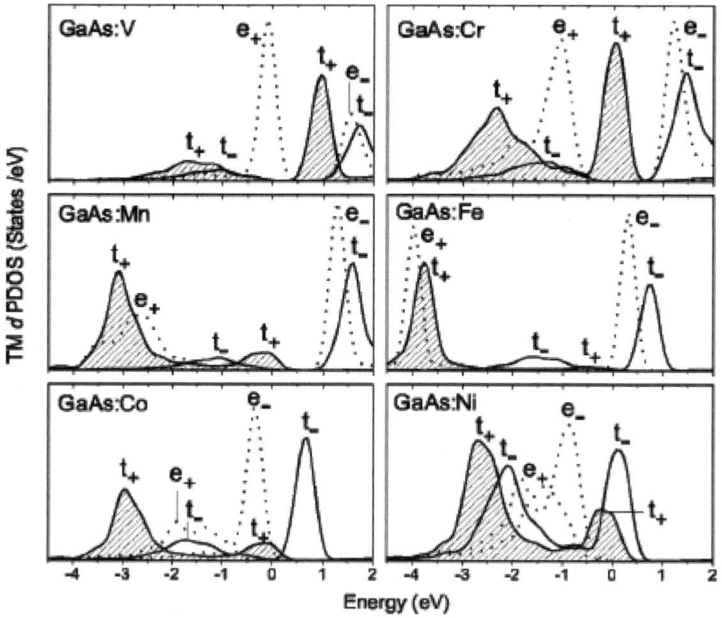

Fig. 3.36. The TM projected density of states for substitute V–Ni impurities in GaAs for spin-up t_+ (shaded region), spin-down t_-. (solid line), and e (dashed line) symmetries. The zero of energy represents the valence band maximum (VBM) of the host.

the TM site are referred to as crystal-field resonances (CFRs), whereas antibonding states with low contribution on the TM which are localized instead on the four nearest As atoms are referred to as the dangling bond hybrid (DBH).

3.7.1. *Simple model of the electronic structure of 3d impurities in GaAs*

Most of the results of the density functional study of the electronic structure of 3d impurities in III–V's as shown in Fig. 3.36 can be captured by a simple model. While the actual supercell calculations are performed at a large set of k points (64) over the entire Brillouin zone, the model uses representative energy levels as an abstraction of the finite-width density of states. The electronic structure of substitute 3d in III–V semiconductors can be understood as arising from the interaction of the host cation vacancy (anion dangling bonds) with the crystal-field and exchange-split orbitals of a 3d ion.

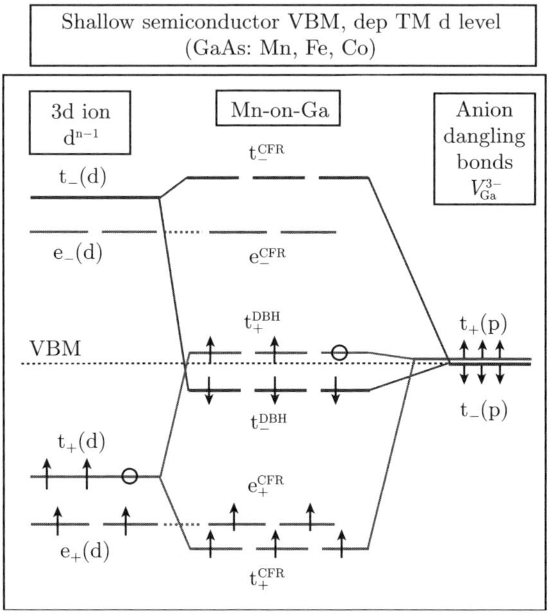

Fig. 3.37. The schematic energy level diagram for the levels (central panel) generated from the interaction between the crystal-field and exchange-split levels on the 3d TM ion (left panel) with the anion dangling bond (right), when the TM d levels are energetically deeper than the dangling bond levels.

Figure 3.37 shows the schematic energy-level diagram for the levels (central panel) generated from the interaction between the crystal-field and exchange-split levels on the 3d TM ion (left panel) with the anion dangling bond (right), when the TM d levels are energetically deeper than the dangling bond levels (Mahadevan and Zunger, 2004). The anion dangling bond gives rise to a fully occupied s-like a_1 level located deep in the host valence band, and a partially occupied p-like t_2 level located just above the host valence band maximum (VBM), with a wave-function amplitude localized primarily on the neighboring atoms. The neutral vacancy V_{III}^0 has a deficiency of three electrons, i.e. the orbital configuration is $a_1^2 t_2^3$ (p), where p denotes its major orbital character. Spin polarization splits this t_2(p) vacancy level into spin-up [t_+(p)] and spin-down [t_-(p)] states, but the splitting is small (90 meV) on account of the rather delocalized nature of these pure host dangling bond orbitals.

The tetrahedral crystal field of the zincblende host splits the TM d levels into e(d) and t_2(d), with e below t_2 in the point-ion limit, which is denoted by $\Delta_{CF}(t_2 - e)$. Exchange interactions further split these levels into spin-up (+) and spin-down (−), with exchange splittings $\Delta_x(e) = [e_-(\mathrm{d}) - e_+(\mathrm{d})]$ and $\Delta_x(t) = [t_-(\mathrm{d}) - t_+(\mathrm{d})]$.

The energy levels of a cation-substituted TM in a III–V semiconductor can be thought of as the result of a hybridization between the anion dangling bonds and the crystal-field and exchange-split d levels of a TM ion as shown in the central panel of Fig. 3.37. There are two cases: when the 3d levels are well below the host anion dangling bonds (Fig. 3.37), or when the 3d levels are well above the host anion dangling bonds (Fig. 3.38). The t_2(p) levels of the anion dangling bond hybridize with the t_2(d) levels of the TM ion. In contrast, the e(d) level of the TM ion remains largely unperturbed since the host does not have localized e states in this energy range. The hybridization in the t_2-channel creates bonding, TM localized CFRs (crystal-field resonance): $t_{\mathrm{CFR}+}$ and $t_{\mathrm{CFR}-}$, as well as the host-anion localized antibonding DBHs (dangling bond hybrid): $t_{\mathrm{DBH}+}$ and $t_{\mathrm{DBH}-}$, whereas the e channel creates the nonbonding $\mathrm{e}_{\mathrm{CFB}+}$ and $\mathrm{e}_{\mathrm{CFB}-}$ states. The available electrons for occupation of these levels are $N = (n-1) + 6$

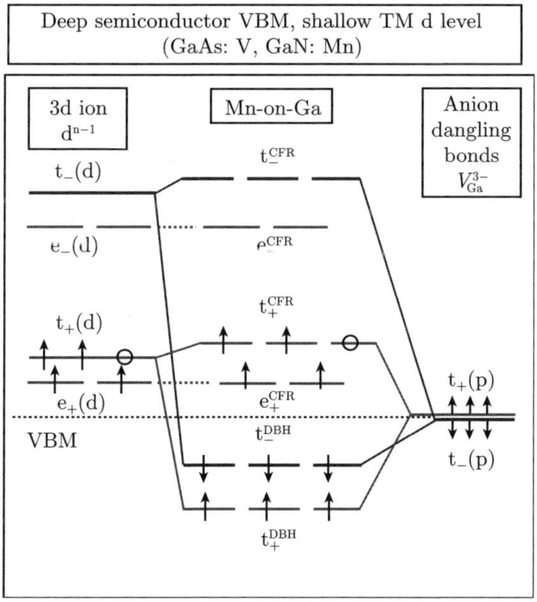

Fig. 3.38. The schematic energy-level diagram for the levels (central panel) generated from the interaction between the crystal-field and exchange-split levels on the 3d TM ion (left panel) with the anion dangling bond (right), when the TM d levels are energetically higher than the dangling bond levels.

for a d^n–s^2 TM atom [three electrons are used to complete the anion dangling bond state to $t^6(p)$, leaving d^{n-1} at the TM ion]. For GaAs: Mn, Fe, Co as shown in Fig. 3.37, the orders of levels are

$$\text{GaAs:Mn} \quad t^3_{\text{CFR}+} < e^2_{\text{CFR}+} < t^3_{\text{DBH}-} < t^2_{\text{DBH}+},$$

$$\text{GaAs:Fe} \quad t^3_{\text{CFR}+} < e^2_{\text{CFR}+} < t^3_{\text{DBH}-} < t^3_{\text{DBH}+},$$

$$\text{GaAs:Co} \quad t^3_{\text{CFR}+} < e^2_{\text{CFR}+} < e^2_{\text{CFR}-} < t^3_{\text{DBH}-} < t^2_{\text{DBH}+}.$$

These orders of levels are in agreement with Fig. 3.36; there are fully filled $t_{\text{CFR}+}$ and $e_{\text{CFR}+}$ levels and 1, 0, and 1 hole in the $t_{\text{DBH}+}$ level, respectively.

Figure 3.38 shows the schematic energy-level diagram for GaAs:V and GaN:Mn, when the TM d levels are energetically higher than the dangling bond levels (Mahadevan and Zunger, 2004). The order of levels is different from that of Fig. 3.37. The orders of levels of GaAs:V and GaN:Mn are

$$\text{GaAs:V} \quad t^3_{\text{DBH}+} < t^3_{\text{DBH}-} < e^2_{\text{CFR}+}$$

$$\text{GaN:Mn} \quad t^3_{\text{DBH}+} < t^3_{\text{DBH}-} < e^2_{\text{CFR}+} < t^2_{\text{CFR}+}.$$

The most difference between GaAs:Mn and GaN:Mn is that for the former the nonfilled state is DBH type, localized at host anions and favored to ferromagnetism; for the latter, the nonfilled state is CFR type, localized at TM ion and not favored to ferromagnetism. As a rule, as the anion dangling bond becomes deeper (GaSb→GaAs→GaP→GaN), or the TM d level becomes shallower (Ni→V), there will be a transform from the DBH to CFR.

The analysis for different TM–host combinations suggests that when the gap level has a t_{DBH} character the spin splitting of the VBM is large (~0.4 eV for GaAs:Mn). If the gap level has a t_{CFR} character, the spin splitting of the VBM is small (~0.1 eV for GaN:Mn) due to the large energy separation between the dangling bond levels and the valence band levels. This VBM spin splitting of the impure system has been traditionally used to estimate the exchange interaction strength J_{pd}.

3.7.2. *Practical rules for ferromagnetism of 3d impurities in semiconductors*

Zhao *et al.* (2005) distilled from the first-principle spin-polarized total-energy calculations some practical rules for predicting the magnetic states (ferromagnetic/antiferromagnetic/paramagnetic) of substitute TM impurity with different charge states in various host crystal group IV, III–V,

II–VI, etc., semiconductors. The basic mechanism is the stabilization of a ferromagnetic bond between two TM ions if the interacting orbitals are partially occupied. The rule is based on the observations that the competition between FM and AFM is decided by the strength of the interaction between partially occupied states (either the hole or the electron) localized on neighboring TM impurities and that FM interaction is strongest for t_2 type (rather than e type) orbital.

Figure 3.39 shows the schematic energy levels for spin-up and spin-down t_2 levels of two identical 3d ions TM_1 and TM_2 with either closed or open shells (Zhao *et al.*, 2005). The energy levels of TM ions split by the exchange interaction, and form a set of bonding and antibonding states for each spin channel. In an FM spin arrangement (Fig. 3.39(a)), both the bonding and antibonding levels of one-spin channel are completely filled, so to first order, there is no gain in energy as a result of ferromagnetic coupling. In an AFM arrangement of fully occupied orbitals (Fig. 3.39(b)), only the bonding states are fully occupied for both spin channels, while the antibonding levels are empty. This will lead to an energy gain for the AFM spin arrangement. Thus, for fully occupied TM orbitals, the AFM arrangement is generally

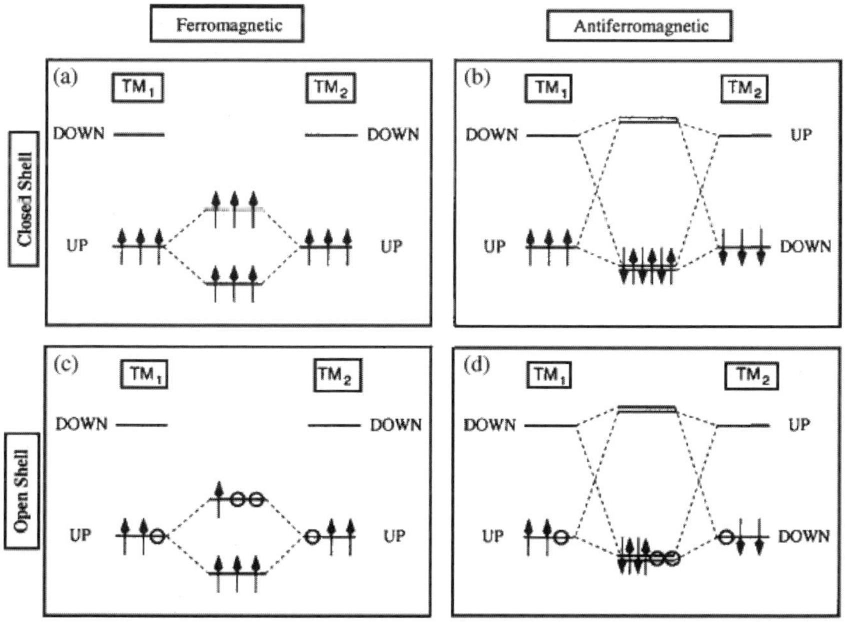

Fig. 3.39. Schematic energy levels for spin-up and spin-down t_2 levels of two identical 3d ions TM_1 and TM_2 with either closed or open shells. FM is generally favored in (c) while AFM is favored in (b).

favored. In an FM spin arrangement with nonfilled orbitals (Fig. 3.39(c)), there is an energy gain as a result of TM–TM coupling since only the bonding orbitals are occupied and the antibonding levels are only partially occupied. Here the TM_1-up and the TM_2-up levels are nearly the same, so the FM energy gain can be large. In turn, in the AFM spin arrangement with nonfilled orbitals (Fig. 3.39(d)), there is also an energy gain, but the energy separation between the TM_1-up and TM_2-up levels are large, so the energy gain would be smaller.

In order to denote the charge state, an "oxidation state" (OS) is defined. If a neutral TM impurity is substituted on a divalent cation (ZnSe:Mn^0), its nominal oxidation is II, whereas if it is substituted on a trivalent cation (GaAs:Mn^0) or tetravalent cation (Ge:Mn^0), its OS is III or IV, respectively. If the system is doped p-type so as to transform the impurity to a positively charged donor, the nominal OS increases. For example, ZnSe:Mn^+ has OS III and GaAs:Mn^+ has OS IV. If the system is doped n-type so the impurity is a negatively charged acceptor, the nominal OS decreases.

Table 3.1 shows the prediction of the type of magnetic behavior for different OSs (roman letters) of various 3d ions in different zinc-blende host semiconductors (Zhao *et al.*, 2005). From Table 3.1, we see that:

(1) Antiferromagnetism is predicted in closed-shell configurations, e.g. group A (t^6e^2) of Fe(I), Co(II), and Ni(III); the HS group C; the LS group D; as well as the HS group F. First-principles calculated examples confirmed most of the results, but still have some exceptions.
(2) Weak FM or AFM is predicted in cases having a hole in an e-type state, e.g. HS group B (t^6e^1) and LS group E (t^3e^1).
(3) Ferromagnetism is predicted in cases with one or two holes in the bonding t_- or antibonding t_+ levels., e.g. LS group B (t^5e^2) of Mn(I), Fe(II), and NI(IV), LS group C (t^4e^2), HS group D (t^5e^0); HS (t^4e^0) and LS (t^2e^2) group E, and LS group F (t^1e^2).

To test the above rules quantitatively, the energy differences of ferromagnetic state and the antiferromagnetic state $\Delta = E_{\rm FM} - E_{\rm AFM}$ are calculated from first principles for Mn in various host crystals while shifting the Fermi level $E_{\rm F}$, thus changing the OS, as shown in Fig. 3.40 (Zhao *et al.*, 2005). The rule of Table 3.1 states that Mn(II) will be weak FM (or even AFM, e.g. HS case C), whereas Mn(III) will be FM (e.g. HS case D), as is Mn(IV) (HS case E). In Fig. 3.40, the Mn(II) is doped in the divalent site of $ZnGeP_2$, $CdGeP_2$, and CdTe; it is seen that in all cases they are

Table 3.1. Prediction of the type of magnetic behavior for different OSs (roman letters) of various 3d ions in different zincblende host semiconductors. LS and HS means low spin and high spin, "weak" means weak FM or AFM. The boxed numbers correspond to the electron numbers in the partially filled orbital.

							Electronic configuration							Prediction	
Case	V	Cr	Mn	Fe	Co	Ni		e_+	t_+^{bond}	t_-^{bond}	t_+^{anti}	e_-	t_-^{anti}	Mag. state	Moment
A $d_+^5(t^6\ e^2)$				I	II	III	HS/LS:	2	3	3	3	2	0	AFM	3
B $d_+^5(t^6\ e^1)$			I	II	III	IV	HS:	2	3	3	3	$\boxed{1}$	0	Weak	4
$d_+^5(t^5\ e^2)$							LS:	2	3	3	$\boxed{2}$	2	0	FM	2
C $d_+^5(t^6\ e^0)$		I	II	III	IV		HS:	2	3	3	3	0	0	AFM	5
$d_+^5(t^4\ e^2)$							LS:	2	3	3	$\boxed{1}$	2	0	FM	2
D $d_+^5(t^5\ e^0)$	I	II	III	IV			HS:	2	3	3	$\boxed{2}$	0	0	FM	4
$d_+^5(t^3\ e^2)$							LS:	2	3	3	0	2	0	PM	0
$d_+^5(t^4\ e^0)$							HS:	2	3	3	$\boxed{1}$	0	0	FM	3
E $d_+^5(t^3\ e^1)$	II	III	IV				LS:	2	3	3	0	$\boxed{1}$	0	Weak	1
$d_+^5(t^2\ e^2)$							LS:	2	3	$\boxed{2}$	0	2	0	FM	1
F $d_+^5(t^3\ e^0)$	III	IV					HS:	2	3	3	0	0	0	AFM	2
$d_+^5(t^1\ e^2)$							LS:	2	3	$\boxed{1}$	0	2	0	FM	2

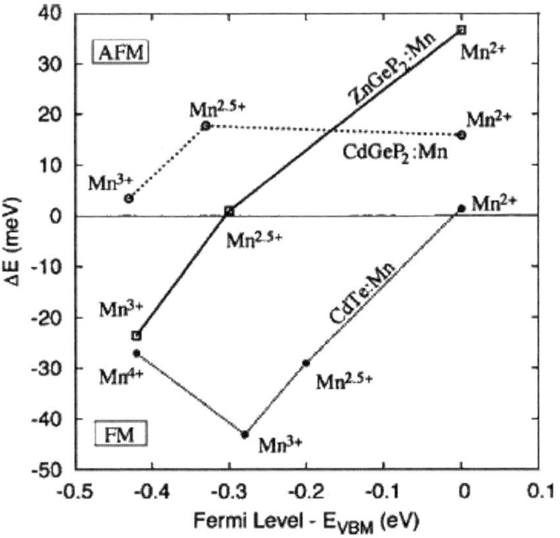

Fig. 3.40. The change of ferromagnetic stability energy of $ZnGeP_2$:Mn, $CdGeP_2$:Mn, and CdTe:Mn by shifting E_F.

AFM as expected of Mn(II). Then the E_F shifts in the calculation toward the valence band, creating higher OS. A transition is found toward FM as expected by the rule: in CdTe it occurs when the OS is ~2.5, and stays for FM at +3 and +4. In $ZnGeP_2$, the transition occurs for OS > 2.5, whereas $CdGeP_2$ also tends to be less AFM as OS increases, but it is still weak AFM for OS $= 3$.

3.8. Magnetic Polaron (MP) — A New Mechanism of Ferromagnetism

The experiments found that the dilute ferromagnetic oxides have Curie temperatures far in excess of 300 K and exceptionally large ordered moments per TM cation. These materials are high-*k* dielectrics with degenerate or thermally activated n-type semiconductivity. Coey *et al.* (2005) proposed that ferromagnetic exchange here and in dilute ferromagnetic nitrides, is mediated by shallow donor electrons that form bound MPs, which overlap to create a spin-split impurity band. High Curie temperatures arise only when empty minority-spin or majority-spin d states lie at the Fermi level in the impurity band. A list of oxides and nitrides where high Curie temperature ferromagnetism has been reported is given in Table 3.2.

Table 3.2. High-temperature ferromagnetism in dilute ferromagnetic oxides and nitrides.

Material	E_g (eV)	Doping X	Moment (μ_B)	T_C (K)	Reference
GaN	3.5	Mn: 9%	0.9	940	34
		Cr	–	>400	35
AIN	4.3	Cr: 7%	1.2	>600	36
		V: 5%	4.2	>400	28
TIO_2	3.2	Co: 1–2%	0.3	>300	4
		Co: 7%	1.4	650–700	37
		Fe: 2%	2.4	>300	38
SnO_2	3.5	Fe: 5%	1.8	610	39
		Co: 5%	7.5	650	6
ZnO	3.3	V: 15%	0.5	>350	27
		Mn: 2.2%	0.16	>300	40
		Fe: 5%, Cu: 1%	0.75	550	41
		Co: 10%	2.0	280–300	5
		Ni: 0.9%	0.06	>300	42
CU_2O	2.0	Co: 5%, Al: 0.5%	0.2	>300	43
$In_{1,8}Sn_{02}O_3$	3.8	Mn: 5%	0.8	>300	44

These compounds are all wide gap semiconductors. When the concentration of magnetic ions x is far smaller than the percolation threshold concentration x_p associated with nearest-neighbor cation coupling, the average moment per dopant cation approaches (or even exceeds) the spin-only moment at low dopant concentration x, and it falls progressively as x increases toward x_p. Their Curie temperature all exceed RT. The ordered magnetic moments per 3d cation at low concentrations are unprecedented, exceeding those in almost all known oxides or alloys, including pure metal. Such high moments per cation are inexplicable in terms of possible known ferromagnetic phases. Dietl's hole-mediated ferromagnetism theory (Dietl *et al.*, 2001) explained the ferromagnetism in III–V DMSs; the Curie temperature in these semiconductors range up to about 170 K. The above oxides and nitrides are all n-type and have high resistance due to compensation, thus Dietl's theory is not suitable.

Coey *et al.* (2005) proposed that the magnetic exchange interaction is through the impurity states or impurity band. The donor electron forms a hydrogen-like orbital around the donor,

$$\begin{aligned} \psi(\boldsymbol{r}) &= (\pi r_H^3)^{-1/2} \exp\left(\frac{-r}{r_H}\right), \\ r_H &= \varepsilon\left(\frac{m_0}{m^*}\right) a_0 = \gamma a_0, \end{aligned} \tag{3.71}$$

where a_0 is the Bohr radius and $r_{\rm H}$ is the effective radius of impurity state. There is a critical donor concentration $n^{\rm crit}$ where the impurity band states become delocalized, and metallic conduction sets in, namely $(n^{\rm crit})^{1/3} r_{\rm H} \approx 0.26$. The corresponding donor concentration $\delta_{\rm p}$ is $n^{\rm crit}/n_0$, where the oxygen density $n_0 = 6 \times 10^{22}\,{\rm cm}^{-3}$ for oxides with a close-packed oxygen lattice. Hence $\gamma^3 \delta_p \approx 4.3$. Suppose that there are several magnetic positive ions such as Fe^{3+} or Mn^{2+} in the region of the donor electron orbital. The donors tend to form bound MPs, coupling the 3d moments of the ions within their orbits. Provided the radius of its orbit is sufficiently large, say $\gamma > 4$, an overlap between a hydrogenic electron and the magnetic cations within its orbit leads to ferromagnetic exchange coupling. This interaction can be written in terms of the s–d exchange parameter $J_{\rm sd}$ as

$$H_{\rm sd} = -J_{\rm sd} \boldsymbol{S} \cdot \boldsymbol{s} |\psi(r)|^2 \Omega, \tag{3.72}$$

where $\boldsymbol{S}$ is the spin of the magnetic ions, Ω is its volume, $\boldsymbol{s}$ is the spin of donor electron, and $\psi(r)$ is its wave function (3.71).

If the donor impurity (for example, oxygen vacancy) concentration is high enough, so that the donor electron orbits overlap each other, i.e. the percolation state and the impurity band are formed, the magnetic momentsy of the bound MPs will tend to one direction, resulting in ferromagnetism, as shown in Fig. 3.41 for the case of overlapping polarons. The related parameters for three compounds are listed in Table 3.3; where $\nu_{\rm c}$ is the number of cation sites within $r_{\rm H}$, $\delta_{\rm p}$ is the polaron percolation threshold, and $x_{\rm p}$ is the cation percolation threshold. Provided the number of the sites within $r_{\rm H}$ occupied by magnetic cations, $\nu = x\nu_{\rm c}$ is large enough (>10), ferromagnetism occurs when $\delta > \delta_{\rm p}$ and $x < x_{\rm p}$.

The effective field of the exchange interaction Eq. (3.72) acting on the spin $\boldsymbol{S}$ of the magnetic ions,

$$H'_W = \frac{N J_{\rm sd} \langle s_z \rangle}{2\mu_0 \mu_{\rm B} \rho^3}, \tag{3.73}$$

where $\rho = r_{\rm H}/r_{\rm c}$ and $r_{\rm c}$ is the radius of Ω. The Curie temperature can be determined by the following transcendental equations,

$$\begin{aligned} \frac{\langle S_z \rangle}{S} &= B_s \left(\frac{N J_{\rm sd} S \langle s_z \rangle}{k_{\rm B} T \rho^3} \right), \\ \frac{\langle s_z \rangle}{s} &= \tanh \left(\frac{\nu J_{\rm sd} s \langle S_z \rangle}{k_{\rm B} T \rho^3} \right). \end{aligned} \tag{3.74}$$

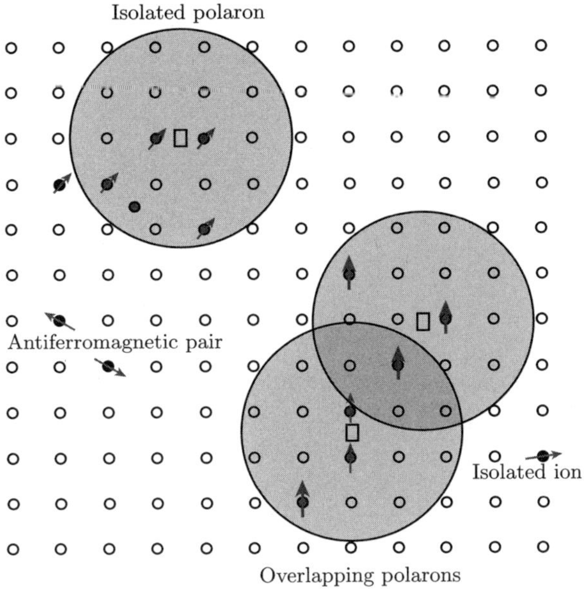

Fig. 3.41. Schematic diagram of MPs for $x = 0.1$, $\gamma = 12$. Cation sites are represented by small circles, oxygen is not shown, and unoccupied oxygen sites are represented by squares. The isolated polaron, antiferromagnetic pair, isolated ion, and overlapping polarons are shown in the figure.

Table 3.3. Parameters for some oxides. n_c, cation density; ε, high-frequency dielectric constant; $r_H = \varepsilon(m/m^*)a_0$ hydrogenic radius; v_c number of cation sites within r_H; δ_p polaron percolation threshold; x_p cation percolation threshold.

Material	Structure	n_c (10^{28}m^{-3})	ε	m^*/m	γ	r_H (nm)	v_c	δ_p (10^{-3})	x_p
ZnO	Wurtzite	3.94	4.0	0.28	14	0.76	64	1.5	0.18
TiO_2	Anatase	2.93	9.0	1.0	9	0.48	14	5.9	0.25
SnO_2	Rutile	2.80	3.9	0.24	16	0.86	75	1.0	0.25

Close to T_C, the functions at the right side of Eq. (3.74) can be linearized using $B_S(y) \approx [(S+1)S/3S]y$ and $\tanh y \approx y$, yielding

$$T_C = \frac{J_{sd}s}{k_B\rho^3}\sqrt{\frac{\nu NS(S+1)}{3}}. \tag{3.75}$$

Taking ZnO:Co as an example: $J_{sd} = 1.5\,\text{eV}$, $s = 1/2$, $S = 3/2$, $n_0 = 6\times10^{22}\,\text{cm}^{-3}$, $x = 0.1$, $r_H = 0.8\,\text{nm}$, $r_c = 0.06\,\text{nm}$, and $\delta = 0.01$. Derived values are $N = 1.3$, $\nu = 13$, $\rho = 13$. The Curie temperature given by

Eq. (3.75) is then $T_C = 18\,\text{K}$. The estimate is too low by one or two orders of magnitude.

From Eq. (3.75), $T_C \sim \rho^{-3}$, where $\rho = r_H/r_c = 13$, and r_c is the radius of ion sphere. The empty 3d states ($3d^{\downarrow}$ in the example of a $3d^7$ impurity, $S = 3/2$) have to hybridize with the donor states at the Fermi level and increase the effective radius r_c^{eff} of the magnetic ions and the electron transfer effect. The electron transfer effect may be represented by introducing a factor into Eq. (3.75),

$$\left(\frac{r_c^{\text{eff}}}{r_c}\right)^3 \frac{1}{\rho^3} = \left(\frac{r_c^{\text{eff}}}{r_c}\right)^3 \left(\frac{r_c}{r_H}\right)^3 = \left(\frac{r_c^{\text{eff}}}{r_H}\right)^3. \tag{3.76}$$

The electronic structure necessary for a high Curie temperature is contrasted with that for a low Curie temperature as shown in Fig. 3.42. On passing along the 3d series from Ti to Cu, the spin-split d levels move down in energy toward the top of valence band. Hence, there are two regions where a high T_C is expected: one near the beginning of the series where the $3d^{\uparrow}$ states cross the Fermi level in the impurity band and one toward the end of the series where the $3d^{\downarrow}$ states cross the Fermi level, as shown in Figs. 3.42(b) and 3.42(c), respectively. For the middle case as shown in Fig. 3.42(a), there is no 3d state crossing the Fermi level, leading to low T_C. The results have been examined by experiments for thin films of ZnO doped with every element in the 3d series from Sc to Cu. Films were deposited from targets of $(Zn_{0.95}M_{0.05})O$ by pulsed laser deposition. The RT moment in μ_B per dopant cation is plotted in Fig. 3.43 (Coey *et al.*, 2005), where the two distinct peaks appear, one at Ti and V corresponding

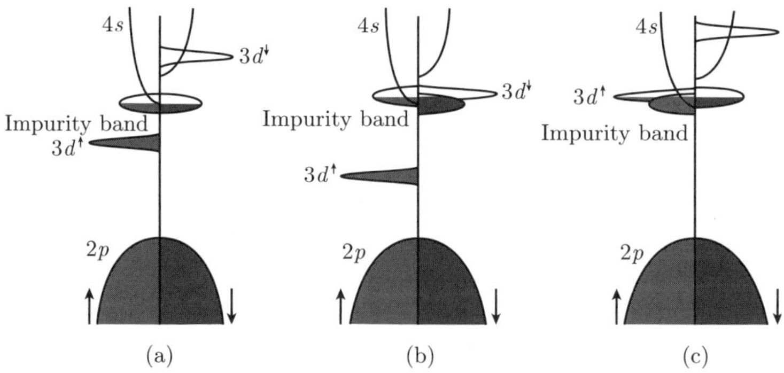

Fig. 3.42. Schematic band structures of an oxide with 3d impurities and a spin-split donor impurity band. (a) The position of the 3d level for low Curie temperature. (b) and (c) The 3d level crosses the impurity band, which leads to high T_C.

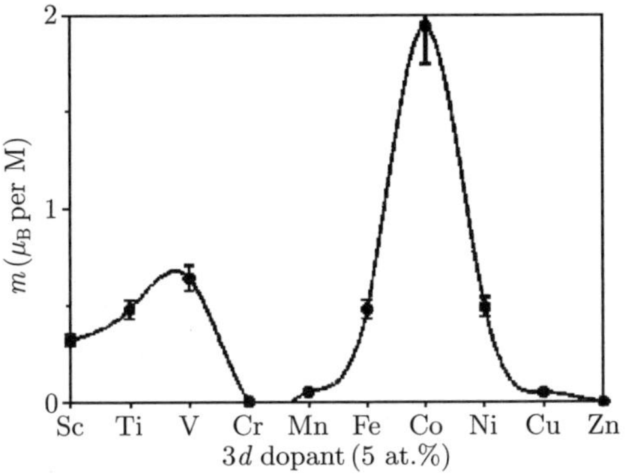

Fig. 3.43. The magnetic moment of thin films produced from ($Zn_{0.95}M_{0.05}O$) (M = Sc–Cu) targets by pulsed laser deposition measured at RT.

to Fig. 3.42(c), the other at Co corresponding to Fig. 3.24(b). Moments are near zero for Cr, Mn, and Cu corresponding to Fig. 3.42(a).

References

Arrot A. (1957) *Phys Rev* **108**: 1394.
Berciu M, Bhatt RN. (2004) *Phys Rev B* **69**: 045202.
Boukari H. (2002) *Phys Rev Lett* **88**: 207204.
Chen YF, Lee WN, Huang JH *et al.* (2005) *J Vac Sci Technol B* **23**: 1376.
Chen YF, Huang JH, Lee WN *et al.* (2007) *Appl Phys Lett* **90**: 022505.
Cocy JMD, Venkatesan M, Fitzgerals CB. (2005) *Nat Mater* **4**: 173.
Dietl T, Ohno H, Matsukura F, Cibert J, Ferrand D. (2000) *Science* **287**: 1019.
Dietl T, Ohno H, Matsukura F. (2001) *Phys Rev B* **63**: 195205.
Fernandez-Rossier J, Brey L. (2004) *Phys Rev Lett* **93**: 117201.
Ferrand D. (2001) *Phys Rev B* **63**: 085201.
Haury A. (1997) *Phys Rev Lett* **79**: 511.
Holub M, Chakrabarti S, Fathpour S *et al.* (2004) *Appl Phys Lett* **85**: 973.
Jeon HC, Jeong YS, Kang TW *et al.* (2002) *Adv Mater* **14**: 1725.
Ku KC, Potashnik SJ, Wang RF, Chun SH *et al.* (2003) *Appl Phys Lett* **82**: 2302.
Mahadevan P, Zunger A. (2004) *Phys Rev B* **69**: 115211.
Matsukura F, Ohno H, Shen A, Sugawara Y. (1998) *Phys Rev B* **57**: R2037.
Moreno M, Trampert A, Jenichen B *et al.* (2002) *J Appl Phys* **92**: 4672.
Norberg NS, Kittilstved KR, Amonette JE *et al.* (2004) *J Am Chem Soc* **126**: 9387.
Ohno H, Munekata H, Penney T, von Molnar S, Chang LL. (1992) *Phys Rev Lett* **68**: 2664.
Ohno H, Shen A, Matsukura F *et al.* (1996) *Appl Phys Lett* **69**: 363.
Ohno H. (1998) *Science* **281**: 951.

Ohno H. (2002) In *Semiconductor Spintrinics and Quantum Computation*, Vol. 1, Awschalom DD, Loss D, Samarth N. (eds). Springer, Berlin.
Overberg ME, Gila BP, Abernathy CR, Pearton SJ *et al.* (2001) *Appl Phys Lett* **79**: 3128.
Ozaki N, Nishizawa N, Nam KT, Kuroda S, Takita K. (2004) *Phys Status Solidi* C**1**: 957.
Ozaki N, Okabayashi I, Kumekawa T *et al.* (2005) *Appl Phys Lett* **87**: 192116.
Reed MJ, Arkun FE, Berkman EA *et al.* (2005) *Appl Phys Lett* **86**: 102504.
Saito H, Zayets V, Yamagata S, Ando K. (2003) *Phys Rev Lett* **90**: 207202.
Sakaki T, Sonoda S, Yamamoto Y *et al.* (2002) *J Appl Phys* **91**: 7911.
Samarth N. (2004) In *Solid State Physics*, Vol. 58, p. 1, Ehrenreich H, Spaepen F. (eds). Elsevier, Amsterdam.
Sato K, Katayama-Yoshida H. (2002) *Semicond Sci Technol* **17**: 367.
Thaler GT, Overberg ME, Gila B *et al.* (2002) *Appl Phys Lett* **80**: 3964.
Theodoropoulou N, Hebard AF, Overberg ME *et al.* (2002) *Phys Rev Lett* **89**: 107203.
Wojtowicz T, Cywinski G, Lim WL *et al.* (2003) *Appl Phys Lett* **82**: 4310.
Yokoyama M, Yamaguchi H, Ogawa T, Tanaka M. (2005) *J Appl Phys* **97**: 100317.
Yosida K. (1957) *Phys Rev* **106**: 893.
Zhang XW, Fan WJ, Zheng YH, Li SS, Xia JB. (2007) *Appl Phys Lett* **90**: 253110.
Zhang XW, Zhu YH, Xia JB. (2008) In *Quantum Dots: Research, Technology, and Applications*, Chap. 3, Knoss RW (eds). Nova Science Publishers, New York.
Zhao YJ, Mahadevan P, Zunger A. (2005) *J Appl Phys* **98**: 113901.
Zheng YH, Xia JB. (2005) *Phys Rev B* **72**: 195204.
Zheng YH, Zhao JH, Bi JF *et al.* (2007) *Chin Phys Lett* **24**: 2118.

Chapter 4

INJECTION OF SPIN-POLARIZED ELECTRONS

4.1. Spin Lifetime and Drift of Electrons in Semiconductors

Now efforts to improve computational power have led the semiconductor industry to push the large-scale integration (LSI) devices to ever-shrinking sizes. As advances in this area have proved the function of today's chip architectures, miniaturization may soon bring additional complications in the form of quantum mechanical effects, such as unpredictable fluctuations in device parameters. These concerns may ultimately force the industry to consider alternative schemes for electronics and to embrace nonclassical electron behavior. Researchers are exploring the possibilities of controlling the electron spin, rather than its charge — the basis for the new field dubbed "spintronics" (Awschalom and Kikkawa, 1999). By discovering regimes in which quantum states based on spin or charge can be carefully prepared and can evolve coherently long enough to be manipulated, researchers hope to provide a new set of tools for the construction of quantum-coherent electronics. Moreover, this ability will make possible new probes of individual quantum state, thereby deepening our understanding of their interactions in electronic structures.

In addition to opening new avenues for the study of fundamental quantum interactions in solids, wave-function engineering can also provide new opportunities for technology development. Coherent approaches to electronics offer two major advantages. First, because interference between two coherently occupied quantum states occurs at a frequency $\Delta E/\hbar$, where ΔE is the energy separation between the states, interference can result in rapidly varying oscillations of classical observables such as charge displacement or spin orientation (magnetization). For instance, a magnetic field can control electron spin precession frequencies in semiconductors at a rate of tens of GHz per tesla. Hence, quantum coherence may be used as a sort of engine to drive the operation of ultrafast devices. The second advantage is that interference may be used as part of a calculation. In the quantum computation, quantum bits are described by wave functions that are well isolated from their environment and whose mutual interactions are also well controlled.

Consider a single-spin system with spin states $\pm 1/2$, its wave function can be written as

$$\psi = \begin{pmatrix} a \\ b \end{pmatrix}, \tag{4.1}$$

where a and b represent the amplitudes of spins $+1/2$ and $-1/2$ states. The Hamiltonian of the spin system in the magnetic field is

$$H = -\mu_0 g \boldsymbol{\sigma} \cdot \boldsymbol{B}, \tag{4.2}$$

where μ_0 is the electron magnetic moment, g is the electron g-factor in the semiconductor, and $\boldsymbol{\sigma}$ is the Pauli matrix. Provided the magnetic field is along the y-direction, then $H = -\mu_0 g B \sigma_y$, insert it into the Schrödinger equation,

$$i\hbar \frac{\partial \psi}{\partial t} = H\psi. \tag{4.3}$$

and obtain

$$i\hbar \frac{\partial}{\partial t} \begin{pmatrix} a \\ b \end{pmatrix} = -\mu_0 g \begin{pmatrix} 0 & -iB \\ iB & 0 \end{pmatrix} \begin{pmatrix} a \\ b \end{pmatrix}. \tag{4.4}$$

Assuming that at the initial time $t = 0$, $a = 1$, $b = 0$, we obtain the solution of Eq. (4.4),

$$a = \cos\left(\frac{\mu_0 g B t}{\hbar}\right), \quad b = \sin\left(\frac{\mu_0 g B t}{\hbar}\right). \tag{4.5}$$

The coefficients a and b are two components of spin along the z-direction. If a and b are looked as two components of a classical vector in the plane perpendicular to the magnetic field, this vector will rotate in the plane with time, which is called the Larmor precession. The frequency of the Larmor precession equals $\mu_0 g B/\hbar$, called the Larmor frequency.

Many single-electron spin states form a system of spin-polarized electrons. The spin coherent state means that all electrons precess with one frequency. Because of the interactions of spin–spin, spin–impurity, spin–phonon, etc., the total spin of the system (magnetization) will decay exponentially with time.

$$M(t) = M(0)\exp\left(-\frac{t}{T_2}\right)\cos\left(\frac{\mu_0 gBt}{\hbar}\right), \tag{4.6}$$

where T_2 is called dephase time, as well as the transverse SRT.

The Larmor precession of electrons in the magnetic field was confirmed in the experiment (Heberle *et al.*, 1994). The sample contains 10 25 nm GaAs quantum wells separated by barriers. The sample is mounted in Voigt configuration with the magnetic field in the y-direction, the growth and observations in the z-direction, the electrons precess in the (x, z) plane. The sample temperature is 10 K. When a pulse laser I_+ of CP σ^+ transmitting in the z-direction irradiates on the sample, it will excite electrons from the $J = -1/2$ light hole (lh) state to the $s = +1/2$ electron state as shown in the inset of Fig. 4.1. When the laser pulse stops, the electrons at the $s = +1/2$ state will transit to the $J = +3/2$ heavy hole (hh) states and emit CP σ^- light I_-. Similarly, when a pulse laser I_- irradiates on the sample, it will excite the electrons from $J = +1/2$ lh state to the $s = -1/2$ electron state, and emit CP σ^+ light I_+. The upper panel of Fig. 4.1 shows the luminescence I_+ and I_- decay at 0 T after CP excitation of the lh exciton (Heberle *et al.*, 1994). They decay with time, and the decay time is about 500 ps. The lower panel shows those at $B = 2$ T. In the magnetic field, except decay, they oscillate with time, the oscillation periods are the same, but the phases are inverse. This reflects the Larmor precession of electrons with spins + and − in the magnetic field, as shown in Eq. (4.5).

Figure 4.2 shows the luminescence decay of I_- for different applied magnetic fields (Heberle *et al.*, 1994). A doubling of the oscillation frequency is obtained when the magnetic field strength is doubled. When $B = 1$ T, the period of the Larmor precession is about $T = 200$ ps, i.e. the oscillation frequency $\nu = 1/T = 5$ GHz. From the expression $\omega = \mu_0 gB/\hbar$, we obtain the g-factor of the electron in GaAs $g \approx 0.357$. The precise value determined from the spin splitting as a function of the magnetic field is $g = -0.390$, where the minus sign means that in the magnetic field the $s = +1/2$ level is lower than the $s = -1/2$ level. The ratio of the intensity I_-/I_+ is therefore equals the ratio of electron number at $s = +1/2$ to at $s = -1/2$. The spin coherent time in the semiconductor T_2 reaches 500 ps (see Figs. 4.1 and 4.2), which is far larger than the charge coherent time (~10 ps); these properties open a path to use the spin coherence.

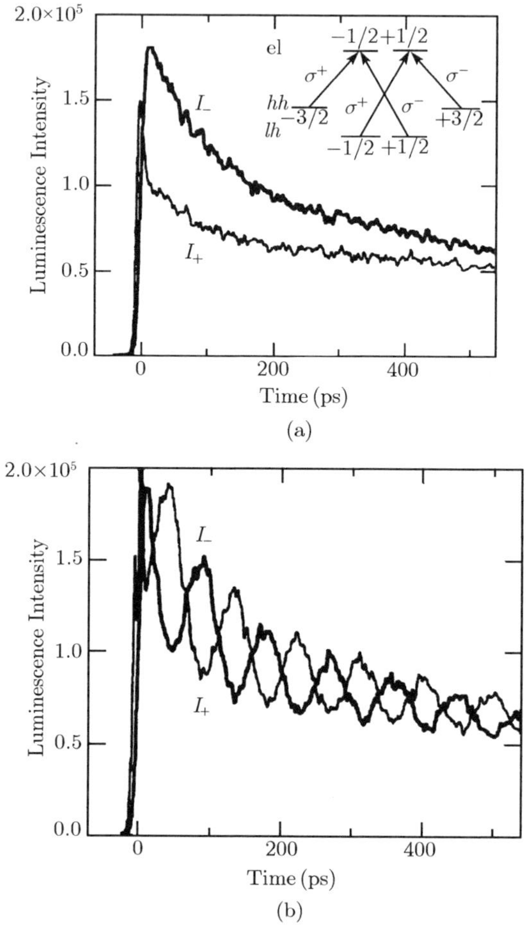

Fig. 4.1. Luminescence decay at (a) 0 T and (b) 2 T after circularly polarized (CP) excitation of the lh exciton. The inset shows the selection rule of the transitions.

The above description of the electron precession is a classical picture. From the point of view of quantum mechanism, the electron spin degenerate states split in the magnetic field; the splitting energy equals $\Delta E = \mu_0 g B$, corresponding to the energy of quantized precession $\hbar\omega$. The interference between the two quantum states is called quantum beats, whose frequency is determined directly by the spin splitting energy ΔE. The electron g-factor can be determined directly by the quantum beat frequency.

The quantum beat behavior has been observed in a vertical cavity surface-emitting laser (VCSEL) (Hallstein *et al.*, 1997). In the experiment,

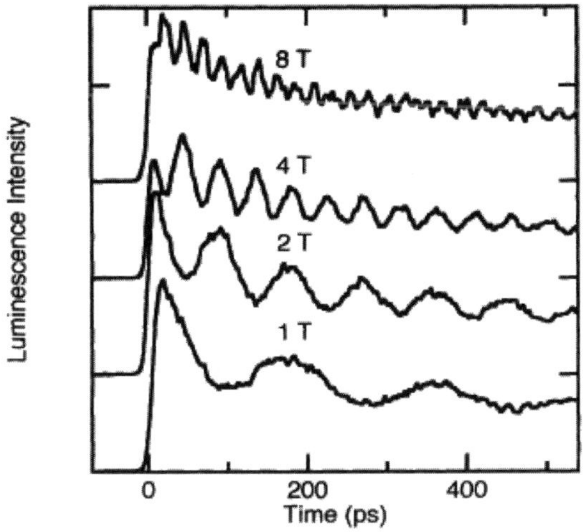

Fig. 4.2. Luminescence decay of I_- for different applied magnetic fields.

they excite a semiconductor microcavity with CP light in Voigt geometry, where the magnetic field is applied perpendicular to both the quantum-well growth direction and the emission direction of the VCSEL. Initially, they create a spin-polarized distribution in the CB which then performs Larmor precession around the magnetic field. Under suitable conditions these coherent spin oscillations are transformed into a pulsing laser emission with a repetition rate determined by the electron Larmor frequency. Figure 4.3(a) shows the time-resolved simulated emission of the quantum-well microcavity laser in $B = 2\,\mathrm{T}$ (Hallstein *et al.*, 1997). Pulsed laser emision with a modulation depth of approximately 96% is clearly seen. The pulse repetition rate is 22 GHz, corresponding to twice the electron Larmor frequency. A comparison in Fig. 4.3(b) shows the emission of a quantum-well reference sample without a cavity. Under otherwise identical excitation conditions, there are no oscillations present. This is the first time that the interband emission of a semiconductor laser has been governed by electron spin dynamics.

To compute the VCSEL emission dynamics, we need the temporal evolution of the electron distributions in the spin-split bands. In a fashion analogous to the semiconductor Bloch equations, they obtained screened Hartree–Fock equations of the electron distribution functions and the Larmor polarization between the electron bands. Physically, it turns out that after the original excitation, first the gain for σ^- polarized light

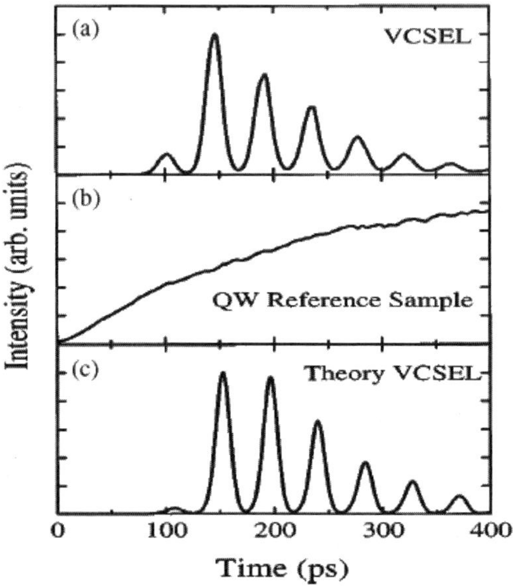

Fig. 4.3. (a) Time-resolved simulated emission of the quantum-well microcavity laser in $B = 2\,\mathrm{T}$. (b) Time-resolved emission of the quantum well reference sample without a cavity. (c) Theoretical result with parameters equal to the conditions of (a).

exceeds the losses at the emission energy and the σ^- emission is therefore strongly amplified. After a quarter precession period ($t = T_{\mathrm{L}}/4$), a sufficient amount of carriers have been transferred to the spin-down band, such that the gain for both light polarizations drops below threshold. Hence, lasing is switched off and the total microcavity emission reaches a minimum. After half a Larmor precession period ($t = T_{\mathrm{L}}/2$), the laser exceeds the threshold again, this time, however, for σ^+ light. The theoretical results for the conditions in Fig. 4.3(a) are shown in Fig. 4.3(c). The comparison between theory and experiment shows a good agreement.

Kikkawa and Awschalom (1998, 1999) obtained large advantages in prolonging the spin coherent time. In the above experiments the existence of the excited holes influences the coherent time of spin-polarized electrons. They used doping the donor impurities, which produce low-density electron gas, absorb light excited holes, and decrease scattering to electron spin. Through all-optical studies of spin precession using TRFR, they showed that a 2DEG can act as a reservoir for optically injected spin polarization, significantly extending spin lifetimes by removing holes which act to efficiently scatter electron spins.

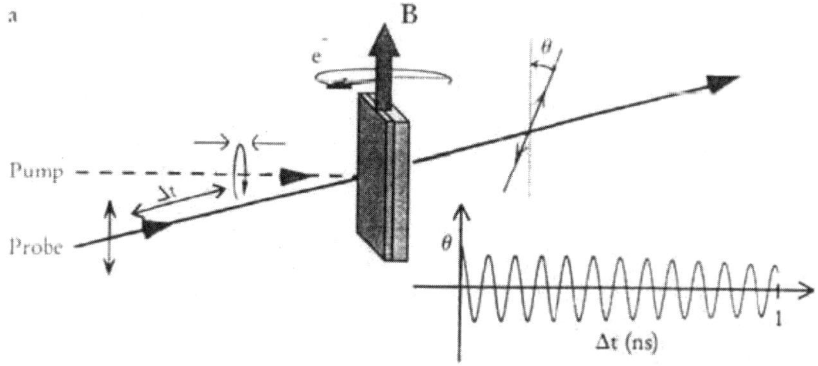

Fig. 4.4. Experimental setting for measuring electron precession using the Faraday rotation method.

They used femtosecond TRFR in the Voigt geometry to measure the electron precession (Kikkawa and Awschalom, 1998), the experimental setting is shown in Fig. 4.4 (Awschalom and Awschalom, 1999). The TRFR measures the Larmor precession of electron spin excited by a normally incident, CP pump pulse. Spin electrons rotate the probe light beam's linear polarization, measured by a polarization bridge technique. This deflection is proportional to the electron moment project on the probe's incident direction, so spin precession about an in-plane magnetic field B causes the Faraday rotation to oscillate in the pump-probe delay, Δt. Scanning Δt then produces a TRFR profile of the form as follows:

$$\theta \sim A e^{-\Delta t/T_2^*} \cos\left(\frac{g\mu_{\rm B} B \Delta t}{\hbar}\right). \tag{4.7}$$

Its frequency is a direct measure of the electron g-factor and its decay yields the transverse spin lifetime T_2^*.

Figure 4.5 shows the TRFR versus decay time Δt for undoped and n-type GaAs at $T = 5\,\text{K}$ and $B = 4\,\text{T}$ (Kikkawa and Awschalom, 1998). The four curves from up to down correspond to the electron densities $n = 0, 10^{16}$, 10^{18}, and $5 \times 10^{18}\,\text{cm}^{-3}$, with the excited electron density $N_{\rm ex} = 2\times10^{14}, 2\times10^{14}, 1.4\times10^{15}$, and $3\times10^{15}\,\text{cm}^{-3}$, respectively. A change in the Larmor frequency with doping reflects an energy dispersion in the GaAs CB g-factor. The free carrier spin lifetime is a nonmonotonic function of carrier density, abruptly increasing between $n \approx 0$ and $n = 1\times10^{16}\,\text{cm}^{-3}$ samples and steadily decreasing thereafter. The inset shows T_2^* versus

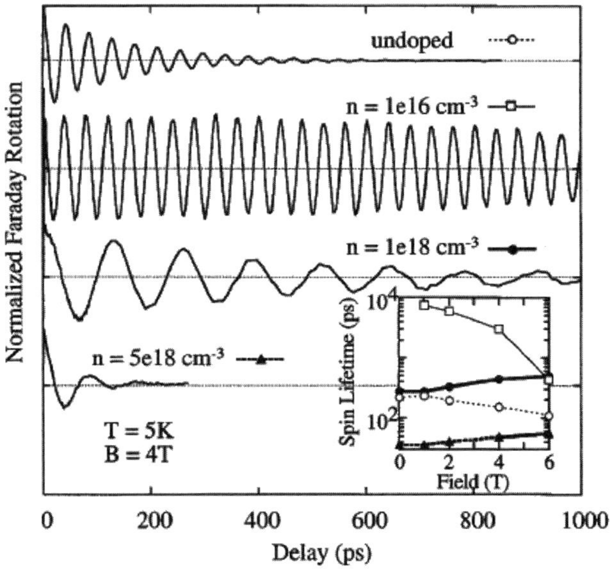

Fig. 4.5. TRFR versus decay time Δt for undoped and n-type GaAs at $T = 5\,\mathrm{K}$ and $B = 4\,\mathrm{T}$. The inset shows T_2^* versus field.

field, spanning over two orders of magnitude and ranging nearly 10 ns for $n = 1 \times 10^{16}\,\mathrm{cm}^{-3}$ to less than 40 ps for $n = 5 \times 10^{18}\,\mathrm{cm}^{-3}$. The spin lifetime has a power dependence. Optical pulse powers are converted to excitation densities N_{ex}. The data indicate a power law $T_2^* \sim N_{\mathrm{ex}}^{-0.4}$, for $B = 0, 0.2$, and 6 T.

Except the decoherence of the single electron spin, the nonuniform factors in a collective electron system, such as the variation of the g-factor and the difference of local magnetic field, will give rise to slightly different precession frequencies. Though all electrons move coherently, consequently the directions of spin will have a distribution. This type of nonuniformity results in an additional delay of spin polarization, which is called dephasing, different from decoherence. Therefore, the observed spin decay time T_2^* is the low limit of T_2, unless the nonuniform effect can be eliminated.

Excite the spins in the sample using the repetitive pumps with the interval t_{rep}, which is smaller than the spin lifetime. When spin precession commensurates with the pulse repetition, the spin polarizations induced by different laser pulses are additional. The observed spin polarization is the summation of various exponentially decayed oscillations,

$$M_s(\Delta t, B) = \sum_n \theta(\Delta t + nt_{\rm rep}) A e^{-(\Delta t + nt_{\rm rep})/T_2^*} \cos\left[\frac{g\mu_{\rm B} B(\Delta t + nt_{\rm rep})}{\hbar}\right], \tag{4.8}$$

where $\theta(t)$ is a step function.

Figure 4.6 shows TRFR versus magnetic field for $n = 10^{16}\,\mathrm{cm}^{-3}$ sample (Kikkawa and Awschalom, 1998). When the magnetic field B equals a definite value, all terms in Eq. (4.8) are additional, the resonance takes place, and the amplitude of the spin polarization is larger than that induced by a single pulse by one order of magnitude. Figure 4.6(c) shows the expanded

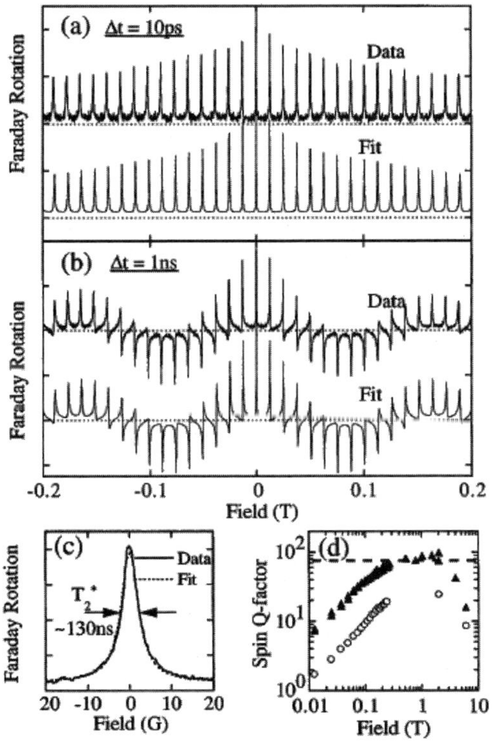

Fig. 4.6. TRFR versus magnetic field for $n = 10^{16}\,\mathrm{cm}^{-3}$ sample with (a) $\Delta t = 10\,\mathrm{ps}$ and (b) $1\,\mathrm{ns}$. (c) An expanded view of the central resonance in (a). The fit given by the dotted line corresponds to a spin lifetime of 130 ns. (d) The spin Q factor versus field on a log–log plot for $N_{\rm ex} = 2\times10^{14}\,\mathrm{cm}^{-3}$ (solid triangles) and $3\times10^{15}\,\mathrm{cm}^{-3}$ (open circles), respectively. The dotted line indicates a thermal ceiling. Data are taken at $T = 5\,\mathrm{K}$.

plot of the central resonance. The fit with Eq. (4.8) gives the spin lifetime $T_2^* = 130\,\text{ns}$ at $B = 0$. Adjusting the parameters T_2^*, A, and g separately for each resonance in Eq. (4.8), the fitting results are shown in Figs. 4.6(a) and 4.6(b), which are in good agreement with the experiments. The T_2^* may be extracted for each resonance and is seen to increase by 10-fold from 1 to 0 T, reaching 130 ns at zero field.

If the quality factor $Q = \mu_0 g B T_2^*/h$ is used to represent the nonuniformity of the electron g-factor, then $Q \propto g/\Delta g$. The Q factor extracted from each resonance in Figs. 4.6(a) and 4.6(b) as a function of field is shown in Fig. 4.6(d). There are three regimes: (a) a low-field region obeying a power law $Q \sim B^{0.8}$, (b) an intermediate regime where Q is independent of B, and (c) a high-field region in which Q decreases. Because dephasing of spin direction produces an effective lifetime which is inversely proportional to B, the resulting $Q \propto g/\Delta g$ is field independent. The plateau at $Q \sim 80$ seen in Fig. 4.6(d) suggests the possibility that spin dephasing limits the Q factors in this region, with a variation of $\Delta g/g = 0.28\%$.

This resonance technique can be modified to produce macroscopic displacement of spin precession — a first step toward manipulating spin coherence in semiconductors. If one applies a lateral electric field to the system, the conduction electrons from each pulse are dragged away after they are excited. One can map the mother of the spins by moving the position of the probe beam. The different packets can be resolved by their phases: older spins will have precessed further in the magnetic field than younger spins. The ability to electrically transport spin through quantum structures provides exciting opportunities for coherent spin transport. Taking into account the drift of electrons in the laterally applied electric field, the total polarization of electron spin equals,

$$M(\Delta x, \Delta t, B) = \sum_n m_n(\Delta x, \Delta t + n t_{\text{rep}}, B). \tag{4.9}$$

Figure 4.7 shows the TRFR versus magnetic field and displacement at an electric field of $\sim 37\,\text{V/cm}$, $\Delta t = 10\,\text{ps}$, for $n = 10^{16}\,\text{cm}^{-3}$ GaAs sample (Kikkawa and Awschalom, 1999). In the presence of an electric field, the spin resonances deduced by pulse laser no longer constructively reinforce each other at the resonance magnetic fields because they drift variable distances that are proportional to their ages. The inset of Fig. 4.7(b) shows the corresponding harmonic power versus position. The data can be used to mark the positions of individual spin packets and indicate that spin drift

(a)

(b)

(c)

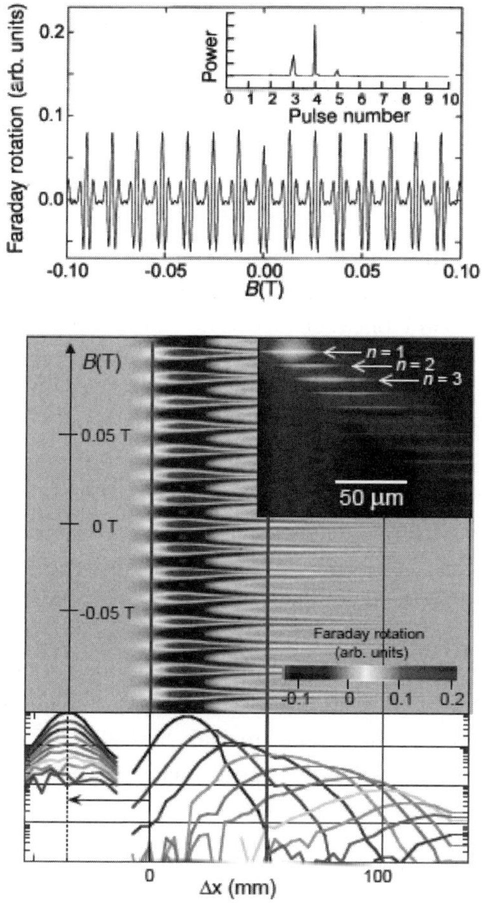

Fig. 4.7. (a) TRFR versus magnetic field at a displacement of $54\,\mu$m and an electric field of $\sim 37\,$V/cm, $\Delta t = 10\,$ps, for $n = 10^{16}\,\text{cm}^{-3}$ GaAs sample. (b) TRFR versus magnetic field and displacement. (c) Amplitudes of the first 10 m_n constituents versus displacement.

is linear in time and corresponds to a drift mobility $\mu_d = 3 \times 10^3\,\text{cm}^2/\text{Vs}$. Despite an exponential time decay of the spin polarization that leads to a spatial decay along the drift direction, spin transport is observable at distance exceeding $100\,\mu$m in rather modest electric fields and mobilities. Figure 4.7(c) shows the amplitudes of the first 10 m_n constituents versus displacement; the left plot is of those at zero bias. From the decay of the 10 constituents, one obtains $T_2^* = 29\,$ns.

4.2. Rashba Effect

4.2.1. *Origin of the Rashba effect*

Rashba (1960) proposed a Hamiltonian about interaction between electron spin and momentum,

$$H_{\mathrm{R}} = \alpha[\boldsymbol{\sigma} \times \boldsymbol{k}] \cdot \boldsymbol{n}, \tag{4.10}$$

where $\boldsymbol{n}$ is the unit vector perpendicular to the surface or interface. As for its origin, de Andrada *et al.* (1994) proposed a general theory from the SOC and interaction between the CBs and valence bands. They studied the magneto-oscillations in asymmetric heterostructures with the aim of clarifying the origin and the effects of the spin–orbit spin splitting in the conduction subband. Consider a modulation-doped heterostructure, suppose that the growth direction is along the z-direction, then the potential of the electron gas along the z-direction is a triangle-type well, in which forms a series of discrete energy levels, each level corresponds a subband in the (x, y) plane. Generally, the electrons fill the ground subband, and the Fermi level lies between the first and second levels.

The semiconductor compound is described by the eight bands $\boldsymbol{k} \cdot \boldsymbol{p}$ Kane model and the solution for the lowest conduction subband is obtained within the multiband effective-mass envelope-function approximation. Start from the Kane model, let S, X, Y, and Z denote the conduction- and the valence bulk Bloch functions at the zone center. Choose the following linear combinations as the basis functions,

$$\begin{aligned}
u_1 &= S\uparrow, \\
u_2 &= \sqrt{\frac{2}{3}}Z\uparrow + \frac{1}{\sqrt{6}}(iX\uparrow + Y\downarrow), \\
u_3 &= -\frac{1}{\sqrt{2}}(-iX\uparrow + Y\downarrow), \\
u_4 &= -\frac{1}{\sqrt{3}}[Y\downarrow - (Z - iX)\uparrow], \\
u_5 &= -S\downarrow, \\
u_6 &= -\sqrt{\frac{2}{3}}Z\downarrow + \frac{1}{\sqrt{6}}(iX\downarrow + Y\uparrow), \\
u_7 &= -\frac{1}{\sqrt{2}}(-iX\downarrow + Y\uparrow), \\
u_8 &= -\frac{1}{\sqrt{3}}[Y\uparrow + (Z + iX)\downarrow],
\end{aligned} \tag{4.11}$$

where the arrows indicate the spin state with respect to the y-axis. Choose the parallel wave vector $\boldsymbol{k}$ along the x-axis, and the electron wave function will be given by

$$\psi(\boldsymbol{r}) = e^{ikx} \sum_{j=1}^{8} f_j(z) u_j(\boldsymbol{r}), \tag{4.12}$$

where $f_j(z)$ are the envelope functions and the effective-mass Hamiltonian is block diagonalized,

$$H = \begin{pmatrix} H_+ & 0 \\ 0 & H_- \end{pmatrix}, \tag{4.13}$$

with

$$H_\pm = \begin{pmatrix} V_1(z) & P\left[\frac{\mathrm{d}}{\mathrm{d}z} \mp \frac{k}{2}\right] & \mp\frac{\sqrt{3}}{2}Pk & \frac{P}{\sqrt{2}}\left[\frac{\mathrm{d}}{\mathrm{d}z} \pm k\right] \\ P\left[-\frac{\mathrm{d}}{\mathrm{d}z} \mp \frac{k}{2}\right] & V_2(z) - E_\mathrm{g} & 0 & 0 \\ \mp\frac{\sqrt{3}}{2}Pk & 0 & V_2(z) - E_\mathrm{g} & 0 \\ \frac{P}{\sqrt{2}}\left[-\frac{\mathrm{d}}{\mathrm{d}z} \pm k\right] & 0 & 0 & V_2(z) - E_\mathrm{g} - \Delta \end{pmatrix}, \tag{4.14}$$

where $P = \sqrt{(2/3)}(\hbar/m_0)\langle iS|P_x|X\rangle$ and $V_1(z)$ and $V_2(z)$ are the confining potential and applied field potential of the CB and the valence band, respectively. The k^2 terms have been neglected in the diagonal matrix elements because the small electron effective mass guarantees that the dispersion is given mainly by the off-diagonal contribution.

Hamiltonian (4.13) leads to two sets of four coupled equations for the two groups of envelope functions. The secular equation is

$$|H_\pm - \varepsilon| = 0. \tag{4.15}$$

Expanding Eq. (4.15) and dividing the obtained equation by the factor $(V_2(z)-E_\mathrm{g}-\varepsilon)^2\cdot(V_2(z)-E_\mathrm{g}-\Delta-\varepsilon)$, one obtains the following Schrödinger equation for the CB envelope functions,

$$\left[-\frac{\hbar^2}{2}\frac{\mathrm{d}}{\mathrm{d}z}\frac{1}{m(z,\varepsilon)}\frac{\mathrm{d}}{\mathrm{d}z} + \frac{\hbar^2 k^2}{2m(z,\varepsilon)} + V_1(z) \mp \alpha(z,\varepsilon)k - \varepsilon\right] f_\pm = 0, \tag{4.16}$$

where $f_\pm$ is the envelope function of the CB.

$$\frac{1}{m(z,\varepsilon)} = \frac{P^2}{\hbar^2}\left[\frac{2}{\varepsilon - V_2(z) + E_\mathrm{g}} + \frac{1}{\varepsilon - V_2(z) + E_\mathrm{g} + \Delta}\right], \tag{4.17}$$

$$\alpha(z,\varepsilon) = \frac{P^2}{2}\frac{\mathrm{d}}{\mathrm{d}z}\left[\frac{1}{\varepsilon - V_2(z) + E_\mathrm{g}} - \frac{1}{\varepsilon - V_2(z) + E_\mathrm{g} + \Delta}\right]. \tag{4.18}$$

The term linear in k in Eq. (4.16) gives the spin splitting. It is zero for the bulk [$V(z) = \text{const.}$] and for the symmetric quantum wells. Only the asymmetric quantum well, such as modulation-doped heterostructure, or in the case of applied electric field ($V(z) = eEz$), the coefficient $\alpha(z,\varepsilon)$ (Eq. (4.18)) does not equal zero. The $\alpha(z,\varepsilon)$ is called the Rashba coefficient. From Eq. (4.18), we see that when the spin–orbit splitting energy of the valence band Δ equals zero, $\alpha = 0$. Thus, the Rashba effect is caused by the SOC of the valence band. The origin of energy ε is at the CB bottom. E_g is the band gap, and if E_g is smaller, then α is larger. Thus, the Rashba splitting is larger for the narrow gap semiconductors.

In Eqs. (4.16)–(4.18), $\varepsilon - V(z)$ is very small relative to E_g and $E_\mathrm{g} + \Delta$ and can be neglected in the zero-order approximation. The zero-order term in Eq. (4.16) gives the parabolic equation,

$$\left[\frac{\hbar^2}{2m^*}\left(-\frac{\mathrm{d}^2}{\mathrm{d}z^2} + k^2\right) + V_1(z) \mp \alpha k - \varepsilon\right] f_\pm = 0, \tag{4.19}$$

$$m^* = \frac{\hbar^2}{P^2}\frac{E_\mathrm{g}(E_\mathrm{g} + \Delta)}{3E_\mathrm{g} + 2\Delta}, \tag{4.20}$$

$$\alpha = \frac{\hbar^2}{2m^*}\frac{\Delta}{E_\mathrm{g}}\frac{2E_\mathrm{g} + \Delta}{(E_\mathrm{g} + \Delta)(3E_\mathrm{g} + 2\Delta)}\frac{\mathrm{d}V}{\mathrm{d}z}. \tag{4.21}$$

Thus, the Rashba coefficient α is proportional to $\Delta, \mathrm{d}V/\mathrm{d}Z$, and inversely proportional to m^*, E_g.

de Andrada *et al.* (1994) assumed that in the triangle potential well of the modulation-doped heterojunction, the trial wave function of the electron ground state is

$$f_\pm(z) = \sqrt{\frac{b^3}{2}}ze^{-bz/2}, \tag{4.22}$$

and taking into account the nonparabolicity of the CB, they calculated the zero-field Rashba splitting of the lowest conduction subband versus wave vectors in the (x,y) plane in InAs- and GaSb-based heterojunctions, as

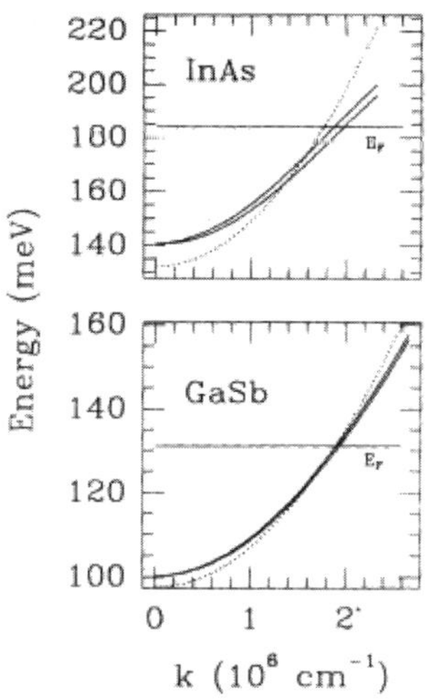

Fig. 4.8. Calculated zero-field Rashba splitting of the lowest conduction subband versus wave vectors in the (x, y) plane in InAs- and GaSb-based heterojunctions, both with $n_s = 6.0 \times 10^{11}$ cm^{-3}. The dashed lines are the results of the parabolic approximation.

shown in Fig. 4.8. The Rashba splitting energy of InAs is obviously larger than that of GaSb.

4.2.2. *Experimental measurement of the Rashba coefficient*

The Rashba coefficient was measured by the beating frequency effect of the SdH oscillations (Engels *et al.*, 1997). The $In_{0.77}Ga_{0.23}As$ modulation-doped heterostructure was grown by the metal organic vapor phase epitaxy on a semi-insulating InP substrate. The fully prepared sample yields a carrier concentration $n = 1.59 \times 10^{12}\,\text{cm}^{-2}$ and a mobility of $\mu = 2 \times 10^5\,\text{cm}^2/\text{V s}$ at zero-voltage gate bias.

Figure 4.9 shows the SdH magnetic resistance oscillations for InGaAs 2DEG at $I = 1\,\mu$A and $T = 0.3$ K with the gate voltage changed from -6 to $+2$ V (Engles *et al.*, 1997). The carrier concentration is lowered due to a negative gate voltage. For a voltage below -6 V, the SdH oscillations vanish between 0 and 1.2 T. As a consequence of the reduction of carrier density and the reduced mobility, the resistance at $B = 0$ increases. The positions of the nodes (as shown by the circles in Fig. 4.9) in the SdH amplitude is shifted, which indicates a change in the beating effect between two closely spaced SdH frequencies.

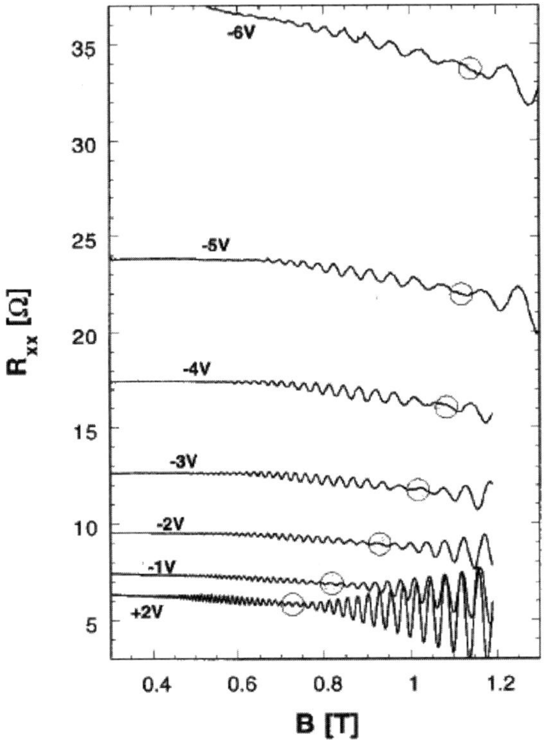

Fig. 4.9. SdH magnetic resistance oscillations for InGaAs 2DEG at $I = 1\,\mu$A and $T = 0.3\,$K with the gate voltage changed from -6 to $+2\,$V.

Figure 4.10 shows the distribution of the electron density with various gate voltages obtained from the $1/B$ Fourier transform taken from SdH oscillations of Fig. 4.9 (Engels *et al.*, 1997). The $1/B$ Fourier transform clearly resolves a double-peak structure, indicating that there are two discrete electron density peaks corresponding one bias gate voltage and the electron subband has split into two nondegenerate subbands. The electron density is larger for the positive gate voltage. The splitting of the electron subband is caused by the Rashba effect, the electron energy equals

$$E = E^{\text{sub}} + \frac{\hbar^2 k_{\parallel}^2}{2m^*} \pm \alpha |k_{\parallel}|, \tag{4.23}$$

where E^{sub} is the energy of the ground state in the heterostructure quantum well, $k_{\parallel}$ is the wave vector in the 2D plane, and α is the Rashba

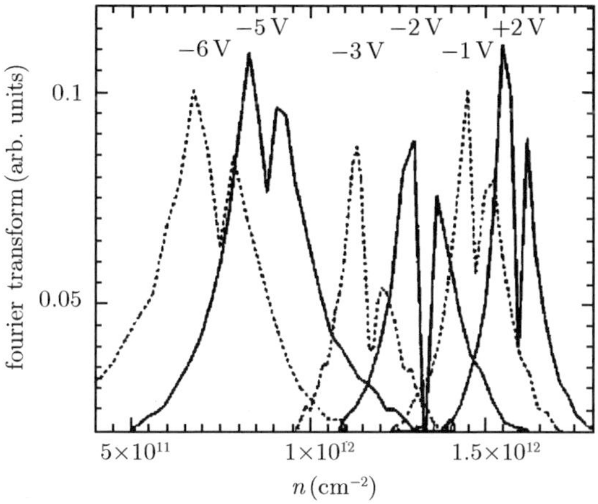

Fig. 4.10. Distribution of the electron density with various gate voltages obtained from the $1/B$ Fourier transform taken from SdH oscillations of Fig. 4.9.

coefficient. The density of states of each spin subband is obtained from Eq. (4.23),

$$Z_{\pm}(E) = \frac{1}{2}\frac{m^*}{\pi\hbar^2}\left(1 \mp \frac{1}{\sqrt{1 + (2E\hbar^2)/(\alpha^2 m^*)}}\right). \tag{4.24}$$

The second term in Eq. (4.24) is caused by the Rashba energy splitting. If $\alpha = 0$, Eq. (4.24) becomes the density of states of general 2D parabolic band — constant.

The difference in carrier concentration Δn of the spin subbands at low temperatures,

$$\Delta n = n_- - n_+ = \int_0^{E_F} Z_-(E)\mathrm{d}E - \int_0^{E_F} Z_+(E)\mathrm{d}E. \tag{4.25}$$

By a simple calculation, the Rashba coefficient is obtained

$$\alpha = \frac{\Delta n\hbar^2}{m^*}\sqrt{\frac{\pi}{2(n - \Delta n)}}. \tag{4.26}$$

From the experimental results of Fig. 4.10 and the theoretical formula (4.26), the values of the Rashba coefficients α as functions of the gate voltage are obtained and shown in Fig. 4.11 (Engels *et al.*, 1997), where the solid and open circles represent the experimental and theoretical values, respectively. When the gate voltage decreases from +2 to −6 V, α increases

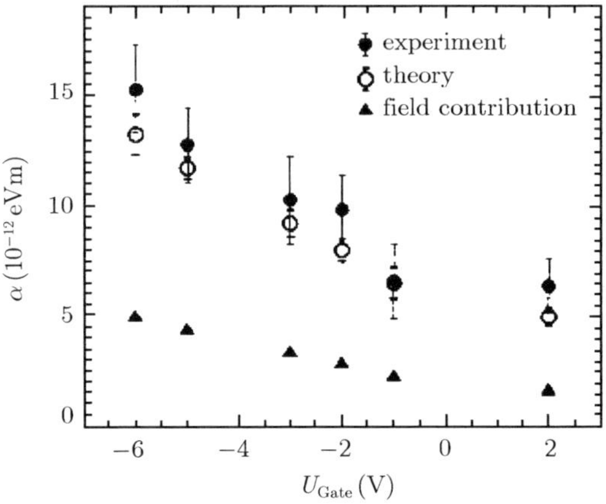

Fig. 4.11. Values of the Rashba coefficients α as functions of the gate voltage given by experiment and theory.

from 6.3×10^{-12} to 15.3×10^{-12} eVm, and the agreement between the experimental and theoretical values is good. The solid triangle in Fig. 4.11 are the electric field contributions; the gate with a more negative voltage leads to an increase in the net electric field in the well and consequently to an increase in the Rashba coefficient.

Next experiment of the Rashba effect was concentrated on the arising and controlling the Rashba coefficient (Grundler, 2000). A square asymmetric quantum well was grown by MBE, which consists of a 4 nm InAs well, 13.5 and 2.5 nm $In_{0.75}Ga_{0.25}As$ barriers on both sides, and outmost layers of higher potential barriers $In_{0.75}Al_{0.25}As$. Two gate electrodes: front gate and back gate are integrated.

Figure 4.12(a) shows the Fourier transform of the low-field SdH oscillation at $T = 0.28$ K, $I = 1\,\mu$A (Grundler, 2000), where the abscissa is the frequency $\nu(T)$, which is related to the electron density by $n^{\pm} = \nu^{\pm} e/h$. Various curves correspond to different electron densities N_s and back gate voltages V_{bg}, the uppermost curve to $N_s = 8.9 \times 10^{11}\,\text{cm}^{-2}$ and $V_{bg} = 150$ V, the below four curves to $N_s = 11.1 \times 10^{11}\,\text{cm}^{-2}$ and V_{bg} equals 80, 110, 140, and 160 V, from up to down respectively. The calculated α as functions of the electron density is shown in Fig. 4.12(b). The α increases with increasing electron density, the maximum α reaches 4×10^{11} eVm, three times the previous value (Engels *et al.*, 1997). The inset in Fig. 4.12(b) shows the α versus the back gate voltage V_{bg} for the $N_s = 11.1 \times 10^{11}\,\text{cm}^{-2}$

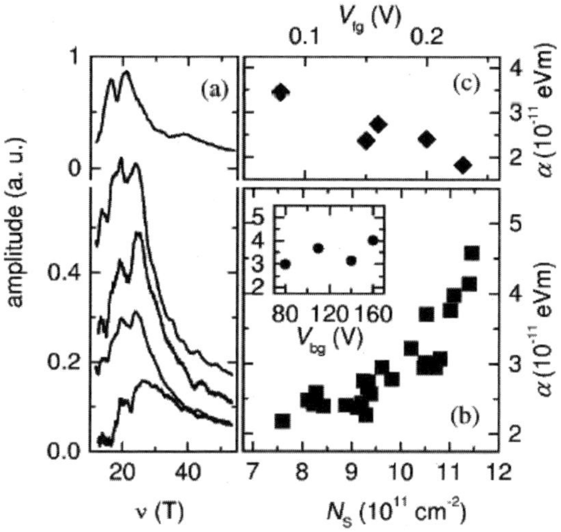

Fig. 4.12. (a) Fourier transform of the low-field SdH oscillation at $T = 0.28$ K, $I = 1\,\mu$A. (b) Calculated α as functions of the electron density. The inset shows the α versus the back gate voltage V_{bg} for the $N_{\text{s}} = 11.1 \times 10^{11}\,\text{cm}^{-2}$ sample. (c) The α versus the front gate voltage V_{fg} for the $N_{\text{s}} = 11.1 \times 10^{11}\,\text{cm}^{-2}$ sample and $V_{\text{bg}} = 160$ V.

sample. The α has little dependence on the V_{bg}. Figure 4.12(c) shows the α versus the front gate voltage V_{fg} for the $N_{\text{s}} = 11.1 \times 10^{11}\,\text{cm}^{-2}$ sample and $V_{\text{bg}} = 160$ V. The α increases with decreasing V_{fg}, same as Fig. 4.11.

According to the author's analysis, such a large Rashba coefficient is mainly caused by the asymmetry of the quantum well. It is proved that the contribution of the barrier is given as the band offset weighted by the probability density of the conduction electrons at the interface, such that in a multiband approach α should scale with $\beta \times (|\psi_b|^2 - |\psi_t|^2)$, where β contains the band structure parameters of the heterostructure and ψ_{b} and ψ_{t} are the envelope functions at the bottom and top barriers, respectively. Applying a gate voltage on the back with respect to the 2DEG, one is able to shift the wave function transverse to the well and thus to change the probability density in the barriers. The electron density inside the barrier and $(|\psi_b|^2 - |\psi_t|^2)$ are enlarged, resulting in a large Rashba coefficient. The gate-controlled position of the wave function can thus be utilized to tune the Rashba interaction at a constant N_{s}. Because the distance between the back gate and the 2DEG is small, the effect of the back gate to increase α is more obvious than the front gate.

4.2.3. *Theoretical calculation of the Rashba coefficient*

Yang and Chang (2006) proposed a theoretical study of the Rashba spin splitting (RSS) in biased quantum wells in the framework of eight bands $k \cdot p$ theory. They found that at large parallel wave vector, RSS is both nonmonotonic and anisotropic as a function of in-plane wave vector, in contrast to the widely used isotropic linear model. Figure 4.13 shows the RSS versus k_x for 20 nm wide (a) $Ga_{0.47}In_{0.53}As/Al_{0.48}In_{0.52}As$ and (b) $Hg_{0.74}Cd_{0.26}Te/CdTe$ quantum wells at the electric field 100 kV/cm (Yang and Chang, 2006). Solid lines denote the full numerical solutions to the eight band Hamiltonian. Dash-dotted lines denote the two contributions $\Delta E^{(1)}$ and $\Delta E^{(2)}$, which are caused by the quantum well offset and applied electric field, respectively. From Eq. (4.18), they can separately be denoted as,

$$\Delta E_n^{(1)}(k_\parallel) = \frac{P^2}{2} k_\parallel \sum_j |F_n(z_j)|^2[\gamma(z_j^+) - \gamma(z_j^-)], \tag{4.27}$$

$$\Delta E_n^{(2)}(k_\parallel) = \frac{P^2}{2} eFk_\parallel \int \mathrm{d}z |F_n(z)|^2 (U_{\mathrm{lh}}^{-2} - U_{\mathrm{so}}^{-2}), \tag{4.28}$$

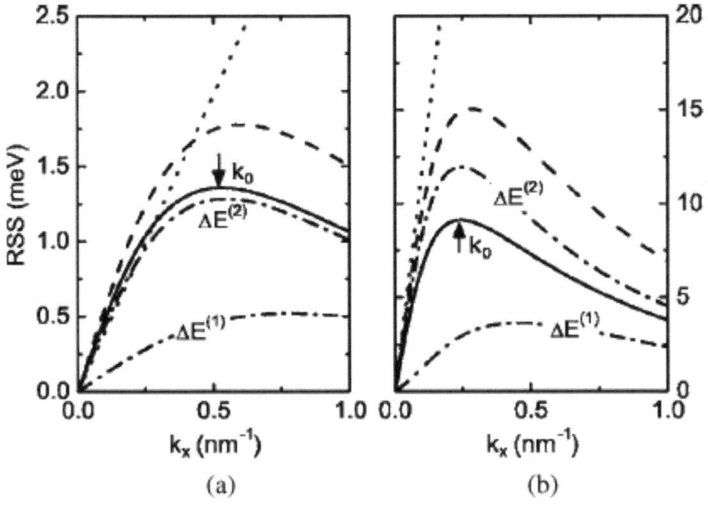

Fig. 4.13. Rashba spin splitting energy versus k_x for 20 nm wide (a) $Ga_{0.47}In_{0.53}As/Al_{0.48}In_{0.52}As$ and (b) $Hg_{0.74}Cd_{0.26}Te/CdTe$ quantum wells at the electric field 100 kV/cm. Solid lines denote the full numerical solutions to the eight-band Hamiltonian. The two contributions $\Delta E^{(1)}$ and $\Delta E^{(2)}$ are denoted by dash-dotted lines.

where $F_n(z)$ is the envelope function of the nth subband in the quantum well,

$$U_{\rm lh} = \varepsilon - V_2(z) + E_{\rm g}, \quad U_{\rm so} = \varepsilon - V_2(z) + E_{\rm g} + \Delta, \tag{4.29}$$

$$\gamma(z) = \frac{1}{U_{\rm lh}(z)} - \frac{1}{U_{\rm so}(z)}, \tag{4.30}$$

z_j is the z coordinate of the interface. Thus, $\Delta E^{(1)}(k_\parallel)$ is caused by the offsets of the valence band and the spin–orbit splitting band, $\Delta E^{(2)}(k_\parallel)$ is proportional to the applied electric field F.

In the case of quantum well, the ε in Eq. (4.29) is the energy of the electron at subband,

$$\varepsilon = E_{n0} + \frac{\hbar^2 k_\parallel^2}{2m^*}, \tag{4.31}$$

where E_{n0} is the nth electron quantum confined energy in the well, the effective energy gap $E_{\rm g}^{\rm eff} = \varepsilon + E_{\rm g}$. When $k_\parallel$ increases to the value that the kinetic energy term $\hbar^2 k_\parallel^2/(2m^*)$ is comparable to the $(E_{\rm g} + E_{n0})$ term, it is called critical wave vector k_0. When $k_\parallel > k_0$, and increase continuously, the effective band gap increases as $E_{\rm g}^{\rm eff} \propto k_\parallel^2$, the RSS starts to decrease.

The critical wave vector k_0 decreases when the band gap $E_{\rm g}$ decrease, since a smaller k_0 is required for the kinetic energy to become comparable to $(E_{\rm g} + E_{n_0})$ term. This can be confirmed from Figs. 4.13(a) and 4.13(b): $E_{\rm g} = 0.816\,{\rm eV}$ for $Ga_{0.47}In_{0.53}As$ in (a), and $E_{\rm g} = 0.2\,{\rm eV}$ for $Hg_{0.74}Cd_{0.26}Te$; the k_0 of (b) is smaller than that of (a).

Define the Rashba coefficient,

$$\alpha = \frac{1}{2}\left[\frac{{\rm d}\Delta E(k_\parallel)}{{\rm d}k_\parallel}\right]_{k_\parallel \to 0}, \tag{4.32}$$

that is, when $k_\parallel \to 0$, the slope of RSS to $k_\parallel$ multiplied by 1/2. Figure 4.14 shows Rashba coefficients for the lowest subband of the $Hg_{0.74}Cd_{0.26}Te$/CdTe quantum well as functions of electric field and well width (Yang and Chang, 2006). The α increases almost linearly with increasing electric field, in agreement with Eq. (4.28), and it increases with increasing well width and saturates at large well width. The latter can be explained through the dependence of the effective band gap on the well width. With increasing well width, the confining energy E_{n0} decreases, such that the effective band gap decreases and approaches a constant value $E_{\rm g} + \hbar^2 k_\parallel^2/(2m^*)(k_\parallel \to 0)$.

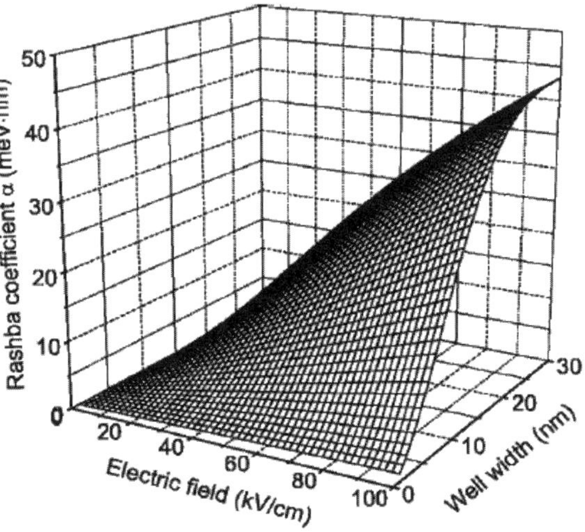

Fig. 4.14. Rashba coefficients for the lowest subband of the $Hg_{0.74}Cd_{0.26}Te/CdTe$ quantum well as functions of electric field and well width.

Zhang and Xia (2006) calculated the Rashba splitting energy of $R = 8\,\text{nm}$ InSb QW at transverse electric field $F = 7.44 \times 10^7\,\text{V/m}$ as a function of longitudinal wave vector k_z, as shown in Fig. 4.15. The solid line is the result of numerical calculation, and the dashed line is the result of the approximate formula (4.33) derived from Eq. (4.18).

$$\alpha(k_z) = \frac{P^2}{2} \frac{\Delta(2E_{\rm g}^{\rm eff} + \Delta)}{(E_{\rm g}^{\rm eff})^2(E_{\rm g}^{\rm eff} + \Delta)^2} \frac{2\varepsilon_0}{\varepsilon_r + \varepsilon_0} eF, \tag{4.33}$$

where the band gap $E_{\rm g}$ is replaced by the effective band gap $E_{\rm g}^{\rm eff}$,

$$E_{\rm g}^{\rm eff} = E_{\rm g} + E_{n0} + \frac{\hbar^2 k_z^2}{2m^*}, \tag{4.34}$$

E_{n0} is the energy of the transverse confined level in the QW, m^* is the electron effective mass moving in the z-direction, and F is the applied transverse electric field. The factor $2\varepsilon_0/(\varepsilon_{\rm r} + \varepsilon_0)$ in Eq. (4.33) is the correction of the inner electric field due to the dielectric effect, $\varepsilon_{\rm r}$ is the dielectric constant of the QW, and ε_0 is the dielectric constant of surrounding air. From Fig. 4.15, we see that the splitting energy has the same sign as the k_z, which is zero when $k_z = 0$. The splitting energy is comparable to the quantum well case (see Fig. 4.13). For positive k_z, the splitting energy increases as k_z increases at first, then decreases when k_z is large, similar to Fig. 4.13. The decrease is caused by the k_z-quadratic term in the effective

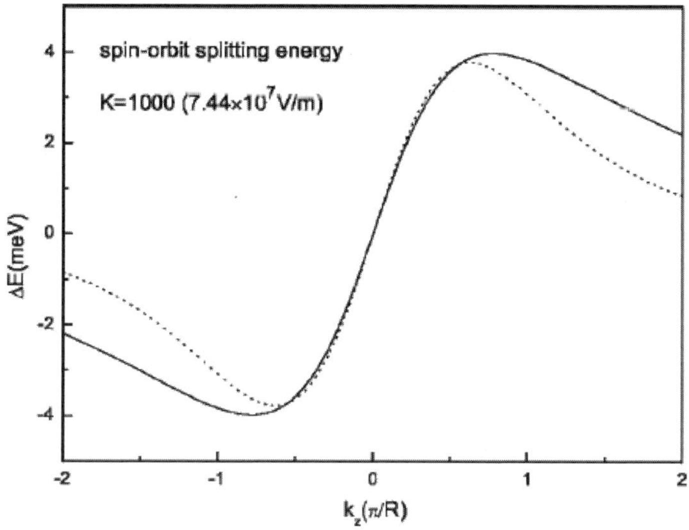

Fig. 4.15. Rashba splitting energy of $R = 8\,\text{nm}$ InSb quantum wire at transverse electric field $F = 7.44 \times 10^7\,\text{V/m}$ as a function of longitudinal wave vector k_z.

band gap Eq. (4.34), which is important in the large k_z range, resulting in a decrease in the Rashba coefficient α (Eq. (4.33)).

Figure 4.16 shows the Rashba coefficients of InSb QWs α versus QW radius R, (a) $F = 0.2 \times 10^7\,\text{V/m}$, (b) $F = 2 \times 10^7\,\text{V/m}$ (Zhang and Xia, 2006). The solid line is the result of the numerical calculation, and the dotted line is the result calculated from Eq. (4.33). The Rashba coefficient α increases with increasing R, and saturates, even decreases. According to Eq. (4.33), when R increases, E_{n0} approaches zero, and α should approach saturation, as shown by the dotted lines in Fig. 4.16. The descent of α in the large R range is due to the Stark effect. As the band gap of the InSb material is quite small, the mixing of the conduction states and the valence band states due to the Stark effect is very large, so the deviations especially in the larger electric field case are quite obvious, as shown in Fig. 4.16(b).

In general case, the Rashba coefficient α is proportional to the applied electric field F, but in InSb QWs it is not true. Figure 4.17 shows the Rashba coefficients of InSb QWs α versus applied transverse electric field F, (a) $R = 8\,\text{nm}$, (b) $R = 24\,\text{nm}$ (Zhang and Xia, 2006). The solid line is the result of numerical calculation, and the dotted line is the result calculated from Eq. (4.33). In the small radius case (Fig. 4.17(a)), α is approximately a linear function of the external electric field. In the large

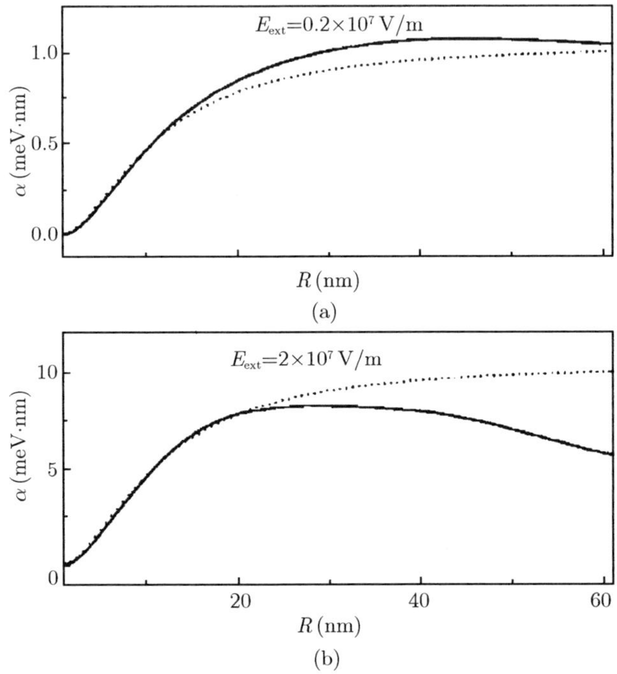

Fig. 4.16. Rashba coefficients of InSb QWs α versus QW radius R: (a) $F = 0.2 \times 10^7$ V/m, (b) $F = 2 \times 10^7$ V/m.

R case (Fig. 4.17(b)), the Rashba coefficient α shows a saturating trend in the large F range. This is also due to the Stark effect which changes the state components, mixing the CB states and the valence band states, resulting in the deviation. If the radius is larger, the Stark effect is more serious.

4.3. Semiconductor Spin Transistor and Quantum Waveguide Theory

4.3.1. *Spin-polarized tunneling transistor*

The spin-polarized tunneling through the superconductor/insulator/ferromagnet device was proposed early in 1971 (Tedrow and Meservy, 1971). The tunneling experiments showed that the quasi-particle energy states in thin superconducting Al films are split in a high magnetic field by the interaction of the field with the quasi-particle spin magnetic moments.

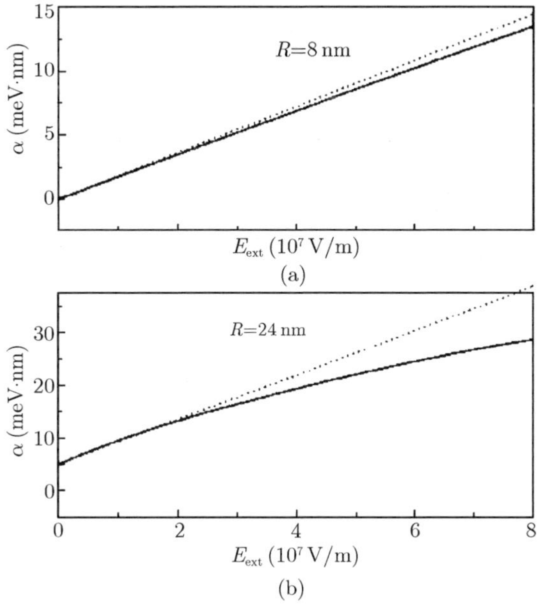

Fig. 4.17. Rashba coefficients of InSb QWs α versus applied transverse electric field F: (a) $R = 8\,\text{nm}$, (b) $R = 24\,\text{nm}$.

Because of the sharpness of the density of states at the edge of the superconducting energy gap, this splitting allows the selection of tunneling electrons of a particular spin direction, so that the superconductor can be used as a spin analyzer. Changing the voltage between the ferromagnet and the superconductor, one obtains a polarized tunneling current which reveals a polarization of the carriers in Ni by causing asymmetry in the tunneling characteristic. For the Al–Al_2O_3–Ni structure, the effective carrier polarization in the Ni is 7.5% with a majority of spin-down (magnetic moment parallel to B) carriers.

Johnson (1993) realized a spin (injection) transistor based upon thin metallic films, as shown in Fig. 4.18(a). The devices formed a sandwich structure, in which the two thin metal film ferromagnets, F1 and F2, were positioned on opposite sides of a 1.6 nm thick disorder, polycrystallic, Au film (labeled P). A second normal metal contact, N, was positioned near the detector ferromagnet F2 to measure the voltage between the ferromagnet contact F2 and P, V_s, and then detect the spin polarization. The current injects in the circuit between the below ferromagnet contacts F1 and P.

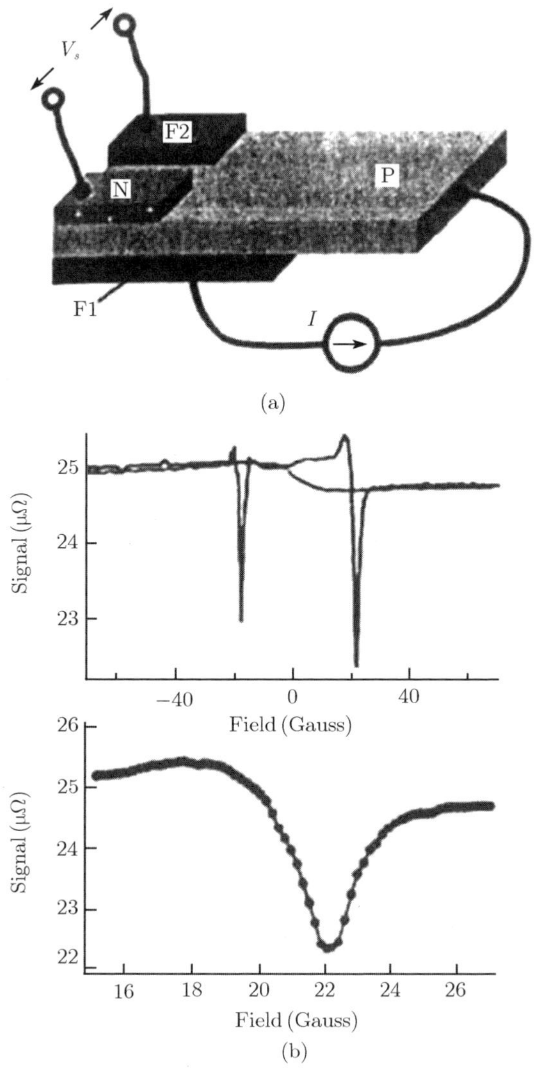

Fig. 4.18. (a) Schematic plot of spin (injection) transistor. (b) Applied magnetic field changes the magnetization directions of the injector and detector from parallel to antiparallel, and the voltage at the detector drops suddenly.

Evidence for spin injection in these devices was based on polarize/analyze experiment. A magnetic field is applied in the plane of the sample, along the easy axis of the ferromagnet. It is swept to cause their relative magnetizations to switch. A voltage jump at the detection terminals, i.e. in

V_s, is expected when the injector and detector magnetizations change from parallel to antiparallel. The obvious dips in the V_s curve in Fig. 4.18(b) correspond to antiparallel orientation of the magnetizations in the ferromagnetic contacts. An injected current of several mA results in a spin-induced voltage at the output of order several μV. Therefore, these devices yield a two-state output depending on the relative orientation of the ferromagnetization in the two ferromagnetic contacts. This result induces expectation to realize spin-based nonvolatile memory and logic elements.

4.3.2. *Semiconductor spin transistor*

Since Johnson (1993) proposed spin transistor based on the ferromagnetic metal, the research develops slowly. At the same time, the experiments found that the SRT in the semiconductors reaches 100 ns, and the spin-polarized drift distance reaches 100 μm. The concepts for novel, all-semiconductor spintronic devices have emerged, and these are generating new enthusiasm for "spin transistor" research. Comparing with the ferromagnetic metal, the semiconductor has the following advantages: (1) Equilibrium carrier densities can be varied through a wide range by doping. (2) The typical carrier densities in semiconductors are low compared with metals and are easily tunable by gate potentials. (3) The material growth technique allows the optimization of interfaces, at the level of atomic-scale control. (4) Advances in materials science of high-quality magnetic semiconductors. All advantages make semiconductor materials perhaps the first and natural choice for future spintronic applications — especially those involving large scale integration of spintronic devices.

Datta and Das (1990) first proposed the principle of transistor controlling electron spin. The title of their paper is "Electronic analog of the electro-optic modulator", whose basic principle is shown in Fig. 4.19, (a) electro-optic modulator and (b) proposed electron wave analog of the electro-optic modulator. In the electro-optic modulator, a polarizer at the input polarizes the light at 45° to the y-axis (in the $y-z$ plane) which can be represented as a linear combination of z- and y-polarized light.

$$\underset{(45^\circ\ \text{pol.})}{\begin{pmatrix}1\\1\end{pmatrix}} = \underset{(z\ \text{pol.})}{\begin{pmatrix}1\\0\end{pmatrix}} + \underset{(y\ \text{pol.})}{\begin{pmatrix}0\\1\end{pmatrix}}.$$

As this light passes through the electro-optic material, the two polarizations suffer different phase shifts k_1L and k_2L, because the electro-optic

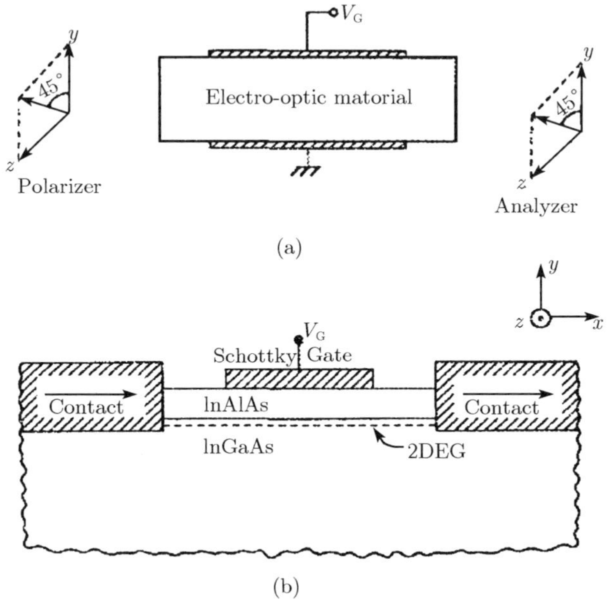

Fig. 4.19. Principle of transistor controlling electron spin: (a) electro-optic modulator, (b) proposed electron wave analog of the electro-optic modulator.

effect makes the dielectric constant ε_{zz} slightly different from ε_{yy}. The light emerging from the electro-optic material is represented as $\begin{pmatrix} e^{ik_1 L} \\ e^{ik_2 L} \end{pmatrix}$. The analyzer at the output lets the component along $\begin{pmatrix} 1 \\ 1 \end{pmatrix}$ to pass through. The output power P_0 is given by

$$P_0 \propto \left| (11) \begin{pmatrix} e^{ik_1 L} \\ e^{ik_2 L} \end{pmatrix} \right|^2 = 4\cos^2 \frac{(k_1 - k_2)L}{2}. \tag{4.35}$$

The light output is modulated by a gate voltage that controls ε_{zz} and ε_{yy}, thus the difference in the phase shift $\Delta\theta = (k_1 - k_2)L$.

The analogous device based on electron waves is shown in Fig. 4.19(b). The polarizer and analyzer can be implemented using contacts made of a ferromagnetic material like iron. A contact magnetized in the x-direction preferentially launches and detects electron spin polarized along positive x. The spin of the injected electron at input is along the x-direction, which can be represented as a linear combination of positive z-spin and negative

z-spin electrons.

$$\underset{(+x \text{ spin})}{\begin{pmatrix}1\\1\end{pmatrix}} = \underset{(+z \text{ spin})}{\begin{pmatrix}1\\0\end{pmatrix}} + \underset{(-z \text{ spin})}{\begin{pmatrix}0\\1\end{pmatrix}}. \tag{4.36}$$

The dominant mechanism for the "zero-field spin splitting" is the Rashba term in the effective mass Hamiltonian (Rashba, 1960),

$$H_R = \alpha(\sigma_z k_x - \sigma_x k_z). \tag{4.37}$$

In no magnetic field case, the Rashba term causes $+z$-spin and $-z$-spin electrons with the same energy to have different wave vectors k_1 and k_2. Consider an electron traveling in the x-direction with $k_x \neq 0$ and $k_z = 0$, the energies of electron with spin $+$ and $-$ equal,

$$\begin{aligned} E(+z \text{ spin}) &= \frac{\hbar^2 k_{x1}^2}{2m^*} + \alpha k_{x1}, \\ E(-z \text{ spin}) &= \frac{\hbar^2 k_{x2}^2}{2m^*} - \alpha k_{x2}. \end{aligned} \tag{4.38}$$

The energies of two spin electrons are equal, so the difference of the wave vectors of two spin electrons are obtained from Eq. (4.38).

$$k_{x1} - k_{x2} = \frac{2m^*\alpha}{\hbar^2}. \tag{4.39}$$

The difference in phase shift through the junction region L is introduced between spin-up and spin-down electrons, which is proportional to the Rashba coefficient α

$$\Delta\theta = (k_{x1} - k_{x2})L = \frac{2m^*\alpha L}{\hbar^2}. \tag{4.40}$$

For InGaAs/InAlAs heterostructures, from the experimentally observed zero-field spin splitting, α was estimated to be $\sim 3.9 \times 10^{-12}$ eVm. Equation (4.40) yields $L(\theta = \pi) = 0.67\,\mu$m, while the mean free paths $\geq 1\,\mu$m can be reached for high-mobility semiconductors at low temperatures. The Rashba coefficient α is proportional to the electric field perpendicular to the heterostructure interface, hence in principle, can be controlled by the application of a gate voltage.

So far, only electrons traveling along the x-direction have been considered. The above effect is reduced as the direction of propagation of the electron turns away from the x-axis. For electrons with nonzero k_x and k_z, the system Hamiltonian is

$$H = H_0 + H_R = \begin{pmatrix} \frac{\hbar^2}{2m^*}(k_x^2 + k_z^2) + \alpha k_x & -\alpha k_z \\ -\alpha k_z & \frac{\hbar^2}{2m^*}(k_x^2 + k_z^2) - \alpha k_x \end{pmatrix}. \tag{4.41}$$

With $k_z = 0$, the eigenstates are $\binom{1}{0}$ and $\binom{0}{1}$, i.e. the above discussed case; if $k_x = 0$, the eigenstates are $\binom{1}{1}$ and $\binom{1}{-1}$, so that the incoming beam is not split at all. It goes entirely into one of the eigenstates. Consequently, no current modulation is expected. As the angle θ of the propagation of the electrons with the x-axis increases, it can be shown that the modulation effect is reduced gradually to zero at $\theta = 90°$.

The spin-injected FET proposed by Datta and Das (1990) has been confirmed experimentally (Koo *et al.*, 2009). They used a high-mobility InAs heterostructure with a strong intrinsic Rashba coefficient α and measured the nonlocal channel conductance rather than the direct channel source–drain conductance suggested by Datta and Das (1990). The observed oscillatory conductance fits to theory with no adjustable parameters. The temperature dependence indicates that the modulation is only observed when the injected electrons have ballistic trajectories to the detector.

The samples with 2DEG channel are shown in Figs. 4.20(a) and 4.20(b), with the orientations of the external magnetic field B_a and the magnetization of the FM electrode in the y and x-directions, respectively. The carrier density and mobility of the 2DEG are $n_S = 1.8 \times 10^{12}$ to $2.8 \times 10^{12}\,\text{cm}^{-2}$ and $\mu = 50{,}000$–$60{,}000\,\text{cm}^2/\text{V s}$ at $T = 1.8\,\text{K}$, respectively. The channel has a width $w = 8\,\text{nm}$. The two FM electrodes were fabricated with spacings of $L = 1.25$ and $1.65\,\mu\text{m}$.

An example of the oscillatory conductance modulation is shown in Fig. 4.20(c). An external magnetic field $B_a = 0.5\,\text{T}$ was applied to fix the magnetization orientations of the FM electrodes in a chosen direction, thereby determining the axis of spin injection and detection. The nonlocal channel conductance was measured as a function of gate voltage for the range $-3 \leq V_G \leq 3\,\text{V}$. The third trace presents data for the configuration A ($B_a \| y$). In this case, the orientation of injected spins along the y-axis is parallel to the Rashba field. There is no spin precession and no modulation of voltage recorded by the detector. The fourth trace presents data for the configuration B ($B_a \| x$), the magnetization orientations of injector

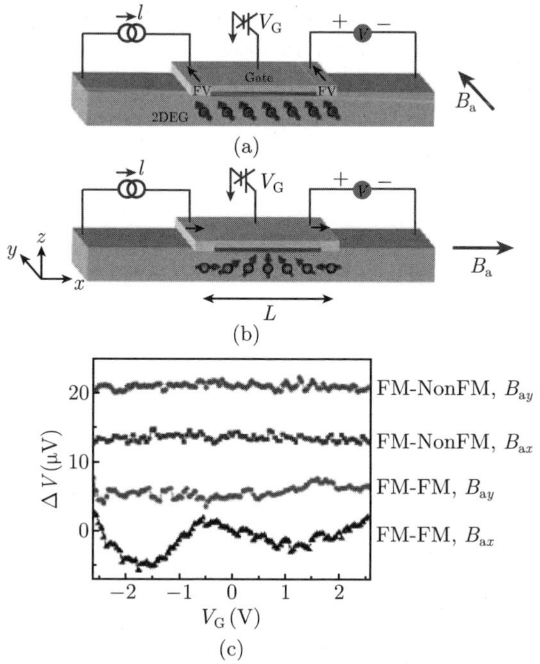

Fig. 4.20. Lateral gated spin valve with an external magnetic field $B_a = 0.5\,\mathrm{T}$: (a) $B_a \| y$, magnetization of FM electrodes $\| y$; (b) $B_a \| x$, magnetization of FM electrodes $\| x$; (c) observation of oscillatory conductance from injector to detector at $T = 1.8\,\mathrm{K}$. The four curves from bottom to up are: data from configuration B, data from configuration A, and data from a control sample similar to B or A, that has a ferromagnetic injector but a nonmagnetic detector, respectively.

and detector are along the $+x$-direction, and the injected spin orientation is perpendicular to the Rashba field (see Fig. 4.19(b)). The spin precession varies as a function of gate voltage, and an oscillation of detected voltage as a function of V_G was observed. The range of gate voltage is sufficiently large that more than one full cycle of voltage was recorded. Thus, the spin transistor proposed by Datta and Das (1990) was realized by the experiment of Koo *et al.* (2009) about 20 years later.

4.4. Quantum Waveguide Theory of Rashba Electrons

Rashba (1960) proposed SOI Hamiltonian Eq. (4.10), where η is the Rashba coefficient. When the potential field is nonsymmetric along the z-direction, $\mathrm{d}V(z)/\mathrm{d}z \neq 0$, $\eta \propto \mathrm{d}V/\mathrm{d}z$. The Rashba effect exists generally in the 2DEG formed by the modulation-doping heterostructure or in the applied electric field. Datta and Das (1990) proposed spin transistor based on the Rashba SOI. In the spin transistor, it demands that the electrons traveling in the channel should be ballistic, i.e. there is no scattering. In this section, we study the quantum waveguide theory of Rashba electron in the 1D circuit.

4.4.1. *1D quantum waveguide theory of Rashba electrons*

Wave function

In a 2D system such as 2DEG, the Hamiltonian taking into account the Rashba interaction is

$$H = \begin{pmatrix} -\dfrac{\hbar^2}{2m^*}\nabla^2 & \dfrac{\alpha}{\hbar}(ip_x + p_y) \\ \dfrac{\alpha}{\hbar}(-ip_x + p_y) & -\dfrac{\hbar^2}{2m^*}\nabla^2 \end{pmatrix}, \tag{4.42}$$

where m^* is the electron effective mass and α is the Rashba coefficient. If the electron moves in a straight line with a polar angle θ, the Hamiltonian (4.12) can be written as

$$H = \begin{pmatrix} -\dfrac{\hbar^2}{2m^*}\dfrac{\partial^2}{\partial l^2} & \alpha e^{-i\theta}\dfrac{\partial}{\partial l} \\ -\alpha e^{i\theta}\dfrac{\partial}{\partial l} & -\dfrac{\hbar^2}{2m^*}\dfrac{\partial^2}{\partial l^2} \end{pmatrix}, \tag{4.43}$$

where l is the coordinate along the line circuit. The electron wave function along a straight line has the plane wave form,

$$\Phi = \phi(\theta)e^{ikl} = \begin{pmatrix} c_1 \\ c_2 \end{pmatrix} e^{ikl}. \tag{4.44}$$

The eigenenergies are determined as

$$E = \frac{\hbar^2}{2m^*}k^2 \pm \alpha k, \tag{4.45}$$

and the wave functions are calculated to be

$$\phi_1(\theta) = \frac{1}{\sqrt{2}}\begin{pmatrix} 1 \\ ie^{i\theta} \end{pmatrix}, \quad \phi_2(\theta) = \frac{1}{\sqrt{2}}\begin{pmatrix} 1 \\ -ie^{i\theta} \end{pmatrix}, \tag{4.46}$$

corresponding to $-\alpha k$ and $+\alpha k$ terms in Eq. (4.45), respectively.

If two electron states have the same energy E, their wave vectors,

$$k_1 = k_0 + k_\delta, \quad k_2 = k_0 - k_\delta,$$
$$k_0 = \frac{m^*}{\hbar^2}\sqrt{\alpha^2 + \frac{2\hbar^2}{m^*}E}, \quad k_\delta = \frac{m^*}{\hbar^2}\alpha. \tag{4.47}$$

The spin orientations of the states ϕ_1 and ϕ_2 can be determined by $\phi_i^+\sigma_x\phi_i$ and $\phi_i^+\sigma_y\phi_i$ $(i = 1, 2)$, where σ_x and σ_y are the Pauli matrices. It is found that the spin orientations of ϕ_1 and ϕ_2 are perpendicular to the circuit l, the angles between the spin and the circuit are $+\pi/2$ and $-\pi/2$, respectively. Afterwards, we call the spin with the angle $+\pi/2$ to the circuit as spin up and that with the angle $-\pi/2$ as spin down.

Therefore, due to the Rashba SOI, two states with the same energy will have different wave vectors, spin orientations, and phases. Note that there is another pair of solutions associated with $-k_0$. These properties are different from that of electron without considering the Rashba SOI.

Now we consider the movement of Rashba electron in several structures as shown in Fig. 4.21 (Liu *et al.*, 2010). When there is a ferromagnetic contact (Fig. 4.21(b)) or a gate (c) at the end of the circuit, then either the spin-up (or -down) electrons or all electrons cannot pass through the circuit. If the electrons of spin up are reflected, then the wave function in the circuit will not be the plane wave as in Eq. (4.44), it has the form of a standing wave as

$$\Phi = \phi_1(\theta)e^{ik_1l} - \phi_2(\theta+\pi)e^{-ik_2l} = \phi_1(\theta)e^{ik_\delta l}\sin(k_0l), \tag{4.48}$$

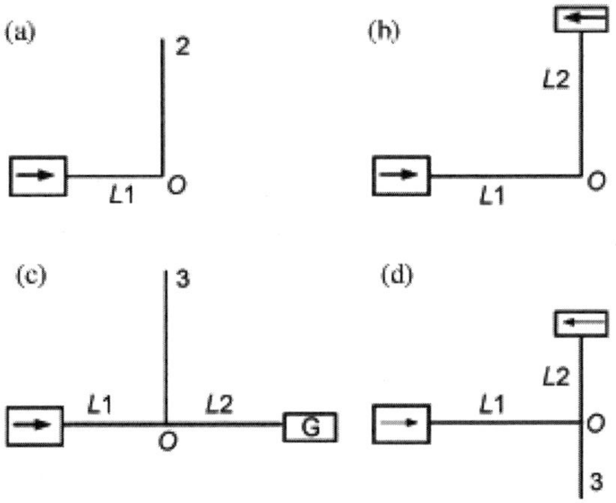

Fig. 4.21. Several structures: (a) turning structure; (b) spin-polarized device; (c) spin-polarized interference device; (d) another spin-polarized interference device.

where the origin of the coordinate l is at the contact or gate, and we have used the relation $\phi_2(\theta+\pi)=\phi_1(\theta)$ and Eq. (4.47). If spin-down electrons are reflected, its wave function equals

$$\Phi = \phi_2(\theta)e^{-ik_\delta l}\sin(k_0 l). \tag{4.49}$$

Boundary conditions at the intersection

Now we know all wave functions in every circuit, another problem is the boundary conditions at the intersection. If the intersection is crossed by n circuits, let Φ_i be the wave function of the ith circuit. At the intersection, there are two boundary conditions: one is the continuity of the wave functions and the other is the conservation of the current density. The first one is presented as

$$\Phi_1 = \Phi_2 = \cdots = \Phi_n. \tag{4.50}$$

The second one can be determined by the current density operator $\hat{L}_i$, which is the derivative of the coordinate l

$$\hat{L} = -\frac{i}{\hbar}[l, H] = \frac{i}{\hbar}\begin{pmatrix} -\dfrac{\hbar}{m^*}\dfrac{\partial}{\partial l} & \alpha e^{-i\theta} \\ -\alpha e^{i\theta} & -\dfrac{\hbar}{m^*}\dfrac{\partial}{\partial l} \end{pmatrix}. \tag{4.51}$$

If we assign all positive directions of l_i to be pointing to the intersection, the second boundary condition has the form,

$$\sum_{i=1}^{n} \hat{L}_i \Phi_i = 0. \tag{4.52}$$

For the plane wave in Eqs. (4.44) and (4.46), we get

$$\hat{L}\Phi = \frac{\hbar k_0}{m^*}\Phi, \tag{4.53}$$

no matter the state is spin up or spin down. For the standing waves in Eqs. (4.48) and (4.49) with the coordinate origin at the intersection, and L being the distance from the contact (gate) to the intersection, we get

$$\hat{L}\Phi = -\frac{\hbar k_0}{m^*} i\cot[k_0(l-L)]\Phi. \tag{4.54}$$

From Eq. (4.53), we see that though the Rashba states have different wave vectors of k_1 and k_2, their travel velocities are the same $\eta k_0/m^*$, this will make the calculation more convenient. Furthermore, the $\eta k_0/m^*$

terms in Eqs. (4.53) and (4.54) of the second boundary condition can be counteracted.

Transport in open circuits

(a) Turning right angle structure (Fig. 4.21(a))

Suppose that an incident electron with energy E enters into circuit 1, there will be transmitted waves in circuit 2 and reflected waves in circuit 1. The wave functions in the circuits can be written as

$$\begin{aligned}\Phi_1 &= a_{10}\phi_1(0)e^{ik_1l_1} + a_{20}\phi_2(0)e^{ik_2l_1}\\ &\quad + a_1\phi_1(\pi)e^{-ik_1l_1} + a_2\phi_2(\pi)e^{-ik_2l_1},\\ \Phi_2 &= c_1\phi_1\left(\frac{\pi}{2}\right)e^{ik_1l_2} + c_2\phi_2\left(\frac{\pi}{2}\right)e^{ik_2l_2},\end{aligned} \tag{4.55}$$

where a_{10}, a_{20} are the coefficients of the known incident wave functions, a_1, a_2 are the coefficients of reflected waves, and c_1, c_2 are the coefficients of transmitting waves. From the two boundary conditions Eqs. (4.50) and (4.52), we obtain the coupled equations about the unknown coefficients a_1, a_2 and c_1, c_2. The solution is

$$\begin{gathered} a_1 = a_2 = 0,\\ c_1 = \frac{1-i}{2}A_{10} + \frac{1+i}{2}A_{20}, \quad c_2 = \frac{1+i}{2}A_{10} + \frac{1-i}{2}A_{20},\\ A_{10} = a_{10}e^{ik_1L_1}, \quad A_{20} = a_{20}e^{ik_2L_1}.\end{gathered} \tag{4.56}$$

From Eq. (4.56), we see that the reflection probabilities are zero, and it is proved that the conclusion is valid for turning structures with any turning angle. If the incident wave has the spin orientation in the x-direction, then $a_{10} = 1/\sqrt{2}, a_{20} = i/\sqrt{2}$, from Eq. (4.56) we obtain $c_1 = e^{ik_0L_1}\sin(k_\delta L_1)$ and $c_2 = e^{ik_0L_1}\cos(k_\delta L_1)$. Thus, the transmission probabilities of two electron waves T_{21}, T_{22} oscillate with the Rashba coefficient α, because two injection waves have different wave vectors, and they interfere with each other when they arrive at the intersection.

(b) Turning structure with a ferromagnetic contact at the output end (Fig. 4.21(b))

The wave function Φ_1 in circuit 1 is same as in Eq. (4.55), the Φ_2 in circuit 2 is written as

$$\Phi_2 = c_1\phi_1\left(\frac{\pi}{2}\right)e^{ik_1l_2} + c_2\phi_2\left(\frac{\pi}{2}\right)e^{-ik_\delta l_2}\sin[k_0(l_2 - L_2)], \tag{4.57}$$

Due to the ferromagnetic contact at the output, the wave of spin down is a standing wave as shown in the second term of Eq. (4.57).

Using the boundary conditions, Eqs. (4.50) and (4.52), the calculated results are

$$\begin{aligned} A_2 &= -iA_1 = -\frac{A_{10} - iA_{20}}{2} e^{2ik_0 L_2}, \\ c_1 &= \frac{1-i}{2} A_{10} + \frac{1+i}{2} A_{20}. \end{aligned} \tag{4.58}$$

Comparing with the simple turning structure Eq. (4.56), the transmission probabilities T_{21} of spin-up electron is the same as the simple turning structure, and oscillate with α. The T_{22} of spin-down electron is zero, and the reflection probabilities are not zero. Thus, this structure is a spin-polarized diode whose current can be modulated by α and L_1. If the polarized direction of the ferromagnetic contact is reversed, then the reflection and the transmission will be upset.

(c) Three-terminal device with a gate (Fig. 4.21(c))

Figure 4.22 shows the T_{31}, T_{32}, R_1 and R_2 of the spin polarized interference device in Fig. 4.21(c) versus α (Liu *et al.*, 2010). The T_{31} and T_{32} oscillate with α. Due to the interference effect of the gate, their amplitudes vary with α, first increase, then decrease to zero at $\alpha = 85\,\text{meV nm}$. Therefore, the interference effect can modulate the amplitudes of T_{31} and T_{32} simultaneously.

We also develop a general quantum waveguide theory of the Rashba electron for a structure with multiple branches (Liu *et al.*, 2010).

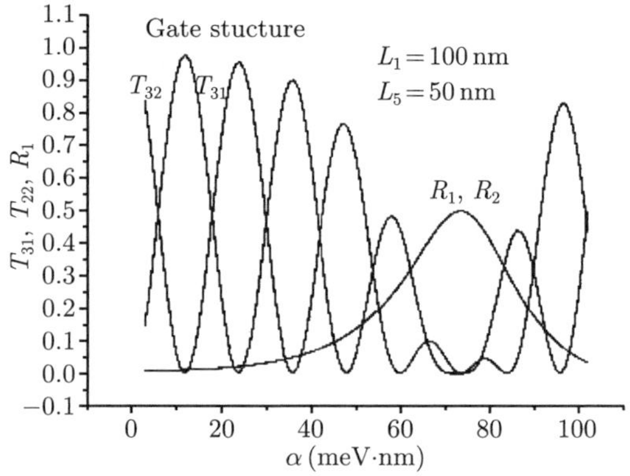

Fig. 4.22. T_{31}, T_{32}, R_1, and R_2 of the spin-polarized interference device in Fig. 4.21. (c) versus α for $m^* = 0.1m_0, E = 0.1\,\text{meV}$, $L_1 = 100\,\text{nm}$, and $L_2 = 50\,\text{nm}$.

4.4.2. *Transport in closed 1D loops*

Koga *et al.* (2006) succeeded in observing gate-controlled electron spin interference in nanolithographically defined square loop (SL) arrays that were fabricated in $In_{0.53}Ga_{0.47}As$ quantum well (QW). They demonstrated electron spin precession in ballistic channels within the QW that is caused by the Rashba effect. The electric sheet conductivities σ_{2D} as a function of magnetic field B for an SL array shows AB-like oscillations with period $h/2e$. The observed oscillations decay rather rapidly with increasing B due to the existence of multiple channels within a finite channel width.

We apply the 1D quantum waveguide theory of Rashba electron in last Section 4.4.1 to the case of closed loops, especially the circular rings (Liu *et al.*, 2009). We study three closed loop structures as shown in Fig. 4.23.

AB ring: closed circular ring

Divide the circle into N segments, each segment can be seen as a straight line segment. Calculate the transfer matrix of every segment and intersection. The total transfer matrix is the product of all matrices in the sequent order of transport.

For one segment, the polar angle of the segment is θ, and the polar angle of next segment is φ, then its transfer matrix is (Liu *et al.*, 2009)

$$M_{\theta\to\varphi} = \frac{1}{2}\begin{pmatrix} 1+e^{-i(\varphi-\theta)} & 1-e^{-i(\varphi-\theta)} \\ 1-e^{-i(\varphi-\theta)} & 1+e^{-i(\varphi-\theta)} \end{pmatrix}, \tag{4.59}$$

which relates the wave function coefficients of the circuit φ with that of the circuit θ. Accordingly, the transfer matrix of one circuit with length L can be written as

$$M_L = \begin{pmatrix} e^{ik_1L} & 0 \\ 0 & e^{ik_2L} \end{pmatrix}. \tag{4.60}$$

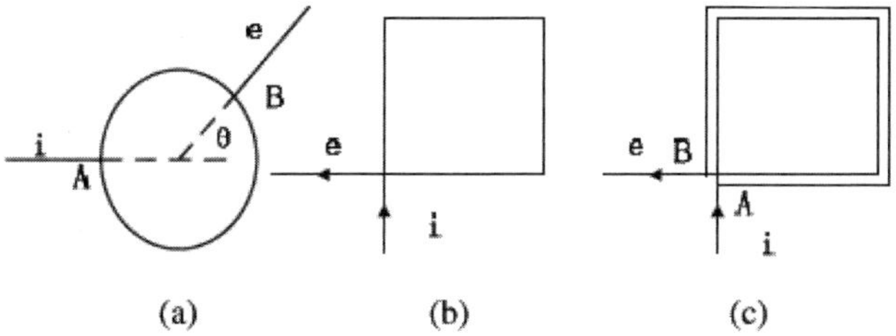

Fig. 4.23. Several closed-loop structures: (a) circular ring, (b) SL, (c) double SL.

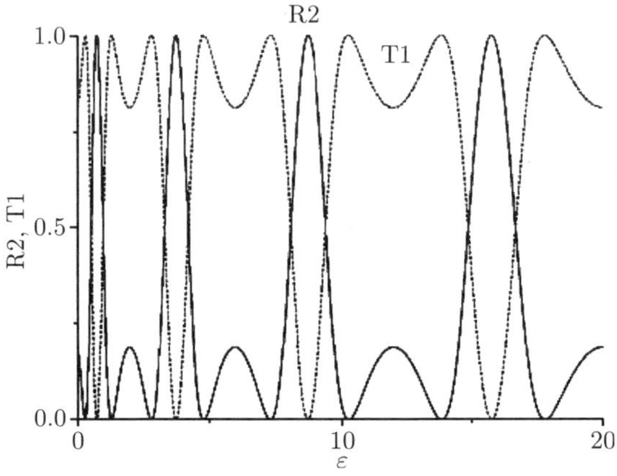

Fig. 4.24. T_1 and R_2 of the AB ring versus electron energy ε for $\alpha = 1, a_{10} = 1$, and $a_{20} = 0$ in zero magnetic field $b = 0$.

For the AB ring with $\theta = 0$ as shown in Fig. 4.23(a), we use the transfer matrices Eqs. (4.59) and (4.60) to calculate the wave function from one end to another end of the circular circuit and the boundary conditions Eqs. (4.50) and (4.52) to calculate the coefficients of the transmitted and reflected waves.

Figure 4.24 shows the T_1 and R_2 of the AB ring versus electron energy ε for $\alpha = 1, b = 0, a_{10} = 1$, and $a_{20} = 0$ (Liu *et al.*, 2009), where T_1 is the transmission probability of the spin-up electron and R_2 is the reflection probability of spin-down electron. ε, α, and b are the dimensionless energy, Rashba coefficient, and magnetic field, their units are $E_0 = \hbar^2/(2m^*R^2), \alpha_0 = \hbar^2/(2m^*R)$, and $B_0 = \hbar c/(2eR^2)$, respectively, R is the radius of ring. It is found that the energies to which the zero points of the R_2 correspond, for example, in Fig. 4.24 ε = 0.3, 1.3, 2.8, 4.8, 7.3, 10.3, 13.8, 17.8, etc., are just the energies of eigenstates in the closed ring. It means that when the energy of the injected electron equals the energy of the eigenstate in the close ring, there will be no reflection completely, and there will occurs resonance.

In the case of an external magnetic field perpendicular to the circuit plane, for a closed loop the wave vector k in the Hamiltonian will be replaced by

$$k_{\text{eff}} = \frac{k \pm eA}{\hbar}, \tag{4.61}$$

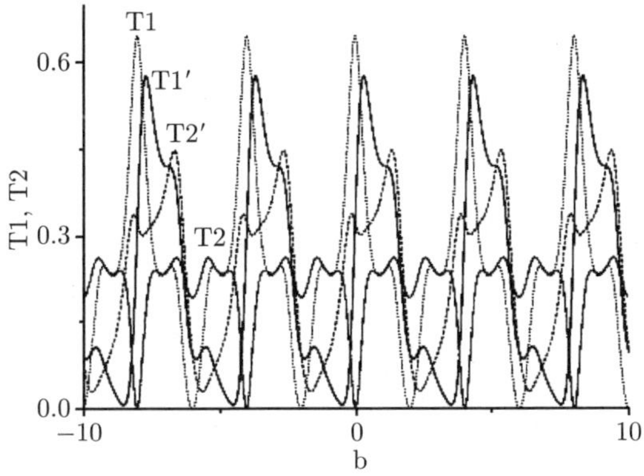

Fig. 4.25. T_1 and T_2 of the AB ring versus magnetic field b for $\alpha = 1$, $\varepsilon = 0.5, a_{10} = 1, a_{20} = 0(T_1, T_2)$ and $a_{10} = 0.707, a_{20} = 0.707(T_1', T_2')$.

where the $\pm$ sign depends on the relative orientation of the magnetic field and the $k, A = \Phi_{\rm m}/L$ is the vector potential, and $\Phi_{\rm m}$ and L are the magnetic flux through the closed loop and its perimeter, respectively.

Figure 4.25 shows the T_1 and T_2 of the AB ring versus magnetic field b for $\alpha = 1, \varepsilon = 0.5, a_{10} = 1, a_{20} = 0(T_1, T_2)$ and $a_{10} = 0.707, a_{20} = 0.707(T_1', T_2')$ (Liu *et al.*, 2009). For both cases, T_1 and T_2 curves all oscillate with the magnetic field b, the oscillating period is $b = 4$, though their zero points are different. From the definition of $b = 2eBR^2/c\hbar$, we obtain the period of the magnetic flux is $\Phi_{\rm m} = hc/e$, independent of the magnitude of the ring. This is the first proof of the AB effect of the Rashba spin current.

The transfer matrix for SL is calculated easily, which includes that of straight line section of length L Eq. (4.60) and that of the turning intersection Eq. (4.59). Figure 4.26 shows the T_1, T_2, R_1 and R_2 of the AB double SL in Fig. 4.23(c) versus magnetic field b for $\alpha = 1, \varepsilon = 0.5, a_{10} = 1$, and $a_{20} = 0$ (Liu *et al.*, 2009). T_1, T_2, R_1, and R_2 all oscillate with the magnetic field, and the oscillating period is $b = 6.28$. According to the definition of b in the SL $b = 2eBL^2/\hbar c$, the oscillating period of the magnetic flux $\Phi_{\rm m} = 3.14\hbar c/e = hc/2e$. It is half of that of the single SL, in agreement with the experimental results (Koga *et al.*, 2006).

4.4.3. *Eigenstates in closed loops*

The transfer matrix can be used to calculate the eigenstates in closed loops. If the transfer matrix around a loop is $M(E)$, which is a function of electron

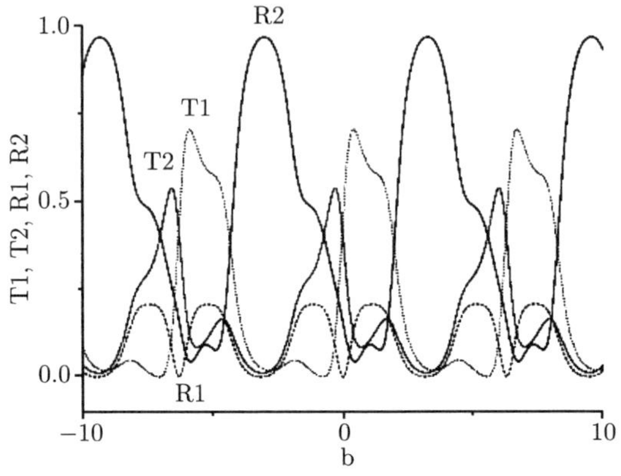

Fig. 4.26. T_1, T_2, R_1 and R_2 of the AB double SL in Fig. 4.23(c) versus magnetic field b for $\alpha = 1, \varepsilon = 0.5, a_{10} = 1$, and $a_{20} = 0$.

energy E, the eigenenergy E satisfies the equation $\det|M(E) - I| = 0$ and the eigenvector can be correspondingly obtained.

First we consider the circular ring and divide it into N small segments. If N is large enough, the length of the segment can be written as $R\delta\theta$, and $\delta\theta = (\varphi - \theta)$ is the turning angle at the intersection in Eq. (4.59). $\delta\theta$ is an infinitesimal quantity and $N\delta\theta = 2\pi$. For one segment, the transfer matrix is

$$M_{\delta\theta} = \frac{1}{2}\begin{pmatrix} 1 + e^{-i\delta\theta} & 1 - e^{-i\delta\theta} \\ 1 - e^{-i\delta\theta} & 1 + e^{-i\delta\theta} \end{pmatrix} \begin{pmatrix} e^{ik_1 R\delta\theta} & 0 \\ 0 & e^{ik_2 R\delta\theta} \end{pmatrix}. \tag{4.62}$$

Because $M_{\delta\theta}$ is very close to I, we assume that its eigenvalue is $e^{im\delta\theta}$, where m is an integral. Ignoring twice and more power terms of $\delta\theta$, and solving the secular equation,

$$\det|M_{\delta\theta} - e^{im\delta\theta}| = 0, \tag{4.63}$$

we obtain the eigenenergy,

$$\varepsilon = \left(m + \frac{1}{2}\right)^2 + \frac{1}{4} \pm \left(m + \frac{1}{2}\right)\sqrt{1 + \alpha^2}. \tag{4.64}$$

In the presence of a magnetic field, the corresponding eigenenergies are

$$\varepsilon = \left(m + \frac{b}{4} + \frac{1}{2}\right)^2 + \frac{1}{4} \pm \left(m + \frac{b}{4} + \frac{1}{2}\right)\sqrt{1 + \alpha^2}. \tag{4.65}$$

The results of Eqs. (4.64) and (4.65) are identical to those obtained from solving the Schrödinger equation with the Hamiltonian in the polar coordinate straightly (Bulgakov and Sadreev, 2002).

We can also get the eigenvectors of the transfer matrix $M_{\delta\theta}$,

$$M_{\delta\theta}\begin{pmatrix}\cos\varphi'\\ \sin\varphi'\end{pmatrix} = e^{im\delta\theta}\begin{pmatrix}\cos\varphi'\\ \sin\varphi'\end{pmatrix}, \tag{4.66}$$

where $\tan\varphi' = \alpha \pm \sqrt{1+\alpha^2}$. The eigen wave function of the state with the polar quantum number m is

$$\Psi_m(\theta) = \phi_1(\theta)\cos\varphi' + \phi_2(\theta)\sin\varphi' = \frac{1}{\sqrt{2\pi}}e^{im\theta}\begin{pmatrix}\cos\varphi\\ \sin\varphi e^{i\theta}\end{pmatrix}, \tag{4.67}$$

where $\tan\varphi = (1+\sqrt{1+\alpha^2})/\alpha$ for "+", and $\tan\varphi = (1-\sqrt{1+\alpha^2})/\alpha$ for "−" in Eq. (4.64) or Eq. (4.65).

Similarly, the eigenenergies and eigenvectors of the states in the closed square single loop and double loop can be calculated using the transfer matrix (Liu *et al.*, 2009).

4.5. Generation and Transport of Spin-polarized Current

4.5.1. *Coherent transport of spin-polarized electrons through an interface of heterostructure*

A critical issue in the development of spintronics is the transmission of classical and quantum information through heterogeneous systems. The spin injection from ferromagnets into semiconductors has proved challenging and promoted all-optical investigations of spin transport in semiconductors. One of the main obstacles is that when spin-polarized electrons travel through a heterointerface how many spin-polarized electrons remain. Similar to the experiment done by Kikkawa and Awschalom (1998) (Fig. 4.4), the GaAs sample is replaced by the GaAs/ZnSe heterostructure and the Faraday rotation is replaced by the KR. The experimental geometry of the TRKR is shown in Fig. 4.27(c) (Malajovich *et al.*, 2000). Figures 4.27(a) and 4.27(b) show the energy bands lineup of ZnSe and GaAs; (d) TRKR angle θ_K versus decay time Δt in the Si GaAs substrate at $T = 5\,\mathrm{K}$ and $B = 0\,\mathrm{T}$; (e) TRKR data versus pump energy $E_p$at $\Delta t = 800\,\mathrm{ps}$ and probe energy $E_{\mathrm{probe}} = 2.80\,\mathrm{eV}$; (f) TRKR date versus E_{probe} at $\Delta t = 800\,\mathrm{ps}$ and $E_p = 1.52\,\mathrm{eV}$ (Malajovoch *et al.*, 2000). The energy gaps of ZnSe and

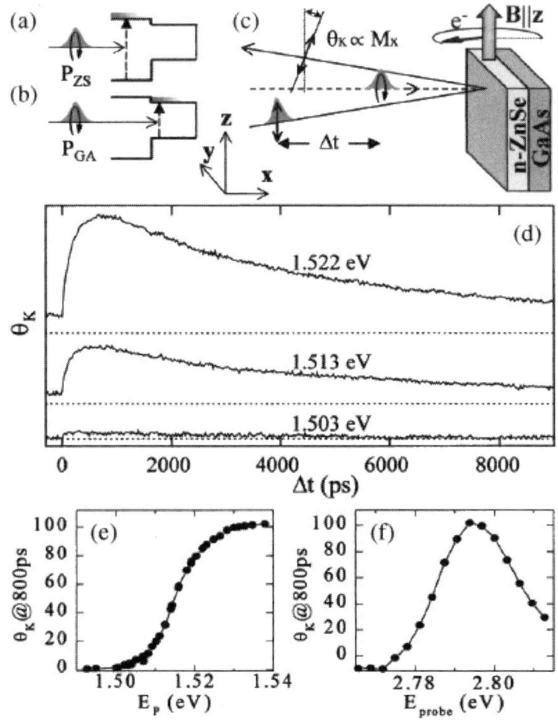

Fig. 4.27. (a) and (b) Energy bands lineup of ZnSe and GaAs. (c) Experimental geometry of TRKR. (d) TRKR angle θ_K versus delay time Δt in the Si GaAs substrate at $T = 5\,K$ and $B = 0\,T$. (e) TRKR data versus pump energy E_P at $\Delta t = 800\,ps$ and probe energy $E_{probe} = 2.80\,eV$; (f) TRKR date versus E_{probe} at $\Delta t = 800\,ps$ and $E_P = 1.52\,eV$.

GaAs are 2.80 and 1.52 eV, respectively, the band offset of the CBs is rather small. A CP pump pulse P creates electron spins oriented along the sample normal, x. P may be tuned to either the ZnSe or GaAs absorption threshold (energy gap) and is denoted P_{ZS} or P_{GA}, accordingly. Whereas primary absorption of P_{ZS} occurs in ZnSe (Fig. 4.27(a)), P_{GA} passes through the ZnSe epilayer, exciting spins in the GaAs substrate (Fig. 4.27(b)).

Figure 4.27(d) shows the TRKR data measured in ZnSe originated from P_{GA}. The pump energy E_P is tuned through the GaAs energy gap. P_{GA} produces a ZnSe magnetization S_x that grows over the first few hundred picoseconds and then decays with a spin lifetime appropriate to the ZnSe host. This suggests spin migration from GaAs to ZnSe, a suspicion confirmed by the sudden growth of S_x as E_p is tuned from below to above the

GaAs absorption threshold (Fig. 4.27(e)). Conversely, the probe's selectivity in measuring only ZnSe spins is seen in Fig. 4.27(f), where θ_K (and hence S_x) is maximized at the ZnSe absorption threshold. An estimate of the fraction of GaAs spins entering ZnSe is 2.5–10% at $B = 0\,\mathrm{T}$, based on the Kerr response obtained by introducing a known number of spins directly into the ZnSe.

The decay of the TRKR data with the delay time Δt can be described by $e^{-\Delta t/\tau}$, where τ is the accumulation time of spin transfer and equals 210 and 440 ps for Si and n-GaAs substrates, respectively. Generally,

$$\frac{1}{\tau} = \left[\frac{1}{\tau_0} + \frac{1}{T_2^*|_{\mathrm{GaAs}}}\right], \tag{4.68}$$

where τ_0 is the accumulation time of GaAs spin transfer, $T_2^*|_{\mathrm{GaAs}}$ is the substrate transverse spin lifetime. Since the spin lifetime in the n-GaAs substrate is nearly three orders of magnitude longer than τ, so that the measured τ is mainly the spin accumulation time.

The application of a transverse magnetic field B (Fig. 4.27(c)) induces coherent spin precession at the Larmor frequency, $\omega = g\mu_B B/\hbar$. Because $g_{\mathrm{GA}} \neq g_{\mathrm{ZS}}$, the phase of electron spins in ZnSe depends on their duration in GaAs. Hence, a distribution of arrival times t_i ($t_i < \Delta t$) can result in inhomogeneous dephasing within the population of transferred electron spins. Assuming that the Larmor angle in each material grows as $\omega\Delta t$, where ω is now material-dependent, the contribution at Δt of all spins that have spent a time t_i in the GaAs is

$$s_i(\Delta t) = \left(\frac{S}{\tau}e^{-t_i/\tau}\right) e^{-\Delta t/T_2^*} \cos[\omega_{\mathrm{GA}} t_i + \omega_{\mathrm{ZS}}(\Delta t - t_i)], \tag{4.69}$$

where S is the total spin transferred, $(S/\tau)e^{-ti}/\tau$ is the spin accumulation at t_i in GaAs. Integration of Eq. (4.69) over $t_i < \Delta t$ yields the total spin contribution,

$$S_x(\Delta t) = Ae^{-\Delta t/T_2^*}[\cos(\omega_{\mathrm{ZS}}\Delta t + \phi) - e^{-\Delta t/\tau}\cos(\omega_{\mathrm{GA}}\Delta t + \phi)], \tag{4.70}$$

where

$$\begin{aligned} \phi &= \tan^{-1}[(\omega_{\mathrm{GA}} - \omega_{\mathrm{ZS}})\tau], \\ A &= \frac{S}{\sqrt{1+\tan^2\phi}}. \end{aligned} \tag{4.71}$$

From Eq. (4.70), we see that due to the difference between the precession frequencies of two materials, the spin amplitude decreases by $\sqrt{1+\tan^2\phi}$. The TRKR angle θ_K is composed of the two oscillations with two different frequencies. If $(\omega_{GA} - \omega_{ZS})\tau \ll 1$, the inhomogeneous dephasing effect can be neglected. When $\Delta t \gg \tau$, the spin electrons transferred from GaAs equal zero, the precession frequency is determined mainly by ω_{ZS}, and the amplitude decays as $e^{-\Delta t/T_2^*}$, where T_2^* is the transverse lifetime of ZnSe.

Using Eq. (4.70) to fit the data over the entire time interval, one obtained T_2^*, A, ϕ, g, and τ as functions of the magnetic field. Figure 4.28(a) shows T_2^* versus B for excitation in ZnSe layer (□), Si GaAs substrate (•), and n-GaAs substrate (×) (Malajovich *et al.*, 2000). The decreases in T_2^* with field for excitations in ZnSe and GaAs are basically the same, and appear to describe properties of the ZnSe host unrelated to the process of

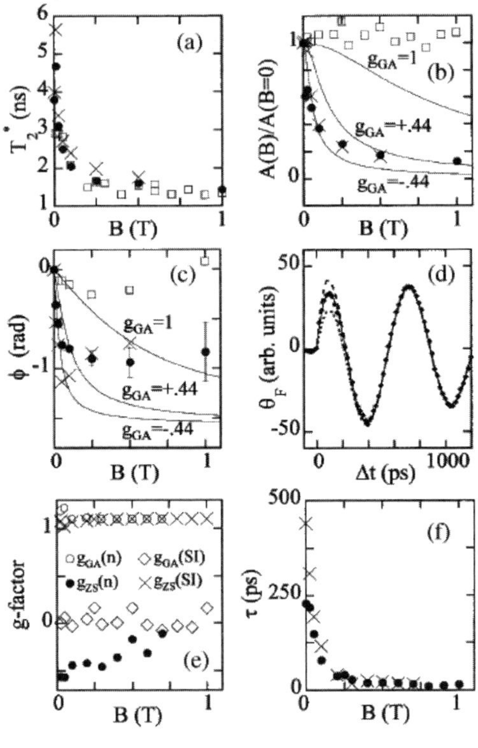

Fig. 4.28. (a) T_2^* versus B for excitation in ZnSe layer (open squares), Si GaAs substrate (solid circles), and n-GaAs substrate (crosses). (b) Spin relative amplitude $A(B)/A(0)$ versus B. (c) Phase shift ϕ versus B. (d) θ_K versus Δt for excitation in GaAs at $B = 100\,\text{mT}$. (e) g_{GA} and g_{ZS} versus B for excitations in Si and n-GaAs. (f) Spin accumulation time τ versus B.

spin transfer. Figure 4.28(b) shows the spin relative amplitude $A(B)/A(0)$ versus B. There is an obvious difference between excitations in ZnSe and GaAs, for former the amplitude does not change with B, while for later it decreases with increasing B. Three curves are calculation results from Eq. (4.70) taking $g_{\mathrm{GA}} = 1, +0.44$ and -0.44. The result of $g_{\mathrm{GA}} = 1 \approx g_{\mathrm{ZS}}$ is close to the experimental result for excitation in ZnSe. The results of $g_{\mathrm{GA}} = +0.44$ and -0.44 are both close to the experimental result for excitation in GaAs, so the + or − sign cannot be determined by this experiment. Figure 4.28(c) shows phase shift ϕ versus B. The phase shift ϕ is almost zero for excitation in ZnSe, and it is not zero for excitation in GaAs, in agreement with Eq. (4.70). But the differences between the experimental values and the theoretical values are rather large. Figure 4.28(d) shows θ_{K} versus Δt for excitation in GaAs at $B = 100\,\mathrm{mT}$, three curves correspond to $g_{\mathrm{GA}} = +0.04$ (solid line), $+0.44$ (dotted line), and -0.44 (dashed line). The results are not sensitive to g_{GA}, the result of $g_{\mathrm{GA}} = +0.04$ is in best agreement with the experimental result, the reason is still not clear. Figure 4.28(e) shows g_{GA} and g_{ZS} versus B for excitations in Si and n-GaAs. For excitation in Si GaAs, $g_{\mathrm{ZS}} \approx 1, g_{\mathrm{GA}} \approx 0$, and for excitation in n-GaAs, $g_{\mathrm{ZS}} \approx -0.25, g_{\mathrm{GA}} \approx 1$. The deviations from the nominal values of g-factor may be due to excess kinetic energy within the incident spins, whose g-factor is known to depend on the wave vector and can cross zero for carrier energies of 10–100 meV. Moreover, as shown in Fig. 4.28(f), the spin accumulation time τ drops sharply with the applied field.

As discussed before, the optical excitation can provide a short (<500 ps) nonequilibrium burst of coherent spin transfer across a GaAs/ZnSe interface, but less than 10% of the total spin crosses into the ZnSe layer, leaving long-lived spins trapped in the GaAs layer. Application of a bias voltage on the heterostructure will increase greatly the number of spin-polarized electrons injected from GaAs to ZnSe (Malajovich *et al.*, 2001). The spin transfer in the CB with (bottom) and without (top) bias is shown in Fig. 4.29(b); in the bias voltage, there are more electrons injected from GaAs to ZnSe. Figure 4.29(c) shows the TRKR resonance scans taken by sweeping the applied magnetic field at $\Delta t = 10\,\mathrm{ps}$, $T = 5\,\mathrm{K}$ for GaAs, ZnSe, and GaAs/ZnSe heterostructures (Malajovich *et al.*, 2001). When the bias voltage increases, the amplitude of resonance peak increases, and the resonance spectrum transforms from that characteristic of ZnSe to one whose spacing more closely resembles that of GaAs. The data show that an electric field can be used to control the magnetic response of the ZnSe layer during spin injection, and vice versa. This behavior results

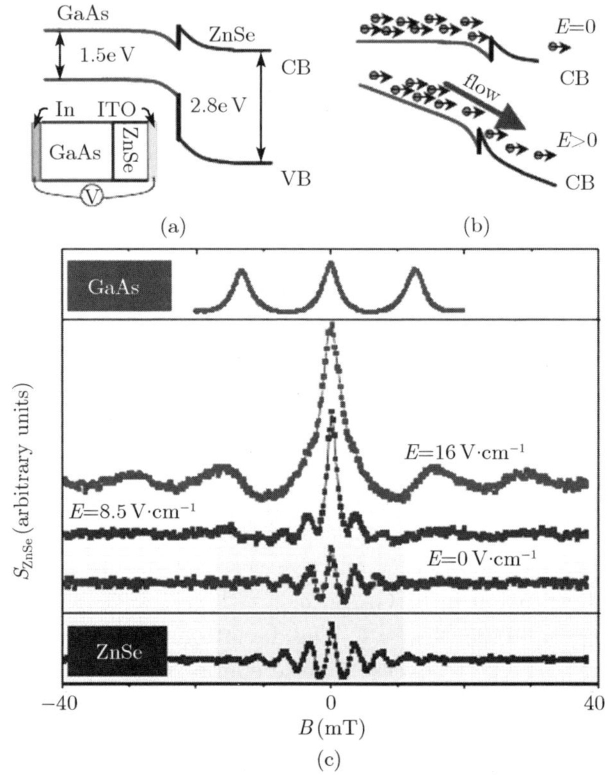

Fig. 4.29. (a) Energy bands of GaAs/ZnSe heterostructure. (b) Spin transfer in the CB with (bottom) and without (top) bias. (c) TRKR resonance scans taken by sweeping the applied magnetic field at $\Delta t = 10\,\text{ps}$, $T = 5\,\text{K}$. The top (bottom) spectrum corresponds to the GaAs (ZnSe) layer. The middle spectra of the GaAs/ZnSe heterostructure, with the bias electric field from 0 to 16 V/cm.

from co-current changes in the amplitude and orientation of the spin current itself.

The analysis of the experimental data indicates that in the bias voltage there are three channels of spin injection: A, B, and C, where A and B are the spin-injected currents in the zero bias, but with different accumulation times $\tau^{\text{A}} \approx 20\,\text{ps}$, $\tau^{\text{B}} \approx 500\,\text{ps}$. While the channel C is a persistent spin current in the bias voltage, its $\tau_0^{\text{C}} \approx \infty$, thus $\tau^{\text{C}} \approx T^*_{2\text{GaAs}}$ (see Eq. (4.68)), which represents not the spin accumulation time but a decay in the persistent spin current. The total spin signal in the epilayer is therefore described by

$$S_{\mathrm{ZnSe}}(\Delta t) = S^{\mathrm{A}} + S^{\mathrm{B}} + S^{\mathrm{C}}, \tag{4.72}$$

where $S^{\mathrm{A}}, S^{\mathrm{B}}, S^{\mathrm{C}}$ are Eq. (4.70) with different τ.

Figure 4.30 shows the TRKR scans versus Δt at $B = 75\,\mathrm{mT}$, $T = 5\,\mathrm{K}$, (a) 300 nm thick, $n = 1.5 \times 10^{18}\,\mathrm{cm}^{-3}$ ZnSe epilayer, (b) 100 nm thick, $n = 1.5 \times 10^{17}\,\mathrm{cm}^{-3}$ ZnSe epilayer. (c) Decomposition of the fit to the biased scan in (b) into three different conduction channels (Malajovich *et al.*, 2001). The following changes are observed in the spin polarization transfer upon biasing: (1) the amplitude of the spin transfer increases, (2) the spin signal survives for longer times, (3) a second precession frequency

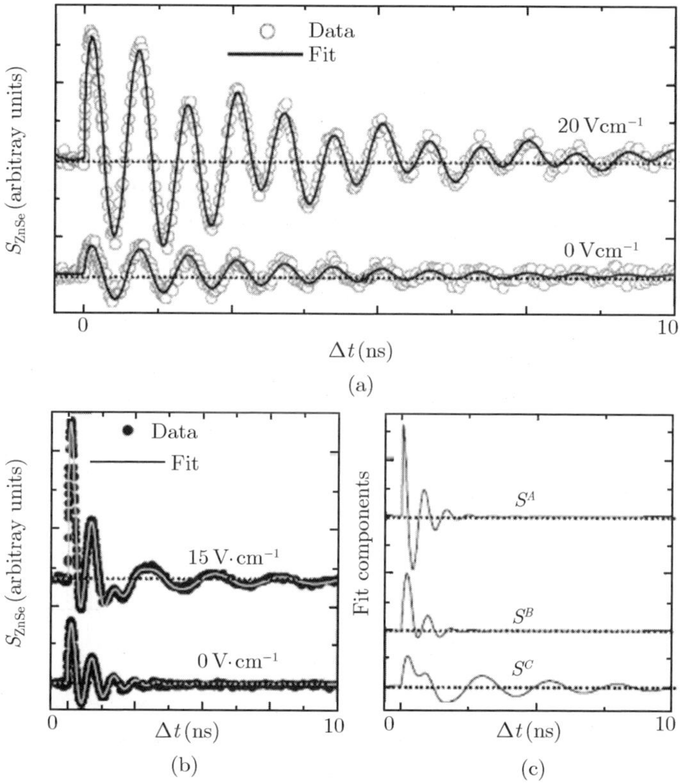

Fig. 4.30. TRKR scans versus Δt at $B = 75\,\mathrm{mT}$, $T = 5\,\mathrm{K}$, (a) 300 nm thick, $n = 1.5 \times 10^{18}\,\mathrm{cm}^{-3}$ ZnSe epilayer, (b) 100 nm thick, $n = 1.5 \times 10^{17}\,\mathrm{cm}^{-3}$ ZnSe epilayer, (c) decomposition of the fit to the biased scan in (b) into three different conduction channels.

corresponding to a different g-factor appears. The appearance of a second g-factor is revealed by the presence of the beating system. A Fourier transform of the biased time scans shows that the second g-factor is within 2% of the nominal -0.44 GaAs value. The increase in the spin lifetime in the biased case is due to the persistent current (Fig. 4.30(c)). The modest electric fields increase spin transfer nearly 500%. About three-fourths of this increase is due to the persistent current, the spin transfers by mechanisms A and B are also enhanced by the bias.

4.5.2. *Injection of spin-polarized electrons (experiment)*

As discussed before, the spin-polarized electrons is generally produced by the optical method, i.e. excitation of light hole excitons by the CP laser pulse (see Fig. 4.1). According to the transition selection rule, the electrons of spin up or spin down can be produced by different polarized light. But this method needs pulse laser, which makes it not convenient to fabricate spin electronic devices. People hope to find a method that the spin-polarized electrons are produced directly from the material itself, e.g. the DMSs and/or FMSs, and ferromagnets are candidates.

In 1999 the successful transfer of spin-polarized electrons from DMS to semiconductor was realized experimentally. Figure 4.31 shows the band structure and the side view of the spin-aligner LED (Fiederling *et al.*, 1999). In DMS BeMnZnSe, the spin degeneracy can be lifted in a magnetic field by the giant Zeeman splitting effect, the electron level of $s = -1/2$ is lower than that of $s = +1/2$, and the splitting energy is about 10 meV. In the bias voltage of the pn junction, most $s = -1/2$ electrons are injected from

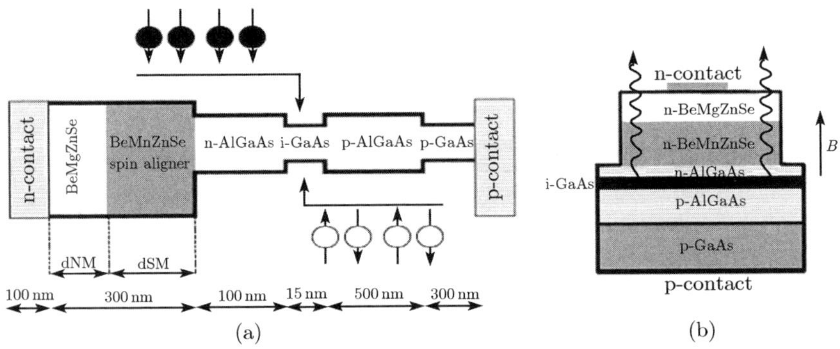

Fig. 4.31. (a) Schematic band structure of the spin-aligner LED. (b) Side view of the device showing the direction of the magnetic field and the emitting light.

the left side to the nonmagnetic active GaAs quantum well. On the other hand, unpolarized holes are injected from the right side. The electrons and holes are recombined to emit light in the well. Because the electrons are spin-polarized, according to the transition selection rule (see the inset of Fig. 4.1), only two transitions are possible: one heavy hole transition (from $-1/2$ to $-3/2$) and one light hole transition (from $-1/2$ to $1/2$). Whereas the heavy hole transition is σ^- CP, and the light hole transition is σ^+ polarized. The heavy hole transition matrix element is a factor of 3 larger than that of the light hole transition, and therefore, the emitted light will be CP, with a degree of polarization as

$$P = \frac{I^+ - I^-}{I^+ + I^-} = \frac{(3n\uparrow + n\downarrow) - (3n\downarrow + n\uparrow)}{(3n\uparrow + n\downarrow) + (3n\downarrow + n\uparrow)} = \frac{n\uparrow - n\downarrow}{2(n\uparrow + n\downarrow)}, \quad (4.73)$$

where $n\uparrow$ and $n\downarrow$ are the respective occupation numbers of the spin $s = +1/2$ and $-1/2$. If $n\uparrow \neq 0$, $n\downarrow = 0$, i.e. there is only one spin-polarized electron, then $P = 0.5$. Conversely, if $n\downarrow \neq 0, n\uparrow = 0$, then $P = -0.5$. The maximum value of $|P|$ is 0.5.

Figure 4.32 shows the degree of circular polarization P of the electroluminescence vs magnetic field (Fiederling *et al.*, 1999). A strongly polarized electroluminescence, with a maximum P of 43% for the structure with a thick dilute magnetic injection layer is detected. P increases with the

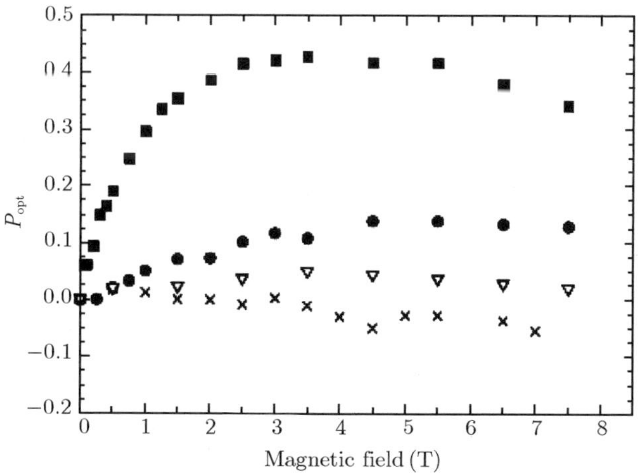

Fig. 4.32. Degree of circular polarization P of the electroluminescence versus magnetic field for n-BeMnZnSe spin aligners ($d_{SM} = 300$ nm, solid squares), ($d_{SM} = 3$ nm, solid circles), nonmagnetic layer ($d_{SM} = 0$, triangles), and GaAs layer (crosses).

magnetic field until it saturates for fields above 3 T. This behavior reflects the spin polarization of the injected carriers achieved in the BeMnZnSe layer, where the Zeeman splitting follows a typical Brillouin function and saturates when all carriers are aligned along the direction of the external magnetic field. Larger $d_{\rm SM}$ means that there are more spin-polarized electron injected to the active region, resulting in larger P.

For the σ^- polarized case, $0 > P > -0.5$. Let $x = n\downarrow/n\uparrow$, from Eq. (4.73), we get

$$x = \frac{0.5 - P}{0.5 + P}. \tag{4.74}$$

Inserting $P = -0.43$ into Eq. (4.74), we obtain $x = 13.3$ and $n\downarrow/(n\downarrow+n\uparrow) = 0.93$. In Fig. 4.32, $T = 3\,\text{K}$, $B = 1.5\,\text{T}$, $d_{\rm SM} = 300\,\text{nm}$, the electrons of spin down injected into the quantum well is about 93% of the total number of electrons. The P decreases with increasing temperature, it is because that the argument x of the Brillouin function decreases when the temperature increases, the splitting of electron energy levels decreases, and the occupation number of electrons at the up level increases.

The above experiment used the giant Zeeman splitting effect in DMS to produce the spin-polarized electrons, there needs strong magnetic field high as several tesla. Ohno *et al.* (1999) demonstrated the injection of spin-polarized holes from an FMS GaMnAs into a GaAs/(In,Ga)As LED. Figure 4.33 shows the (a) schematic structure of spin-polarized hole-injected LED and (b) electroluminescence intensity (EL) and polarization of the device versus photon energy at $T = 6\,\text{K}$, $B = 1000\,\text{Oe}$ (Ohno *et al.*, 1999). A peak in the EL intensity (FWHM $= 13\,\text{meV}$) and polarization is seen at the QW ground-state transition energy ($E = 1.34\,\text{eV}$). At $T = 6\,\text{K}$, $B = 1000\,\text{Oe}$, the hole magnetization approaches saturation. The spin-polarized holes are injected into the InGaAs quantum well, recombined with the unpolarized electrons to emit the light. The polarization P reaches $\sim$0.09 from Fig. 4.33(b).

Figure 4.34 shows the relative changes of polarization $\Delta P = P - P_{\rm bk}$ versus in-plane magnetic field at $T = 6$–$52\,\text{K}$, $d = 140\,\text{nm}$, $E = 1.34\,\text{eV}$, and $I = 2.8\,\text{mA}$ (Ohno *et al.*, 1999), where $P_{\rm bk}$ is the polarization of the background. When the temperature is lower than 52 K, ΔP shows a hysteretic loop. The coercive field increases from 0 to 40 Oe as the temperature decreases from 52 to 4 K. The inset shows relative remanent polarization (ΔP at $B = 0\,\text{Oe}$) (solid squares) and magnetic moment measured by SQUID (solid line) versus T. The two independently measured results are in good agreement, indicating that ΔP is originated from the magnetic

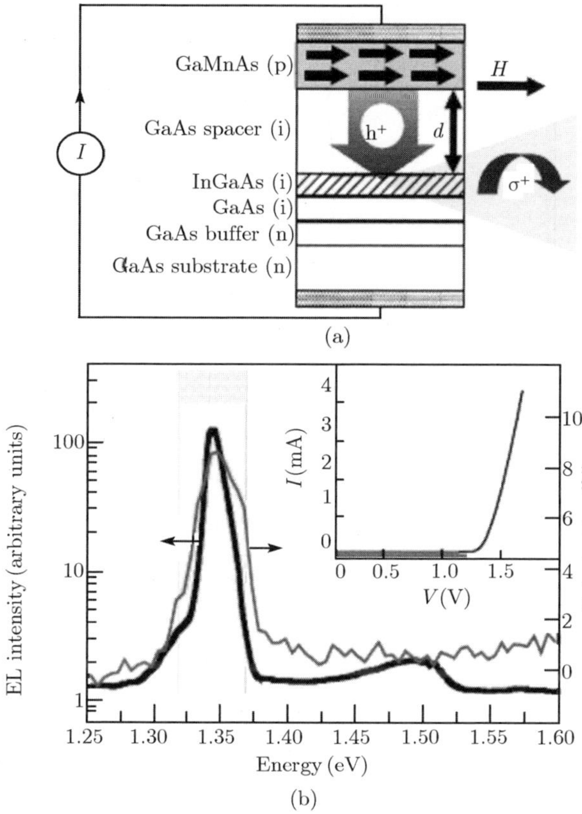

Fig. 4.33. (a) Schematic structure of spin-polarized hole-injected LED. (b) Electroluminescence intensity and polarization of the device versus photon energy at $T = 6\,\mathrm{K}$, $B = 1000\,\mathrm{Oe}$. The inset shows the current–voltage I–V characteristic.

moment of the FMS. No polarization hysteresis is observed for $T \geq 52\,\mathrm{K}$, when (Ga,Mn)As becomes paramagnetic. Therefore, in this case, the Curie temperature of (Ga,Mn)As is 52 K.

Though the generation of spin-polarized holes by the FMS has the advantage: there is no need for large magnetic fields, the circular polarization P of the emitting light is only several percent. The reason is that the SRT of hole is much smaller than that of electron, and the hole spin flips at a short distance.

Injection of spin-polarized electrons from a metal into semiconductor is demonstrated for a GaAs/(In,Ga)As LED covered with Fe (Zhu *et al.*, 2001). The circular polarization degree of the observed electroluminescence

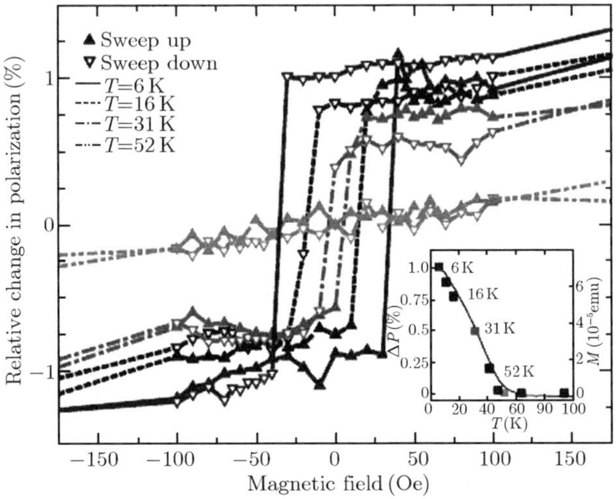

Fig. 4.34. Relative changes of polarization ΔP versus in-plane magnetic field at $T = 6{-}52\,\text{K}$. The inset shows relative remanent polarization (ΔP at $B = 0\,\text{Oe}$) (solid squares) and magnetic moment measured by SQUID (solid line) versus T.

reveals a spin injection efficiency of 2%. The reason is that the metal/semiconductor interface has a rather large destructive effect on the spin polarization, unlike the semiconductor/semiconductor interface. The LED structure is similar to those shown in Figs. 4.31 and 4.33, but the DMS or FMS is replaced by Fe, at the n-side of the p–n junction. The experiments were done in the Faraday geometry, i.e. with the magnetic field direction parallel to the light propagation. The EL signal was collected from the wafer back side. The intensity component $I^+(I^-)$ of right (left) CP light has been determined by integrating over the low-energy part of the spectrum which can be safely attributed to heavy-hole transition alone. In this case, instead of Eq. (4.73), the circular polarization degree is identical to the spin polarization of recombining electrons,

$$P = \frac{I^+ - I^-}{I^+ + I^-} = \frac{n\uparrow - n\downarrow}{n\uparrow + n\downarrow}. \tag{4.75}$$

Figure 4.35 shows the circular polarization degree P versus magnetic field for LEDs (Zhu *et al.*, 2001). For magnetic field $|B| < 4\text{T}$, the polarization degree P follows very closely the magnetization curve of thin Fe layers. This suggests that the spin-polarized electrons comes from Fe and

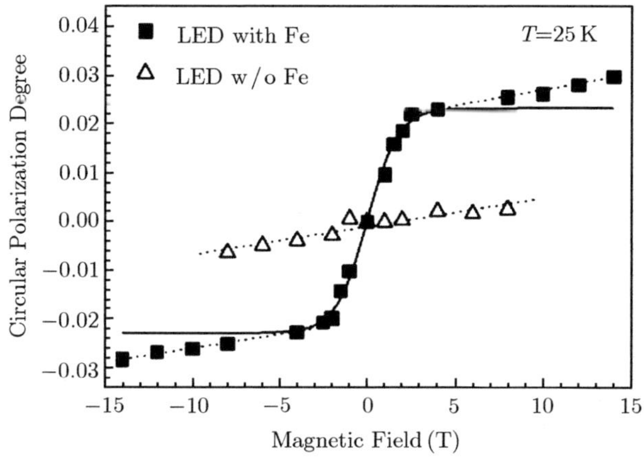

Fig. 4.35. Circular polarization degree P versus magnetic field for LEDs with (full squares) and without (open triangles) Fe cap layer at $T = 25\,\text{K}$. The solid line is the magnetization curve of a thin Fe layer in arbitrary units for comparison.

are injected into GaAs with an efficiency of about 2%. For $|B| > 4\,\text{T}$, the polarization degree P deviates from the magnetization curve and has no saturation. Because in the strong magnetic field, the electron level splits into spin-up and spin-down levels due to the Zeeman splitting effect. The occupation numbers of electrons at the two levels are different for different magnetic fields. They have different contributions to the spin-polarized electrons, making the circular polarization curve unsaturated. The LED without the Fe cap layer reveals a clearly smaller circular polarization degree P with a linear dependence on the magnetic fields as shown in Fig. 4.32 (open triangles). This magnetic field dependence can be explained simply by the Zeeman splitting-induced spin alignment in GaAs, without invoking any spin injection. The slope of the linear dependence is almost identical to that found for the LED with the Fe cap layer at $|B| > 4\,\text{T}$. The room temperature results provide further evidence for the injection of spin-polarized electrons from Fe into GaAs with an efficiency of 2%, which is enabled by the large Curie temperature of Fe.

The injection of spin-polarized electrons can also be obtained by use of the resonant tunneling effect (Gruber *et al.*, 2001). The energy band of the tunneling junction is schematically shown in Fig. 4.36; the potential well material is DMS ZnMnSe and the barrier material is nonmagnetic BeTe; underneath the resonant tunneling diode (RTD) (right side in Fig. 4.36),

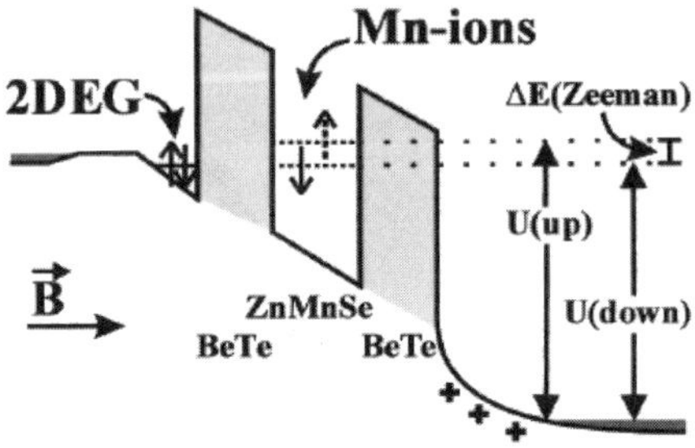

Fig. 4.36. Spin-dependent tunneling in a semimagnetic resonant tunneling diode (RTD).

there is an AlGaAs/GaAs p–i–n LED, which is used to measure the circular polarization degree of the emitting light. In the Faraday geometry (magnetic field parallel to the current direction), the circular polarization degree reflects directly the spin polarization of the injected electrons. It has been shown that the spin-dephasing times for electrons in GaAs are as long as 10 ns, and the spin orientation survives for a diffusion length of up to 100 μm. Hence, the depolarization during the diffusion into GaAs quantum well can be assumed to be small. The magnetic field causes a giant Zeeman splitting of the electron level in ZnMnSe well, the splitting energy is about 20 meV. When the applied bias voltage makes the spin-down level in the well coincided with the Fermi level of 2DEG at the input end, there will only be spin-down electrons tunneling out, which forms polarized current of spin down.

Figure 4.37 shows the electroluminescence polarization degree versus magnetic field at bias $V = 1.9$ V, $T = 1.6$ K (the solid circles) (Gruber *et al.*, 2001). The optical circular polarization degree increases with the magnetic field until it saturates at 80% for fields above 2–3 T, coinciding very well with the saturation of the giant Zeeman splitting in DMSs. This indicates that the injected electrons are almost completely spin-polarized, as expected from Fig. 4.36. The measurements with optical excitation of nonpolarized carriers in the GaAs quantum well in the same structure as well as with nonmagnetic BeTe–ZnSe tunneling structures reveal only a small polarization degree below 15%. This demonstrates that the giant Zeeman splitting in the semimagnetic ZnMnSe well and the subsequent large spin polarization of the injected electrons is responsible for the circular polarization of emission. When the bias voltage increases from 1.8 to 2.3 V, the circular polarization degree at a constant drive current of

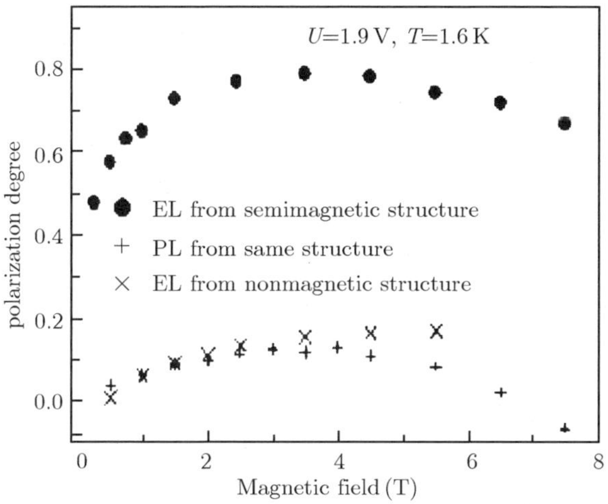

Fig. 4.37. Electroluminescence polarization degree versus magnetic field at $V = 1.9\,\text{V}$, $T = 1.6\,\text{K}$. Test measurements are on a nonmagnetic reference structure and PL on the same structure.

$50\,\mu\text{A}$ decreases from 80% to 38%. This can be explained by the onset of transport through the second spin-up subband.

4.5.3. *Injection of spin-polarized electrons (theory)*

Chang *et al.* (2002) studied theoretically the longitudinal spin transport in DMS superlattices taking into account the giant Zeeman splitting effect. Figure 4.38 shows the schematic CB lineup of a ZnSe/$Zn_{0.96}Mn_{0.04}Se$ (shaded region) DMS superlattice in a perpendicular magnetic field. (a) magnetic field $B = 0$, (b) $B \neq 0$, CB of spin up, (c) $B \neq 0$, CB of spin down (Chang *et al.*, 2002). When $B = 0$, the $Zn_{0.96}Mn_{0.04}Se$ layer is the potential well, the band offset is small. When $B \neq 0$, the giant Zeeman splitting induced by the s–d exchange interaction is larger than the band offset. Consequently, the spin-up band shifts upward, and exceeds the CB bottom of ZnSe, and the $Zn_{0.96}Mn_{0.04}Se$ layer becomes barrier as shown in Fig. 4.38(b). The spin-down band shifts downward, the $Zn_{0.96}Mn_{0.04}Se$ well becomes deeper as shown in Fig. 4.38(c). Thus, when electrons travel longitudinally through the superlattice, the electrons with spin up and spin down will meet different potential wells and barriers and have different transport properties.

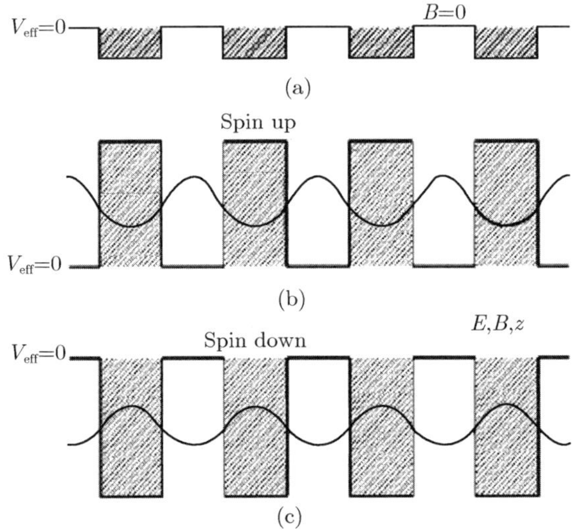

Fig. 4.38. Schematic CB lineup of a ZnSe/$Zn_{0.96}Mn_{0.04}Se$ (shaded region) DMS superlattice in a perpendicular magnetic field. (a) Magnetic field $B = 0$, (b) $B \neq 0$, CB of spin up, and (c) $B \neq 0$, CB of spin down.

In the mean-field approximation, the electron Hamiltonian can be written as

$$H = \frac{p_x^2}{2m^*} + \frac{(p_y + eBx)^2}{2m^*} + \frac{p_z^2}{2m^*} + V_{\text{conf}}(z) + J_{\text{s-d}}\langle S_z \rangle \sigma_z, \quad (4.76)$$

where $J_{\text{s-d}} = -N_0 \alpha x$, $\langle S_z \rangle$ is the average magnetic moment of magnetic ions (see Eq. (2.36)), and σ_z is the electron spin operator.

The electron wave function in the superlattice can be calculated from Eq. (4.76) as

$$\Psi = \frac{1}{\sqrt{L_y}} \phi_n(x - X_0) e^{ik_y y} \psi_{k_z}(z), \quad (4.77)$$

where $\phi_n(x)$ is the wave function of the harmonic oscillator, $X_0 = -k_y l_B^2$, $l_B = \sqrt{\hbar/eB}$ is the magnetic length, and $\psi_{kz}(z)$ is the envelope function along the z-axis. The eigenenergies are

$$E_{n,s}(k_z) = \left(n + \frac{1}{2}\right) \hbar\omega_c + \varepsilon_s(k_z), \quad (4.78)$$

where $\varepsilon_s(k_z)$ is the energy spectrum of the miniband in the superlattice (only consider the transport in the lowest miniband), which depends on the spin s, $(n + 1/2)\hbar\omega_c$ is the energy of the Landau level, $\omega_c = eB/m^*c$.

The group velocity of the electron along the z-direction is

$$v_s = \frac{1}{\hbar}\frac{\partial \varepsilon_s(k_z)}{\partial k_z}, \quad s = \uparrow, \downarrow . \tag{4.79}$$

The ballistic current density J_s is the sum of the contributions from each Landau level with different spins.

$$J_s = -ne\langle v_s \rangle = -\frac{eF}{2\pi l_B^2}\sum_n \int_{-\pi/L}^{\pi/L} \frac{\mathrm{d}k_z}{2\pi} v_s(n,k_z) f_s(n,k_z), \tag{4.80}$$

where $1/2\pi l_B^2$ is the degeneracy of the Landau level for each spin, $f_s(n,k_z)$ is the electron distribution function in the state (n,k_z,s), and F is the electric field strength along the z-direction.

In the relaxation time approximation,

$$f = f_0 + f_1 = f_0 - ev\tau F\frac{\partial f_0}{\partial E}, \tag{4.81}$$

where f_0 is the equilibrium distribution function.

$$f_{0,ns} = \frac{1}{\exp[(E_{n,s}(k_z) - E_\mathrm{F})/k_\mathrm{B}T] + 1}. \tag{4.82}$$

The Fermi energy E_F can be determined by the electron density n_e,

$$n_\mathrm{e} = \frac{e^2}{4\pi^2 l_B^2}\sum_{s,n}\int_{-\pi/L}^{\pi/L} f_0[E_{n,s}(k_z)]\mathrm{d}k_z. \tag{4.83}$$

Inserting Eq. (4.81) into Eq. (4.80), we obtain

$$J_s = \frac{e^2 F\tau}{4\pi^2 l_B^2}\sum_n \int_{-\pi/L}^{\pi/L} \mathrm{d}k_z \left(-\frac{\partial f_{0,ns}}{\partial E}\right) v_s^2. \tag{4.84}$$

The conductivity $\sigma = J/F$, and the spin polarization degree of the current density,

$$P = \frac{J_\downarrow - J_\uparrow}{J_\downarrow + J_\uparrow}. \tag{4.85}$$

The parameters in the calculation are: $m^* = 0.16m_0$, $x = 0.04$, $T_0 = 1.4\,\mathrm{K}$, electron density $n_e = 2 \times 10^{17}\,\mathrm{cm}^{-3}$, width of DMS layer $L_\mathrm{D} = 10\,\mathrm{nm}$, width of nonmagnetic layer $L_\mathrm{N} = 10\,\mathrm{nm}$, in the DMS layer the band offset $V_\mathrm{conf} = -3\,\mathrm{meV}$, $g_\mathrm{e} = 2$, $N_0\alpha = 0.27\,\mathrm{eV}$. Figure 4.39 shows the fan plot of Landau levels of the lowest subband in DMS superlattice versus magnetic filed for spin up (thin curves) and spin down (thick curves) (Chang *et al.*, 2002). When the magnetic field is smaller, parts of levels do not increase linearly with the magnetic field, quite different from that in

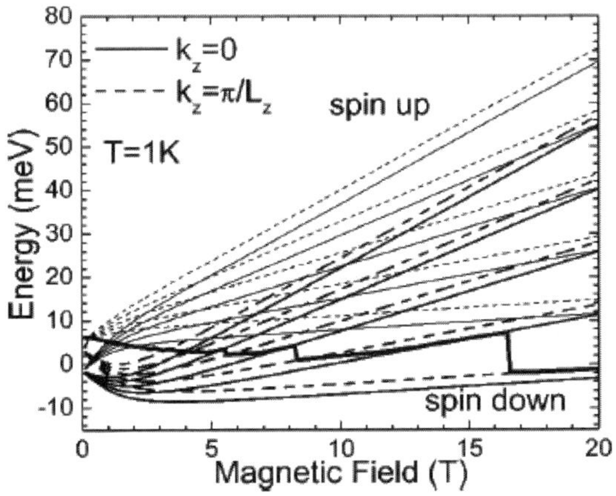

Fig. 4.39. Fan plot of Landau levels of lowest subband in DMS superlattice versus magnetic filed for spin up (thin curves) and spin down (thick curves). The thickest solid curve shows the Fermi energy versus magnetic field.

a nonmagnetic semiconductor. Due to the giant Zeeman splitting in the DMS, the Landau levels of the spin-down electron exhibit minima with increasing magnetic field, while those of the spin-up electron increase nonlinearly with increasing magnetic field. When the magnetic field is smaller, the increase in the giant Zeeman splitting energy exceeds the increase in the Landau level $\hbar\omega_c$; when $B \geq 5\,\mathrm{T}$, the splitting energy $\langle S_z \rangle$ approaches saturation, but the Landau level increases linearly as $\hbar\omega_c$. It is also noticed that below the Fermi level, most levels are levels of spin down.

Figure 4.40 shows the spin polarization of the current in a DMS superlattice vs magnetic field for two temperatures (Chang *et al.*, 2002). The inset shows the spin-up and spin-down conductivity components versus magnetic field at the same conditions same as Fig. 4.40. In the small magnetic field, the spin-up conductivity is larger than that of spin down. When the magnetic field increases, the spin-up conductivity decreases rapidly to zero and mainly the spin-down conductivity remains. The oscillation of the conductivity is associated with the noncontinuous variation of the Fermi level with the magnetic field (see Fig. 4.39), there are some correlations between them. The spin polarization of the current P (Fig. 4.40) approaches 1 when $B > 2\,\mathrm{T}$, which means that the carriers in the current are all spin down.

Xia *et al.* (2003) calculated the spin polarized currents through a DMS RTD. For comparison with experimental results (Gruber *et al.*, 2001), the

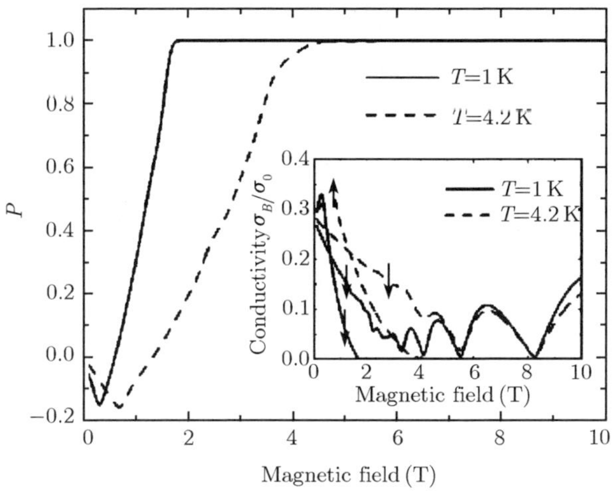

Fig. 4.40. Spin polarization of the current in a DMS superlattice versus magnetic field for two temperatures. The inset shows the spin-up and spin-down conductivity components versus magnetic field.

RTD is a BeTe/$ZnMn_xSe_{1-x}$/BeTe double potential barrier structure as shown in Fig. 4.36, in the applied magnetic field the CB, i.e. the confined states in the DMS well split into spin-up and spin-down levels.

The tunneling current is given by

$$J = \frac{em^*}{2\pi^2\hbar^3}\int \mathrm{d}E_z \int \mathrm{d}E_\parallel T(E_\parallel, E_z)[f(E, E_\mathrm{F}) - f(E, E_\mathrm{F} - eV_\mathrm{b})], \quad (4.86)$$

where E_z and $E_\parallel$ are energies of electron in the perpendicular and parallel directions to the junction plane, respectively, T is the tunneling probability, f is the electron distribution function, and V_b is the bias voltage to the junction. The terms in the square brackets of Eq. (4.86) represent the difference in the electron distribution functions at the input end and output end.

Assuming that the effective masses in the well and barrier regions are equal, the electron movements in the z-direction (tunneling direction) and the xy plane are independent, T depends only on E_z. Integrating Eq. (4.86) over $\mathrm{d}E_\parallel$, we obtain

$$J = \frac{em^* k_\mathrm{B}T}{2\pi^2\hbar^3}\int_0^\infty \mathrm{d}E_z T(E_Z)\ln\left\{\frac{1+\exp[(E_\mathrm{F}-E_z)/k_\mathrm{B}T]}{1+\exp[(E_\mathrm{F}-E_z-eV_\mathrm{b})/k_\mathrm{B}T]}\right\}. \quad (4.87)$$

At $T = 0\,\mathrm{K}$, for $eV_\mathrm{b} > E_\mathrm{F}$, Eq. (4.87) becomes

$$J = \frac{em^*}{2\pi^2\hbar^3}\int_0^{E_\mathrm{F}} \mathrm{d}E_z(E_F - E_z)T(E_z). \tag{4.88}$$

In the case of applied perpendicular magnetic field, the 2D density of states is replaced by density of states of Landau levels, and Eq. (4.88) is written as

$$J = \frac{e}{2\pi\hbar}\left(\frac{eB}{\pi\hbar c}\right)\int_0^{E_\mathrm{F}} dE_z N(E_\mathrm{F} - E_z), \tag{4.89}$$

where $(eB/\pi\hbar c)$ is the degeneracy of each Landau level and $N(E_\mathrm{F} - E_z)$ is the number of Landau levels in the range of $(E_F - E_z)$, which equals

$$N(E_F - E_z) = \left(\frac{E_F - E_z}{\hbar\omega_c} - \frac{1}{2}\right) \text{ maximum integer.} \tag{4.90}$$

The tunneling probability $T(E_z)$ can be calculated using the transfer matrix method (Xia, 1988).

The parameters are taken coincided with the experiment (Gruber *et al.*, 2001): $B = 2\,\mathrm{T}$, $T = 1.6\,\mathrm{K}$, $T_0 = 1.4\,\mathrm{K}$, $x = 0.06$, the splitting energy of the CB $\Delta E_s = 17.68\,\mathrm{meV}$, $m^* = 0.16m_0$, well width 3.5 nm, and barrier width 3 nm. The barrier height 1890 meV, and the energies of the confined states in well are 132, 521, and 1135 meV. The injected electron density is $10^{19}\,\mathrm{cm}^{-3}$, and the Fermi energy $E_\mathrm{F} = 105.8\,\mathrm{meV}$ is calculated by use of $E_\mathrm{F} = (\hbar^2/2m^*)(3\pi^2 n)^{2/3}$.

Figure 4.41 shows the tunneling current with spin-down (triangle down) and spin-up (triangle up) versus applied electric field at $B = 2.5\,\mathrm{T}$ (Xia *et al.*, 2003). At different electric fields, the ratio of the spin-up and spin-down current is different leading to a nonzero polarization. Especially, when the electric field is such that the Fermi level is between two quasi-confined states of spin up and spin down, there is only current of spin down, and the polarization degree reaches 1. As the electric field increases, the current of other spin, i.e. spin up increases, and the polarization degree decreases gradually. These results are basically consistent with the experimental results Fig. 4.37 (Gruber *et al.*, 2001). When the electric field continuously increases such that the bottom of the CB at the input end is close to the quasi-confined states of spin up, there will be only current of spin up, and the polarization degree reaches -1.

The results of tunneling through the second quasi-confined state close to 521 meV are shown in Fig. 4.41(b). In this case, the applied electric field

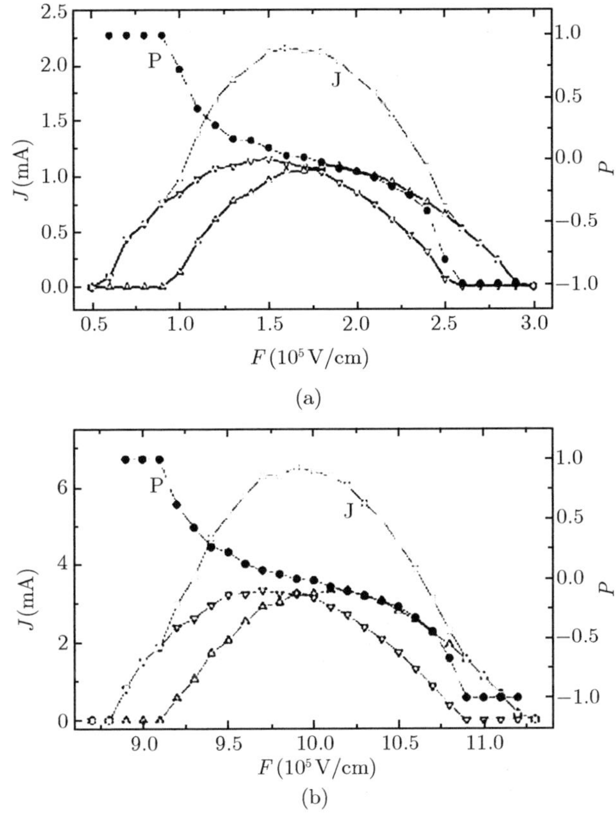

Fig. 4.41. Tunneling current with spin down (triangle down) and spin up (triangle up) versus applied electric field at $B = 2.5\,\mathrm{T}$ when the Fermi level crosses (a) the first and (b) the second quasi- confined state. The total current and the spin polarization degree are indicated by the squares and the dots, respectively.

is larger than that of the first one and the current is about three times larger. However, the behavior of the spin-polarized currents as well as the polarization degree is very similar to Fig. 4.41(a). Therefore, using the tunneling through the second quasi-confined state, we can obtain larger polarized current.

Li *et al.* (2003) studied the spin-dependent transport through CdMnTe DMS QDs in both the electric field and the magnetic field using the recursion method. Assume that there is a DMS $Cd_{1-x}Mn_xTe$ quantum sphere embedded in the nonmagnetic semiconductor $Cd_{1-y}Mg_yTe$, an external magnetic field and an electric field are applied to the system along the

z-direction. The electron effective-mass Hamiltonian can be written as

$$H = \frac{(\boldsymbol{p} + e\boldsymbol{A})^2}{2m^*} \pm g_{\mathrm{e}}\mu_{\mathrm{B}}B - eFz + V_{\mathrm{exch}} + V(\boldsymbol{r}). \tag{4.91}$$

The vector potential of the magnetic field is taken as

$$\boldsymbol{A} = \frac{1}{2}\boldsymbol{B} \times \boldsymbol{r} = \frac{B}{2}(-y, x, 0). \tag{4.92}$$

$V(\boldsymbol{r})$ is the confined potential of the QD, which is zero in the QD, E_{c} otherwise. V_{exch} is the s–d exchange potential between the electron and the magnetic ions.

$$V_{\mathrm{exch}} = J_{\mathrm{s-d}}\langle S_z \rangle \sigma_z. \tag{4.93}$$

In the xy plane, we construct a 2D superlattice structure. Each unit cell of the superlattice includes a QD, the unit cell is taken to be large enough so that the coupling between dots can be neglected. According to the Bloch theorem, electron states can be expanded in terms of plane waves with discrete reciprocal-lattice vectors $(k_x + nK_x, k_y + mK_y)$, where k_x and k_y are the wave vectors of the electron in the x and y directions, respectively. The wave function can be written as

$$\Psi(\boldsymbol{r}) = \frac{1}{\sqrt{L_x L_y}} \sum_{mn} C_{mn}(z) \exp\{i[(k_x + mK_x)x + (k_y + nK_y)y]\}, \tag{4.94}$$

where $K_x = 2\pi/L_x, K_y = 2\pi/L_y$, and L_x and L_y are the periods of the superlattice in the x and y-directions, respectively. Calculating the Hamiltonian matrix elements between states mn and $m'n'$ and employing the diagonalization process, we obtain the effective potential at each z point $V_{\mathrm{eff}}^i(z)$, where i is the index of the transverse quantum energy level. Then the Schrödinger equation becomes 1D,

$$\left[\frac{p_z^2}{2m^*} + V_{\mathrm{eff}}^i(z)\right]\psi(z) = E_z\psi(z), \tag{4.95}$$

where $E_z = \sqrt{E^2 - (E_{\parallel}^i)^2}$, E is the electron total energy and $E_{\parallel}^i$ is the energy of the transverse ith sublevel ($k_x = k_y = 0$).

Equation (4.95) is not a definite state equation, the energy E_z and the wave function at the input end are known, and the transmitted wave function at the output end and the reflection wave function at the input end are to be calculated. Using the Numerov method, Eq. (4.95) can be transformed into the three-term differential equation with the mesh size h_z,

$$a(z_p)\Psi(z_{p+1}) - b(z_p)\Psi(z_p) + c(z_p)\Psi(z_{p-1}) = 0, \tag{4.96}$$

where

$$a(z_p) = 1 - \frac{1}{12}\frac{2m^*h^2}{\hbar^2}[V^i_{\text{eff}}(z_{p+1}) - E], \tag{4.97}$$

$$b(z_p) = 2 + \frac{5}{6}\frac{2m^*h^2}{\hbar^2}[V^i_{\text{eff}}(z_p) - E], \tag{4.98}$$

$$c(z_p) = a(z_{p-2}). \tag{4.99}$$

Equation (4.96) can be changed into the form,

$$S(z_{p-1}) = \frac{c(z_p)}{b(z_p) - a(z_p)S(z_p)}, \tag{4.100}$$

where

$$S(z_p) = \frac{\Psi(z_{p+1})}{\Psi(z_p)}. \tag{4.101}$$

Let the wave function at the entrance be $\Psi(z_0)$, and the wave function at the exit equals

$$\Psi(z_N) = \prod_{p=0}^{N-1} S(z_p)\Psi(z_0). \tag{4.102}$$

The spin-dependent transmission coefficients of the system can be calculated by

$$T = \sum_i t^i = \sum_i \left[\frac{\Psi^i(z_N)}{\Psi^i(z_0)}\right]. \tag{4.103}$$

For the $Cd_{1-x}Mn_xTe/Cd_{1-y}Mg_yTe$ QD system, the CB offset $E_c = 0.8(E_g^2 - E_g^1)$, where $E_g^1 = (1.586 + 1.51x)\,\text{eV}$ and $E_g^2 = (1.586 + 1.705y)\,\text{eV}$ are the energy gaps of $Cd_{1-x}Mn_xTe$ and $Cd_{1-y}Mg_yTe$, respectively. If $x = 0.07$ and $y = 0.06$, then E_c is very small, so that can be neglected. The band offset of two materials are determined mainly by the giant Zeeman splitting. For spin-up electrons, the QD is the barrier; for spin-down electrons, the QD is the well. Thus, the spin-up and spin-down electrons have different transport properties.

Taking $x_{\text{eff}} = 0.045$, $g_{\text{Mn}} = 2, N_0\alpha = 0.22\,\text{eV}$, $S_0 = 1.32$, $T_0 = 3.1\,\text{K}$, $g_e^* = -0.7, L_x = L_y = 20\,\text{nm}$, $B = 10\,\text{T}\ F = 0$, the calculated spin-dependent transmission coefficients versus electron Fermi energy for different QD radii

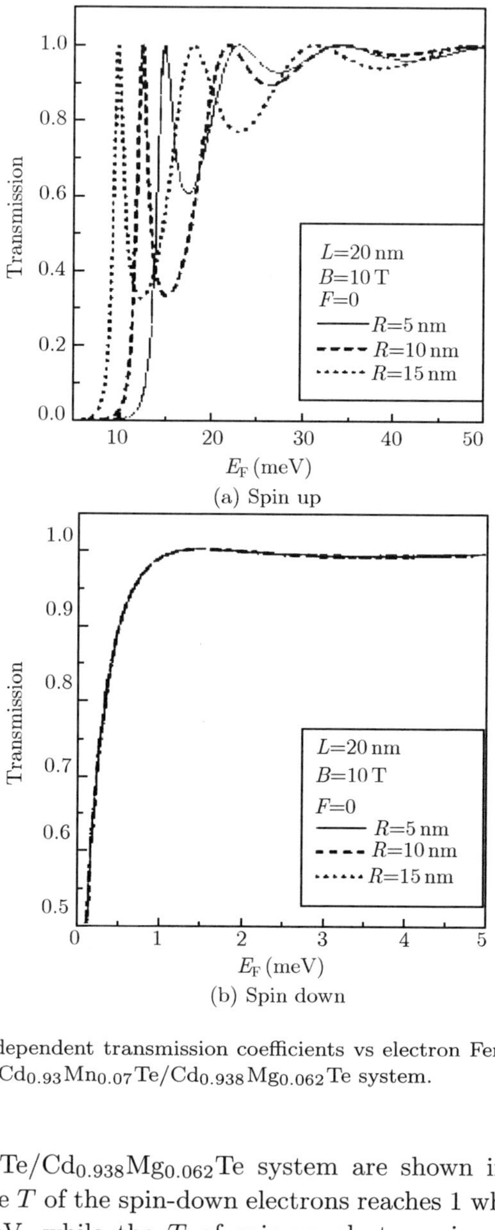

Fig. 4.42. Spin-dependent transmission coefficients vs electron Fermi energy for different QD radii in $Cd_{0.93}Mn_{0.07}Te/Cd_{0.938}Mg_{0.062}Te$ system.

in $Cd_{0.93}Mn_{0.07}Te/Cd_{0.938}Mg_{0.062}Te$ system are shown in Fig. 4.42 (Li *et al.*, 2003). The T of the spin-down electrons reaches 1 when the electron energy $E = 1$ meV, while the T of spin-up electrons increases gradually at $E \geq 10$ meV and oscillates with the electron energy. The transmission peaks of spin-up electrons move to lower Fermi energy with increasing QD radius. For the spin-up electrons, the QD is the barrier, the resonance energy levels along the z-direction will decrease for a larger QD. For the spin-down electrons, the transmission is that of the quantum well, and the transmission is affected slightly by the QD radius. Using this property, the spin-polarized current can be obtained by choosing the electron energy in the range of 1–10 meV.

4.6. Magnetic Semiconductor Tunneling Junction

The discovery of the GMR effect (Baibich *et al.*, 1988; Barnas *et al.*, 1990) open a rapidly developing field — magnetic electronics. GMR effect is the spin-related transport across an FM/M/FM sandwich structure arising from spin-relevant interlayer scattering and coupling. The resistance approaches the minimum (maximum) when the magnetic moments of two FM layers are parallel (antiparallel) for the electric current perpendicular or parallel to the interface. Replacing the nonmagnetic metal layer with an insulating metal oxide, one can obtain a MTJ (Moodera *et al.*, 1995). GMR and MTJ have very wide applications, e.g. spin valve based on GMR (Dieny *et al.*, 1991a,b; Tsang *et al.*, 1994) used in the head of the hard disk, nonvolatile MRAM utilizing MTJ (Gallagher and Parkin, 2006). The state-of-the-art growth and fabrication techniques make it possible to fabricate these devices in FMSs, for instance, an FMS/NMS/FMS tunneling junction. These semiconductor devices are more easily to be integrated with semiconductor integrated circuit to realize a new type of spintronic devices with both magnetic and electric functionalities.

4.6.1. *Property of GaAs/GaMnAs heterostructures*

Figure 4.43 shows the XRD spectrum of GaMnAs/GaAs superlattice with 20 periods (Shen *et al.*, 1997); the structure parameters are: the thicknesses of GaMnAs and GaAs layers are 11.4 and 12.1 nm, respectively. The concentration of Mn is 0.054, and the fitting parameters are given in the figure. In this figure, one can see that the numerical results agree very well with the experiment, indicating that the thin film sample and the interfaces have a high quality.

The experimental measurement of the transport property of GaMnAs multilayer shows that the GaMnAs layer is still ferromagnetic when the thickness approaches to 5 nm, but becomes paramagnetic when the

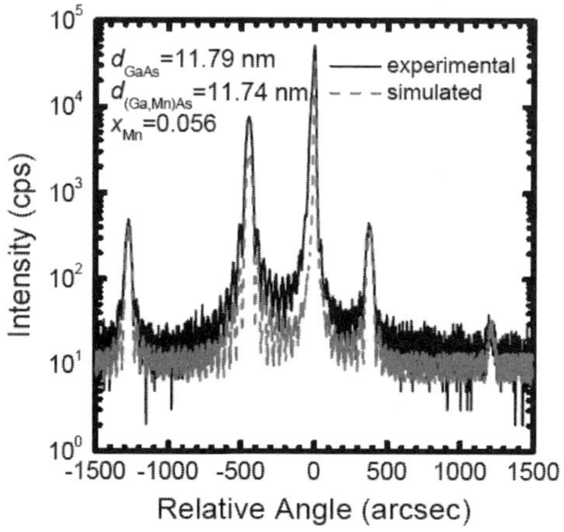

Fig. 4.43. XRD spectrum of a GaMnAs/GaAs superlattice (Shen *et al.*, 1997).

thickness decreases further. A recent experiment reports that a superlattice sample consisting of the m single layer of GaMnAs and n single layer GaAs still displays the ferromagnetism even when $8 \leq m \leq 12, 4 \leq n \leq 8$ (Kepa *et al.*, 2001). Figure 4.44 describes the relation between the thickness of GaMnAs sample and magnetic moment (Matsukura *et al.*, 1998, 1999) (notice that $R_{\text{Hall}}/R_{\text{sheet}} \propto M$); the Curie temperature as a function of the sample thickness is plotted in the inset.

The band offset between GaMnAs and GaAs can be extracted from the temperature dependence of the $I-V$ character and the Fermi energy in the GaMnAs layer (see Fig. 4.45). Figure 4.45 shows the dependence of the Mn concentration of the barrier height Δ and the effective Richard constant (A/A^*) (Ohno *et al.*, 2002), indicating that the band offset of holes across the GaMnAs/GaAs interface is 100 meV, which is almost independent of the concentration x.

4.6.2. *FM/NM/FM trilayer*

FMS/nonmagnetic semiconductor/FMS (FS/NS/FS) sandwiched structure is an important building block of magnetic sensors and magnetic storage devices. It can be used to detect various spin transport processes, especially for spin-related scattering processes, which lead to the GMR. These spin-relevant scattering processes play an important role in tunneling junctions consisting of magnetic metal and insulators and result in the TMR. Similar TMR behaviors were also reported in various

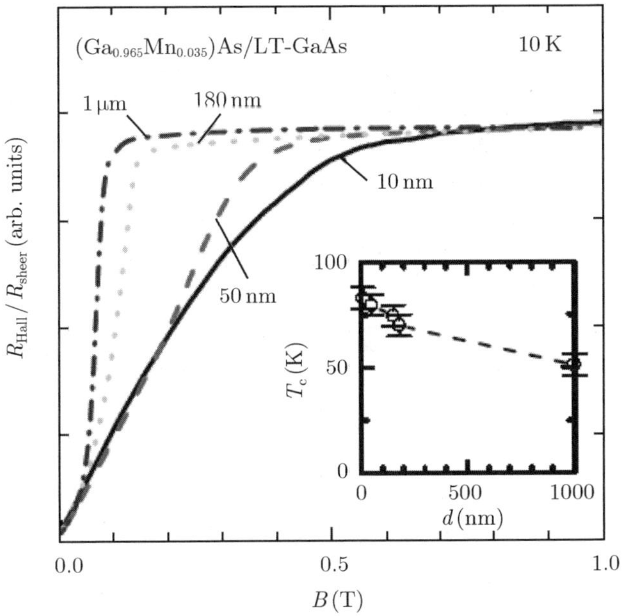

Fig. 4.44. Magnetic moment in GaMnAs versus the sample thickness (Matsukura *et al.*, 1998, 1999).

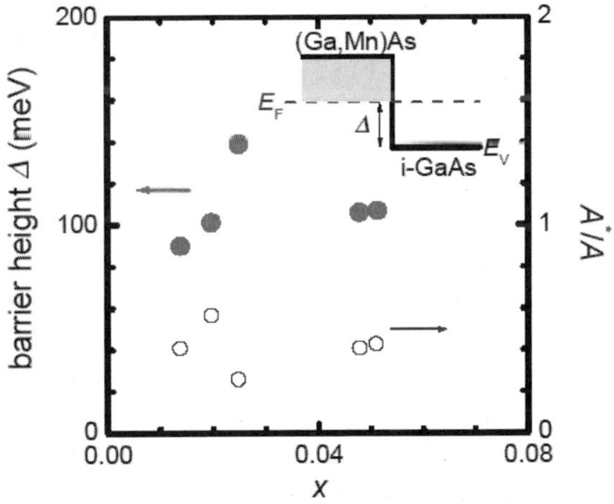

Fig. 4.45. The barrier height versus effective Richard constant (A/A^*) versus the concentration x of Mn (Ohno *et al.*, 2002).

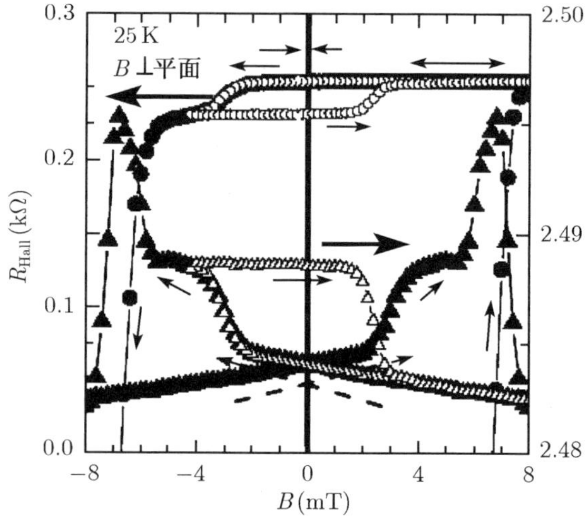

Fig. 4.46. Hall resistance and MR of GaMnAs/AlGaAs/GaMnAs trilayer structure against magnetic fields (Akiba *et al.*, 2000).

GaMnAs/AlGaAs/GaMnAs trilayer tunneling junctions (Akiba *et al.*, 1998, 2000; Chiba *et al.*, 2000; Higo *et al.*, 2001; Tanaka and Higo, 2001).

Figure 4.46 shows the magnetic field dependence of the Hall resistance and MR of 30 nm GaMnAs/2.8 nm AlGaAs/30 nm GaMnAs trilayer structure at temperature $T = 30\,$K (Akiba *et al.*, 2000). The samples are grown by low temperature MBE onto 1 μm $In_{0.15}Ga_{0.85}As$ buffer layer grown at 500°C on semi-insulating GaAs (100) substrates. The solid and open scatters in the figure denote the major and minor loops. The in-plane MR ratio $(R_{\rm sheet} - R_0)/R_0$ in the FS/NS/FS trilayer structure increases clearly, where R_0 is the MR in the absence of external magnetic fields. This clear increase in the MR indicates the presence of spin-dependent scattering process in the trilayer structure. The MR ratio decreases with increasing temperature T and/or the Al composition y. The T dependence is believed to reflect the dependence of the MR ratio on M and thus on the spin polarization of the holes in the (GaMn)As layer, and the y dependence due in the decrease in the number of holes that travel across the (AlGa)As layer. The magnitude of the coupling J can be determined from the magnetic field shift $B_s(T)$ of the minor loop with respect to B =0, i.e. $J = M_s(T)B_s(T)d$, where $M_s(T)$ is the spontaneous magnetization and $d$the thickness of the magnetic layer. The magnitude of the coupling strength $J\,(<1\,\mu{\rm J/m^2})$ extracted from the experiments indicates that the exchange coupling is always ferromagnetic and orders of magnitude smaller than those reported in metallic systems and becomes smaller with increasing the temperature of T, i.e. reduction in M, and decreases with an increase in the Al composition y i.e. increase in hole barrier, which suggests that the interlayer coupling is mediated by itinerant and/or tunneling holes.

By decreasing the thickness and the Al composition in the barrier, the interlayer coupling can be enhanced. Figure 4.47 shows the results of vertical magnetotransport measurements (Akiba *et al.*, 2000). In this figure the magntization M and MR are plotted as a function of magnetic fields B at 20 K. The temperature T dependence of the MR ratio is shown in the inset. The different coercive forces produce the plateau structure in Fig. 4.47, where M of the two ferromagnetic layers are antiparallel. The resistance increase is observed between 8 and 16 mT, precisely in the field region of antiparallel configuration of M. This semiconducting (Ga,Mn)As/(Al,Ga)As/(Ga,Mn)As trilayer structures having the easy axis of magnetization perpendicular to the plane. The AlAs barrier is very high (550 meV), indicating that the measured MR ratio is actually the TMR

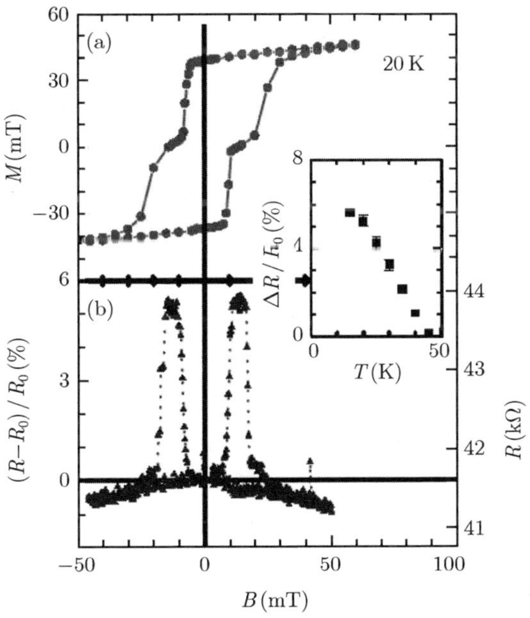

Fig. 4.47. (a) Magnetization and (b) TMR of a $Ga_{0.95}Mn_{0.05}As$/AlAs(3 nm)/ $Ga_{0.97}Mn_{0.03}As$ tunnel junction at 20 K. The inset shows the temperature dependence of TMR ratio (Chiba *et al.*, 2000).

since the holes have to tunnel across the barrier. The measured TMR ratio is about 5.5% at 20 K. With increasing the temperature, the magnetization of the GaMnAs layer decreases, leading to a rapid decrease in the TMR. When the thickness of the AlAs barrier layer decreases to 1.6 nm, one can obtain a very high TMR which approaches to 70% (Tanaka and Higo, 2001) (see Fig. 4.48). This high TMR indicates that the spin polarization of holes in GaMnAs is very high.

Lyu and Moon (2001) studied theoretically the spin-polarized tunneling process in the GaMnAs/AlAs/GaMnAs trilayer structure, showing that the TMR ratio has a strong dependence on the hole density and magnetic ion concentration.

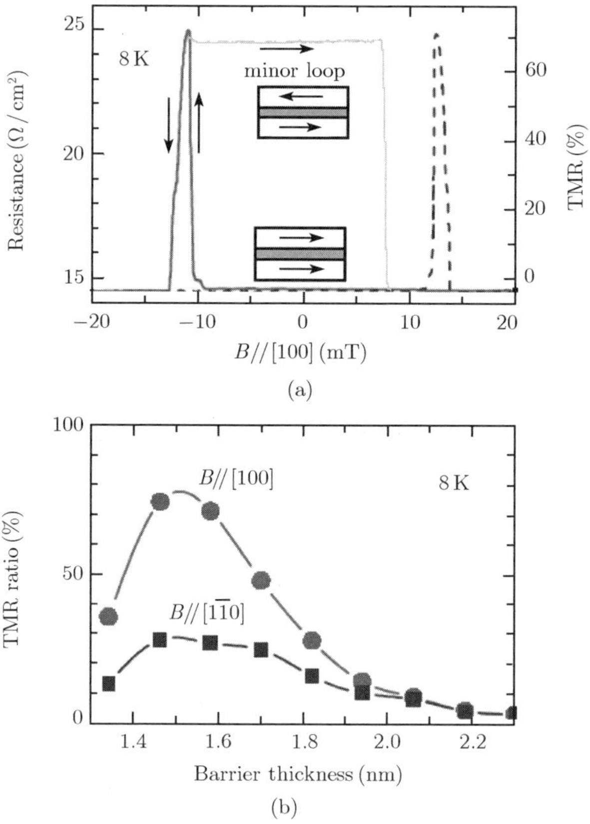

Fig. 4.48. (a) Magnetoresistance and (b) TMR ratio for vertical transport through 1nm AlAs barrier layer (Tanaka and Higo, 2001).

4.6.3. *Hybridized structure of ferromagnetic metals and semiconductors*

There are four essential issues in semiconductor-based spintronic devices: (1) how to realize high efficient spin injection into semiconductors; (2) how to control spin coherent transport in semiconductors and enhance SRT; (3) how to manipulate electron spin coherently and construct spintronic devices with novel functionality; (4) detecting spin state and spin polarization efficiently. So far spin injection can be realized via FM/I/SC and DMS/SC tunneling junctions (Fiederling *et al.*, 1999; Ohno *et al.*, 1999). For the DMS/SC system, the magnetic semiconductors are FMS GaMnAs or paramagnetic semiconductor ZnMnSe. FMS GaMnAs is p-type and the Curie temperature can approach 200 K. ZnMnSe is an n-type paramagnetic DMS, so that one have to obtain the high spin injection efficiency (>80%) at low temperatures (<10 K) and high magnetic fields (>1 Te). The FM/SC systems have the following advantages compared with DMS/SC systems: high Curie temperature; electron spin injection; low coercive forces, rapid switching, and well-developed technique.

Unfortunately, the direct spin injection from ferromagnetic metals to semiconductors is quite low, only 0.1–1% (Hammar *et al.*, 1999). Some theory groups extended the theory developed by van Son *et al.* (1987) and Valet and Fert (1993) to study spin-polarized transport from ferromagnetic metal to semiconductors (Schmidt *et al.*, 2000; Fert and Jaffres, 2001; Smith and Silver, 2001).

Schmidt *et al.* (2000) calculated the spin polarizations of injected electrons from FMS for different three SC/FM configurations as a function of the polarization of FM as shown in Fig. 4.49. From this figure, one can see that there are two criteria for achieving high injection efficiency in the diffusion regime: (1) the matched conductivities of ferromagnetic metals and semiconductors; (2) fully spin-polarized ferromagnetic metals. If these two requirements cannot be met, the spin injection into semiconductors should be very low (<1%). There are no existing ferromagnetic metals which can meet these two criteria except CrO_2, and so far there is no report about the growth of CrO_2 above the semiconductor. Interestingly, ZnMnSe can satisfy these requirements simultaneously.

Rashba (2000) recently proposed a scheme to solve this problem by inserting a tunneling barrier between the ferromagnetic metal and semiconductor layers, forming spin-dependent chemical potentials, to solve the conductivity mismatching between the ferromagnetic metal and semiconductor layers. Al_2O_3 is a conventional insulator and widely used as the

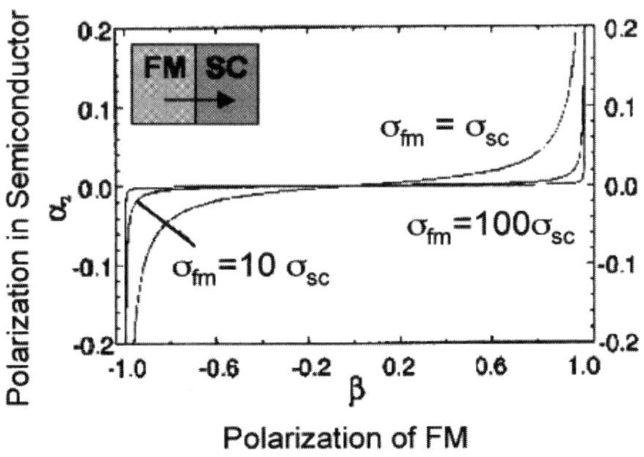

Fig. 4.49. Spin polarizations of injected electrons from FMS for different three SC/FM configurations against the polarization of FM (Schmidt *et al.*, 2000).

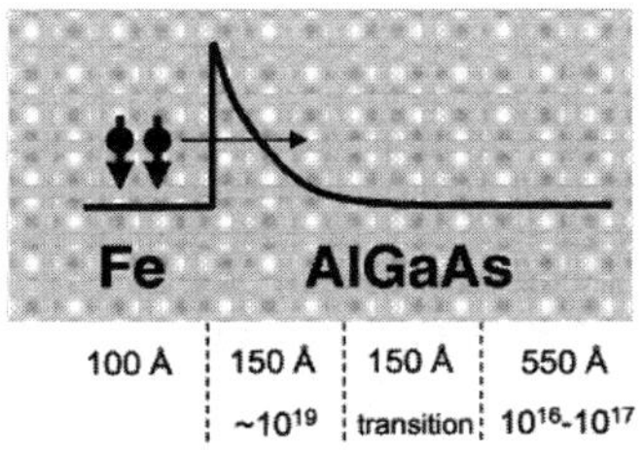

Fig. 4.50. Schematic of FM/SC contact, and the band bending of semiconductors forms the Schotky barrier (Hanbicki *et al.*, 2002).

tunneling barrier in MTJs, for instance, CoFe–Al_2O_3–NiFe (Moodera *et al.*, 1995). But for FM/SC contacts, the Schtoky barrier caused by band bending forms a barrier, as shown in Fig. 4.50. In an n-type semiconductor, electrons are transferred in the metal layer, and electrons near the FM/SC interface are depleted, resulting in the band bending up and forming a triangular-shaped barrier. Usually, the doping concentration in conventional semiconductors is about $n \sim 10^{16}$–$10^{18}\,\text{cm}^{-3}$, the thickness of depletion layer approaches several hundred angstroms. The thickness can be adjusted by tuning the doping concentration, for example, heavily doped depletion layer can be reduced to several nanometers, which make electrons in ferromagnetic metals tunnel into semiconductors more easily.

Hanbicki *et al.* (2002) fabricated Fe/AlGaAs/GaAs semiconductor LED structure and realized spin injection from Fe to AlGaAs; the spin injection efficiency can approach 30%. The LED structure is used to measure the

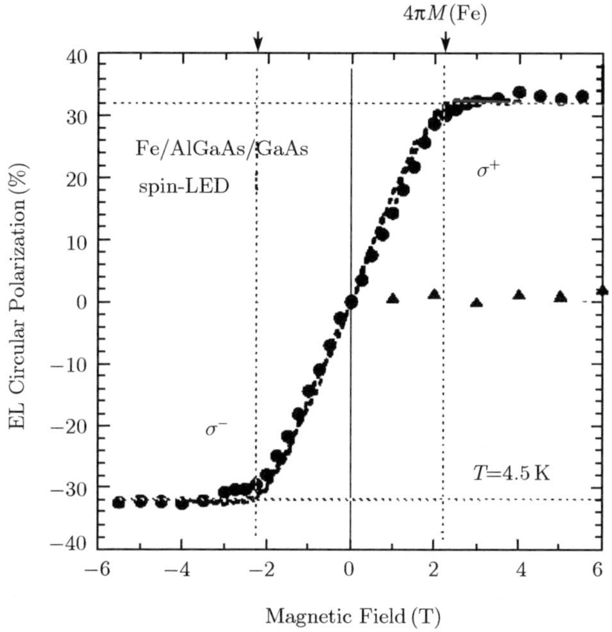

Fig. 4.51. The circular polarization of the PL spectrum in the LED structure versus magnetic fields (Hanbicki *et al.*, 2002).

spin injection efficiency. The circular polarization $P_{\text{cir}} = (I(\sigma^+) - I(\sigma^-))/(I(\sigma^+)+I(\sigma^-))$, here $I(\sigma^+)$ and $I(\sigma^-)$ are the intensities of photoluminence peaks. Figure 4.51 depicts the circular polarization as a function of the magnetic field. The circular polarization P_{cir} increases rapidly as magnetic fields increase, until the magnetic moments of Fe atoms are fully aligned. At high magnetic fields, the circular polarization P_{cir} increases continuously at a slow rate (1%/T) and saturates at 32%. This feature indicates that the spin-polarization injection approaches 32% in this injection process from Fe to the AlGaAs layer.

Comparing with DMSs, the ferromagnetic metals show high Curie temperatures, which are crucial for practical applications. To estimate the temperature effect on the spin injection efficiency, Fig. 4.52 displays the temperature dependence of circular polarization. At 4.5 K, the circular polarization is 13% and decreases with increasing temperature. At RT 240 K, the circular polarization approaches 4%.

Above we focus on the spin injection into the III–VI group semiconductor GaAs material. Now several groups study the spin injection from

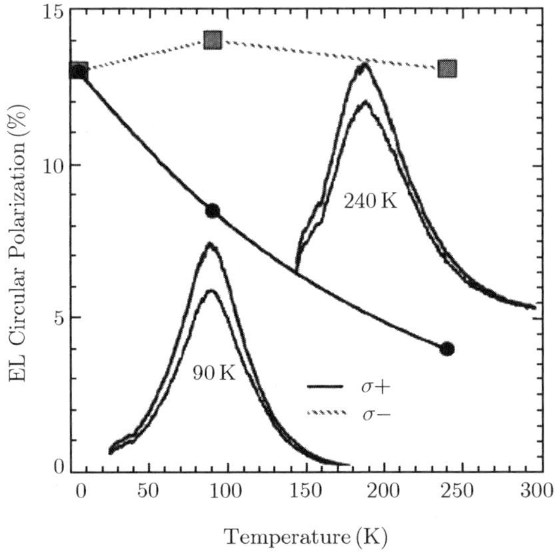

Fig. 4.52. The temperature dependence of the circular polarization of the PL spectrum in the LED structure (Hanbicki *et al.*, 2002).

ferromagnetic metal into IV group semiconductor Si, aiming at the integration with the current microelectronics technique (Jonker *et al.*, 2007). Comparing with the GaAs material, Si displays the following advantages: (i) weak SOI in Si coming from a small mass of a Si atom and the spatial inversion symmetry in Si, leading to long SRT; (ii) isotope of Si^{28} shows a vanishing nuclear spin, and consequently enhances the SRT further; (iii) the well-developed technique in the Si-based semiconductor integrated circuit makes it possible to become a very good candidate for realizing practical spintronic devices. Jonker *et al.* (2007) demonstrated experimentally the spin injection from thin Fe layer into a n-type Si through a Al_2O_3 tunneling barrier. Figures 4.53(a) and 4.53(b) show two sample structures, and Figs. 4.53(c) and 4.53(d) are the schematics of the band structure and the LED device.

Figure 4.54 depicts the circular polarization of the PL spectrum in AlGaAs/GaAs quantum well LED structure for spin injection process from Fe into Si against the magnetic fields. The inset shows the electro PL spectrum at 20 K and 3 T. The PL peak at 1.54 eV corresponds to the free exaction in GaAs quantum well. From this figure, one can see that the circular polarization saturation is 5.6% at 20 K and 2.8% at 125 K. These values are significantly smaller than that of the spin injection from Fe into GaAs quantum wells directly (Hanbicki *et al.*, 2002). This reduction

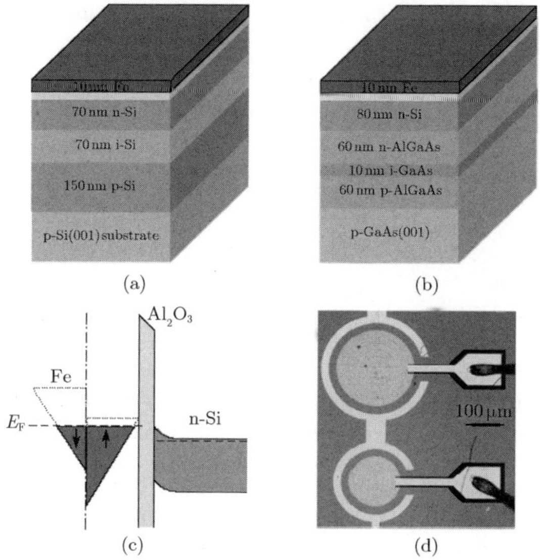

Fig. 4.53. (a) and (b) The schematics of two Fe/Al_2O_3/Si sample structures. (c) and (d) The schematic of the band structure and the LED devices (Jonker *et al.*, 2007).

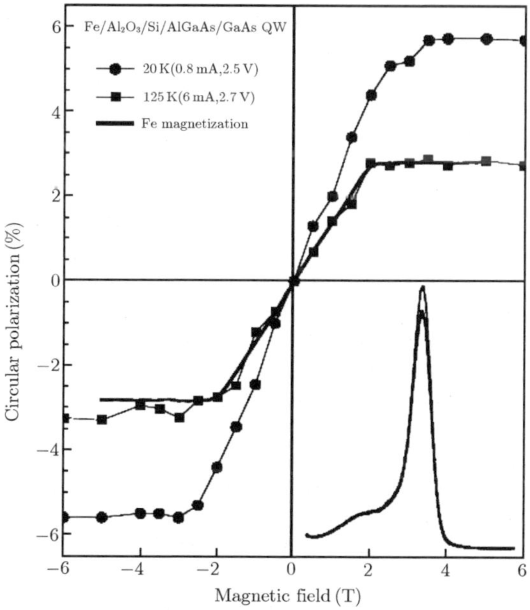

Fig. 4.54. The circular polarization of the PL spectrum in the AlGaAs/GaAs quantum well LED structure versus magnetic fields; the inset is the electro PL spectrum at 20 K when $B = 3\,\mathrm{T}$ (Jonker *et al.*, 2007).

arises from the recombination process in indirect bandgap semiconductor Si involving acoustic phonons.

References

Akiba N, Matsukura F, Shen A, Ohno Y, Ohno H, Oiwa A, Katsumoto S, Iye Y. (1998) *Appl Phys Lett* **73**: 2122.

Akiba N, Chiba D, Nakata K, Matsukura F, Ohno Y, Ohno H. (2000) *J Appl Phys* **87**: 6436.

Baibich MN, Broto JM, Fert A, *et al.* (1988) *Phys Rev Lett* **61**: 2472.

Baibich MN, Broto JM, Fert A, Van Dau FN, Petroff F, Eitenne P, Creuzet G, Friederich A, Chazelas J. (1988) *Phys Rev Lett* **61**: 2472.

de Andrada EA, La Rocca GC, Bassani F. (1994) *Phys Rev B* **50**: 8523.

Awschalom DD, Kikkawa JM. (1999) *Phys Today* **52**: 33.

Barnas J, Fuss A, Camley RE, Gruberg P, Zinn W. (1990) *Phys Rev B* **42**: 8110.

Bulgakov EN, Sadreev AF. (2002) *Phys Rev B* **66**: 075331.

Chang K, Xia JB, Peeters FM. (2002) *Phys Rev B* **65**: 155211.

Chiba D, Akiba N, Matsukura F, Ohno Y, Ohno H. (2000) *Appl Phys Lett* **77**: 1873.

Datta S, Das B. (1990) *Appl Phys Lett* **56**: 665.

Dieny B, Speriosu VS, Metin S, Parkin SSP, Gurney BA, Baumgart P, Wilhoit DR. (1991a) *J Appl Phys* **69**: 4774.

Dieny B, Speriosu VS, Parkin SSP, Gurney BA, Wilhoit DR, Mauri D. (1991b) *Phys Rev B* **43**: 1297.

Engels G, Lange J, Shäpers Th, Lüth H. (1997) *Phys Rev B* **55**: R1958.

Fert A, Jaffres H. (2001) *Phys Rev B* **64**: 184420.

Fiederling R, Keim M, Reuscher G, Ossau W, Schmidt G, Waag A, Molenkamp LW. (1999) *Nature* **402**: 787.

Gallagher WJ, Parkin SSP. (2006) *IBM J Res Dev* **50**: 5.

Gruber Th, Keim M, Fiederling R, Reuscher G, Ossau W, Schmidt G, Molenkamp LW, Waag A. (2001) *Appl Phys Lett* **78**: 1101.

Grundler D. (2000) *Phys Rev Lett* **84**: 6074.

Hallstein S, Berger JD, Hilpert M, Schneider HC, Rühle WW, Jahnke F, Koch SW, Gibbs HM, Khitrova G, Oestreich M. (1997) *Phys Rev B* **56**: R7076.

Hammar PR, Bennett BR, Yang MJ, Johnson M. (1999) *Phys Rev Lett* **83**: 203.

Hanbicki AT, Jonker BT, Itskos G, Kioseoglou G, Petrou A. (2002) *Appl Phys Lett* **80**: 1240.

Heberle AP, Rühle WW, Ploog K. (1994) *Phys Rev Lett* **72**: 3887.

Higo Y, Shimizu H, Tanaka MJ. (2001) *Appl Phys* **89**: 6745.

Johnson M. (1993) *Science* **260**: 320.

Jonker BT, Kioseoglou G, Hanbicki AT, Li CH, Thompson PE. (2007) *Nature Phys* **3**: 542.

Kepa H, Kutner-Pielaszek J, Twardowski A, Majkrzak CF, Sadowski J, Story T, Giebultowicz TM. (2001) *Phys Rev B* **64**: 121302.

Kikkawa JM, Awschalom DD. (1998) *Phys Rev Lett* **80**: 4313.

Kikkawa JM, Awschalom DD. (1999) *Nature* **397**: 139.
Koga T, Sekine Y, Nitta J. (2006) *Phys Rev* B **74**: 041302(R).
Koo HC, Kwon JH, Eom J, Chang J, Han SH, Johnson M. (2009) *Science* **325**: 1515.
Li SS, Chang K, Xia JB. (2003) *Phys Rev B* **68**: 245306.
Liu DY, Xia JB, Chang YC. (2009) *J Appl Phys* **106**: 093705.
Liu DY, Xia JB, Chang YC. (2010) *Sci China G* **53**: 16.
Lyu P, Moon K. (2001) *Phys Rev B* **64**: 035201.
Malajovich I, Kikkawa JM, Awschalom DD, Berry JJ, Samarth N. (2000) *Phys Rev Lett* **84**: 1015.
Malajovich I, Berry JJ, Samarth N, Awschalom DD. (2001) *Nature* **411**: 770.
Matsukura F, Shen A, Sugawara Y, Omiya T, Ohno Y, Ohno H. (1998) In *Proc 25th Int Symp Compound Semiconductors*, Nara, Japan (12–16 October 1998).
Matsukura F, Shen A, Sugawara Y, Omiya T, Ohno Y, Ohno H. (1999) In *Proc 25th Int Symp Compound Semiconductors*, Institute of Physics Conference Series, No. 162, IOP Publishing Ltd., Bistrol, p. 547.
Moodera JS, Kinder LR, Wong TM, Meservey R. (1995) *Phys Rev Lett* **74**: 3273.
Ohno Y, Young DK, Beschoten B, Matsukura F, Ohno H, Awschalom DD. (1999) *Nature* **402**: 790.
Ohno Y, Arata I, Matsukura F, Ohno H. (2002) *Physica E* **13**: 521.
Rashba EI. (1960) *Sov Phys Solid State* **2**: 1109.
Rashba EI. (2000) *Phys Rev B* **62**: R16267.
Schmidt G, Ferrand D, Molenkamp LW, Filip AT, van Wees BJ. (2000) *Phys Rev B* **62**: R4790.
Shen AD, Ohno H, Matsukura F, Sugawara Y, Ohno Y, Akiba N, Kuroiwa T. (1997) *Jpn J Appl Phys* **36**: L73.
Smith DL, Silver RN. (2001) *Phys Rev B* **64**: 045323.
van Son PC, van Kempen H, Wyder P. (1987) *Phys Rev Lett* **58**: 2271.
Tanaka M, Higo Y. (2001) *Phys Rev Lett* **87**: 026602.
Tedrow PM, Meservy R. (1971) *Phys Rev Lett* **26**: 192.
Tsang C, Fontana RE, Lin T, Heim DE, Speriosu VS, Gurney BA, Mason ML. (1994) *IEEE Trans Magn* **30**: 3801.
Valet T, Fert A. (1993) *Phys Rev B* **48**: 7099.
Xia JB. (1988) *Phys Rev* B **38**:8365.
Xia JB, Hai GQ, Wang JN. (2003) *Solid State Commun* **127**: 489.
Yang W, Chang K. (2006) *Phys Rev B* **73**: 113303.
Zhang XW, Xia JB. (2006) *Phys Rev B* **74**: 075304.
Zhu HJ, Ramsteiner M, Kostial H, Wassermeier M, Schönherr HP, Ploog KH. (2001) *Phys Rev Lett* **87**: 016601.

Chapter 5

SPIN RELAXATION

5.1. SRTs T_1 and T_2

Originally, the SRTs T_1 and T_2 were proposed in the nuclear spin investigation, this concept is very general and suitable for the spin of the conduction electrons. The coupling between the nuclear spin and the perturbation decides the properties of spin relaxation. For example, the coupling with phonons (nonelastic spin–lattice relaxation) determines T_1, while the direct interaction between magnetic dipole moments determines T_2. The electron spin relaxation can also be described by T_1 and T_2, but has a rather large difference from the nuclear spin relaxation. Because electrons can move freely in the space, the spin is generally coupled with the orbital movement (SOC). Consequently, the electron spin relaxation mechanisms are various, and the intensity is larger.

The simplest spin system is the two-level system, corresponding to the spin of $\pm 1/2$. External nonequilibrium perturbation has two effects: (1) change relative populations of two states when the energies of two states are different; (2) change the coherent relation (phase) of two states. These two perturbations can be described by relaxation times T_1 and T_2, T_1 is called spin lifetime and T_2 is called coherent time. When the two-level system is placed in the magnetic field, the T_1 corresponds to the decay time of the longitudinal magnetization and T_2 to the decay time of transverse magnetization.

When the energies of two states are different, T_1 and T_2 have essential difference. The state occupation is relaxed from nonequilibrium to equilibrium, i.e. electrons are transferred from high-energy state to low-energy state, it needs energy exchange between the spin system and other systems. While the dephasing of the phase relation between two states does not need exchange with the external systems, the dephasing is deduced by the elastic magnetic moment interaction between spins. Generally, the magnetic moment interaction is much stronger than the spin–lattice interaction, so that $T_1 \gg T_2$.

When a solid is put in the magnetic field, the magnetic moments of conduction electrons (including spin and orbit) tend to arrange along the magnetic field direction. This process is not performed immediately, because the magnetic energy of the random arranged magnetic moments should be transferred to other freedom degree, and then to the thermal energy. The energy transfer is due to the interaction between electron spin and other freedom degree; thus, the speed of the magnetization is determined by the strongest interaction of all these interactions. For an isotropic solid, the longitudinal relaxation time T_1 is defined as the time needed for the magnetization component M_z along the magnetic field direction to reach its equilibrium value M_0. T_1 is determined by the following equation:

$$\frac{dM_z}{dt} = -\frac{M_z - M_0}{T_1}. \tag{5.1}$$

T_1 can be measured by the direct method. It is also called spin–lattice relaxation time, because the lattice is generally assumed to be a thermal reservoir. The relaxation time T_2 is determined experimentally by the reciprocal of the FWHM of the spin resonant peak.

The scattering processes in most nonmagnetic semiconductors are not associated with the spin freedom degree of electrons or holes, i.e. the Hamiltonian describing the spin scattering process does not include the spin operator. If the eigenstates of the system are spin eigenstates, this scattering event does not influence the spin state. At the neighborhood of the CB bottom, the electron state is approximately a spin eigenstate. In the changing process of the electron momentum, the probability of the spin quantum number changing is very small, but is not zero. Because the electron Hamiltonian consists of the SOC term, the electron eigenstate consists of different spin components. Though the interaction of the electron–phonon and electron-impurity scatterings do not consist of spin operator, there still exists the scattering of spin flip.

5.2. Elliot–Yafet Relaxation Mechanism

As was first pointed out by Elliot (1954) and later by Yafet (1961), SOI in crystals causes the mixing of electron wave functions with opposite spin orientations. As a result, in the process of momentum scattering, the disorientation of electron spin also becomes possible. The spin relaxation rate increases with increasing momentum scattering rates. This mechanism is called the Elliot–Yafet (EY) spin relaxation mechanism.

Taking into account the SOC effect, the effective mass equation of the conduction electron in semiconductors is written as

$$\left[\frac{p^2}{2m^*} + \frac{\hbar}{2m_0^2c^2}(\nabla V \times \boldsymbol{p}) \cdot \boldsymbol{\sigma}\right] \psi = E\psi. \tag{5.2}$$

Because the Hamiltonian consists of a term with spin operator, the electron eigenstate should be the linear combination of different spin functions,

$$\psi_1 = [a_{\boldsymbol{k}}|+\rangle + b_{\boldsymbol{k}}|-\rangle]e^{i\boldsymbol{k}\cdot\boldsymbol{r}}, \tag{5.3}$$

where $|+\rangle$ and $|-\rangle$ are spin states with components of $\pm 1/2$. If the crystal has a space inversion center, the electron state is always two-degenerate due to the time inversion. Operating the space inversion operator to Eq. (5.3), we obtain

$$[a_{-\boldsymbol{k}}|+\rangle + b_{-\boldsymbol{k}}|-\rangle]e^{-i\boldsymbol{k}\cdot\boldsymbol{r}}. \tag{5.4}$$

Inversing the time axis, we obtain

$$\psi_2 = [a^*_{-\boldsymbol{k}}|-\rangle + b^*_{-\boldsymbol{k}}|+\rangle]e^{i\boldsymbol{k}\cdot\boldsymbol{r}}. \tag{5.5}$$

ψ_2 has the same $\boldsymbol{k}$ and energy as ψ_1 in Eq. (5.3), and this is Kramers' theorem.

In general case, the coefficients $a_{\boldsymbol{k}} \gg b_{\boldsymbol{k}}$, or $b_{\boldsymbol{k}} \gg a_{\boldsymbol{k}}$, for convenience the states are still represented as (+) and (−) states according to their main components. Suppose that the interaction Hamiltonian H_{int} scatters the electron $\boldsymbol{k}$ state to $\boldsymbol{k}'$ state and does not change the spin. The scattering matrix element is the integration

$$\int a^*_{\boldsymbol{k}'} H_{\text{int}} a_{\boldsymbol{k}} \mathrm{e}^{\mathrm{i}(\boldsymbol{k}-\boldsymbol{k}')\cdot\boldsymbol{r}} \mathrm{d}\boldsymbol{r}. \tag{5.6}$$

Besides, there is a scattering matrix element from one spin state to another spin state,

$$\int (a_{-\boldsymbol{k}'} H_{\text{int}} b_{\boldsymbol{k}} - b_{-\boldsymbol{k}'} H_{\text{int}} a_{\boldsymbol{k}}) \mathrm{e}^{\mathrm{i}(\boldsymbol{k}-\boldsymbol{k}')\cdot\boldsymbol{r}} \mathrm{d}\boldsymbol{r}. \tag{5.7}$$

The ratio of Eqs. (5.7) and (5.6) is $c = b_{\boldsymbol{k}}/a_{\boldsymbol{k}}$ ($\ll 1$); thus, the spin flip scattering probability is c^2 times of the momentum scattering probability.

Assuming that the matrix element of the spin flip is known, we calculate the spin relaxation time T_1. Let N_+ and N_- be the numbers of electrons with spin up and spin down in a solid of unit volume, respectively. When a small magnetic field is applied, the electron magnetization is proportional to the difference of the population $D = N_- - N_+$. The relaxation time T_1 is defined by the variation rate of D,

$$\frac{\mathrm{d}D}{\mathrm{d}t} = -\frac{D - D_0}{T_1}, \tag{5.8}$$

where D_0 is the equilibrium value.

In statistical physics, the two system of spin up and spin down can be described by two Fermi functions $f_\uparrow(k)$ and $f_\downarrow(k)$, respectively, whose Fermi energies have a slight difference. Let $W_{\uparrow\downarrow}$ be the transition number of electrons from the spin-up system to spin-down system in unit time ($W_{\downarrow\uparrow}$ inverse), then

$$\frac{\mathrm{d}D}{\mathrm{d}t} = 2(W_{\uparrow\downarrow} - W_{\downarrow\uparrow}). \tag{5.9}$$

$W_{\uparrow\downarrow}$ equals $-W_{\downarrow\uparrow}$, which is given by

$$W_{\uparrow\downarrow} = (2\pi)^{-6} \int \mathrm{d}\boldsymbol{k}\,\mathrm{d}\boldsymbol{k'} \{f_\downarrow(\boldsymbol{k})[1 - f_\uparrow(\boldsymbol{k'})]W_{\boldsymbol{k}\downarrow,\boldsymbol{k'}\uparrow} - f_\uparrow(\boldsymbol{k'})[1 - f_\downarrow(\boldsymbol{k})]W_{\boldsymbol{k'}\uparrow,\boldsymbol{k}\downarrow}\}, \tag{5.10}$$

where $W_{\boldsymbol{k}\downarrow,\boldsymbol{k'}\uparrow}$ is the transition probability from ($\boldsymbol{k}\downarrow$) state to ($\boldsymbol{k'}\uparrow$) state.

Suppose that the interaction is the electron–phonon interaction, the scattering process consists of the absorption and emission of a phonon, the phonon wave vector $\pm\boldsymbol{q} = \boldsymbol{k'} - \boldsymbol{k} + \boldsymbol{k}_n$ ($\boldsymbol{k}_n$ is the reciprocal lattice vector, generally can be neglected). By use of the standard formula of the transition probability, $W_{\boldsymbol{k}\downarrow,\boldsymbol{k'}\uparrow}$ can be written as

$$W_{\boldsymbol{k}\downarrow,\boldsymbol{k'}\uparrow} = \frac{2\pi}{\hbar} \left|M_{\boldsymbol{k}\downarrow,\boldsymbol{k'}\uparrow}\right|^2 \frac{\hbar\omega_{\boldsymbol{q}}}{2NM\omega_{\boldsymbol{q}}^2} \big\{n_{\boldsymbol{q}}\delta(E_{\boldsymbol{k}\downarrow} - E_{\boldsymbol{k'}\uparrow} + \hbar\omega_{\boldsymbol{q}}) + (n_{-\boldsymbol{q}} + 1)\delta(E_{\boldsymbol{k}\downarrow} - E_{\boldsymbol{k'}\uparrow} - \hbar\omega_{-\boldsymbol{q}})\big\}, \tag{5.11}$$

where $M_{\boldsymbol{k}\downarrow,\boldsymbol{k'}\uparrow}$ is the transition matrix element, N is the atom number in unit volume, M is the atom mass, $\omega_{\boldsymbol{q}}$ is the frequency of the phonon $\boldsymbol{q}$, and $n_{\boldsymbol{q}} = [\exp(\hbar\omega_{\boldsymbol{q}}/k_BT) - 1]^{-1}$ is the equilibrium distribution number of phonon $\boldsymbol{q}$.

By use of the invariabilities of space inversion and time inversion, we obtain (see Eq. (5.5)),

$$M_{\boldsymbol{k}\downarrow,\boldsymbol{k'}\uparrow} = -(M_{\boldsymbol{k}\uparrow,\boldsymbol{k'}\downarrow})^*. \tag{5.12}$$

Expanding the Fermi distribution functions $f_\downarrow$ and $f_\uparrow$ with respect to the difference in the Fermi energy and the equilibrium Fermi energy to first order, expressing $D-D_0$ by this difference, and using Eqs. (5.11) and (5.12), we obtain

$$\begin{aligned}\frac{1}{T_1} = \frac{4\pi}{\hbar}\frac{1}{(2\pi)^3}\Big\langle \int \mathrm{d}\boldsymbol{k}' \frac{\hbar}{2NM\omega_{\boldsymbol{q}}} \left|M_{\boldsymbol{k}\uparrow,\boldsymbol{k}'\downarrow}\right|^2 \\ \times\{[1-f(\boldsymbol{k}')][n_{\boldsymbol{q}}\delta(E_{\boldsymbol{k}}-E_{\boldsymbol{k}'}+\hbar\omega_{\boldsymbol{q}})+(n_{-\boldsymbol{q}}+1)\delta(E_{\boldsymbol{k}}-E_{\boldsymbol{k}'}-\hbar\omega_{\boldsymbol{q}})] \\ + f(\boldsymbol{k}')[n_{-\boldsymbol{q}}\delta(E_{\boldsymbol{k}}-E_{\boldsymbol{k}'}-\hbar\omega_{\boldsymbol{q}})+(n_{\boldsymbol{q}}+1)\delta(E_{\boldsymbol{k}}-E_{\boldsymbol{k}'}+\hbar\omega_{\boldsymbol{q}})]\}\Big\rangle_{\boldsymbol{k}},\end{aligned} \tag{5.13}$$

where $f(\boldsymbol{k}')$ is the equilibrium Fermi distribution function. The angle bracket represent the average of $\boldsymbol{k}$ over the Fermi surface,

$$\langle F(\boldsymbol{k})\rangle_{\boldsymbol{k}} = \frac{\int F(\boldsymbol{k})(\partial f/\partial E)\mathrm{d}\boldsymbol{k}}{\int(\partial f/\partial E)\mathrm{d}\boldsymbol{k}}. \tag{5.14}$$

Under the stress, the eight bands (including CB and valence band) effective-mass Hamiltonian for the direct gap III–V semiconductors without inverse center is written as (Pikus and Titkov, 1984),

$$H=\begin{vmatrix} \frac{\hbar^2k}{2m_e'}+C_1\varepsilon & 0 & -K_+ & \sqrt{\frac{2}{3}}K_z & \sqrt{\frac{1}{3}}K_- & 0 & -\sqrt{\frac{1}{3}}K_z & -\sqrt{\frac{2}{3}}K_- \\ 0 & \frac{\hbar^2k}{2m_e}+C_1\varepsilon & 0 & -\sqrt{\frac{1}{3}}K_+ & \sqrt{\frac{2}{3}}K_z & K_- & -\sqrt{\frac{2}{3}}K_+ & \sqrt{\frac{1}{3}}K_z \\ -K_+^* & 0 & F-E_g & H & I & 0 & -\sqrt{\frac{1}{2}}H & -\sqrt{2}I \\ \sqrt{\frac{2}{3}}K_z^* & -\sqrt{\frac{1}{3}}K_+^* & H^* & G-E_g & 0 & I & -\frac{G-F}{\sqrt{2}} & \sqrt{\frac{2}{3}}H \\ \sqrt{\frac{1}{3}}K_-^* & \sqrt{\frac{2}{3}}K_z^* & I^* & 0 & G-E_g & -H & \sqrt{\frac{2}{3}}H^* & \frac{G-F}{\sqrt{2}} \\ 0 & K_-^* & 0 & I^* & -H^* & F-E_g & \sqrt{2}I^* & -\sqrt{\frac{1}{2}}H^* \\ -\sqrt{\frac{1}{3}}K_z^* & -\sqrt{\frac{2}{3}}K_+^* & -\sqrt{\frac{1}{2}}H^* & -\frac{G-F}{\sqrt{2}} & \sqrt{\frac{3}{2}}H^* & \sqrt{2}I & \begin{matrix}\frac{1}{2}(G-F)\\-E_g-\Delta\end{matrix} & 0 \\ -\sqrt{\frac{2}{3}}K_-^* & \sqrt{\frac{1}{3}}K_z^* & -\sqrt{2}I^* & \sqrt{\frac{3}{2}}H^* & \frac{G-F}{\sqrt{2}} & -\sqrt{\frac{1}{2}}H & 0 & \begin{matrix}\frac{1}{2}(G-F)\\-E_g-\Delta\end{matrix} \end{vmatrix} \tag{5.15}$$

where

$$K_z = \frac{\hbar}{m_0} P k_z - \frac{i}{m_{\text{cv}}} k_x k_y - i C_2 \varepsilon_{xy}. \tag{5.16}$$

Other components are obtained by cyclic permutation of indices.

$$\begin{aligned}
K_\pm &= \sqrt{\frac{1}{2}}(K_x \pm i K_y), \quad P = \langle S|p_z|Z\rangle, \\
F &= Ak^2 + \frac{1}{2}B(k^2 - 3k_z^2) + a\varepsilon + \frac{1}{2}b(\varepsilon - 3\varepsilon_{zz}), \\
G &= Ak^2 - \frac{1}{2}B(k^2 - 3k_z^2) + a\varepsilon - \frac{1}{2}b(\varepsilon - 3\varepsilon_{zz}), \\
H &= -Dk_z(k_x - ik_y) - d(\varepsilon_{xz} - i\varepsilon_{yz}), \\
I &= -\frac{\sqrt{3}}{2}Bt(k_x^2 - k_y^2) + iDk_xk_y - \frac{\sqrt{3}}{2}b(\varepsilon_{xx} - \varepsilon_{yy}) + id\varepsilon_{xy}
\end{aligned} \tag{5.17}$$

m_e' and A, B, D are effective mass parameters of CB and valence band, respectively. P and $1/m_{\text{cv}}$ are interaction parameters between CB and valence band. C_2 and a, b, d are deformation potential parameters of CB and valence band, respectively, $\varepsilon = \varepsilon_{xx} + \varepsilon_{yy} + \varepsilon_{zz}, \varepsilon_{xy}, \ldots$ are strain components. Δ is the spin–orbit splitting energy. The basis functions for the Hamiltonian (5.17) are in following sequence: CB, $c_1 = |1/2, 1/2\rangle$, $c_2 = |1/2, -1/2\rangle$, valence band, $u_1 = |3/2, 3/2\rangle$, $u_2 = |3/2, 1/2\rangle$, $u_3 = |3/2, -1/2\rangle$, $u_4 = |3/2, -3/2\rangle$, and spin–orbit splitting band, $v_1 = |1/2, 1/2\rangle$, $v_2 = |1/2, -1/2\rangle$.

In the framework of perturbation theory, taking into account the non-diagonal terms describing the interaction between c (CB) and u (Γ_8) and v (Γ_6) valence bands, we can write the CB Hamiltonian in the form,

$$H_c = \frac{\hbar^2 k^2}{2m_e} + \frac{1}{2}\alpha_c \hbar^3 (2m_e^3 E_g)^{-1/2} (\boldsymbol{\kappa} \cdot \boldsymbol{\sigma}) + \frac{1}{2} C_3 (\boldsymbol{\sigma} \cdot \boldsymbol{\varphi}), \tag{5.18}$$

where σ is the Pauli matrix, $\kappa_z = k_z(k_x^2 - k_y^2)$, $\varphi_z = \varepsilon_{zx}k_x - \varepsilon_{zy}k_y$, and other components are obtained by cyclic permutation of indices,

$$\begin{aligned}
\frac{1}{m_e} &= \frac{2P^2}{m_0^2 E_g}\left(1 - \frac{\eta}{3}\right) + \frac{1}{m_e'}, \quad \eta = \frac{\Delta}{E_g + \Delta}, \\
\alpha_c &= \frac{4}{3}\frac{m_e}{m_{\text{cv}}}\eta\left(1 - \frac{\eta}{3}\right)^{-1/2}, \\
C_3 &= \frac{4}{3}\hbar C_2 \eta \left[2E_g m_e\left(1 - \frac{\eta}{3}\right)\right]^{-1/2}.
\end{aligned} \tag{5.19}$$

The first term in the right-hand side of Eq. (5.18) is the general effective mass kinetic energy term; the second term is the k^3 term caused by the lake of space inversion symmetry (III–V compounds), resulting in splitting of CB at $k \neq 0$; the third term describes the conduction band splitting caused by the strain, which depends only on the shear strain $\varepsilon_{xz}, \varepsilon_{xy}$, etc.

The wave functions of the CB spin-up state and spin-down states can also be obtained by the perturbation theory from Hamiltonian (5.15),

$$
\begin{aligned}
\psi_{c\uparrow} &= c_1 + \frac{1}{E_g}\left(-K_- u_1 + \sqrt{\frac{2}{3}} K_z u_2 + \sqrt{\frac{1}{3}} K_+ u_3\right) \\
&\quad + \frac{1}{E_g + \Delta}\left(-\sqrt{\frac{1}{3}} K_z v_1 - \sqrt{\frac{2}{3}} K_+ v_2\right), \\
\psi_{c\downarrow} &= c_2 + \frac{1}{E_g}\left(-\sqrt{\frac{1}{3}} K_- u_2 + \sqrt{\frac{2}{3}} K_z u_3 + K_+ u_4\right) \\
&\quad + \frac{1}{E_g + \Delta}\left(-\sqrt{\frac{2}{3}} K_- v_1 + \sqrt{\frac{1}{3}} K_z v_2\right).
\end{aligned}
\tag{5.20}
$$

Thus, $\psi_{c\uparrow}$ consists of spin-down state and $\psi_{c\downarrow}$ consists of spin-up state.

Let H_s be the electron scattering Hamiltonian, such as ionized impurity scattering, acoustic and optical phonon scatterings, etc. They are only the interactions with electron orbit. But for the electron wave functions in Eq. (5.20), H_s can cause the transition between spin inverse states. For example,

$$
\begin{aligned}
&\langle \psi_{c\uparrow}(\boldsymbol{k}) | H_s | \psi_{c\downarrow}(\boldsymbol{k}') \rangle \\
&= \frac{\sqrt{2}}{3}(-K_z K'_- + K_- K'_z)\left(\frac{1}{E_g^2} - \frac{1}{(E_g + \delta)^2}\right)\langle \phi_c(\boldsymbol{k}) | H_s | \phi_c(\boldsymbol{k}'),
\end{aligned}
\tag{5.21}
$$

where $\phi_c(\boldsymbol{k})$ and $\phi_c(\boldsymbol{k}')$ are envelope functions of the conduction states in the effective mass approximation, having the plane wave form. The EY mechanism does not consider the nonsymmetry of space inversion and strain effect, so that the $\boldsymbol{k}$ in Eq. (5.21) is just the wave vector $\boldsymbol{k}$ (see Eq. (5.16)),

$$
\boldsymbol{K} = \frac{\hbar}{m_0} P \boldsymbol{k}. \tag{5.22}
$$

Similarly, we can calculate the matrix elements of H_s between $\psi_{c\uparrow}(\boldsymbol{k})$ and $\psi_{c\uparrow}(\boldsymbol{k}')$, $\psi_{c\downarrow}(\boldsymbol{k})$ and $\psi_{c\uparrow}(\boldsymbol{k}')$, $\psi_{c\downarrow}(\boldsymbol{k})$, and $\psi_{c\downarrow}(\boldsymbol{k}')$. Writing in the matrix form, the results can be denoted by a simple formula,

$$H_{\boldsymbol{kk}'} = H^{\text{int}}_{\boldsymbol{kk}'} \frac{i\hbar^2[\boldsymbol{\sigma}\cdot(\boldsymbol{k}\times\boldsymbol{k}')]\eta(1-1/2\eta)}{2m_e E_g\left(1-\frac{1}{3}\eta\right)}, \tag{5.23}$$

where

$$H^{\text{int}}_{\boldsymbol{kk}'} = \langle\phi_c(\boldsymbol{k})|H_s|\phi_c(\boldsymbol{k}')\rangle. \tag{5.24}$$

For ionized impurity scattering, which is elastic scattering, $k' = k$; thus,

$$\left(\frac{\hbar^2}{m_e}[\boldsymbol{k}\times\boldsymbol{k}']\right)^2 = 4E^2_{\boldsymbol{k}}\sin^2\theta, \tag{5.25}$$

where θ is the angle between $\boldsymbol{k}$ and $\boldsymbol{k}'$. In the isotropic case, the transition probability should be averaged over all direction of $\boldsymbol{k}$. Then the spin relaxation rate equals

$$\frac{1}{\tau_s(\boldsymbol{k})} = 2\frac{2\pi}{\hbar}\sum_{\boldsymbol{k}'}|H_{\boldsymbol{k}m,\boldsymbol{k}'m'}|^2\delta(E_{\boldsymbol{k}}-E_{\boldsymbol{k}'}), \quad m\neq m', \tag{5.26}$$

where m denotes the spin states $\uparrow$ and $\downarrow$.

The momentum relaxation rate is (Seeger, 1982)

$$\frac{1}{\tau_p(\boldsymbol{k})} = \frac{2\pi}{\hbar}\sum_{\boldsymbol{k}'}|H^{\text{int}}_{\boldsymbol{kk}'}|^2(1-\cos\theta)\delta(E_{\boldsymbol{k}}-E_{\boldsymbol{k}'}). \tag{5.27}$$

Then

$$\frac{1}{\tau_s(\boldsymbol{k})} = \frac{16}{27}\frac{1}{\tau_p(\boldsymbol{k})}\left(\frac{E_{\boldsymbol{k}}}{E_g}\right)^2\eta^2\left(\frac{1-\eta/2}{1-\eta/3}\right)^2\Phi, \tag{5.28}$$

where

$$\Phi = \frac{\int_{-1}^{1}(1-\mu^2)\sigma(\mu)\mathrm{d}\mu}{\int_{-1}^{1}(1-\mu)\,\sigma(\mu)\mathrm{d}\mu}, \tag{5.29}$$

$\mu = \cos\theta$, and $\sigma(\mu)$ is the scattering cross-section. For ionized impurity scattering,

$$\sigma(\mu)\sim\left[(1-\mu)+\frac{2}{\beta}\right]^{-2}, \quad \beta=(2kL_D)^2, \tag{5.30}$$

where L_D is the Debye length. For scattering by polar optical phonons, if the electron kinetic energy $E_{\mathbf{k}}$ is much higher than the LO phonon energy $\hbar\omega_q$, the scattering can be looked approximately as elastic scattering, $\sigma(\mu)\sim(1-\mu)^{-1}$, $\Phi = 1$.

After averaging over the Maxwell distribution of electrons, the general expression for the spin relaxation rate due to the EY mechanism can be written as

$$\frac{1}{\tau_{\rm s}} = r\frac{1}{\tau_{\rm p}}\left(\frac{k_{\rm B}T}{E_{\rm g}}\right)^2 \eta^2 \left(\frac{1-\eta/2}{1-\eta/3}\right)^2, \tag{5.31}$$

where

$$r = \frac{32}{81}\Phi\frac{\langle E_{\boldsymbol{k}}^2 \tau_{\rm p}^{-1}(\boldsymbol{k})\rangle\langle E_{\boldsymbol{k}}\tau_{\rm p}(\boldsymbol{k})\rangle}{(k_{\rm B}T)^3}. \tag{5.32}$$

The angle brackets denote the averaging over the Maxwell distribution. The momentum relaxation time is given by

$$\tau_{\rm p} = \frac{\langle E_{\boldsymbol{k}}\tau_{\rm p}(\boldsymbol{k})\rangle}{\langle E_{\boldsymbol{k}}\rangle} = \frac{2\langle E_{\boldsymbol{k}}\tau_{\rm p}(\boldsymbol{k})\rangle}{3k_{\rm B}T}. \tag{5.33}$$

From Eq. (5.31), we see that the spin relaxation time $\tau_{\rm s}$ depends on η (see Eq. (5.19)). If $E_{\rm g}$ is small, Δ large, then $\tau_{\rm s}$ is small, i.e. the EY mechanism is the main spin relaxation mechanism in the semiconductors of narrow energy gap and large spin–orbit splitting energy.

5.3. Dyakonov–Perel Relaxation Mechanism

The Dyakonov–Perel (DP) mechanism is caused by the spin splitting of energy bands in the crystals without inversion centre. The degeneracy of the CB is completely lifted for $k \neq 0$. The Hamiltonian of the CB is given by Eq. (5.18), where the second term describes the splitting of the CB. The spin splitting of the CB is equivalent to the presence in the crystal of an effective internal magnetic field with magnitude and orientation dependent on magnitude and orientation of $\boldsymbol{k}$. The spin electrons precess in this internal magnetic field with the Larmor frequency Ω. For $\Omega > 1/\tau_{\rm p}$, the transverse component of the electron spin vanishes before the first momentum scattering occurs. In the opposite limiting case, i.e. at $\Omega < 1/\tau_{\rm p}$, the angular rotation of electron spin over time $\tau_{\rm p}$ is small and spin relaxation occurs as a result of a number of accidental small rotations. In III–V compounds for thermalized electrons, the latter condition $\Omega < 1/\tau_{\rm p}$ is usually valid.

The conduction electron density matrix $\rho(\boldsymbol{k})$ is a 2×2 matrix, including two components of spin, which satisfies the following equation:

$$\frac{\rho(\boldsymbol{k})}{\tau} + \frac{i}{\hbar}[H_1(\boldsymbol{k}), \rho(\boldsymbol{k})] + \sum_{\boldsymbol{k}'} W_{\boldsymbol{k}\boldsymbol{k}'}(\rho(\boldsymbol{k}) - \rho(\boldsymbol{k}')) = G, \tag{5.34}$$

where G is the generation matrix, τ is the lifetime, $W_{\boldsymbol{kk}'}$ is the momentum scattering probability, supposed to be diagonal with spin, and $H_1(\boldsymbol{k})$ is the spin-splitting operator as the second term in Eq. (5.18). Let $\rho(\boldsymbol{k}) = \bar{\rho} + \rho_1(\boldsymbol{k})$, where $\bar{\rho}$ is the result of averaging $\rho(\boldsymbol{k})$ over all directions of $\boldsymbol{k}$. The equation of $\bar{\rho}$ can be obtained from Eq. (5.34),

$$\frac{\bar{\rho}}{\tau} + \overline{\frac{i}{\hbar}[H_1(\boldsymbol{k}), \rho_1(\boldsymbol{k})]} = \bar{G}. \tag{5.35}$$

Here the overbar denotes averaging over all directions of $\boldsymbol{k}$.

Inserting Eq. (5.35) into Eq. (5.34), and taking into account the case of $\Omega\tau_{\rm p} \ll 1$, $\rho_1(\boldsymbol{k}) \ll \bar{\rho}$, $\tau_{\rm p} \ll \tau$, we obtain the equation of $\rho_1(\boldsymbol{k})$,

$$\frac{i}{\hbar}[H_1(\boldsymbol{k}), \bar{\rho}] + \int \frac{\mathrm{d}\Omega}{2\pi}\sigma(\theta)(\rho_1(\boldsymbol{k}) - \rho_1(\boldsymbol{k}')) = 0. \tag{5.36}$$

Suppose that

$$\rho_1(\boldsymbol{k}) = -\frac{i\tau^*}{\hbar}[H_1(\boldsymbol{k}), \bar{\rho}], \tag{5.37}$$

where τ^* is a parameter which should be determined from Eq. (5.36). In order to find the solution, we have to expand $\rho_1(\boldsymbol{k})$ in Eq. (5.37) over the spherical harmonic functions (lth order)

$$\rho_1^l(\boldsymbol{k}) = \sum_m C_{l,m} k^l Y_m^l(\vartheta, \varphi). \tag{5.38}$$

Insert Eq. (5.38) into Eq. (5.36), and using the well-known relation for spherical harmonic functions,

$$\int \frac{\mathrm{d}\Omega'}{2\pi} Y_m^l(\vartheta', \varphi')\sigma(\theta) = Y_m^l(\vartheta, \varphi) \int_0^\pi \sigma(\theta) P_l(\cos\theta) \sin\theta \,\mathrm{d}\theta, \tag{5.39}$$

where ϑ, φ are the angular coordinates of $\boldsymbol{k}, \vartheta'$, and φ' are the angular coordinates of $\boldsymbol{k}'$, θ is the angle between $\boldsymbol{k}$ and $\boldsymbol{k}'$, and $P_l(\cos\theta)$ is the Lagendre polynomial, then we get

$$\frac{1}{\tau^*} = \int_{-1}^{1} \sigma(\theta)(1 - P_l(\mu))\mathrm{d}\mu, \quad \mu = \cos\theta. \tag{5.40}$$

The momentum scattering time is given by

$$\frac{1}{\tau_{\rm p}} = \int_{-1}^{+1} \sigma(\mu)(1 - \mu)\mathrm{d}\mu.$$

Thus, $1/\tau^*$ can be rewritten as

$$\frac{1}{\tau^*} = \gamma_l \frac{1}{\tau_{\rm p}}, \quad \gamma_l = \frac{\int_{-1}^{1} \sigma(\mu)(1 - P_l(\mu))\mathrm{d}\mu}{\int_{-1}^{1} \sigma(\mu)(1 - \mu)\mathrm{d}\mu}. \tag{5.41}$$

Substituting Eq. (5.37) into Eq. (5.35), we obtain the equation of $\bar{\rho}$,

$$\frac{\bar{\rho}}{\tau} + \frac{\tau^*}{\hbar}\overline{[H_1(\boldsymbol{k})[H_1(\boldsymbol{k}), \bar{\rho}]]} = \bar{G}. \tag{5.42}$$

The second term in Eq. (5.42) describes the spin relaxation due to the DP mechanism,

$$\left(\frac{\partial \rho}{\partial t}\right)_{\rm DP} = -\frac{\tau^*}{\hbar}\overline{[H_1(\boldsymbol{k})[H_1(\boldsymbol{k}), \bar{\rho}]]}. \tag{5.43}$$

From Eq. (5.18), for $s = 1/2$, the matrix $H_1(\boldsymbol{k})$ can be written as

$$H_1(\boldsymbol{k}) = \frac{\hbar}{2}[\boldsymbol{\sigma} \cdot \boldsymbol{\Omega}(\boldsymbol{k})], \tag{5.44}$$

where Ω is a vector proportional to κ or φ, and

$$\bar{\rho} = \frac{1}{2}(\boldsymbol{s} \cdot \boldsymbol{\sigma}), \tag{5.45}$$

s_i is the average of electron spin projection on the x_i direction,

$$s_i = \frac{1}{2}\mathrm{Tr}(\rho \cdot \sigma_i). \tag{5.46}$$

Substituting Eqs. (5.44) and (5.45) into Eq. (5.43), and using the properties of Pauli matrix (Merzbacher, 1970): if the vectors $\boldsymbol{A}$ and $\boldsymbol{B}$ commute with σ, then we have the equality,

$$(\boldsymbol{\sigma} \cdot \boldsymbol{A})(\boldsymbol{\sigma} \cdot \boldsymbol{B}) = \boldsymbol{A} \cdot \boldsymbol{B} + i\boldsymbol{\sigma} \cdot (\boldsymbol{A} \times \boldsymbol{B}). \tag{5.47}$$

We prove that

$$\begin{aligned} [H_1(\boldsymbol{k}), \bar{\rho}] &= \frac{\hbar}{4}[(\boldsymbol{\sigma} \cdot \boldsymbol{\Omega}), (\boldsymbol{s} \cdot \boldsymbol{\sigma})] \\ &= \frac{\hbar}{2} i\boldsymbol{\sigma} \cdot (\boldsymbol{\Omega} \times \boldsymbol{s}), \\ [H_1(\boldsymbol{k})[H_1(\boldsymbol{k}), \bar{\rho}]] &= \left(\frac{\hbar}{2}\right)^2 i[(\boldsymbol{\sigma} \cdot \boldsymbol{\Omega}), \boldsymbol{\sigma} \cdot (\boldsymbol{\Omega} \times \boldsymbol{s})] \\ &= -\frac{\hbar^2}{2}[(\boldsymbol{\Omega} \cdot \boldsymbol{s})\boldsymbol{\sigma} \cdot \boldsymbol{\Omega} - \Omega^2 \boldsymbol{\sigma} \cdot \boldsymbol{s}]. \end{aligned} \tag{5.48}$$

From Eqs. (5.43) and (5.48), we obtain

$$\left(\frac{\partial s_z}{\partial t}\right)_{\mathrm{DP}} = -\tau^*[s_z(\bar{\Omega}_x^2 + \bar{\Omega}_y^2) - s_x(\overline{\Omega_x\Omega_z}) - s_y(\overline{\Omega_y\Omega_z})]. \tag{5.49}$$

$(\partial s_x/\partial t), (\partial s_y/\partial t)$ can be obtained by cyclic permutation of indices. From $\partial s_z/\partial t = s_z/\tau_{\mathrm{s}}(k)$ and Eq. (5.49), it follows that the electron spin relaxation time is generally a tensor, with components

$$\begin{aligned} \frac{1}{[\tau_{\mathrm{s}}(k)]_{ii}} &= \frac{1}{\gamma_l}\tau_{\mathrm{p}}(k)(\Omega^2 - \overline{\Omega_i^2}), \\ \frac{1}{[\tau_{\mathrm{s}}(k)]_{ij}} &= \frac{1}{\gamma_l}\tau_{\mathrm{p}}(k)(\overline{\Omega_i\Omega_j}). \end{aligned} \tag{5.50}$$

For III–V compounds, from Eq. (5.18), one obtains

$$\boldsymbol{\Omega}(\boldsymbol{k}) = \alpha_{\mathrm{c}}\hbar^2(2m_{\mathrm{e}}^3E_{\mathrm{g}})^{-1/2}\boldsymbol{\kappa}. \tag{5.51}$$

And $\overline{\kappa_i^2} = (4/105)k^6, \overline{\kappa_i\kappa_j} = 0$, so that the $1/\tau_{\mathrm{s}}(k)$ of conduction electrons in Eq. (5.50) is a scalar, which is in accordance with general symmetry consideration

$$\frac{1}{\tau_{\mathrm{s}}(k)} = \frac{32}{105\gamma_3}\tau_{\mathrm{p}}(k)\alpha_{\mathrm{c}}^2\frac{E_{\boldsymbol{k}}^3}{\hbar^2E_{\mathrm{g}}}. \tag{5.52}$$

After averaging over the Maxwell distribution, we obtain the spin relaxation rate,

$$\frac{1}{\tau_{\mathrm{s}}} = Q\tau_{\mathrm{p}}\alpha_{\mathrm{c}}^2\frac{(k_{\mathrm{B}}T)^3}{\hbar^2E_{\mathrm{g}}}, \tag{5.53}$$

where

$$Q = \frac{32}{105\gamma_3}\frac{\langle\tau_{\mathrm{p}}(\boldsymbol{k})E_{\boldsymbol{k}}^3\rangle\langle E_{\boldsymbol{k}}\rangle}{\langle\tau_{\mathrm{p}}(\boldsymbol{k})E_{\boldsymbol{k}}\rangle(k_BT)^3} = \frac{16}{35}\frac{\langle\tau_{\mathrm{p}}(\boldsymbol{k})E_{\boldsymbol{k}}^3\rangle}{\langle\tau_{\mathrm{p}}(\boldsymbol{k})E_{\boldsymbol{k}}\rangle(k_BT)^2}. \tag{5.54}$$

The parameter Q depends on the scattering process. For $\tau_{\mathrm{p}}(k) \sim E_{\boldsymbol{k}}^{\nu}$,

$$Q = \frac{16}{35\gamma_3}\left(\nu + \frac{7}{2}\right)\left(\nu + \frac{5}{2}\right). \tag{5.55}$$

For scattering by polar optical phonons, $\sigma(\mu) = (1-\mu)^{-1}$, substitution of $P_3(\mu) = \mu(5\mu^2 - 3)/2$ in Eq. (5.41) gives $\gamma_3 = 11/6$. In this case, $\nu = 0.5$ and it follows from Eq. (5.55) that $Q = 3.0$. For scattering by ionized impurities, the main contribution to $1/\tau^*$ is given by the scattering for small angle θ. Then $1 - \mu = \beta^{-1} = (2kL_{\mathrm{D}})^2$, $\sigma(\mu)$ does not depend on μ. $\gamma_3 = 6$, $\nu = 3/2$, $Q = 1.5$.

DP mechanism in uniaxially deformed crystals

According to Eq. (5.18), in uniaxially deformed crystals the electron Hamiltonian $H_1(\boldsymbol{k})$ in addition to terms cubic in k also equals

$$\boldsymbol{\Omega} = \frac{C_3}{\hbar}\varphi. \tag{5.56}$$

Deformation splitting of the CB should cause an additional contribution to the electron spin relaxation rate. In this case,

$$\begin{aligned}\overline{\Omega_z^2} &= \frac{1}{3\hbar^2}C_3^2k^2(\varepsilon_{xz}^2+\varepsilon_{yz}^2),\\ \overline{\Omega_x\Omega_y} &= \frac{1}{3\hbar^2}C_3^2k^2\varepsilon_{xz}\varepsilon_{yz}.\end{aligned} \tag{5.57}$$

Other components are obtained by cyclic permutation of coordinates x, y, and z. From Eq. (5.57), we see that only the shear strain (ε_{ij}, $i \neq j$) can induce splitting of the CB and spin relaxation. The axial strain (ε_{ii}) cannot induce splitting of the CB and has no contribution to the spin relaxation.

If a compressive stress is applies to the crystal along the $\langle 111\rangle$ axis, then $\varepsilon_{xy} = \varepsilon_{xz} = \varepsilon_{yz} = \varepsilon'_{111}/3$, and the corresponding contribution to $1/\tau^*$ is given by Eq. (5.50),

$$\frac{1}{[\tau_s(k)]_{ii}} = \frac{4}{[\tau_s(k)]_{ij}} = \frac{4}{3}\tau_p(k)\frac{C_3^2k^2}{\hbar^2}\left(\frac{\varepsilon'_{111}}{3}\right)^2. \tag{5.58}$$

where it is taken $\gamma_1 = 1$. ε'_{111} is the uniaxial strain along the $\langle 111\rangle$ axis,

$$\varepsilon'_{111} = \varepsilon_{111} - \varepsilon_{1\bar{1}0} = \frac{S_{44}P_{111}}{2}, \tag{5.59}$$

where S_{44} and P_{111} are the compliance coefficient and the applied stress, respectively. The principal axis of the tensor $1/\tau_s(k)$ lies in the $\langle 111\rangle$ direction. In this coordinate system, according to Eq. (5.58),

$$\frac{1}{[\tau_s(k)]_{||}} = \frac{2}{[\tau_s(k)]_{\perp}} = \frac{32}{81}\tau_p(k)\frac{\eta^2E_k}{\hbar^2E_g(1-\eta/3)}(C_2\varepsilon'_{111})^2, \tag{5.60}$$

Here we have used the expression of C_3 Eq. (5.19). After averaging over the Maxwell distribution, we obtain

$$\frac{1}{\tau_{s||}} = \frac{2}{\tau_{s\perp}} = \frac{16}{27}\frac{\tau_pk_BT\eta^2}{\hbar^2E_g(1-\eta/3)}(C_2\varepsilon'_{111})^2. \tag{5.61}$$

For compression along the $\langle 110\rangle$ axis, when $\varepsilon_{xy} = \varepsilon'_{110}/2, \varepsilon_{xz} = \varepsilon_{yz} = 0$,

$$\varepsilon'_{110} = \varepsilon_{110} - \varepsilon_{1\bar{1}0} = \frac{1}{2}S_{44}P_{110}. \tag{5.62}$$

From Eqs. (5.50) and (5.57), we obtain

$$\frac{1}{[\tau_s(k)]_{zz}} = \frac{2}{[\tau_s(k)]_{xx}} = \frac{2}{[\tau_s(k)]_{yy}} = \frac{2}{3}\tau_p(k)\frac{k^2}{\hbar^2}(C_3\varepsilon_{xy})^2. \tag{5.63}$$

In this case, the principal axis of the tensor $1/\tau_s(k)$ lies in the $\langle 001\rangle$ direction, which is perpendicular to the stress direction. After averaging over the Maxwell distribution, we obtain

$$\frac{1}{\tau_{s||}} = \frac{2}{\tau_{s\perp}} = \frac{4}{9}\frac{\tau_p k_B T}{\hbar^2 E_g}\frac{\eta^2}{(1-\eta/3)}(C_2\varepsilon'_{110})^2. \tag{5.64}$$

5.4. Bir–Aronov–Pikus Mechanism

In this mechanism, the spin flip is caused by the electron scattering by the free or bound holes; the similar effect occurs on scattering by the paramagnetic impurities, which is an exchange interaction. In the Bohr approximation, the electron spin relaxation time due to scattering on holes is defined by

$$\frac{1}{\tau_s(\boldsymbol{k})} = 4\pi^3 a_B^6 \sum_{\substack{\alpha\alpha' \\ \boldsymbol{pq}}} |\psi(0)|^4 |D_{m'\alpha',m\alpha}(\boldsymbol{p}+\boldsymbol{q},\boldsymbol{p})|^2 f_{\boldsymbol{p}}(1-f_{\boldsymbol{p}+\boldsymbol{q}})$$
$$\times\, \delta(E_{\boldsymbol{k}} + \varepsilon_{\boldsymbol{p}} - E_{\boldsymbol{k}-\boldsymbol{q}} - c\varepsilon_{\boldsymbol{p}+\boldsymbol{q}}), \quad (m \neq m'). \tag{5.65}$$

Here m and m' are the spin indices of an electron, α, and α' are those of a hole in the initial and final states, respectively. $\boldsymbol{p}$ and ε_p are the hole wave vector and energy, $f_{\boldsymbol{p}}(\varepsilon_{\boldsymbol{p}})$ is the hole distribution function, and $\hbar\boldsymbol{q}$ is the momentum transferred in the course of scattering. $\psi(\boldsymbol{r})$ is the wave function of electron–hole relative motion (exciton), the variable is the relative coordinate $\boldsymbol{r} = \boldsymbol{r}_e - \boldsymbol{r}_h$. In III–V compounds as a rule $m_e \ll m_h$, so that both the transferred momentum $\hbar\boldsymbol{q}$ and the electron momentum $\hbar\boldsymbol{k}$ are much smaller than that of the hole $\hbar\boldsymbol{p}$. As a result, the total momentum of an electron–hole pair, $\hbar\boldsymbol{k}$ is practically equal to $\hbar\boldsymbol{p}$, and operator $D(\boldsymbol{p},\boldsymbol{p}+\boldsymbol{q})$ depends only on the direction of $\boldsymbol{p}$. The scattering process in Eq. (5.65) corresponds to an exciton transition from the (m,α) state to the (m',α') state, where the electron spin is inversed, and the transferred momentum is $\hbar\boldsymbol{q}$, i.e. the electron wave vector changes from $\boldsymbol{k}$ to $\boldsymbol{k}-\boldsymbol{q}$, and that of hole from $\boldsymbol{p}$ to $\boldsymbol{p}+\boldsymbol{q}$.

In the following, we give the spin relaxation rates of Bir–Aronov–Pikus (BAP) mechanism for three cases (Bir *et al.*, 1975).

1. Holes are nondegenerate

$$\frac{1}{\tau_{\rm s}} = \frac{2}{\tau_0}\frac{\nu_k}{\nu_{\rm B}} N_{\rm p} a_{\rm B}^3, \tag{5.66}$$

where

$$\begin{aligned} \frac{1}{\tau_0} &= \frac{\pi}{2\hbar}\frac{\overline{D_{\rm s}^2}}{E_{\rm B}}|\psi(0)|^4, \\ \overline{D_{\rm s}^2} &= \int \frac{{\rm d}\Omega_{\rm p}}{4\pi}\frac{1}{2}\sum_{\alpha\alpha'}|D_{m'\alpha',m\alpha}|^2, \quad (m \neq m') \\ E_{\rm B} &= \frac{\hbar^2}{2m_{\rm e}a_{\rm B}^2}, \quad \nu_{\rm B} = \frac{\hbar}{m_{\rm e}a_{\rm B}}, \quad a_{\rm B} = \frac{\hbar^2\varepsilon_0}{{\rm e}^2 m_{\rm e}}, \end{aligned} \tag{5.67}$$

$a_{\rm B}$ is the effective Bohr radius in semiconductors, $N_{\rm p}$ is the hole concentration, and ν_k is the electron velocity.

2. Holes are degenerate with Fermi energy $\varepsilon_{\rm F} \gg E_{\boldsymbol{k}}$, k_BT

$$\frac{1}{\tau_{\rm s}} = \frac{3}{\tau_0}\frac{\nu_k}{\nu_{\rm B}}\frac{k_BT}{\varepsilon_{\rm F}} N_{\rm p} a_{\rm B}^3. \tag{5.68}$$

For hot electrons, $E_{\boldsymbol{k}} \gg k_BT$,

$$\frac{1}{\tau_{\rm s}} = \frac{2}{\tau_0}\frac{\nu_{\rm F}}{\nu_{\rm B}}\frac{E_k}{\varepsilon_{\rm F}} N_{\rm p} a_{\rm B}^3, \tag{5.69}$$

where $\nu_{\rm F}$ is the Fermi velocity of heavy holes.

3. Spin relaxation due to scattering on bound holes at low temperature

$$\frac{1}{\tau_{\rm s}} = \frac{2}{\tau_0}\frac{\nu_k}{\nu_{\rm B}} N_{\rm A} a_{\rm B}^3, \tag{5.70}$$

where $N_{\rm A}$ is the acceptor concentration.

Comparison of EY, DP, and BAP mechanisms

1. The EY and DP mechanisms principally differ by their opposite dependences on $\tau_{\rm p}$: for the EY mechanism, $1/\tau_{\rm s} \sim 1/\tau_{\rm p}$ (Eq. (5.31)), whereas for the DP one $1/\tau_{\rm s} \sim \tau_{\rm p}$ (Eq. (5.53)). Therefore, when scattering on impurity predominates, the efficiency of the EY mechanism increases and that of the DP mechanism decreases with increasing impurity concentration.
2. The efficiency of the BAP mechanism, as in the case of the EY mechanism, also increases with acceptor concentration. However, when the

acceptor concentration is higher than a critical concentration N_0, the increase in $1/\tau_s$ slows down mainly due to the screening of $|\psi(0)|^2$ by the delocalized holes.

3. For electron scattering by phonons, the spin relaxation rate $1/\tau_s$ for both the EY and DP mechanisms increases rapidly with temperature (Eqs. (5.31) and (5.53)). The dependences of $1/\tau_s$ on temperature for these mechanisms are different which gives a possibility to distinguish the EY and DP mechanisms. For the BAP mechanism the, temperature dependence of $1/\tau_s$ is more complicated (Eqs. (5.66) and (5.67)), which is determined mainly by the variation with temperature of the electron velocity ν_k, ionized degree N_p/N_A, and $|\psi(0)|^4$.
4. Both the EY and DP mechanisms are related to the spin–orbit splitting of the valence band. In the case when scattering on polar optical or piezo-electrical phonons predominates, according to Eqs. (5.31) and (5.53),

$$\frac{\tau_{s,DP}}{\tau_{s,EY}} = \frac{9}{16}\frac{\hbar^2}{\tau_p^2 k_B T E_g}\left(\frac{m_{cv}}{m_e}\right)^2 \frac{(1-\eta/2)^2}{1-\eta/3}. \tag{5.71}$$

Since $\hbar/\tau_p < k_B T \ll E_g$, it follows from Eq. (5.71) that $1/\tau_{s,EY}$ can exceed $1/\tau_{s,DP}$ only in narrow gap semiconductors with $m_e \ll m_{cv}$ (m_{cv} is the reduced mass of that of the electron and the heavy hole). The η term has small influence (Eq. (5.19)).

5.5. Experimental Studies of Spin Relaxation in III V Compounds

5.5.1. *Optical orientation method*

When a semiconductor is illuminated by a circular polarization light, the electrons will be excited from the valence band to the CB. According to the basis functions of the heavy hole states (hh), light hole states (lh), and spin–orbit splitting states (sh), the transition selection rules and probabilities can be obtained, as shown in Fig. 5.1 (D'yakonov and Perel, 1984).

For example, for the transition from the hh state $|3/2, 3/2\rangle$ to electron $|1/2, 1/2\rangle$ state, the basis function of the hh state is

$$\left|\frac{3}{2},\frac{3}{2}\right\rangle = \frac{1}{\sqrt{2}}(X+iY)\uparrow.$$

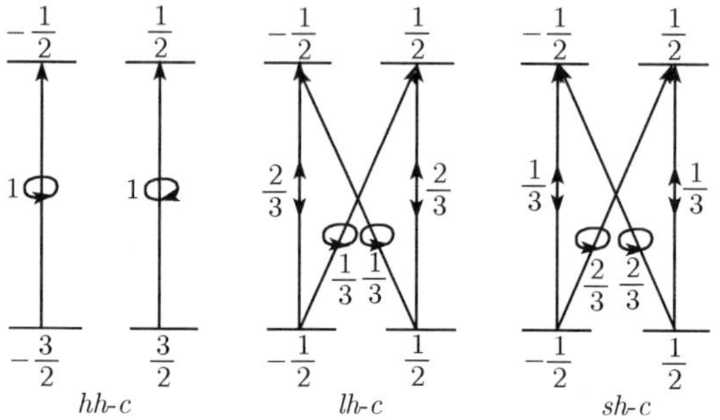

Fig. 5.1. Possible hh-c, lh-c, and sh-c transitions. The straight and circular arrows represent the linear and circular polarization lights, respectively. The numbers near the arrows represent the relative transition intensities. The number near the levels correspond to the projection of the total angular momentum on the quantization axis.

The basis function of the electron state is

$$\left|\frac{1}{2},\frac{1}{2}\right\rangle = S\uparrow .$$

The dipole operator of the circular polarization light is

$$p_{\pm} = \frac{1}{\sqrt{2}}(p_x \pm ip_y).$$

The nonzero transition matrix element is

$$\left\langle S\uparrow\left|\frac{1}{\sqrt{2}}(p_x - ip_y)\right|\frac{1}{\sqrt{2}}\right\rangle (X+iY)\uparrow$$
$$= \frac{1}{2}[\langle S|p_x|X\rangle + \langle S|p_y|Y\rangle] = \langle S|p_x|X\rangle = P.$$

Therefore, when a left circular polarization light illuminates on the sample, the electrons are excited from the hh $|3/2,3/2\rangle$ state to the electron $|1/2,1/2\rangle$ state with the transition probability equal to P^2. Other transition selection rules can be similarly obtained. The dipole operator of the linear polarization transition is p_z. The transition selection rules can be summarized as follows: the differences of the angular momentum components of the final state (electron) and the initial state (hole) are $\Delta m = -1$, $+1$, and 0 for left, right, and linear polarization lights, respectively, as shown clearly in Fig. 5.1.

In the optical orientation experiment, when a circular polarization light (suppose left circular polarization) illuminates on the semiconductor, the electrons are excited from the valence band to the CB. The electrons in the CB have a lifetime, then recombine with the holes in the valence band,

and emit left and right circular polarization lights. Let the intensities of the left and right circular polarization lights be I_- and I_+, respectively, and the degree of circular polarization of luminescence is defined as

$$\rho = \frac{I_+ - I_-}{I_+ + I_-}. \tag{5.72}$$

It can be proved that in bulk materials the measured degree of circular polarization does not exceed 0.25 (ideal value). In bulk materials, the energies of the heavy hole band and light hole band are equal, that of the spin–orbit splitting band is lower than them by Δ, which will be not considered in following. When a left circular polarization light illuminates on the semiconductor, the electrons are excited from the hh $|3/2, 3/2\rangle$ state to the electron (e) $|1/2, 1/2\rangle$ state with the probability 1, and from the lh $|3/2, 1/2\rangle$ state to the e $|1/2, -1/2\rangle$ state with the probability 1/3. Thus, there are one part of $|1/2, 1/2\rangle$ electron and 1/3 part of $|1/2, -1/2\rangle$ electron in the CB. In the recombination process, the $|1/2, 1/2\rangle$ state electron transits to the hh $|3/2, 3/2\rangle$ state and emits right circular polarization light ($\Delta m = +1$) with the relative intensity 1; at the same time, the $|1/2, 1/2\rangle$ state electron also transits to the lh $|3/2, -1/2\rangle$ state and emits left circular polarization light with the intensity 1/3. Similarly, the $|1/2, -1/2\rangle$ state electron transits to the hh $|3/2, -3/2\rangle$ and lh $|3/2, 1/2\rangle$ states, and emits left and right circular polarization lights with intensities 1 and 1/3, respectively. The intensity of the right circular polarization light is $I_+ = 1 \times 1 + (1/3) \times (1/3) = 10/9$, whereas that of the left polarization is $I_- = 1 \times (1/3) + (1/3) \times 1 = 2/3$, so that the degree of the circular polarization is $\rho = (10 - 6)/(10 + 6) = 0.25$. This is the ideal value, after taking into account other factors, ρ is generally smaller than 0.25. For quantum wells and superlattices, the quantum confinement effect splits the heavy and light hole bands and lowers the light hole band, ρ may be larger than 0.25.

Due to the spin relaxation, the degree of the spin polarization of electrons in the CB will decrease, the measured degree of circular polarization of luminescence ρ will be

$$\rho = \frac{\rho_0}{1 + \tau/\tau_s}, \tag{5.73}$$

where τ_s is the spin relaxation time, and τ is the lifetime, and ρ_0 is the ideal ρ (0.25).

A magnetic field B transverse to the direction of light propagation causes depolarization of the luminescence, because of the procession of electron spins in the magnetic field (the Hanle effect). The decrease in the degree of polarization is given by

$$\rho(B) = \frac{\rho}{1 + \Omega_L^2 T_s^2}, \tag{5.74}$$

where Ω_L is the Larmor frequency in the transverse field B, $\Omega_L = \mu_B g B/\hbar$, and $T_s^{-1} = \tau_s^{-1} + \tau^{-1}$. If $\Omega_L T_s \gg 1$, which means that in the time T_s the spin has rotated many circles around the magnetic field. The average spin along the initial spin direction is very small, and the degree of the circular polarization $\rho(B)$ will decrease as shown in Eq. (5.74).

After optical excitation, the spin component varies with time,

$$S_z = S_0 \cos \Omega_L t \exp\left(\frac{-t}{\tau_s}\right). \tag{5.75}$$

The number of electrons in the CB varies with time,

$$W(t) = \frac{1}{\tau} e^{-t/\tau}. \tag{5.76}$$

Combining Eqs. (5.75) and (5.76), and averaging over time, we obtain

$$S_z(B) = S_0 \int_0^\infty W(t) e^{-t/\tau_s} \cos \Omega_L t \, dt = \frac{S_z(0)}{1 + \Omega_L^2 T_s^2}. \tag{5.77}$$

$$S_z(0) = \frac{S_0}{1 + \tau/\tau_s}. \tag{5.78}$$

Rewriting them in degree of circular polarization, Eqs. (5.77) and (5.78) are just Eqs. (5.74) and (5.73).

The τ_s and τ can be determined independently by measuring the degree of luminescence polarization ρ and observing its decrease in the transverse magnetic field $\rho(B)$. The time T_s is usually determined from the field value B_0 for which the initial polarization is halved: $\rho(B_0) = \rho(0)/2$. In this B_0,

$$\frac{1}{T_s} = \Omega_L = \frac{\mu_B g B_0}{\hbar} = \frac{1}{\tau} \frac{\rho_0}{\rho(0)}.$$

The electron spin relaxation time τ_s and lifetime τ are then,

$$\begin{aligned} \tau &= \frac{\rho_0}{\rho(0)} \frac{\hbar}{\mu_B g B_0}, \\ \tau_s &= \frac{\rho_0}{\rho_0 - \rho(0)} \frac{\hbar}{\mu_B g B_0}. \end{aligned} \tag{5.79}$$

For GaAs, $|g_e| = 0.44$, $\rho(0) \ll \rho_0$, from Eq. (5.79), we obtain $\tau_s = 2.58 \times 10^{-11} B_0^{-1}$, where the unit of τ_s is s, unit of B_0 is tesla. Thus, if B_0 is the order of tesla, τ_s is that of 10^{-11} s.

5.5.2. *Spin relaxation in InSb (EY mechanism)*

Figure 5.2 shows the $\tau_s^{-1/2}$ versus electron Fermi level in p-InSb at 1.3 K (Benoit *et al.*, 1974). Excitation with a CO laser made it possible to achieve high concentration of photoelectrons, n. The electron distribution was found to be degenerate, with the Fermi level E_F. The E_F varies with the change of excitation intensity in the range of 2–7 meV, i.e. corresponded to the increase of photoelectron concentration n from 0.7×10^{15} to $5 \times 10^{15}\,\text{cm}^{-3}$. Because of the strong degeneracy of electrons at $T = 1.3\,\text{K}$, only electrons with an energy near E_F could experience spin relaxation. From Eq. (5.28), we obtain

$$\left(\frac{1}{\tau_s}\right)_{EY} = \frac{16}{27\tau_p}\left(\frac{E_F}{E_g}\right)^2 \eta^2 \left(\frac{1-\eta/2}{1-\eta/3}\right)^2 \Phi. \tag{5.80}$$

In these conditions, $1/\tau_p \sim N_p E_k^{-3/2} = \text{const.}$, since the free hole concentration N_p, which equals the electron concentration n, is proportional to $E_F^{3/2}$. Therefore, from Eq. (5.80), we get $1/\tau_s \sim E_F^2$ in agreement with the experimental data (Fig. 5.2). It is noticed that the rate of spin relaxation via the DP mechanism at $\tau_p = \text{const.}$, according to Eq. (5.52), should be proportional to E_F^3. Thus, the DP mechanism does not explain the observed dependence of $1/\tau_s$ on E_F.

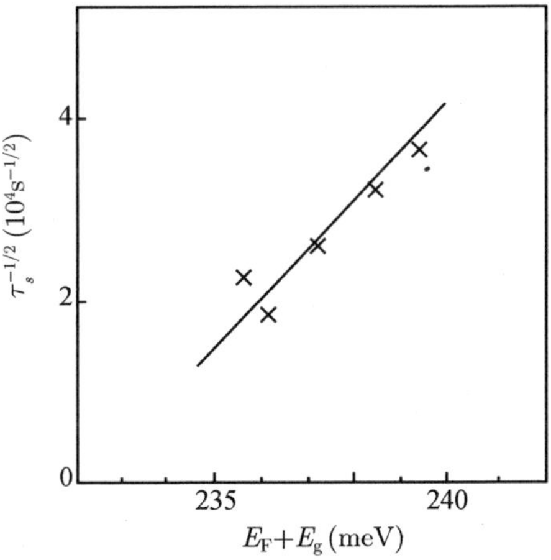

Fig. 5.2. Square root of spin relaxation rate of degenerate electrons versus electron Fermi level in InSb at $T = 1.3\,\text{K}$.

5.5.3. *Spin relaxation in GaAs (DP mechanism)*

Figure 5.3 shows the spin relaxation rate $1/\tau_s$ and hole mobility μ_h versus temperature for two GaAs samples (Maruschak *et al.*, 1983). The two samples have the same acceptor concentrations $N_A = 4 \times 10^{16}\,\text{cm}^{-3}$, but different degrees of compensation. The compensation of acceptors affords the possibility of studying crystals with substantially different carrier mobilities. The onset of the DP mechanism shows a sharp increase in the spin relaxation rate at $T \geq 30$–$40\,\text{K}$. According to Eq. (5.53), the $1/\tau_s$ should rise in proportion to $\tau_p T^3$. When the power dependence of τ_p is taken into account, we obtain $1/\tau_s \sim T^{3+n}$, where n varies with temperature. The temperature dependence of n is caused by the variation with temperature of both the relative efficiencies of electron scattering by impurities and phonons and the number of thermally ionized acceptors. The data in Fig. 5.3 show that in the temperature range $40 < T < 77\,\text{K}$,

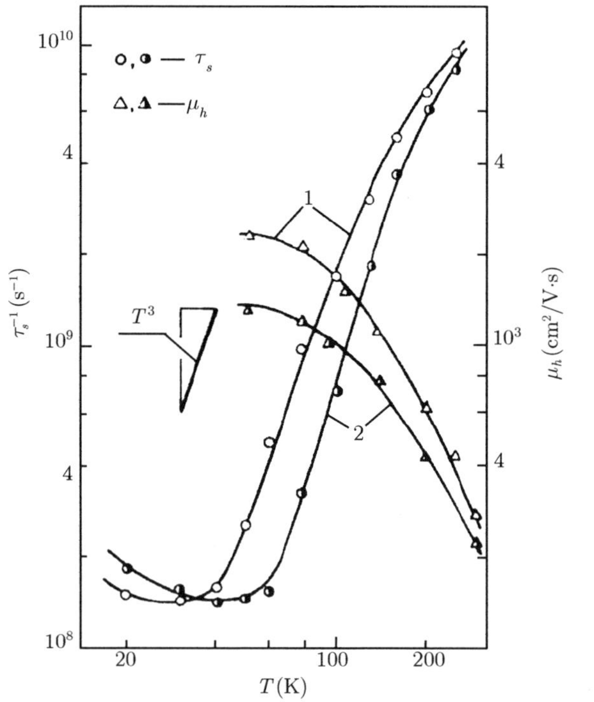

Fig. 5.3. Spin relaxation rate $1/\tau_s$ and hole mobility μ_h versus temperature for two GaAs samples.

where τ_p increases, i.e. $n > 0$, the slope of the experimental curves exceeds 3, whereas at $T > 77\,\text{K}$, where τ_p decreases, the slope becomes smaller than 3.

From Eq. (5.53), we see that $1/\tau_s$ is proportional to τ_p, this is a special characteristic of DP mechanism. In Fig. 5.3, sample 2 has smaller mobility μ_h (larger acceptor compensation); hence, smaller τ_p, so that the $1/\tau_s$ is also smaller. The dependence of $1/\tau_s$ on the degree of acceptor compensation provides unambiguous evidence of the domination of the DP mechanism at $T > 30{-}40\,\text{K}$.

At $T < 30{-}40\,\text{K}$, the DP mechanism no longer explains the experimental data. The increase in $1/\tau_s$ in this temperature range is the consequence of the localization of electrons at low temperatures and of the manifestation of other spin relaxation mechanisms in new conditions, such as hyperfine interaction of electron spins with magnetic momenta of lattice nuclei.

Kikkawa and Awschalom (1998) studied T_2^* as a function of temperature by the resonant spin amplification (RSA) technique (see Fig. 8.5(b)). They found that when $T \geq 30\,\text{K}$, the measured T_2^* versus T is in agreement with the prediction of the DP mechanism. When $T < 30\,\text{K}$, the T_2^* has a strong dependence on the magnetic field, the magnetic field made T_2^* to decrease by two orders of magnitude. In the zero-field case, at low temperatures, the mechanism of spin relaxation is changed from the DP to the EY mechanism. Lau *et al.* (2001) calculated T_2 as functions of temperature using the DP mechanism for three n-type semiconductors: GaAs ($n = 1.0 \times 10^{16}\,\text{cm}^{-3}$), InAs ($n = 1.7 \times 10^{16}\,\text{cm}^{-3}$), and GaSb ($n = 1.49 \times 10^{18}\,\text{cm}^{-3}$). Their results and corresponding experimental results are shown in Fig. 5.4. The agreement with experimental measurements for GaAs at higher temperatures is good, whereas for low temperatures other spin relaxation mechanisms may dominate. The smaller T_2's in InAs and GaSb are due partly to the larger conduction spin splitting, which originates from a larger ratio of the SOC to the band gap.

5.5.4. *Spin relaxation in GaAs (BAP mechanism)*

In GaAs crystals, the BAP mechanism reveals itself at the doping level $N_A \geq 10^{17}\,\text{cm}^{-3}$, at first in the low-temperature range. As the acceptor concentration increases the BAP mechanism begins to dominate at progressively higher temperatures. Figure 5.5 shows the spin relaxation rate $1/\tau_s$ versus temperature for three GaAs samples: (1) $N_A = 3.5 \times 10^{17}\,\text{cm}^{-3}$, (2, 3) $N_A = 2.2 \times 10^{17}\,\text{cm}^{-3}$ (Pikus and Titkov, 1984). The DP mechanism

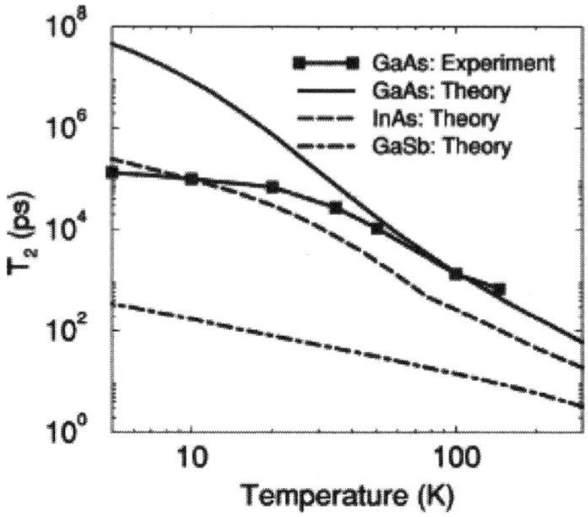

Fig. 5.4. Calculated T_2 with the DP mechanism versus temperature for three n-semiconductors: GaAs ($n = 1.0 \times 10^{16}\,\text{cm}^{-3}$), InAs ($n = 1.7 \times 10^{16}\,\text{cm}^{-3}$), and GaSb ($n = 1.49 \times 10^{18}\,\text{cm}^{-3}$). The squares represent the experimental results of GaAs.

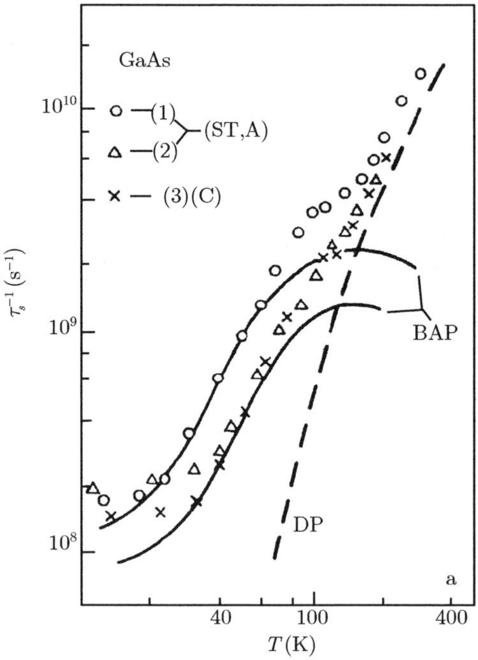

Fig. 5.5. Spin relaxation rate $1/\tau_s$ versus temperature for three GaAs samples: (1) $N_A = 3.5 \times 10^{17}\,\text{cm}^{-3}$, (2, 3) $N_A = 2.2 \times 10^{17}\,\text{cm}^{-3}$. The ST, A, and C denote experimental dots taken from different references.

gives a good description of experimental data only at high temperature, $T \geq 150\,\text{K}$. In the region of lower temperatures, the BAP mechanism is seen to exert its influence. This becomes stronger with increasing acceptor concentration. Taking into account the electron scattering by both free holes and holes bound on acceptors, adding Eqs. (5.66) and (5.70), we get

$$\frac{1}{\tau_\text{s}} = \frac{2}{\tau_0}(N_\text{A} a_\text{B}^3)\left[\frac{N_\text{p}}{N_\text{A}} + 1.66\frac{N_\text{A} - N_\text{p}}{N_\text{A}}\right]\frac{\nu_k}{\nu_\text{B}}. \tag{5.81}$$

The solid lines in Fig. 5.5 are obtained from Eq. (5.81).

The temperature variation of the spin relaxation rate in Fig. 5.5 is a combined effect of three factors: (1) increase in electron velocity ν_k with temperature, (2) increase in the degree of acceptor ionization N_p/N_A with temperature, and (3) decrease in the Sommerfeld factor $|\psi(0)|^4$ with temperature (see Eq. (5.67)). A specific feature of the temperature dependence of $1/\tau_\text{s}$ is its nonmonotonic character: in the region of acceptor depletion, the rapid decrease in the Sommerfild factor with rising temperature leads to a slower spin relaxation and even to a decrease in $1/\tau_\text{s}$. The experimental dependences of $1/\tau_\text{s}$ in Fig. 5.5 exhibit only saturation at $T > 100\,\text{K}$, the expected decrease in $1/\tau_\text{s}$ at higher temperatures is smeared by the DP mechanism.

For degenerate semiconductor samples with higher acceptor concentrations, the temperature dependence of spin relaxation rate $1/\tau_\text{s}$ is simpler. Figure 5.6 shows the $1/\tau_\text{s}$ versus temperature for degenerate GaSb sample: (1) $N_\text{A} = 1.8 \times 10^{18}\,\text{cm}^{-3}$, (2) $6.5 \times 10^{18}\,\text{cm}^{-3}$ and GaAs sample: (1, 3) $N_\text{A} = 5 \times 10^{18}\,\text{cm}^{-3}$, (2) $4 \times 10^{19}\,\text{cm}^{-3}$ (Pikus and Titkov, 1984). In a wide temperature range, $1/\tau_\text{s} \sim T^{3/2}$, which is consistent with the BAP mechanism in degenerate semiconductor, Eq. (5.68), where the electron velocity $\nu_k \sim T^{1/2}$.

It is noticed in Fig. 5.6 that the $1/\tau_\text{s}$ approach saturated at low temperatures, especially for GaAs samples, the saturation starts at $40 \sim 60\,\text{K}$. This is due to the incomplete themolization in the CB during the electron lifetime. The crystals studied have rather short lifetimes of $2–3 \times 10^{-10}\,\text{s}$ for GaAs and $4–5 \times 10^{-10}\,\text{s}$ for GaSb. The scattering by acoustic phonons cannot provide the effective thermolization channel as the corresponding energy relaxation times of the order $10^{-9} \sim 10^{-8}\,\text{s}$ are far in excess of the lifetime. Thus, at $T = 10\,\text{K}$, the electrons in degenerate GaAs and GaSb crystals are certain not to reach the lattice temperature, so this is the reason of saturation of $1/\tau_\text{s}$ at low temperatures.

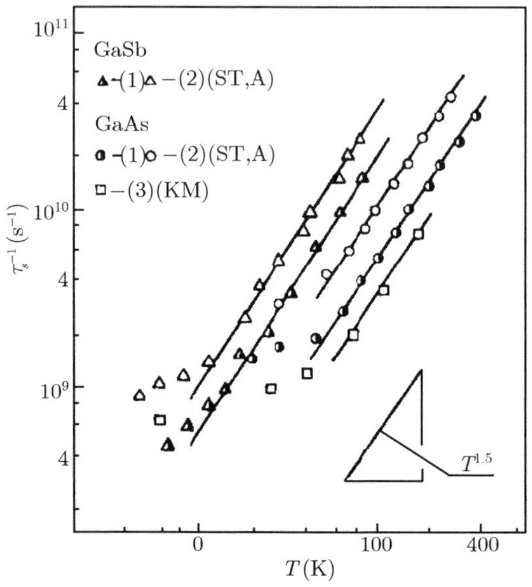

Fig. 5.6. $1/\tau_s$ versus temperature for degenerate GaSb sample: (1) $N_A = 1.8 \times 10^{18}\,cm^{-3}$, (2) $6.5 \times 10^{18}\,cm^{-3}$ and GaAs sample: (1, 3) $N_A = 5 \times 10^{18}\,cm^{-3}$, (2) $4 \times 10^{19}\,cm^{-3}$. ST, A, and KM denote experimental dots taken from different references.

For thermo-electrons $E_k \gg k_B T$, the $1/\tau_s$ of the BAP mechanism is given by Eq. (5.69),

$$\frac{1}{\tau_s(T_e)} = \frac{2}{\tau_0}\frac{\nu_F}{\nu_B}\frac{k_B T_e}{\varepsilon_F} N_p a_B^3, \tag{5.82}$$

where T_e is the "temperature" of the thermo-electrons. Whereas for thermolized electrons from Eq. (5.68),

$$\frac{1}{\tau_s(T)} = \frac{\sqrt{3}}{\tau_0}\frac{(k_B T)^{3/2}}{\sqrt{m_e}\nu_B}\frac{1}{\varepsilon_F} N_p a_B^3. \tag{5.83}$$

Comparing Eqs. (5.82) and (5.83), we get

$$\frac{\tau_s(T)}{\tau_s(T_e)} = \frac{T_e}{T}\left(\frac{2m_e\varepsilon_F}{3m_h T}\right)^{1/2}, \tag{5.84}$$

Here $\tau_s(T)$ is the spin relaxation time for thermolized electrons defined by Eq. (5.68), which can be obtained from the extrapolation of the experimental curves at $T > 60\,K$. $\tau_s(T_e)$ is the spin relaxation time for hot electrons in the region of the plateau in Fig. 5.6. At $T = 10\,K$, it is estimated from

Eq. (5.84) that $T_e = 40\,\text{K}$ for GaAs sample with $N_A = 3.8 \times 10^{19}\,\text{cm}^{-3}$; $T_e = 29\,\text{K}$ at $N_A = 5 \times 10^{18}\,\text{cm}^{-3}$.

5.5.5. *Dependence of spin relaxation rate on acceptor concentration*

Figure 5.7 shows the spin relaxation rate $1/\tau_s$ versus acceptor concentration for GaAs (Aronov *et al.*, 1983). When the acceptor concentration N_A is smaller than a critical concentration N_0, $1/\tau_s \sim N_A^1$; when $N_A > N_0$, $1/\tau_s$ is nearly a constant; when N_A increases continuously $N_A > 10^{19}\,\text{cm}^{-3}$, $1/\tau_s \sim N_A^{1/3}$. At $N_A < N_0$, the electron relaxation rate in the BAP mechanism is determined by Eq. (5.81). When $N_A > N_0$, the saturation of the dependence of $1/\tau_s$ on N_A is due to rapid decrease in the Sommerfild factor $|\psi(0)|^4$ as a result of screening, which almost totally compensates the increase in $1/\tau_s$ with rising acceptor concentration. For instance, in GaAs at 77 K and $N_A = 5 \times 10^{17}\,\text{cm}^{-3}$ when the screening is not very important, $|\psi(0)|^4 \approx 20$. At concentration $N_A = 10^{19}\,\text{cm}^{-3}$, $|\psi(0)|^4$ is approximately 1, and hence in this concentration range, the product $|\psi(0)|^4 \times N_A$ is practically constant. When $N_A > 10^{19}\,\text{cm}^{-3}$, the semiconductor is degenerate, $1/\tau_s$ is determined by Eq. (5.68). In this case, the $|\psi(0)|^4$ is a constant, $\varepsilon_F = (3\pi^2\hbar^3 N_p)^{2/3}/2m_h$, $N_p \approx N_A$, thus $1/\tau_s \sim N_A^{1/3}$. At high temperature $T = 300\,\text{K}$, the range of $1/\tau_s \sim N_A$ is small, and only the latter two cases appear.

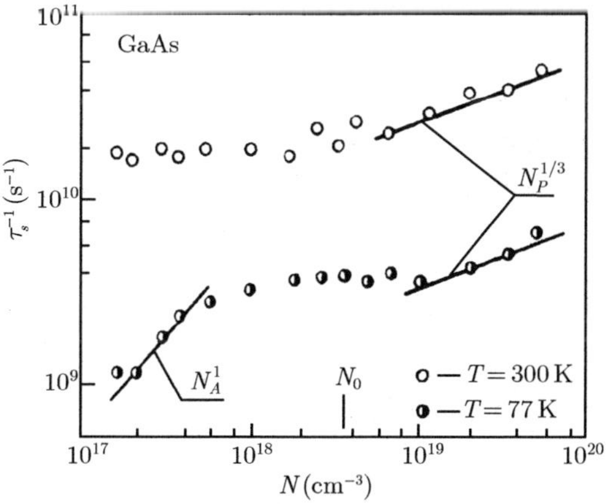

Fig. 5.7. $1/\tau_s$ versus acceptor concentration N_A in doped GaAs at $T = 77$ and 300 K.

5.6. Spin Relaxation in Quantum Wells

Consider the DY mechanism, from Eq. (5.49), we obtain (Lau *et al.*, 2001)

$$T_1^{-1} = \langle \Omega_\perp^2 \rangle \tau,$$
$$T_2^{-1} = \left(\frac{\Omega_\perp^2}{2 + \Omega_{||}^2} \right) \tau, \tag{5.85}$$

where τ is the momentum relaxation time, and $\Omega(\boldsymbol{k})$ is the inner magnetic field equivalent to the spin splitting of the CB. In III–V compounds, it is given by Eq. (5.51),

$$\boldsymbol{\Omega}(\boldsymbol{k}) \propto k_x(k_y^2 - k_z^2)\hat{\boldsymbol{x}} + k_y(k_z^2 - k_x^2)\hat{\boldsymbol{y}} + k_z(k_x^2 - k_y^2)\hat{\boldsymbol{z}}. \tag{5.86}$$

From Eq. (5.86), we see that when $\boldsymbol{k}\|(111), \boldsymbol{\Omega}(\boldsymbol{k}) = 0$ (the lowest order of perturbation). If the crystal is cubic, $\Omega_x = \Omega_y = \Omega_z$, so $T_1 = T_2$. The T_1 and T_2 differ, however, for systems of lower symmetry.

$\boldsymbol{\Omega}(\boldsymbol{k})$ is induced by the bulk inversion asymmetry (BIA) of the crystal, another effect caused by the structure inversion asymmetry (SIA) is the Rashba effect, whose Hamiltonian is

$$H_{\mathrm{R}} = [\boldsymbol{\alpha}(\boldsymbol{r}) \times \boldsymbol{p}] \cdot \boldsymbol{\sigma}, \tag{5.87}$$

where $\boldsymbol{\alpha}$ is a vector, whose direction is along the symmetry-breaking direction. If an electric field is applied in the direction perpendicular to the quantum well plane (z-direction), the electric potential will produce asymmetry in the z-direction, so $\boldsymbol{\alpha}$ is along the z-direction. If the electron momentum $\boldsymbol{p}$ is along the x-direction, then the effective inner magnetic field Ω is along the y-direction.

In the quantum well grown in the (001) direction, because k_z is no longer a good quantum number. From Eq. (5.86), Ω is limited in the quantum well plane, in the x- or y-direction. When the quantum well is in the electric field along the (001) direction, or the quantum well is a modulation doped structure, the Ω caused by the Rashba effect is also in the quantum well plane, perpendicular to the electron wave vector $\boldsymbol{k}$.

In order to calculate T_1, we expand $\boldsymbol{\Omega}(\boldsymbol{k})$ with spherical harmonic functions (three-dimension) or triangle functions (two-dimension),

$$\Omega(\boldsymbol{k}) = \sum_{lm} \Omega_{lm}(E) Y_{lm}(\theta, \varphi), \quad \text{(3D)}$$
$$\Omega(\boldsymbol{k}) = \sum_{l} [\Omega_{l1}(E)\cos(l\theta) + \Omega_{l2}(E)\sin(l\theta)], \quad \text{(2D)}. \tag{5.88}$$

Time reversal invariance requires $\boldsymbol{\Omega}(\boldsymbol{k}) = -\boldsymbol{\Omega}(-\boldsymbol{k})$, so $\Omega_l = 0$ for even l, $\Omega_{lm} = 0$. Each component can be looked as an independent perturbation field.

The time for effective field reversal (τ_l) depends on the angular index l of the field component (Ω_l). For example, an $l = 1$ component (Ω_1) requires a 180° change in the angle of $\boldsymbol{k}$ to change sign, whereas an $l = 3$ component (Ω_3) only requires a 60° change, so typically $\tau_3 < \tau_1$.

Quasi-elastic scattering will simplify greatly the calculation, in this case

$$\frac{1}{T_1} = \frac{1}{n} \int D(E) f(E)[1 - f(E)] \sum_l \tau_l(E) \Omega_l^2(E) \mathrm{d}E, \tag{5.89}$$

where $D(E)$ is the density of states, $f(E)$ is the Fermi distribution function, and n is the electron density.

$$\Omega_l^2(E) = \sum_{lm} \Omega_{\alpha,lm}^2(E), \tag{5.90}$$

α represents a certain scattering mechanism. $\tau_l(E)$ is the momentum scattering time of angular momentum l,

$$\begin{aligned} \tau_l^{-1}(E) &= \int_{-1}^{1} \sigma(\theta, E)[1 - P_l(\cos\theta)] \mathrm{d}\cos\theta, \quad \text{(3D)} \\ \tau_l^{-1}(E) &= \int_{0}^{2\pi} \sigma(\theta, E)[1 - \cos(l\theta)] \mathrm{d}\theta, \quad \text{(2D)} \end{aligned} \tag{5.91}$$

where $\sigma(\theta, E)$ is the scattering cross section. The electron mobility

$$\mu = \left(\frac{e}{m^* n}\right) \int D(E) f(E)[1 - f(E)] \tau_1(E) E \, \mathrm{d}E. \tag{5.92}$$

D'yakonov and Kachorovskii (1986) applied the spin relaxation DP mechanism in bulk semiconductors to the quantum wells. They made the following approximations: (1) negligible penetration of states into the barriers is assumed, so $E_1 \ll \Delta E_c$, where E_1 is the confinement energy of the first quantum well state and ΔE_c is the CB offset. (2) The perturbative expressions $\Omega_1^2(E) \propto E(4E_1 - E)^2$, $\Omega_3^2(E) \propto E^3$ are used. Assume that the electron energy $E \ll E_1$, then $\Omega_1^2(E) \propto E$, $\Omega_3^2(E)$ is ignored. In these approximations,

$$\begin{aligned} \frac{1}{T_1} &= \frac{1}{n} \int D(E) f(E)[1 - f(E)] \tau_1(E) \Omega_1^2(E) \mathrm{d}E \\ &\propto \int D(E) f(E)[1 - f(E)] \tau_1(E) E \, \mathrm{d}E = \mu. \end{aligned} \tag{5.93}$$

The spin relaxation time $T_1(\tau_s)$ is obtained,

$$\tau_s = \frac{E_g \hbar^2}{2\alpha^2 E_1^2 k_B T} \frac{1}{\tau_v}, \tag{5.94}$$

where $\alpha = \gamma\hbar^{-3}(E_g)^{1/2}(2m^*)^{3/2}$, E_1 is the energy of the quantum level in quantum well, and τ_v is the electron momentum relaxation time, which determines the electron mobility $\mu = e\tau_v/m^*$. Because in quantum wells the relation between τ_v and temperature is unclear, so it is difficult to explain the experimental results of $\tau_s(T)$ with Eq. (5.94).

Terauchi *et al.* (1999) measured the spin relaxation times τ_s as functions of the electron mobility μ and the electron quantized energy E_1 in n-type GaAs/$Al_{0.4}Ga_{0.6}$As multiple quantum well (MQW) at 300 K by CP pump-probe technique, the results are shown in Figs. 5.8 and 5.9, respectively. From Fig. 5.8, it can be seen that the data can be well represented by $\tau_s \propto \mu^{-1}$, the relation predicted by the DP theory. In Fig. 5.8, the thick solid line is the result of DP theory calculated based on Eq. (5.94). In the calculation, τ_v is calculated from the measured mobility, and taking $E_1 = 47$ meV, $\alpha = 0.07$, which is the value for bulk GaAs. Thus, the

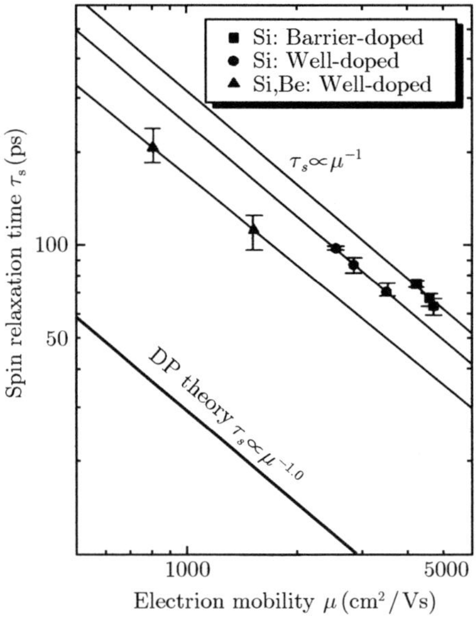

Fig. 5.8. Electron mobility μ dependence of the electron spin relaxation time τ_s in n-type GaAs/$Al_{0.4}Ga_{0.6}$As multi-quantum well. The thick solid line is the result of DP theory calculated based on Eq. (5.94).

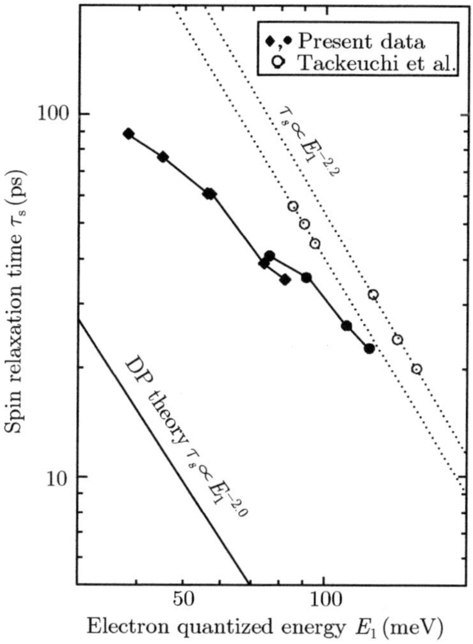

Fig. 5.9. Electron quantized energy E_1 dependence of the spin relaxation time τ_s. The thick solid line is the result of the DP theory.

calculated τ_s is about one order of magnitude smaller than that determined from the experimental results.

Figure 5.9 shows the electron quantized energy E_1 dependence of the spin relaxation time τ_s. The thick solid line is the result of the DP theory (Terauchi *et al.*, 1999). The experimental results are plotted as closed dots, and the open dots are results of another group (Tackeuchi *et al.*, 1996). Both sets of experimental data reveal a similar trend in that τ_s decreases as E_1 increases. However, these results cannot be fitted by the relation $\tau_s \propto E_1^{-2}$ (Eq. (5.94)) of the DP theory, especially in the low E_1 region. These results of Figs. 5.8 and 5.9 suggest that some correction must be made in the simple theory in order to explain the mechanism of spin relaxation in quantum wells.

Lau *et al.* (2001) calculated the spin lifetime $T_1(\tau_s)$ as a function of (a) mobility μ, (b) confinement energy E_1, (c) temperature T, for 7.5 nm $GaAs/Al_{0.4}Ga_{0.6}As$ MQW at RT by using a more strict theory, taking into account the optical phonon (OP) scattering and neutral impurity (NI) scattering. The results are shown in Fig. 5.10. The theoretical results are

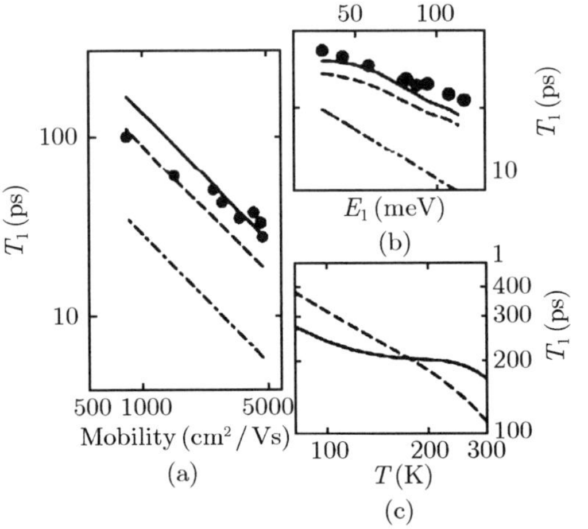

Fig. 5.10. $T_1(\tau_s)$ as a function of (a) mobility μ, (b) confinement energy E_1, and (c) temperature T, for 7.5 nm GaAs/$Al_{0.4}Ga_{0.6}As$ MQW at RT. Closed circles represent the experimental results (Terauchi *et al.*, 1999). The nonperturbative theory results with OP scattering (solid lines) and NI scattering (dashed lines) are shown, as well as the DP theory results (dot-dashed lines).

in agreement with the experimental results. The T_1 is larger than that predicted by the DP theory by about one order of magnitude. When the mobility μ drops, T_1 shifts from OP scattering to NI scattering, resulting in the deviation of $T_1 \propto \mu^{-1}$. In Fig. 5.10(b), the weaker dependence of T_1 on E_1 is also in agreement with the experimental results, which is due to wave-function penetration into the barriers and nonperturbative effects (the spin splitting for damped waves in the barrier is of the opposite sign as that for propagating wave in the wells). In Fig. 5.10(c), the OP results appear insensitive to temperature from 100 to 250 K, this is due to the rapid decrease in the mobility from OP scattering with increasing temperature.

In order to check the DP mechanism further more, Karimov *et al.* (2003) prepared (110)-oriented GaAs/AlGaAs quantum wells and found a 10-fold increase in the spin relaxation time T_1, and variation of T_1 with the applied electric field from 20 to 80 kV/cm at 170 K. It can be proved from Eq. (5.86) that in the (110)-oriented QWs, as shown in Fig. 5.11, $\Omega^{\rm BIA}$ lies approximately along the growth axis for all $\boldsymbol{k}$, $\Omega_\perp^2 = 0$, $T_1^{-1} \sim 0$, causing only weak spin relaxation along the growth axis and thereby leading to greatly enhanced spin memory, as predicted by D'yakonov and Kachorovskii (1986). Spin relaxation along the growth axis therefore

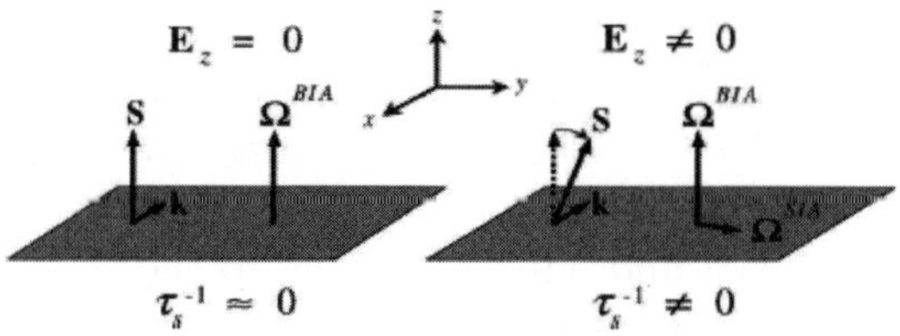

Fig. 5.11. $\Omega^{\rm BIA}$, $\Omega^{\rm SIA}$, and $\boldsymbol{S}$, $\boldsymbol{k}$ in a (110)-oriented QW (shown shaded) in the cases of $E_z = 0$ and $E_z \neq 0$.

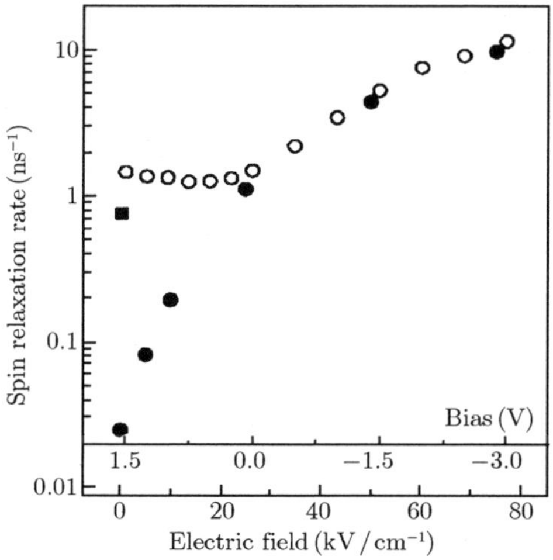

Fig. 5.12. Measured spin relaxation rate versus bias and corresponding electric field (open circles) compared with calculation for a symmetrical QW (solid circles) and for a QW with one two-monolayer graded interface (solid square) assuming electron mobility $0.6\,{\rm m}^2/{\rm V}^{-1}{\rm s}$.

becomes more sensitive to $\Omega^{\rm SIA}$ and should be significantly changed by the application of an external electric field.

If an external electric field is applied to the quantum well in the z-direction, the $\boldsymbol{\Omega}^{\rm SIA}$ will be in the quantum well plane perpendicular to $\boldsymbol{k}$ due to the Rashba effect (Eq. (5.87)), as shown in the right figure of Fig. 5.10. The $\boldsymbol{\Omega}^{\rm SIA}$ induces precession of the electron spin away from the z-axis and therefore contributes a term to DP spin relaxation along the axis; thus, $\Omega_\perp^2 \neq 0$, $T_1^{-1} \neq 0$, which can be varied with applied electric field. Figure 5.12 shows the measured spin relaxation rate versus bias and corresponding electric field (open circles) compared with calculation for a symmetrical QW (solid circles) and for a QW with one two-monolayer graded interface (solid square) (Karimov *et al.*, 2003). When $E > 20\,\text{kV/cm}$, the theoretical results are in agreement with the experimental results. When $E < 20\,\text{kV/cm}$, the T_1^{-1} remains basically a constant, and $T_1 \sim 1\,\text{ns}$ as $E = 0$, which is larger than that in (001)-oriented QW by one order of magnitude (see Figs. 5.8 and 5.9). According to the theory, when $E = 0$, $T_1^{-1} \sim 0$, but actually it is not the case. The authors proposed that a possible cause of the additional spin relaxation is random built-in asymmetry in the QWs, which would give a field-independent SIA contribution. Such asymmetry could result from alloy fluctuations or from differences of top and bottom interface morphology for the GaAs/AlGaAs QWs and for a given $\boldsymbol{k}$ would generate an $\boldsymbol{\Omega}^{\text{SIA}}$ component in the QW plane. Considering these factors, the theoretical calculation gave the T_1^{-1} at $E = 0$ as shown by the solid square in Fig. 5.10, which is in agreement with the experimental results.

People hope to use a smaller electric field to modulate the spin lifetime T_1, and design high-speed spin switch modulated only by electric field. The band gap of GaAs is wider, so the Rashba coefficient α is smaller. Hence, Hall *et al.* (2005) used (110)-oriented narrow gap InAs/AlSb quantum wells to study the spin lifetime. Because the gap of InAs is small, the Rashba coefficient is large, it is expected that the spin lifetime T_1 can be modulated by a smaller electric field. The InAs/AlSb quantum wells with 10 or 20 periods (InAs: 21.5 nm, AlSb: 20 nm) were grown by MBE on a n-type (110)-oriented GaSb substrate. The thick AlSb barrier layers and the large CB offset between InAs and AlSb (2.1 eV) ensure that electrons remain confined in the quantum wells under the applied bias range.

The electron spin dynamics were measured using the pump-probe femtosecond polarization-resolved differential transmission experiment. The measured bias per QW dependence of the spin relaxation time T_1 in a (110) InAs/AlSb MQW at 300 K is shown in Fig. 5.13(a) (Hall *et al.*, 2005). When a single quantum well bias is 140 mV, corresponding to a modest electric field of $50\,\text{kV/cm}$, T_1 decreases 70% from 25 to 10 ps. This indicates that we can use the Rashba effect to tune the spin lifetime by a small applied electric field at RT. As the InAs/AlSb MQW structure is nominally symmetric, the spin lifetime is expected to be reduced at both positive and negative voltages symmetrically. However, the asymmetric shape of the spin lifetime versus applied bias as shown in Fig. 5.13(a) indicates that there are unintentional built-in fields within the structure.

Theoretical calculated spin lifetime T_1 versus bias is shown in Fig. 5.13(b). The electric field required to switch the T_1 decreases below

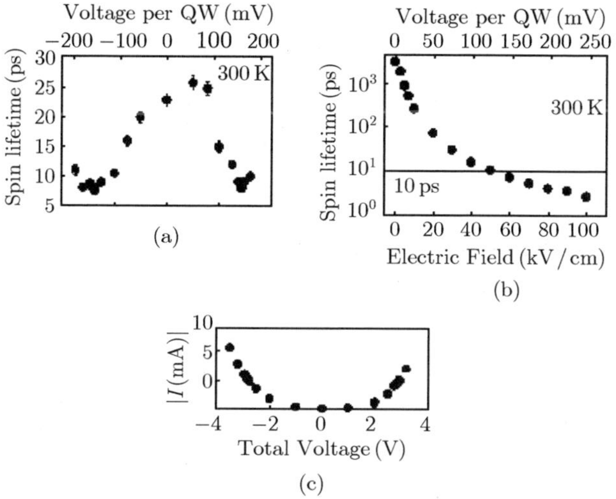

Fig. 5.13. (a) Measured bias dependence of the spin relaxation time in a (110) InAs/AlSb MQW at 300 K; (b) theoretical spin lifetime versus bias; (c) hole photocurrent versus applied bias under the same condition as (a).

10 ps is in agreement with the experimental value. A further reduction in spin lifetime is predicted for larger bias values, which is desirable for spin FET applications. The theory predicts a much longer lifetime at aero bias voltage than the maximum observed value in Fig. 5.13(a) of ~30 ps. This can be attributed to nonunifrom built-in electric fields within the thick (~400 nm) multiple quantum well (MQW) structure. As shown in Fig. 5.13(b), a nonuniform built-in electric field with an average of 70 mV per quantum well would be sufficient to account for the experimentally observed lifetime. This value is also consistent with the asymmetry in the data of Fig. 5.13(a), the maximum of T_1 is at 70 mV. Figure 5.13(c) shows the hole photocurrent versus applied bias. When the bias voltage is smaller than 140 mV, the hole photocurrent is zero, which indicates that sample heating does not play an important role in the experiments, the device can work in this bias range. Due to the large Rashba effect in the InAs/AlSb quantum well, which arises from the small band gap and strong SOC, a large in-plane pseudomagnetic field may be induced with a small applied voltage. The electronic structure calculations indicate that the pseudomagnetic field induced by a gate voltage of 140 mV per quantum well for carriers at the Fermi energy (30 meV) is 1.1 T. For comparison with the (110) GaAs quantum wells (Karimov *et al.*, 2003), approximately twice

the applied electric field is required to reduce the spin lifetime to 80 ps, reflecting the much weaker spin–orbit effects in GaAs quantum wells. The large gate-controlled Rashba effect is promising for the high-speed spin transistor operating at a much lower threshold voltage than conventional CMOS technology.

5.7. Electron Spin Relaxation Studied by a Kinetic Spin Bloch Equation

Jiang and Wu (2009) studied the electron spin relaxation in bulk III–V semiconductors from a fully microscopic kinetic spin Bloch equation (KSBE) approach, where almost all electron scatterings are explicitly included. The dependences of the spin relaxation time on electron density, temperature, initial spin polarization, photo-excitation density, and hole density are studied thoroughly. They found many new results in contrast to the previous investigations. The effect of electric field on spin relaxation in n-type III–V semiconductors is also studied.

5.7.1. *Kinetic spin Bloch equation*

The microscopic KSBE is derived by the nonequilibrium Green function method,

$$\partial_t \hat{\rho}_{\boldsymbol{k}} = \partial_t \hat{\rho}_{\boldsymbol{k}}|_{\text{coh}} + \partial_t \hat{\rho}_{\boldsymbol{k}}|_{\text{drift}} + \partial_t \hat{\rho}_{\boldsymbol{k}}|_{\text{scat}}, \tag{5.95}$$

where $\hat{\rho}_{\boldsymbol{k}}$ is the single particle density matrix with the diagonal terms representing the distributions of each spin band and the off-diagonal terms denoting the correlations of the two spin bands. The coherent term is given by

$$\partial_t \hat{\rho}_{\boldsymbol{k}}|_{\text{coh}} = -i\left[\boldsymbol{\Omega}(\boldsymbol{k}) \cdot \frac{\hat{\sigma}}{2} + \hat{\Sigma}_{\text{HF}}(\boldsymbol{k}), \hat{\rho}_{\boldsymbol{k}}\right], \tag{5.96}$$

where [,] is the commutator and $\boldsymbol{\Omega}(\boldsymbol{k}) = \boldsymbol{\Omega}_{\text{D}}(\boldsymbol{k}) + \boldsymbol{\Omega}_{\text{S}}(\boldsymbol{k})$ with

$$\begin{aligned} \boldsymbol{\Omega}_{\text{D}}(\boldsymbol{k}) &= 2\gamma_{\text{D}}[k_x(k_y^2 - k_z^2), k_y(k_z^2 - k_x^2), k_z(k_x^2 - k_y^2)], \\ \boldsymbol{\Omega}_{\text{S}}(\boldsymbol{k}) &= 2\beta(k_x, -k_y, 0), \end{aligned} \tag{5.97}$$

due to the Dresselhaus SOC and the strain-induced SOC, respectively. $\hat{\Sigma}_{\text{HF}}(\boldsymbol{k}) = -\sum_{\boldsymbol{k}'} V_{\boldsymbol{k}-\boldsymbol{k}'} \hat{\rho}_{\boldsymbol{k}'}$ is the Coulomb HF term of the electron–electron interaction.

The drift term is given by

$$\partial_t \hat{\rho}_{\boldsymbol{k}}|_{\text{drift}} = -e\boldsymbol{F} \cdot \nabla_{\boldsymbol{k}} \hat{\rho}_{\boldsymbol{k}}, \tag{5.98}$$

where e is the electron charge ($e < 0$) and $\boldsymbol{F}$ is the electric field. As the hole spin relaxation is very fast (~100 fs), the authors assumed that the hole system is kept in the thermal equilibrium state where the hole distribution is described by the Fermi distribution. The scattering term $\partial_t \hat{\rho}_{\boldsymbol{k}}|_{\text{scat}}$ contains the contributions from the electron–impurity scattering (ei), electron–phonon scattering (ep), electron–electron scattering (ee), electron–hole scattering (eh), and electron–hole exchange scattering (ex) (Jiang and Wu, 2009).

All relevant spin relaxation mechanisms have been fully incorporated in the KSBE approach. By solving the KABEs (Eq. (5.95)), one obtains the temporal evolution of the spin density matrix $\hat{\rho}_{\boldsymbol{k}}$. After that, the time evolution of macroscopic quantities, such as the electron density along the z-axis, $s_z(t) = \sum_{\boldsymbol{k}} \mathrm{Tr}[\hat{s}_z \hat{\rho}_{\boldsymbol{k}}(t)]$, is obtained. By fitting the decay of s_z with an exponential decay, one obtains the SRT.

5.7.2. *Comparison of different spin relaxation mechanisms*

In n-type III–V semiconductors, the BAP mechanism is ineffective due to the lack of holes. The ratio of the SRT of DP mechanism and EY mechanism is given by (see Eq. (5.71)),

$$\frac{\tau_{EY}}{\tau_{\rm DP}} = k_{\rm B}T\frac{Q}{A}\tau_{\rm p}^2\Theta, \quad \Theta = 8\gamma_{\rm D}^2 m_{\rm e}^3 E_{\rm g}^2 \frac{(1-\eta/3)^2}{(1-\eta/2)^2\eta^2}, \tag{5.99}$$

where Q and A are numerical factors of order of 1, $\tau_{\rm p}$ is the momentum scattering time, $\gamma_{\rm D}$ is the SOC parameter, and η is defined in Eq. (5.19). From Eq. (5.99), one arrives at the conclusion that the EY mechanism is dominant at low temperatures and/or when the momentum scattering is strong (such as, for heavy doped samples). However, the electron system may enter into the degenerate regime at low temperature and/or in heavily doped samples, in which the above conclusion fails.

In Eq. (5.99), the only un-definite parameter is $\gamma_{\rm D}$. As mentioned in the explanatory note [66] of Jiang and Wu (2009): "Lists of the SOC parameters $\gamma_{\rm D}$ in GaAs calculated and measured via various methods are given in the supplementary information of J. J. Krich and B. I. Halperin, [*Phys. Rev. Lett.* **98**: 226802 (2007)] and in A. N. Chantis, M. van Schilfgaarde, and T. Kotani [*ibid.* **96**: 086405 (2006)]. Our fitting value of $\gamma_{\rm D}$ is in the reasonable range in these lists." The calculation of the DP mechanism for GaAs in

Table 5.1. Values of γ_D and Θ of several III–V semiconductors.

	GaAs	GaSb	InAs	InSb	InP
γ_D (eV Å^3)	23.9	168	43.2	389	
Θ (eV)	0.23	0.12	3.6×10^{-4}	9.2×10^{-4}	0.27

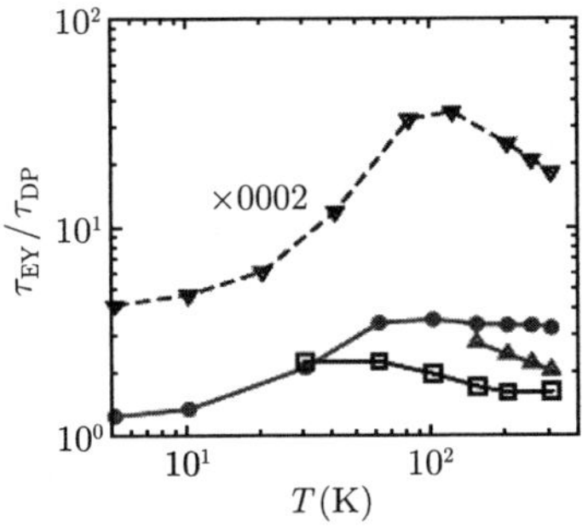

Fig. 5.14. Ratio of the SRT τ_{EY}/τ_{DP} for n-InAs (solid curves) and n-GaAs (dashed curve) as functions of temperature for various electron density: $n_e = 10^{16}\,\text{cm}^{-3}$ (•), $2 \times 10^{17}\,\text{cm}^{-3}$ (□), $10^{18}\,\text{cm}^{-3}$ (△) for InAs, and $n_e = 10^{16}\,\text{cm}^{-3}$ (▽) for GaAs (note that the value in the figure has been rescaled by a factor of 0.02).

Jiang and Wu (2009) gave a fit of the SOC and parameter $\gamma_D = 8.2\,\text{eV Å}^3$, which is different from the value $\gamma_D = 23.9\,\text{eV Å}^3$ calculated from the tight binding method (Fu *et al.*, 2008) or $\boldsymbol{k} \cdot \boldsymbol{p}$ theory (Jancu *et al.*, 2005). From Eq. (5.99), it can be seen that the γ_D affects strongly $\tau_{EY}/\tau_{DP} \sim \gamma_D^2$. Table 5.1 gives the values of γ_D and Θ of several III–V semiconductors taken in Jiang and Wu (2009).

Figure 5.14 shows the ratio of the SRT τ_{EY}/τ_{DP} for n-InAs (solid curves) and n-GaAs (dashed curve) as functions of temperature (Jiang and Wu, 2009). For GaAs, the ratio τ_{EY}/τ_{DP} is much larger than 1 no matter what the γ_D is, so in GaAs and other similar semiconductors, the DP mechanism is dominant, while the EY mechanism is not important. For InAs, the ratio τ_{EY}/τ_{DP} is nearly 1, so in narrow gap semiconductors, we cannot say which mechanism is more important due to the uncertainty of γ_D, it remains to be solved by further experiments.

5.7.3. *DP spin relaxation in n-type GaAs*

The DP mechanism has been introduced in Sec. 5.7.2, the KSBE approach includes the electron–electron (ee) Coulomb scattering, which is important

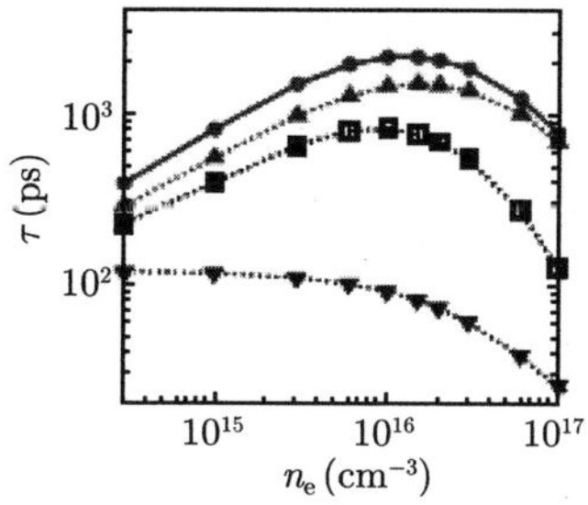

Fig. 5.15. Electron spin relaxation time τ of n-GaAs at $T = 40\,\text{K}$ versus electron density n_e from full calculation (•), with only ee scattering (□), with only electron–impurity scattering (Δ), and with only electron–phonon scattering (∇).

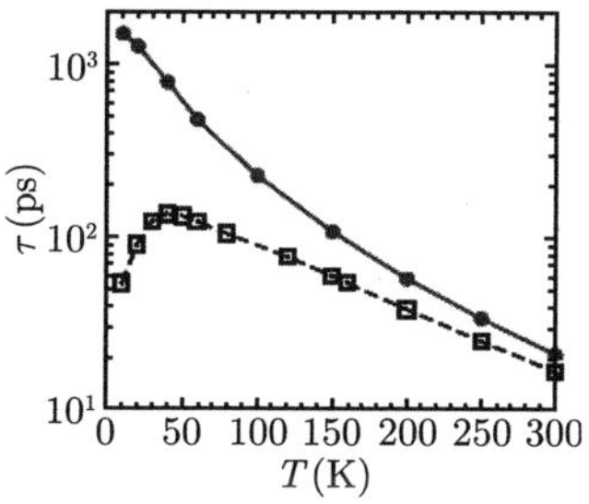

Fig. 5.16. SRT τ of n-GaAs versus temperature T for $n_i = n_e = 10^{17}\,\text{cm}^{-3}$ (•) and $n_i = 0.01\ n_e$ (□).

in the high electron density and low-temperature region (degenerate case). Thus, the temperature and electron density dependences of SRT will appear nonmonotonic characteristics.

Figure 5.15 shows the electron spin relaxation time τ of n-GaAs at $T = 40\,\text{K}$ versus electron density n_e from various scattering mechanisms (Jiang and Wu, 2009). The density dependence of the SRT is nonmonotonic and there is a peak in the $\tau - n_e$ curve. It has been proved further that the appearance of the peak is a universal behavior in the metallic regime in n-type bulk III–V semiconductors. It is noted that the electron–impurity and electron–electron scattering are the relevant scattering mechanisms, while the electron–phonon scattering is less important at low temperatures.

Figure 5.16 shows the SRT τ of n-GaAs versus temperature T for $n_i = n_e = 10^{17}\,\text{cm}^{-3}$ and $n_i = 0.01\ n_e$ (Jiang and Wu, 2009). For $n_i = n_e$, the SRT decreases with temperature monotonically, in agreement with previous experimental results (see Figs. 5.3 and 5.4). For $n_i = 0.01\ n_e$ (like the modulation doped case), the SRT shows a nonmonotonic behavior with a peak around 40 K.

As shown in Fig. 5.15, one of the main scattering mechanisms is the electron–electron scattering. The asymptotic density and temperature dependences of the electron–electron scattering time $\tau_{\rm p}^{\rm ee}$ in the degenerate and nondegenerate regimes are given by

$$\begin{aligned} \tau_{\rm p}^{\rm ee} &\propto \frac{n_{\rm e}^{2/3}}{T^2}, \quad T \ll T_{\rm F}, \\ \tau_{\rm p}^{\rm ee} &\propto \frac{T^{3/2}}{n_{\rm e}}, \quad T \gg T_{\rm F}, \end{aligned} \tag{5.100}$$

where $T_{\rm F}$ is the Fermi temperature. In the nondegenerate (low density, high temperature) regime, the $\tau_{\rm p}^{\rm ee}$ decreases with increasing density and decreasing temperature. The inhomogeneous broadening $\sim \langle(|\boldsymbol{\Omega}(\boldsymbol{k})|^2 - \Omega_{\rm z}^2(\boldsymbol{k}))\rangle$ changes slowly as the distribution function is close to the Boltzmann distribution. In degenerate (high density, low temperature) regime, both $\tau_{\rm p}^{\rm ee}$ and inhomogeneous broadening increase with increasing density and decreasing temperature. Thus, the SRT, which can be estimated as $\tau \sim 1/[\langle|\boldsymbol{\Omega}(\boldsymbol{k})|^2 - \Omega_{\rm z}^2(\boldsymbol{k})\rangle\tau_{\rm p}^{\rm ee}]$, shows the nonmonotonic behaviors with density or temperature. For the electron–impurity scattering, the scenario is similar (Jiang and Wu, 2009).

5.7.4. *Spin relaxation in intrinsic GaAs*

For intrinsic semiconductors, the electrons and holes are created by optical excitation, and their numbers are equal. Compared with n-type semiconductors, there are two additional scattering mechanisms: the electron–hole Coulomb scattering and electron–hole exchange scattering, where the latter corresponds to the BAP mechanism. The impurity density in intrinsic semiconductors is very low, which offers a good platform for demonstrating the effect of the many-body carrier–carrier scattering on spin relaxation.

Figure 5.17(a) shows the SRT τ versus temperature T due to BAP and DP mechanism in intrinsic GaAs with $N_{\rm ex} = 10^{17}\,{\rm cm}^{-3}$ (Jiang and Wu, 2009). The SRT due to the BAP mechanism is larger than that due to the DP mechanism by more than one order of magnitude, which indicates that the BAP mechanism is negligible for intrinsic GaAs. There is a same problem about the value of $\gamma_{\rm D}$, which decides the $\tau_{\rm DP}$. The authors indicated in the explanatory note [91] of Jiang and Wu (2009): "the SOC

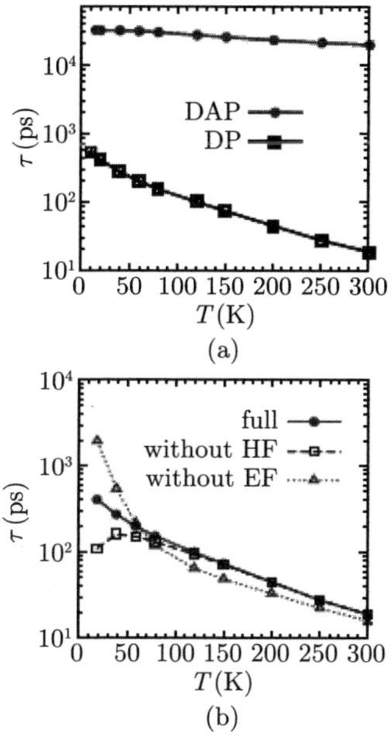

Fig. 5.17. Intrinsic GaAs with $N_{\text{ex}} = 10^{17}\,\text{cm}^{-3}$. (a) SRT τ versus temperature T due to BAP and DP mechanism. (b) SRT τ versus T due to the DP mechanism from full calculation (•), from calculation without the Coulomb HF term (□), and without the electron–hole Coulomb scattering (△).

parameter we used in this section ($\gamma_{\text{D}} = 23.9\,\text{eV}\,\text{Å}^3$) is larger than the value ($\gamma_{\text{D}} = 8.2\,\text{eV}\,\text{Å}^3$) fitted from experimental data. However, as the DP spin relaxation is more than two orders of magnitude larger than the BAP spin relaxation, the choice of the τ_{DP} does not change the conclusion that BAP mechanism is unimportant."

Figure 5.17(b) shows the SRT τ versus T due to the DP mechanism from full calculation, from calculation without the Coulomb HF term, and without the electron–hole Coulomb scattering (Jiang and Wu, 2009). The Coulomb HF term has an important effect on spin relaxation only for low temperature case ($T < 60\,\text{K}$). Without the electron–hole Coulomb scattering, the SRT is larger for $T < 60\,\text{K}$ but smaller for $T > 60\,\text{K}$ compared with that from the full calculation. The results indicate that the electron–hole scattering is comparable with the electron–electron and electron–LO–phonon scatterings. In other words, besides the screening from holes, the main contribution of the hole system to electron relaxation comes from the electron–hole Coulomb scattering in intrinsic semiconductors. Without the electron–hole Coulomb scattering, the momentum scattering time is elongated, which increases the SRT.

The SRT also shows a nonmonotonic photoexcitation density dependence with a peak at some density n_c, which resembles that in n-type semiconductors. Further calculation gives $n_c = 0.8 \times 10^{16}\,\text{cm}^{-3}$ ($T_F = 25\,\text{K}$) for $T = 40\,\text{K}$ case and $n_c = 0.9 \times 10^{17}\,\text{cm}^{-3}$ ($T_F = 127\,\text{K}$) for $T = 300\,\text{K}$ case. The experiments observed that the SRT decreases with N_{ex} when $N_{ex} > 10^{17}\,\text{cm}^{-3}$, and the SRT increases with N_{ex} when it is lower; however, the peak has not been reported.

5.7.5. *Electron spin relaxation in p-type GaAs*

The main sources of spin relaxation are the BAP mechanism and the DP mechanism. In order to avoid exaggerating the DP mechanism, the value of γ_D fitted from the experimental data $8.2\,\text{eV}\,\text{Å}^3$ is used in the calculation instead of $23.9\,\text{eV}\,\text{Å}^3$.

Figure 5.18(a) shows the ratio of SRTs τ_{BAP}/τ_{DP} versus temperature T in p-GaAs with $N_{ex} = 10^{14}\,\text{cm}^{-3}$ and $n_i = n_h$ for various hole densities, and Fig 5.18(b) shows the SRT due to the BAP and DP mechanisms, the total SRT, and the ratio τ_{BAP}/τ_{DP} (Δ) versus temperature for $n_h = 3 \times 10^{18}\,\text{cm}^{-3}$ (Jiang and Wu, 2009). The DP mechanism dominates at high temperatures, whereas the BAP mechanism dominates at low temperatures, which is consistent with previous investigations. A typical case is shown in Fig. 5.18(b) for $n_h = 3 \times 10^{18}\,\text{cm}^{-3}$. It is noted that the "plateau"

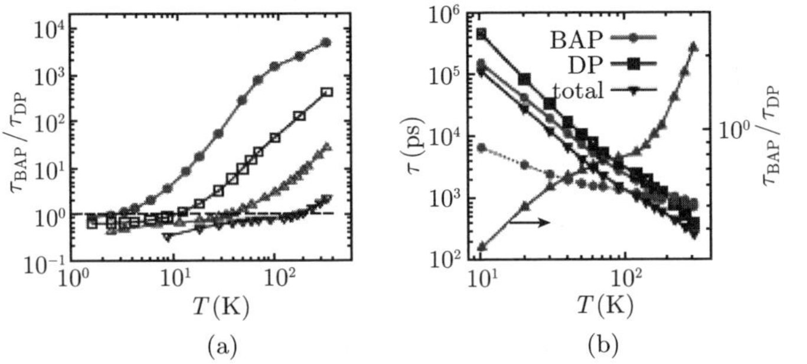

Fig. 5.18. (a) Ratio of SRTs τ_{BAP}/τ_{DP} versus temperature T in p-GaAs with $N_{ex} = 10^{14}\,\text{cm}^{-3}$ and $n_i = n_h$ for various hole densities $n_h = 3 \times 10^{15}\,\text{cm}^{-3}$ (•), $3 \times 10^{16}\,\text{cm}^{-3}$ (□), $3 \times 10^{17}\,\text{cm}^{-3}$ (Δ), $3 \times 10^{18}\,\text{cm}^{-3}$ (∇). (b) The SRT due to the BAP and DP mechanisms, the total SRT, and the ratio τ_{BAP}/τ_{DP} (Δ) versus temperature for $n_h = 3 \times 10^{18}\,\text{cm}^{-3}$. The dotted curve represents the SRT due to the BAP mechanism without the Pauli blocking of holes.

is around the hole Fermi temperature $T_F^h = 156\,K$ which is given by

$$T_F^h = \frac{(3\pi^2 n_h)^{2/3}}{2k_B m_0[(\gamma_1 - 2\gamma_2)^{-3/2} + (\gamma_1 + 2\gamma_2)^{-3/2}]^{2/3}}. \tag{5.101}$$

The underlying physics is that: on the one hand, the Pauli blocking of holes becomes important when $T \leq T_F^h$, which slows down the BAP spin relaxation effectively; on the other hand, the increase in the screening (mainly from holes) with decreasing temperature weakens the electron–impurity and carrier–carrier scatterings and thus enhances the DP spin relaxation. Consequently, the decrease in the ratio with decreasing temperature slows down and the "plateau" is formed around T_F^h.

Figure 5.19 is same as Fig. 5.18, but with high photoexcitation $N_{ex} = 0.1\,n_h$ (Jiang and Wu, 2009). The ratio is nonmonotonic and has a minimum roughly around the Fermi temperature of electrons, $T \sim T_F$. The BAP mechanism is comparable with the DP mechanism only in the moderate temperature regime around T_F, whereas for higher or lower temperatures it becomes unimportant. From Fig. 5.19(b), it can be seen that the Pauli blocking of electrons and holes largely suppresses the BAP spin relaxation in the low-temperature regime and hence makes τ_{BAP} always increase with decreasing temperature.

At $n_h = 3 \times 10^{17}\,cm^{-3}$ and $T = 50\,K$, the SRT due to the DP mechanism as a function of the photoexcitation density first decreases slowly, then ($N_{ex} > 10^{16}\,cm^{-3}$) rapidly with the photoexcitation density. The scenario is as following: in p-type semiconductors at low temperatures,

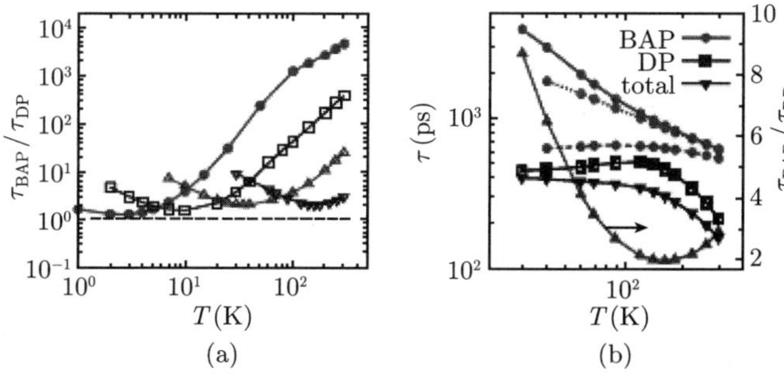

Fig. 5.19. (a) Ratio of SRTs τ_{BAP}/τ_{DP} versus temperature T in p-GaAs with $N_{ex} = 0.1n_h$ and $n_i = n_h$ for various hole densities $n_h = 3\times10^{15}\,cm^{-3}$ (•), $3\times10^{16}\,cm^{-3}$ (□), $3\times10^{17}\,cm^{-3}$ (Δ), $3\times10^{18}\,cm^{-3}$ (∇). (b) The SRT due to the BAP and DP mechanisms, the total SRT, and the ratio τ_{BAP}/τ_{DP} (Δ) versus temperature for $n_h = 3\times10^{18}\,cm^{-3}$. The dotted (dashed) curve represents the SRT due to the BAP mechanism without the Pauli blocking of electrons (holes).

the dominant scattering mechanisms are the electron–hole and electron–impurity scatterings. The momentum scattering due to these two mechanisms changes little with photoexcitation (electron) density ($N_{\rm ex} = n_{\rm e}$) for $N_{\rm ex} < n_{\rm h}$. In the low-density regime ($n_{\rm e} < 3 \times 10^{15}\,{\rm cm}^{-3}$), where the electron system is nondegenerate, the increase in density affects the inhomogeneous broadening very little. Thus, the SRT changes slowly with the photoexcitation density. In the high-density regime ($n_{\rm e} > 3 \times 10^{16}\,{\rm cm}^{-3}$), the electron is degenerate where the inhomogeneous broadening increases fast with density. Consequently, the SRT decreases rapidly with the photoexcitation density.

Wu *et al.* (2010) also investigated the electron spin relaxations in QWs and quantum wires, and their results are published in a long review paper.

References

Aronov AG, Pikus GE, Titkov AN. (1983) *Zh Eksp Teor Fiz* **83**: 1170.

Benoit La G, Lavallard CP, Richard R. (1974) In *Proc. 12th Int Conf on the Physics of Semiconductors*, Stuttgart p. 766, Pilkuhn M (ed).

Bir GL, Aronov AG, Pikus GE. (1975) *Sov Phys JETP* **42**: 705.

D'yakonov MI, Perel VI. (1984) In *Optical Orientation*, p. 1, Meier M, Zakharchenya BP (eds). Elsevier Science Publisher.

D'yakonov MI, Kachorovskii VY. (1986) *Sov Phys Semicond* **20**: 110.

Elliot RJ. (1954) *Phys Rev* **96**: 266.

Fu JY, Weng MQ, Wu MW. (2008) *Physica E* **40**: 2890.

Hall KC, Gündoğdu K, Hicks JL, *et al.* (2005) *Appl Phys Lett* **86**: 202114.

Jancu JM, Scholz R, de Andrada e Silva EA. (2005) *Phys Rev B* **72**, 193201.

Jiang JH, Wu MW. (2009) *Phys Rev B* **79**, 125206.

Karimov OZ, John GH, Harley RT, Lau WH, Flatté ME, Henini M, Airey R. (2003) *Phys Rev Lett* **91**: 246601.

Kikkawa JM, Awschalom DD. (1998) *Phys Rev Lett* **80**: 4314.

Lau WH, Olesberg JT, Flatté ME. (2001) *Phys Rev B* **64**: 161301R.

Maruschak VI, Lagunova TV, Stepanova NN, Titkov AN. (1983) *Fiz Tverd Tela* **25**: 2140.

Merzbacher E. (1970) *Quantum Mechanics*, p. 271, John Wiley & Sons, New York.

Pikus GE, Titkov AN. (1984) in *Optical Orientation*, p. 73, Meier M, Zakharchenya BP (eds). Elsevier Science Publisher, North-Holland, Amsterdam.

Seeger K. (1982) *Semiconductor Physics*, p. 159, Springer-Verlag, Berlin.

Tackeuchi A, Nishikawa Y, Wada O. (1996) *Appl Phys Lett* **68**: 797.

Terauchi R, Ohno Y, Adachi T, *et al.* (1999) *Jpn J Appl Phys* **38**: 2549.

Wu MW, Jiang JH, Weng MQ. (2010) *Phys Rep* **493**: 61.

Yafet Y. (1963) in *Solid State Physics*, Vol. 14, p. 1, Seitz F, Turnbull D (eds). Academic, New York.

Chapter 6

RASHBA AND DRESSELHAUS EFFECTS

6.1. Spin Splitting Induced by Spin–Orbit Interaction (SOI) in Inversion Asymmetrical Semiconductors — Rashba and Dresselhaus Effects

Spin splitting means that due to the linear term of momentum appearing in the system Hamiltonian, the parabolic dispersion curve of the electron energy will split into two curves centered symmetrically away from $k_x = 0$. Such splitting is induced by Rashba effect (Bychkov and Rashba, 1984) resulted from SIA and Dresselhaus effect (Dresselhaus, 1955) resulted from BIA. Under the EMA and based on crystal symmetry, the two effects can be derived by using the $\boldsymbol{k} \cdot \boldsymbol{p}$ method. Here we will mainly introduce this approach. Meanwhile, it is also possible to derive the Rashba effect via relativistic quantum mechanics (Dirac equation), implying that this effect is essentially a relativistic quantum mechanical effect.

6.1.1. *Effective mass approximation (Bir and Pikus, 1974; Foreman, 2005)*

We will first introduce the $\boldsymbol{k} \cdot \boldsymbol{p}$ method in EMA, but the tedious derivation is not included, rather a relatively simple theory of invariant will be adopted to reach the same result.

We first consider semiconductors with the wurtzite structure as an example. The space group of the wurtzite structure is C_{6v}^4 and the point group (i.e. only includes the rotation part) is C_{6v}. The basis functions of the irreducible representation table are listed in Table 6.1.

Table 6.1. The basis functions of the irreducible representation (spin not included) of point group C_{6v}.

Irreducible representations		Basis functions
KDWS	Tinkham	
Γ_1	A_1	$1, z, z^2, x^2+y^2$
Γ_2	A_2	J_z
Γ_3	B_2	x^3-3xy^2
Γ_4	B_1	y^3-3yx^2
Γ_5	E_1	(x,y); (xz,yz); $(J_y,-J_x)$
Γ_6	E_2	(x^2-y^2,xy); $((x+iy)^2, (x-iy)^2)$

For Γ_5, we can also take the basis functions as $(u^5_{+1}, u^5_{-1}) = (x+iy, x-iy)$. Note that

$$\begin{aligned}(x+iy, x-iy) \to (J_y - iJ_x, J_y + iJ_x)\\ = i(-J_x - iJ_y, J_x - iJ_y),\\ = -i(J_x + iJ_y, -J_x + iJ_y),\end{aligned}$$

thus we may ignore i in front of the brackets to compose a pair of functions which transfer to each other under time reversal.

When there is no SOI, the symmetry of the conduction band (CB) of a wurtzite crystal is Γ_1, while the valence band splits into a doublet state Γ_5 and a singlet state Γ_1. Taking spin into account, the symmetry and basis functions of the CB and valence band become:

CB

$$\begin{aligned}\Gamma_1 \otimes D_{1/2} = \Gamma_7\\ |u^7_{+1/2}\rangle = |u^7_+\rangle = |u_7,+\rangle = |u_1\rangle|+\rangle = |s\rangle|+\rangle,\\ |u^7_{-1/2}\rangle = |u^7_-\rangle = |u_7,-\rangle = |u_1\rangle|-\rangle = |s\rangle|-\rangle,\end{aligned} \tag{6.1}$$

where $D_{1/2}$ is the representation of a particle with spin of 1/2.

Valence band

$$\Gamma_1 \otimes D_{1/2} = \Gamma_7$$

$$\begin{aligned}|v^7_{+1/2}\rangle = |v^7_+\rangle = |z\rangle|+\rangle,\\ |v^7_{-1/2}\rangle = |v^7_-\rangle = |z\rangle|-\rangle,\end{aligned}$$

$$\Gamma_5 \otimes D_{1/2} = \Gamma_7 \oplus \Gamma_9,$$

$$\begin{aligned}|v^7_{+1/2}\rangle = |v^7_+\rangle = -|u^5_{+1}|-\rangle,\\ |v^7_{-1/2}\rangle = |v^7_-\rangle = -|u^5_{-1}|+\rangle,\end{aligned} \quad \text{and} \quad \begin{aligned}|v^9_{+3/2}\rangle = |u^5_{+1}\rangle|+\rangle,\\ |v^9_{-3/2}\rangle = |u^5_{-1}\rangle|-\rangle.\end{aligned}$$

The basis functions of the valence band state Γ_7 can be generally written as

$$
\begin{aligned}
|v_+^7\rangle &= \alpha|z\rangle|+\rangle - \beta|u_{+1}^5|-\rangle = \alpha|z\rangle|+\rangle - \beta\frac{1}{\sqrt{2}}|x+iy\rangle|-\rangle \\
|v_-^7\rangle &= \alpha|z\rangle|-\rangle + \beta|u_{-1}^5|+\rangle = \alpha|z\rangle|-\rangle + \beta\frac{1}{\sqrt{2}}|x-iy\rangle|+\rangle,
\end{aligned} \tag{6.2a}
$$

while the basis functions for state Γ_9 are

$$
\begin{aligned}
|v_{+3/2}^9\rangle &= |u_{+1}^5|+\rangle = \frac{1}{\sqrt{2}}|x+iy\rangle|+\rangle \\
|v_{-3/2}^9\rangle &= |u_{-1}^5|-\rangle = \frac{1}{\sqrt{2}}|x-iy\rangle|+\rangle.
\end{aligned} \tag{6.2b}
$$

Due to time reversal, it is required that:

$$\Theta|j,m\rangle = (-1)^{j-m}|j,-m\rangle,$$

where Θ is the time-reversion operator, i.e.

$$
\begin{aligned}
\Theta\left|\frac{1}{2},\pm\frac{1}{2}\right\rangle &= \pm\left|\frac{1}{2},\mp\frac{1}{2}\right\rangle, \\
\Theta|1,\pm1\rangle &= |1,\pm1\rangle, \\
\Theta\left|\frac{3}{2},\pm\frac{3}{2}\right\rangle &= \pm\left|\frac{3}{2},\mp\frac{3}{2}\right\rangle.
\end{aligned} \tag{6.3}
$$

One may therefore write the state $|v_{\pm3/2}^9\rangle$ as $|3/2,\pm3/2\rangle$. Meanwhile, although the state $|v_\pm^7\rangle$ is a linear combination of $|3/2,\pm1/2\rangle$ and $|1/2,\pm1/2\rangle$ (if the general T_d marks are taken), based on time reversal, they are only written as $|1/2,+1/2\rangle$.

In the invariants approach, we simply observe that since the Γ_7 representation is 2D, its effective-mass Hamiltonian must be some linear combination of 2×2 matrices multiplied by quadratic polynomials in $\boldsymbol{k}$. There are four linearly independent 2×2 matrices, which we can choose to be

$$1, \quad \sigma_x, \quad \sigma_y, \quad \sigma_z.$$

We also have the following quadratic functions of $\boldsymbol{k}$:

$$1, \quad k_x, \quad k_y, \quad k_z, \quad k_x^2, \quad k_y^2, \quad k_z^2, \quad k_yk_z, \quad k_zk_x, \quad k_xk_y.$$

Now the Hamiltonian must be invariant under all symmetry operations of group C_{6v}, i.e. for all operations R belonging to C_{6v}, we have

$$D(R)HD^+(R) = H, \quad (R\in C_{6v}). \tag{6.4}$$

Therefore, we wish to construct linear combinations of 2×2 matrices multiplied by quadratic polynomials in $\boldsymbol{k}$ that are invariant under all rotations of C_{6v}. But we already know one way of constructing invariants from irreducible tensor operators:

$$S \equiv \sum_g (T_g^{(k)})^+ U_g^{(k)} \to D(R) S D^+(R) = S, \tag{6.5}$$

we therefore should first construct tensor operators from the given matrices and functions of $\boldsymbol{k}$. But these may simply be read directly from the basis functions given in the character tables:

$$\begin{aligned}
&\Gamma_1(\mathrm{A}_1): && 1, \quad k_z, \quad k_x^2 + k_y^2, \quad k_z^2 \\
&\Gamma_2(\mathrm{A}_2): && \sigma_z \\
&\Gamma_3(\mathrm{B}_2): && \\
&\Gamma_4(\mathrm{B}_1): && \\
&\Gamma_5(\mathrm{E}_1): && (k_x, k_y) \quad \text{or} \quad (k_x k_z, k_y k_z) \quad \text{or} \quad (\sigma_y, \sigma_x) \\
&\Gamma_6(\mathrm{E}_2): && (k_x^2 - k_y^2, k_x k_y).
\end{aligned}$$

Here we have utilized a fact that the Γ_7 basis kets transform in the same way as the spin $-1/2$ basis kets $|1/2, \pm 1/2\rangle \equiv |\pm\rangle$ under all rotations of C_{6v} (and under time reversal). Therefore, the Pauli $\boldsymbol{\sigma}$ matrices represent the spin operator $\boldsymbol{S}$ in this basis:

$$\boldsymbol{S} \cong \frac{\hbar}{2} \boldsymbol{\sigma},$$

which means that they transform as pseudovectors.

The most general invariant of order k^2 that can be constructed from these functions is

$$\begin{aligned}
H = {} & C_1^{(1)} 1 + C_2^{(1)} 1 k_z + C_3^{(1)} 1 (k_x^2 + k_y^2) + C_4^{(1)} 1 k_z^2 \\
& + C_1^{(5)} (k_x \sigma_y - k_y \sigma_x) + C_2^{(5)} (k_x k_z \sigma_y - k_y k_z \sigma_x).
\end{aligned} \tag{6.6}$$

Since $\boldsymbol{k}$ and $\boldsymbol{\sigma}$ are Hermitian, all of the coefficients $C_i^{(j)}$ in this equation must be real if H is to be Hermitian. Also, $\boldsymbol{k}$ and $\boldsymbol{\sigma}$ (like $\boldsymbol{p}$ and $\boldsymbol{S}$) are both odd under time reversal, whereas H must be even:

$$\Theta H \Theta^{-1} = H;$$

therefore

$$C_2^{(1)} = C_2^{(5)} = 0,$$

and the most general Hermitian invariant of order k^2, which is also time-reversal-invariant, is

$$H = C_1^{(1)}1 + C_3^{(1)}1(k_x^2 + k_y^2) + C_4^{(1)}1k_z^2 + C_1^{(5)}(k_x\sigma_y - k_y\sigma_x). \tag{6.7}$$

This is just the Hamiltonian derived from the $\boldsymbol{k} \cdot \boldsymbol{p}$ theory; the only difference is that the method of invariants does not tell us the values of the $C_i^{(j)}$ coefficients.

It should be noted that, in general, the Pauli $\boldsymbol{\sigma}$ matrices do not form a matrix representation for the spin operator $\boldsymbol{S}$ (because in a general 2D basis, the basis kets are not of the form $|j = 1/2, m = \pm 1/2\rangle$). This means that $\boldsymbol{\sigma}$ is generally not a pseudovector. Therefore, to apply the method of invariants in other situations, we need a more general way of constructing matrices that transform as irreducible tensor operators.

Here we can simply use the Wignar–Eckar theorem:

$$\langle \alpha' j' m' | T_g^{(k)} | a j m \rangle = \sum_{\sigma} \langle \alpha' j' \| T_g^{(k)} \| a_j \rangle_\sigma \langle k j g m | \sigma j' m' \rangle, \tag{6.8}$$

where $\langle kjgm|\sigma j'm'\rangle$ is given in the vector coupling coefficient (VCC) tables. In the table, m' represents the row and m represents the line. To construct a matrix representing that portion of the operator $T_g^{(k)}$ which couples the j' subspace to the j subspace, we simply hold k and g fixed and write down the complex conjugate of the array $\langle kjgm|\sigma j'm'\rangle$ with m' labeling the rows and m labeling the columns. If there is more than one value of σ, there will be more than one array, and we can choose an arbitrary linear combination of the given arrays.

For example, if j' and j are both Γ_7, we obtain the following results from the VCC tables for the basis $\{|u_7, +\rangle, |u_7, -\rangle\}$:

$$\begin{gathered} T^{(1)} \cong \begin{bmatrix} 1 & 0 \\ 0 & 1 \end{bmatrix} = 1, \quad T^{(2)} \cong \begin{bmatrix} -i & 0 \\ 0 & i \end{bmatrix}^* = \begin{bmatrix} i & 0 \\ 0 & -i \end{bmatrix} = i\sigma_z \\ T_{+1}^{(5)} \cong \begin{bmatrix} 0 & -1 \\ 0 & 0 \end{bmatrix}^*, \quad T_{-1}^{(5)} \cong \begin{bmatrix} 0 & 0 \\ 1 & 0 \end{bmatrix}^*. \end{gathered} \tag{6.9}$$

For the pseudovectors $\boldsymbol{S}$, we have the standard basis

$$(u_{+1}^5, u_{-1}^5) = (-(S_x + iS_y), S_x - iS_y). \tag{6.10}$$

Thus,

$$S_x = \frac{1}{2}(u_{-1}^5 - u_{+1}^5), \quad S_y = \frac{i}{2}(u_{-1}^5 + u_{+1}^5). \tag{6.11}$$

So that we can define

$$T_x^{(5)} \equiv \frac{1}{2}(T_{-1}^{(5)} - T_{-1}^{(5)}) \cong \frac{1}{2}\begin{bmatrix} 0 & 1 \\ 1 & 0 \end{bmatrix} = \frac{1}{2}\sigma_x, \tag{6.12a}$$

$$T_y^{(5)} \equiv \frac{1}{2}\left(T_{-1}^{(5)} + T_{-1}^{(5)}\right) \cong \frac{1}{2}\begin{bmatrix} 0 & -i \\ i & 0 \end{bmatrix} = \frac{1}{2}\sigma_y. \tag{6.12b}$$

Therefore, from the VCC tables, we have obtained the same results:

$$\begin{aligned} &\Gamma_1: \quad 1 \\ &\Gamma_2: \quad \sigma_z \\ &\Gamma_5: \quad (\sigma_y - \sigma_x) \end{aligned}$$

that we obtained earlier.

Next, let us consider a different physical system: a symmetric GaAs/AlAs quantum well or superlattice. If the superlattice is grown along the crystallographic [001] axis, then the point group is D_{2d}. For simplicity, we shall construct the Hamiltonian by the method of invariants. For a wide quantum well, the lowest CB corresponds to Γ_1 (without spin) or Γ_6 (with spin), Γ_6 is a 2D spinor representation transforming in the same way as $|j = 1/2, m = \pm 1/2\rangle$. As usual, our first step is to find matrices and quadratic polynomials in $\boldsymbol{k}$ that transform as tensor operators under the rotations of the D_{2d} group.

From the D_{2d} character tables, we obtain the basis functions:

$$\begin{aligned} &\Gamma_1(\mathrm{A}_1): \quad 1 \quad \text{or} \quad k_x^2 + k_y^2 \quad \text{or} \quad k_z^2 \\ &\Gamma_2(\mathrm{A}_2): \quad \sigma_z \\ &\Gamma_3(\mathrm{B}_1): \quad k_x^2 - k_y^2 \\ &\Gamma_4(\mathrm{B}_2): \quad k_z \quad \text{or} \quad k_x k_y \\ &\Gamma_5(E): \quad (k_x, k_y) \quad \text{or} \quad (k_y k_z, k_z k_x) \quad \text{or} \quad (\sigma_x, -\sigma_y), \end{aligned}$$

where the sign of σ_y in the last term was chosen to obtain a standard basis. The most general invariant we can construct from these is

$$\begin{aligned} H = C_1^{(1)} 1 + C_2^{(1)} 1(k_x^2 + k_y^2) + C_3^{(1)} 1 k_z^2 + C_1^{(5)}(\sigma_x k_x - \sigma_y k_y) \\ + C_2^{(5)}(\sigma_x k_y k_z - \sigma_y k_z k_x), \end{aligned} \tag{6.13}$$

where $C_i^{(j)}$ must be real if it is to be Hermitian. But if H is to be time-reversal-invariant, $C_2^{(5)}$ must be imaginary, hence

$$C_2^{(5)} = 0,$$

and we obtain

$$H = E_n 1 + \gamma_1 1(k_x^2 + k_y^2) + \gamma_2 1 k_z^2 + \alpha_D(\sigma_x k_x - \sigma_y k_y). \quad (6.14)$$

This is similar to the Rashba Hamiltonian for wutzite semiconductors, except that the linear $-k$ spin-splitting term

$$\alpha_D(\sigma_x k_x - \sigma_y k_y) \quad (6.15)$$

is not quite the same. It is worth noticing that this term is called the Dresselhaus term (Dresselhaus, 1955), although it was not discovered by Dresselhaus.

The true Dresselhaus term is actually a third-order invariant that occurs in the Γ_6 CB Hamiltonian for a bulk crystal of GaAs (T_d group):

$$\sigma_x k_x(k_y^2 - k_z^2) + \sigma_y k_y(k_z^2 - k_x^2) + \sigma_z k_z(k_x^2 - k_y^2). \quad (6.16)$$

In a GaAs/AlAs quantum well, the k_z component is quantized, so we may (roughly speaking) replace k_z and k_z^2 by their mean values:

$$\sigma_x k_x(k_y^2 - \langle k_z^2\rangle) + \sigma_y k_y(\langle k_z^2\rangle - kx^2) + \sigma_z\langle k_z\rangle(k_x^2 - k_y^2). \quad (6.17)$$

The only terms which are linear in $\boldsymbol{k} = (k_x, k_y, 0)$ in this expression are combined to be:

$$-\underbrace{\langle k_z^2\rangle}_{\text{const}}(\sigma_x k_x - \sigma_y k_y) \quad (6.18)$$

that is just the "Dresselhaus" term from above, as $\langle k_z^2\rangle$ can be taken as a constant. Thus, this rough argument illustrates how this term originates in the k^3 terms of the bulk $\boldsymbol{k}\cdot\boldsymbol{p}$ Hamiltonian.

Finally, let us consider an asymmetric GaAs/AlAs quantum well or superlattice, which has symmetry of point group C_{2v}. In this case, the lowest CB state belongs to $\Gamma_1(\mathrm{A}_1)$, without spin, or Γ_5 with spin. The relevant basis functions for C_{2v} are:

$\Gamma_1(\mathrm{A}_1)$: 1 or k_z or k_x^2 or k_y^2 or k_z^2

$\Gamma_2(\mathrm{B}_1)$: k_x or $k_x k_z$ or σ_y

$\Gamma_3(\mathrm{A}_2)$: $k_x k_y$ or σ_z

$\Gamma_4(\mathrm{B}_2)$: k_y or $k_y k_z$ or σ_x

This yields the Hamiltonian (to $0(k^2)$)

$$H = C_1^{(1)}1 + C_2^{(1)}1k_z + C_3^{(1)}1k_x^2 + C_4^{(1)}1k_y^2 + C_5^{(1)}1k_z^2 + C_1^{(2)}\sigma_y k_x + C_2^{(2)}\sigma_y k_x k_z + C_1^{(3)}\sigma_z k_x k_y + C_1^{(4)}\sigma_x k_y + C_2^{(4)}\sigma_x k_y k_z z, \quad (6.19)$$

in which all coefficients are real. Excluding all terms that are odd under time reversal, it leaves

$$H = 1(C_1^{(1)} + C_3^{(1)}k_x^2 + C_4^{(1)}k_y^2 + C_5^{(1)}k_z^2) + C_1^{(2)}\sigma_y k_x + C_1^{(4)}\sigma_x k_y. \quad (6.20)$$

In the D_{2d} case, the x, y, and z-axes correspond to the crystallographic [100], [010], and [001] axes. However, for C_{2v}, we have $x//[110]$, $y//[\bar{1}10]$, $z//[001]$, i.e. rotating the x–y plane by 45°. Thus, in order to facilitate a direct comparison with D_{2d}, we should relate the coordinates by (Fig. 6.1):

$$x \to x' = \frac{1}{\sqrt{2}}(x+y); \quad y \to y' = \frac{1}{\sqrt{2}}(-x+y)$$

that yields

$$\begin{aligned} H &= 1\left[C_1^{(1)} + \frac{1}{2}C_3^{(1)}(k_x - k_y)^2 + \frac{1}{2}C_4^{(1)}(k_x + k_y)^2 + C_5^{(1)}k_z^2\right] \\ &\quad + \frac{1}{2}C_1^{(2)}(\sigma_x + \sigma_y)(k_x - k_y) + \frac{1}{2}C_1^{(4)}(\sigma_x - \sigma_y)(k_x + k_y) \\ &= 1\left[C_1^{(1)} + \frac{1}{2}C_3^{(1)}(k_x - k_y)^2 + \frac{1}{2}C_4^{(1)}(k_x + k_y)^2 + C_5^{(1)}k_z^2\right] \\ &\quad + \frac{1}{2}(C_1^{(2)} + C_1^{(4)})(\sigma_x k_x - \sigma_y k_y) \\ &\quad + \frac{1}{2}(C_1^{(2)} - C_1^{(4)})(k_x\sigma_y - k_y\sigma_x) \end{aligned} \quad (6.21)$$

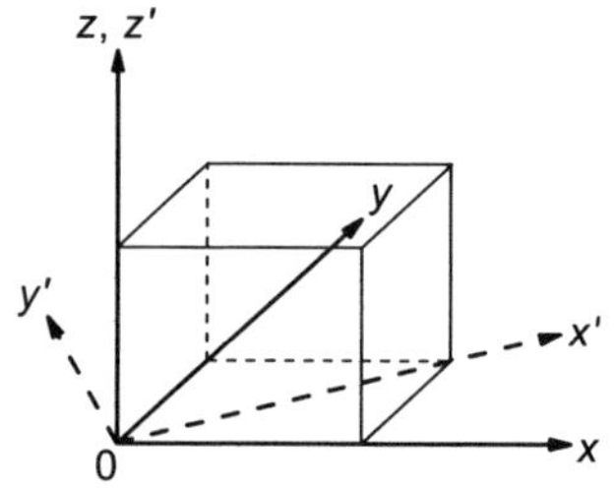

Fig. 6.1. Coordination systems in D_{2d} (solid line) and C_{2v} (dash line) symmetry.

or

$$H = 1[E_n + \gamma_1(k_x^2 + k_y^2)^2 + \gamma_2 k_z^2 + \gamma_3 k_x k_y] \\ + \alpha_D(\sigma_x k_x - \sigma_y k_y) + \alpha_R(k_x\sigma_y - k_y\sigma_x), \tag{6.22}$$

where we have identified

$$\alpha_D \equiv \frac{1}{2}(C_1^{(2)} + C_1^{(4)}) \tag{6.23}$$

as the Dresselhaus coefficient and

$$\alpha_R = \frac{1}{2}(C_1^{(2)} - C_1^{(4)}) \tag{6.24}$$

as the Rashba coefficient.

Therefore, in an asymmetric GaAs/AlAs quantum well with C_{2v} symmetry, the linear-k spin splitting contains contributions from both the Rashba and Dresselhaus terms. The asymmetry of the quantum well can be changed by applying an electric field along the z-axis, which allows one to tune the value of α_R, the Rashba coefficient.

6.1.2. *General description of the Dresselhaus effect*

An article by D'yakonov and Kachorovskiĭ (1986) has been commonly viewed as one of the best descriptions of the Dresselhaus effect, in which the Rashba effect was also introduced. That paper carefully discussed the spin splitting in a 2DEG of BIA systems such as zincblende structured GaAs. In a 3D structure, when a uniaxial stress exists, the CB Hamiltonian of a semiconductor in GaAs-like structure is

$$H = \frac{p^2}{2m} + \frac{\hbar}{2}(\boldsymbol{\sigma} \cdot \boldsymbol{\Omega}), \tag{6.25}$$

where

$$\boldsymbol{\Omega} = \frac{\beta \boldsymbol{\kappa}}{\hbar m^{3/2}(2E_g)^{1/2}} \tag{6.26}$$

with

$$\kappa_x = p_x(p_y^2 - p_z^2); \quad \kappa_y = p_y(p_z^2 - p_x^2); \quad \kappa_z = p_z(p_x^2 - p_y^2). \tag{6.27}$$

In Eq. (6.26) and the following description, in order to be consistent with the general expressions, we have reversed the coefficients α and β in

D'yakonov and Kachorovskiĭ's paper in 1986, m is the effective mass, $\boldsymbol{p}$ the electron momentum, $\boldsymbol{\sigma}$ the Pauli matrices, E_g the band gap, and β is a numerical coefficient representing the magnitude of the spin splitting in the CB (with β taken as 0.07 in GaAs).

In the 2D case, Eq. (6.25) can be written as (D'yakonov and Perel', 1972)

$$H = \frac{k^2}{2m} + \frac{\hbar}{2}(\sigma \cdot \omega), \tag{6.28}$$

where $\boldsymbol{k}$ is the in-plane 2D momentum. As the spin–orbit term is small, so that if we take an average of $\boldsymbol{\Omega}$, for the electron movement perpendicular to the 2DEG plane, we can express $\boldsymbol{\omega} = \langle \boldsymbol{\Omega} \rangle \cdots$. For that reason, we write the momentum as $\boldsymbol{p} = \boldsymbol{q} + \boldsymbol{k}$, with $\boldsymbol{q} = \boldsymbol{n}(\boldsymbol{p} \cdot \boldsymbol{n}), \boldsymbol{k} = \boldsymbol{n} \times (\boldsymbol{p} \times \boldsymbol{n})$, $\boldsymbol{n}$ is the unit vector perpendicular to the plane. Obviously,

$$\langle q_i \rangle = 0, \quad \langle q_i q_j \rangle = \langle q^2 \rangle n_i n_j, \quad \langle q_i q_j q_l \rangle = 0. \tag{6.29}$$

The magnitude of the square of the vertical wave vector $\langle q^2 \rangle$ is determined by the potential well that constrains the electron vertical movement. For an infinite square well with a width of a, $\langle q^2 \rangle = (\pi\hbar/a)^2$. Then from Eqs. (6.27) and (6.29), we get

$$\langle \kappa_x \rangle = \langle q^2 \rangle [2n_x(n_y k_y - n_z k_z) + (n_y^2 - n_z^2)k_x]. \tag{6.30}$$

Similarly, $\langle \kappa_y \rangle$ and $\langle \kappa_z \rangle$ can be obtained from the permutation of the subscripts in the above equation. In the procedure of deriving Eq. (6.30), the k^3 term has been ignored, due to the reason explained before: $\langle q^2 \rangle \gg k^2$. Therefore, the effective magnetic field applied to the electron spin which is proportional to vector $\boldsymbol{\omega} \approx \langle \boldsymbol{\kappa} \rangle$ is linear to the 2D wave vector $\boldsymbol{k}$ and strongly depends on the orientation of the plane against the crystal axes.

For spin relaxation, similar to the case of 3D, we have

$$\frac{\mathrm{d}S_i}{\mathrm{d}t} = -\sum_j \Gamma_{ij} S_j, \tag{6.31}$$

where S is the average spin, and the expression of the relaxation tensor Γ_{ij} is

$$\Gamma_{ij} = \frac{\int_0^a \mathrm{d}z(\partial F/\partial \varepsilon)\tau(\varepsilon)(\overline{\omega^2}\delta_{ij} - \overline{\omega_i \omega_j})}{\int_0^a \mathrm{d}z(\partial F/\partial \varepsilon)}, \tag{6.32}$$

where $F(\varepsilon)$ is the electron distribution function against energy, $\tau(\varepsilon)$ is the momentum relaxation time, and the bar on top of the variables means to take average against the 2D momentum $\boldsymbol{k}$. In the procedure of taking average, we use:

$$\overline{k_i k_j} = \frac{(\delta_{ij} - n_i n_j)k^2}{2}. \tag{6.33}$$

In the derivation of Eqs. (6.31) and (6.32), it is assumed that $\omega\tau \ll 1$, and the energy relaxation time is much longer than the momentum relaxation time, but much smaller than the spin relaxation time. Furthermore, it is assumed that the electrons are degenerate with small average spin.

Using Eqs. (6.30), (6.32), and (6.33), we get

$$\Gamma_{ij} = \tau_s^{-1}(\varepsilon_{ij} T_r \hat{B} - B_{ij}), \tag{6.34}$$

where

$$\frac{1}{\tau_s} = \frac{a^2 \langle q^2 \rangle^2}{2\hbar^2 m^2 E_g} \frac{\int_0^\infty d\varepsilon \varepsilon \tau(\varepsilon)(\partial F/\partial \varepsilon)}{\int_0^\infty d\varepsilon (\partial F/\partial \varepsilon)}. \tag{6.35}$$

The symmetric tensor $\hat{B}$ is determined by the relative orientation of the plane against crystal axes. The elements of $\hat{B}$ in the coordinates relative to crystal axes are

$$B_{xx} = (n_y^2 - n_z^2)^2 + 4n_x^2(n_y^2 + n_z^2) - 9n_x^2(n_y^2 - n_z^2). \tag{6.36a}$$

$$B_{xy} = -2n_x n_y(n_x^2 + n_y^2) + -9n_x n_y(n_y^2 - n_z^2)(n_z^2 - n_x^2). \tag{6.36b}$$

The other components of $\hat{B}$ can be obtained by permutation of the coordinates.

In the case of no degeneracy, the expression of Eq. (6.35) is

$$\frac{1}{\tau_s} = \frac{a^2 \langle q^2 \rangle^2 \tau T}{2\hbar^2 m^2 E_g}, \tag{6.37}$$

where τ is the average relaxation time which is used in the equation of mobility: $\mu = e\tau/m$. In the degenerate case, temperature T has to be replaced by Fermi energy ε_F/k.

The above equations, namely Eqs. (6.31), (6.34), (6.36), and (6.37), are for random orientations of the 2DEG plane. Now we discuss some particularly useful orientations:

(a) *⟨001⟩ orientation*

In this case, $n_x = n_y = 0$, and $n_z = 1$; thus, the nonzero elements of tensor $\hat{\Gamma}$ can be derived from Eqs. (6.34) and (6.36) as $\Gamma_{xx} = \Gamma_{yy} = \Gamma_{zz}/2 = \tau_s^{-1}$; thus, the in-plane components of the spin relaxation rates are determined by $\tau_{\parallel} = \tau_s$, while the component normal to the plane is twice of the in-plane one: $\tau_{\perp} = \tau_s/2$. The physical reason for this difference is that for this orientation $\langle \kappa_z \rangle = 0$, i.e. the effective magnet field is essentially varied in the plane.

(b) *⟨111⟩ orientation*

In this case, $n_x = n_y = n_z = 3^{-1/2}, \Gamma_{xx} = \Gamma_{yy} = \Gamma_{zz} = 16/(9\tau_{\rm s})$, and $\Gamma_{xy} = \Gamma_{xz} = \Gamma_{yz} = 4/(9\tau_{\rm s})$. Tensor $\hat{\Gamma}$ is reduced to the major axis, and therefore, still the normal component is twice in magnitude of the in-plane one, and also the effective magnetic field is varied in the 2DEG plane.

(c) *⟨110⟩ orientation*

In this case, $n_x = n_y = 2^{-1/2}, n_z = 0, \Gamma_{xx} = \Gamma_{yy} = \Gamma_{zz}/2 = -\Gamma_{xy} = (8\tau_{\rm s})^{-1}$, and $\Gamma_{xz} = \Gamma_{yz} = 0$. Therefore, $\tau_{\|} = 4\tau_{\rm s}, \tau_{\perp} = \infty$. That means that there is no relaxation for the spin component normal to the plane. It seemed that the surprising result is not hard to understand, as it can be seen from Eq. (6.30) that the effective magnetic field is always perpendicular to the plane. When electrons are scattered, only the strength and sign of the effective magnetic field will be affected. In reality, of course, the perpendicular component of spin will be slowly relaxed. If the higher order terms are not ignored in Eq. (6.30), then the value of $\tau_{\perp}$ would be limited, i.e. $\tau_{\perp}/\tau_{\rm s} \approx (\hbar^2/ma^2\varepsilon)^2 \gg 1$.

If the quantum confinement induced by potential barrier is asymmetric, then there must be an additional term in the Hamiltonian (6.28) (Bychkov and Rashba, 1984), which is the Rashba term $V = \alpha\boldsymbol{\sigma} \cdot (\boldsymbol{k} \times \boldsymbol{n})$. This term can also induce spin splitting, and thus spin relaxation. If the Rashba interaction V is predominant, the tensor $\hat{\Gamma}$ should be the same as that in the case of ⟨001⟩ orientation, but Eqs. (6.35) and (6.37) have to be corrected by the following replacement:

$$\frac{a^2\langle q^2\rangle^2}{2\hbar^2 m^2 E_{\rm g}} \rightarrow \frac{4m\alpha^2}{\hbar^2}. \tag{6.38}$$

6.1.3. *Relativistic quantum mechanical understanding*

(1) Dirac equation — the relativistic effect in solid:

(a) *Relativistic de Broglie relationship*: It is well known that the Lorentz transforms are

$$\begin{aligned} t' &= \frac{t - vt}{(1 - v^2/c^2)^{1/2}} \\ x' &= \frac{x - x/v}{(1 - v^2/c^2)^{1/2}} \\ y' &= y, \quad z' = z. \end{aligned} \tag{6.39}$$

For the energy and momentum of a particle, there are following invariants:

$$E = \frac{mc^2}{(1 - v^2/c^2)^{1/2}}$$
$$p = \frac{mv}{(1 - v^2/c^2)^{1/2}} \quad (6.40)$$

as well as the Lorentz invariant:

$$E^2 - c^2(p_x^2 - p_y^2 - p_z^2) = E^2 - c^2p^2 = m^2c^4. \quad (6.41)$$

For a nonrelativistic particle, $E = \hbar\omega, \boldsymbol{p} = \hbar\boldsymbol{k}$, which is consistent with the de Broglie relationship and can lead to the nonrelativistic Schrödinger equation. But in a relativistic system,

$$\hbar^2(\omega^2 - c^2k^2) = m^2c^4. \quad (6.42)$$

Therefore the wavefunction of a free particle

$$\psi \propto e^{i(\boldsymbol{k}\cdot\boldsymbol{r} - \omega t)}$$

satisfies the following equation

$$-\hbar^2\frac{\partial^2\psi}{\partial t^2} + \hbar^2c^2\nabla^2\psi = m^2c^2\psi. \quad (6.43)$$

If only the particle number is conserved, i.e.

$$\frac{\partial\rho}{\partial t} + \nabla\cdot j = 0. \quad (6.44)$$

Equation (6.43) can be taken as the relativistic Schrödinger equation, here

$$\rho = \frac{i\hbar}{2mc^2}\left(\psi^*\frac{\partial\psi}{\partial t} - \psi\frac{\partial\psi^*}{\partial t}\right) \quad (6.45)$$

$$j = -\frac{i\hbar}{2m}(\psi^*\nabla\psi - \psi\nabla\psi^*). \quad (6.46)$$

In order to make the equation be kept in the first-order differentiation, we take the first-order expression of energy,

$$E = (p^2c^2 + m^2c^4)^{1/2}. \quad (6.47)$$

(b) *Dirac equation*: Dirac introduced a Hamiltonian:

$$H = \boldsymbol{\alpha}\cdot c\boldsymbol{p} + \beta mc^2 \quad (6.48)$$

and applied that into the Schrödinger equation

$$i\hbar\frac{\partial\psi}{\partial t} = H\psi. \tag{6.49}$$

The Hamiltonian is expressed as

$$H^2 = c^2\alpha_a\alpha_b p_a p_b + mc^3(\alpha_a\beta + \beta\alpha_a)p_a + m^2c^4\beta^2. \tag{6.50}$$

The subscripts in the equation represent summation from 1 to 3, and the numbers represent x, y, and z components. For example, the dot product of $\boldsymbol{A}$ and $\boldsymbol{B}$ is written as

$$\boldsymbol{A}\cdot\boldsymbol{B} = A_a B_a. \tag{6.51}$$

To be consistent with Eq. (6.47), we should have

$$\beta^2 = 1 \tag{6.52}$$

$$\alpha_a\beta + \beta\alpha_a = \{\alpha_a, \beta\} = 0 \tag{6.53}$$

$$\alpha_a\alpha_b + \alpha_b\alpha_a = \{\alpha_a, \alpha_b\} = 2\delta_{ab}, \tag{6.54}$$

where

$$\{A, B\} = AB + BA. \tag{6.55}$$

That is to say that A and B anticommute. We then have

$$\frac{1}{2}(\alpha_a\alpha_b + \alpha_b\alpha_a)p_a p_b = p_a p_b \delta_{ab} = p^2. \tag{6.56}$$

Equation (6.47) is thus satisfied.

But what are the parameters α and β? A hint is that the Pauli spin matrices are

$$\sigma_x = \begin{pmatrix} 0 & 1 \\ 1 & 0 \end{pmatrix}, \quad \sigma_y = \begin{pmatrix} 0 & -i \\ i & 0 \end{pmatrix}, \quad \sigma_z = \begin{pmatrix} 1 & 0 \\ 0 & -1 \end{pmatrix} \tag{6.57}$$

and

$$\sigma_a^2 = 1, \quad \sigma_x\sigma_y = i\sigma_z, \quad \sigma_y\sigma_x = -i\sigma_z. \tag{6.58}$$

Therefore,

$$\{\sigma_a, \sigma_b\} = 2\delta_{ab}. \tag{6.59}$$

Based on the time reversal requirement, Dirac, with great talent, chose

$$\alpha_a = \begin{pmatrix} 0 & \sigma_a \\ \sigma_a & 0 \end{pmatrix}, \quad \beta = \begin{pmatrix} I & 0 \\ 0 & -I \end{pmatrix},$$

i.e.

$$\alpha_x = \begin{pmatrix} 0 & \sigma_x \\ \sigma_x & 0 \end{pmatrix} = \begin{bmatrix} 0 & 0 & 0 & 1 \\ 0 & 0 & 1 & 0 \\ 0 & 1 & 0 & 0 \\ 1 & 0 & 0 & 0 \end{bmatrix}. \tag{6.60}$$

Two 2×2 matrices makes a 4×4 matrix:

$$\begin{aligned} A \otimes B &= \begin{pmatrix} A_{11}B & A_{12}B \\ A_{21}B & A_{22}B \end{pmatrix} \\ &= \begin{bmatrix} A_{11}B_{11} & A_{11}B_{12} & A_{12}B_{11} & A_{12}B_{11} \\ A_{11}B_{21} & A_{11}B_{22} & A_{12}B_{21} & A_{12}B_{22} \\ A_{21}B_{11} & A_{21}B_{12} & A_{22}B_{11} & A_{22}B_{12} \\ A_{21}B_{21} & A_{21}B_{22} & A_{22}B_{21} & A_{22}B_{22} \end{bmatrix}. \end{aligned} \tag{6.61}$$

As

$$(A \otimes B)(C \otimes D) = AC \otimes BD \tag{6.62}$$

so that

$$\alpha_a = \sigma_1 \otimes \sigma_a, \quad \beta = \sigma_3 \otimes I; \tag{6.63}$$

thus

$$\alpha_x\beta = (\sigma_1 \otimes \sigma_x)(\sigma_3 \otimes I) = \sigma_1\sigma_3 \otimes \sigma_x I = -i\sigma_2 \otimes \sigma_x$$

and

$$\beta\alpha_x = \sigma_3\sigma_1 \otimes I\sigma_x = i\sigma_2 \otimes \sigma_x.$$

Therefore,

$$\alpha_x\beta + \beta\alpha_x = \{\alpha_x, \beta\} = 0. \tag{6.64}$$

The corresponding wavefunction is also 4D:

$$\psi = \begin{bmatrix} \psi_1(\boldsymbol{r}, t) \\ \psi_2(\boldsymbol{r}, t) \\ \psi_3(\boldsymbol{r}, t) \\ \psi_4(\boldsymbol{r}, t) \end{bmatrix} = \begin{bmatrix} \psi_+ \\ \psi_- \end{bmatrix}. \tag{6.65}$$

The Dirac equation for a free particle is thus obtained as:

$$i\hbar\frac{\partial}{\partial t}\begin{bmatrix}\psi_+\\ \psi_-\end{bmatrix} = c\boldsymbol{p}\cdot\begin{bmatrix}0 & \boldsymbol{\sigma}\\ \boldsymbol{\sigma} & 0\end{bmatrix}\begin{bmatrix}\psi_+\\ \psi_-\end{bmatrix} + mc^2\begin{bmatrix}I & 0\\ 0 & -I\end{bmatrix}\begin{bmatrix}\psi_+\\ \psi_-\end{bmatrix} \tag{6.66}$$

or in the form of simultaneous equations:

$$i\hbar\frac{\partial}{\partial t}\psi_+ = \boldsymbol{\sigma}\cdot c\boldsymbol{p}\psi_- + mc^2\psi_+ \tag{6.67a}$$

$$i\hbar\frac{\partial}{\partial t}\psi_- = 0\boldsymbol{\sigma}\cdot c\boldsymbol{p}\psi_+ - mc^2\psi_-, \tag{6.67b}$$

where $\boldsymbol{\sigma}$ is the Pauli spin operator, the subscript of $\psi_\pm$ represents the two eigenstates with different energy eigenvalues, due to SOI.

(2) **Electron spin**: An electron is a Fermion with spin of 1/2. The general expression of the spin operators are

$$S_x = \frac{\hbar}{2}\begin{pmatrix}0 & 1\\ 1 & 0\end{pmatrix}, \quad S_y = \frac{\hbar}{2}\begin{pmatrix}0 & -i\\ i & 0\end{pmatrix}, \quad S_z = \frac{\hbar}{2}\begin{pmatrix}1 & 0\\ 0 & -1\end{pmatrix}$$

$$[S_i, S_j] = i\varepsilon_{ijk}\hbar S_k, \quad \{S_i, S_j\} = \frac{\hbar^2}{2}\delta_{ij}, \quad [S^2, S_i] = 0$$

$$S^2 \equiv S_x^2 + S_y^2 + S_z^2 = \frac{3}{4}\hbar^2,$$

where $\varepsilon_{123} = 1$ is a replacement mark, when the subscripts exchange, its sign will change.

Let us discuss the spin in details. In a spherically symmetrical potential, the Hamiltonian is

$$H = \boldsymbol{\alpha}\cdot c\boldsymbol{p} + \beta mc^2 + V(\boldsymbol{r}). \tag{6.68}$$

The angular momentum $\boldsymbol{L} = \boldsymbol{r}\times\boldsymbol{p}$ can be expressed as

$$L_a = \varepsilon_{abc} r_b p_c, \tag{6.69}$$

thus

$$[H, L_\mathrm{a}] = \varepsilon_{abc} c\alpha_a [p_a, r_b p_c] = -i\varepsilon_{abc} c\alpha_b p_c. \tag{6.70}$$

This shows that the angular momentum is not conserved.

Introducing

$$\Sigma_a = I \otimes \sigma_a = \begin{pmatrix} \sigma_a & 0 \\ 0 & \sigma_a \end{pmatrix}, \tag{6.71}$$

we have

$$[H, \Sigma_a] = 2i\varepsilon_{abc} c\alpha_b p_c. \tag{6.72}$$

Therefore, the sum of

$$J_a = L_a + \frac{\hbar}{2}\Sigma_a. \tag{6.73}$$

will commute with the Hamiltonian, i.e. being a conserved quantity. The spin of a Dirac particle is thus introduced as

$$S_a = \frac{\hbar}{2}\Sigma_a. \tag{6.74}$$

It is easy to verify that

$$[S_a, S_b] = i\hbar\varepsilon_{abc} S_c \tag{6.75}$$

and

$$S^2 \equiv S_x^2 + S_y^2 + S_z^2 = \frac{3}{4}\hbar^2. \tag{6.76}$$

Equations (6.75) and (6.76) indicate that the eigenvalue of the spin is $\hbar/2$. As for

$$S_z = \frac{\hbar}{2}\begin{pmatrix} \sigma_z & 0 \\ 0 & \sigma_z \end{pmatrix}, \tag{6.77}$$

the eigenfunction corresponding to its eigenvalue $S_z = \hbar/2$ is

$$\begin{bmatrix} \psi_+(\uparrow) \\ 0 \end{bmatrix} \quad \text{and} \quad \begin{bmatrix} 0 \\ \psi_-(\uparrow) \end{bmatrix} \tag{6.78}$$

(3) **Magnetic moment**: In a magnetic field $\boldsymbol{B} = \nabla \times \boldsymbol{A}$, $c\boldsymbol{p} \rightarrow c\boldsymbol{p} - q\boldsymbol{A}$, so that

$$H = \boldsymbol{\alpha} \cdot (c\boldsymbol{p} - q\boldsymbol{A}) + \beta mc^2. \tag{6.79}$$

This leads to the following simultaneous equations:

$$i\hbar\frac{\partial}{\partial t}\psi_+ = \boldsymbol{\sigma} \cdot (c\boldsymbol{p} - q\boldsymbol{A})\psi_- + mc^2\psi_+ \tag{6.80a}$$

$$i\hbar\frac{\partial}{\partial t}\psi_- = \boldsymbol{\sigma} \cdot (c\boldsymbol{p} - q\boldsymbol{A})\psi_+ + mc^2\psi_-. \tag{6.80b}$$

For a relativistic system, $E \sim mc^2$, therefore the time-related function can be written as

$$\psi_\pm = \phi_\pm e^{-mc^2 t/\hbar}, \tag{6.81}$$

and thus

$$i\hbar\frac{\partial}{\partial t}\phi_+ = \boldsymbol{\sigma}\cdot(c\boldsymbol{p} - q\boldsymbol{A})\phi_- \tag{6.82a}$$

$$i\hbar\frac{\partial}{\partial t}\phi_- = \boldsymbol{\sigma}\cdot(c\boldsymbol{p} - q\boldsymbol{A})\phi_+ - 2mc^2\phi_-. \tag{6.82b}$$

If the left-hand side of Eq. (6.82b) is ignored (as it is much smaller than the mc^2 term), we get

$$\phi_- = \frac{1}{2mc^2}\boldsymbol{\sigma}\cdot(c\boldsymbol{p} - q\boldsymbol{A})\phi_+. \tag{6.83}$$

Substituting Eq. (6.83) into Eq. (6.82a), we get

$$i\hbar\frac{\partial}{\partial t}\phi_+ = \frac{1}{2mc^2}\sigma_a\sigma_b(cp_a - qA_a)(cp_b - qA_b)\phi_+,$$

and applying

$$\sigma_a\sigma_b = \delta_{ab} + i\varepsilon_{abc}\sigma_c,$$

we get

$$\begin{aligned}
&\frac{1}{2mc^2}\sigma_c\varepsilon_{abc}(cp_a - qA_a)(cp_b - qA_b)\phi_+ \\
&\quad = \frac{iq}{2mc^2}\sigma_c\varepsilon_{abc}(cp_aA_b + cA_ap_b)\phi_+ \\
&\quad = \frac{q\hbar}{2mc^2}\sigma_c\varepsilon_{abc}(\nabla_aA_b + A_b\nabla_a)\phi_+ \\
&\quad = \frac{q\hbar}{2mc^2}\sigma_c\varepsilon_{abc}\frac{\partial A_b}{\partial r_a}\phi_+ = -\frac{q\hbar}{2mc}\boldsymbol{\sigma}\cdot\boldsymbol{B}\phi_+,
\end{aligned} \tag{6.84}$$

therefore

$$i\hbar\frac{\partial}{\partial t}\phi_+ = \frac{(\boldsymbol{p} - (q/c)\boldsymbol{A})^2}{2mc}\phi_+ - \frac{q}{mc}\boldsymbol{s}\cdot\boldsymbol{B}\phi_+, \tag{6.85}$$

where $\boldsymbol{s} = \hbar\boldsymbol{\sigma}/2$, and the second term automatically implies that the Landé factor of the particle is $g = 2$.

(4) **SOC**: We remember that in the case of no magnetic field,

$$i\hbar\frac{\partial}{\partial t}\psi_+ = \boldsymbol{\sigma}\cdot c\boldsymbol{p}\psi_- + mc^2\psi_+$$
$$i\hbar\frac{\partial}{\partial t}\psi_- = \boldsymbol{\sigma}\cdot c\boldsymbol{p}\psi_+ - mc^2\psi_-.$$

If the total energy of the particle is mc^2+E, with E represents the contributions of the potential and kinetic energies, the energy eigenvalue equation is then:

$$\boldsymbol{\sigma}\cdot c\boldsymbol{p}\psi_- + mc^2\psi_+ + V\psi_+ = (E+mc^2)\psi_+ \tag{6.86}$$
$$\boldsymbol{\sigma}\cdot c\boldsymbol{p}\psi_+ - mc^2\psi_- + V\psi_- = (E+mc^2)\psi_-, \tag{6.87}$$

therefore,

$$\psi_- = \frac{1}{2mc^2+E-V}\boldsymbol{\sigma}\cdot c\boldsymbol{p}\psi_+ \tag{6.88}$$

and so that

$$\boldsymbol{\sigma}\cdot c\boldsymbol{p}\frac{1}{2mc^2+E-V}\boldsymbol{\sigma}\cdot c\boldsymbol{p}\psi_+ + V\psi_+ = E\psi_+. \tag{6.89}$$

This is just an energy eigenvalue equation, and the first term can be written as

$$\delta_{ab} + i\varepsilon_{abc}\sigma_c c^2 p_a F p_b \psi_+,$$

where F takes the following form:

$$F = \frac{1}{2mc^2+E-V} = \frac{1}{2mc^2}\left(1+\frac{V-E}{2mc^2}\right), \tag{6.90}$$

thus

$$p_a F p_b = [p_a, F]p_b + Fp_a p_b = -i\hbar\frac{\partial F}{\partial r_a}p_b + Fp_a p_b. \tag{6.91}$$

Being multiplied by ε_{abc}, the second term will disappear, and meanwhile,

$$\frac{\partial F}{\partial r_a} = \frac{1}{4m^2c^4}\frac{\partial V}{\partial r_a} = \frac{1}{4m^2c^4}\frac{\mathrm{d}V}{\mathrm{d}r}\frac{r_a}{r}, \tag{6.92}$$

so that

$$\frac{1}{2m}\left(p^2 + \boldsymbol{p}\cdot\frac{V-E}{2mc^2}\boldsymbol{p}\right)\psi_+ + \frac{\hbar}{4m^2c^2r}\frac{\mathrm{d}V}{\mathrm{d}r}\varepsilon_{abc}\sigma_c r_a p_b\psi_+ = E\psi_+. \tag{6.93}$$

Using the angular momentum operator, we finally get

$$\frac{1}{2m}\left(p^2 + \boldsymbol{p}\cdot\frac{V-E}{2mc^2}\boldsymbol{p}\right)\psi_+ + \frac{1}{2m^2c^2r}\frac{\mathrm{d}V}{\mathrm{d}r}\boldsymbol{L}\cdot\boldsymbol{s}\psi_+ = E\psi_+, \tag{6.94}$$

where $V = q\phi = -e\phi$ and ϕ is the static potential, and SOI is included in the second term, while the first term includes the first-order relativistic correction for the dynamic energy.

6.2. The SOI Hamiltonian in a Rashba System

The last term for SOI in the Dirac Hamiltonian is

$$\begin{aligned}\frac{1}{2m^2c^2r}\frac{\mathrm{d}V}{\mathrm{d}r}\boldsymbol{L}\cdot\boldsymbol{s} &= \frac{1}{2m^2c^2r}\frac{\mathrm{d}V}{\mathrm{d}r}(\boldsymbol{r}\times\boldsymbol{p})\cdot\boldsymbol{s}\\ &= \frac{1}{2m^2c^2r}\frac{\mathrm{d}V}{\mathrm{d}r}(\boldsymbol{p}\times\boldsymbol{s})\cdot\boldsymbol{r}\\ &= \frac{\hbar}{4m^2c^2}(-\nabla V)\cdot(\boldsymbol{\sigma}\times\boldsymbol{p}),\end{aligned} \tag{6.95}$$

where $-\nabla V$ is the electric field. Assuming z is the quantum confinement direction in the 2D system, which is perpendicular to the 2D plane of the electron movement, then the spin–orbit Hamiltonian is written as

$$H = \frac{\alpha}{\hbar}(\boldsymbol{\sigma}\times\boldsymbol{p})_z = i\alpha\left(\sigma_y\frac{\partial}{\partial x} - \sigma_x\frac{\partial}{\partial y}\right), \tag{6.96}$$

where α is named as the Rashba coefficient, representing the integrated strength of various SOI related to SIA, and generally being $1\text{–}10\times 10^{-10}$ eVcm.

Under the EMA, the Hamiltonian is written as

$$H = \frac{p_x^2 + p_y^2}{2m} - \frac{\alpha}{\hbar}(\sigma_y p_x - \sigma_x p_y). \tag{6.97}$$

As the operators p_x and p_y commute with H, there are eigenstates:

$$|\psi\rangle = e^{i(k_x x + k_y y)}(c_\uparrow|\uparrow\rangle + c_\downarrow|\downarrow\rangle), \tag{6.98}$$

$|\uparrow\rangle$ and $|\downarrow\rangle$ are the two spin states with spin up and down, respectively. We know that

$$\sigma_x = \begin{pmatrix} 0 & 1 \\ 1 & 0 \end{pmatrix}; \quad \sigma_y = \begin{pmatrix} 0 & -i \\ i & 0 \end{pmatrix}.$$

The Hamiltonian can then be written as

$$H = \begin{pmatrix} \dfrac{\hbar^2 k^2}{2m} & \alpha k_y + i\alpha k_x \\ \alpha k_y - i\alpha k_x & \dfrac{\hbar^2 k^2}{2m} \end{pmatrix}. \tag{6.99}$$

Diagonalizing the Hamiltonian in the spin space, we get the eigenvalues as

$$E_\pm(k) = \left(\frac{\hbar^2 k^2}{2m}\right) \pm \alpha k \tag{6.100}$$

and eigenfunctions as

$$|\psi_{E_+}\rangle = \frac{e^{i(k_x x + k_y y)}}{\sqrt{2}}(e^{i\theta_\gamma/2}|\uparrow\rangle + e^{i\theta_\gamma/2}|\downarrow\rangle) \tag{6.101a}$$

$$|\psi_{E_-}\rangle = \frac{e^{i(k_x x + k_y y)}}{\sqrt{2}}(e^{i\theta_\gamma/2}|\uparrow\rangle + e^{i\theta_\gamma/2}|\downarrow\rangle), \tag{6.101b}$$

where the meaning of θ_γ is shown in Fig. 6.2. It can be seen from the figure that the electrons are under the action of an effective magnetic field perpendicular to the plane of its movement. In the steady state, the spin

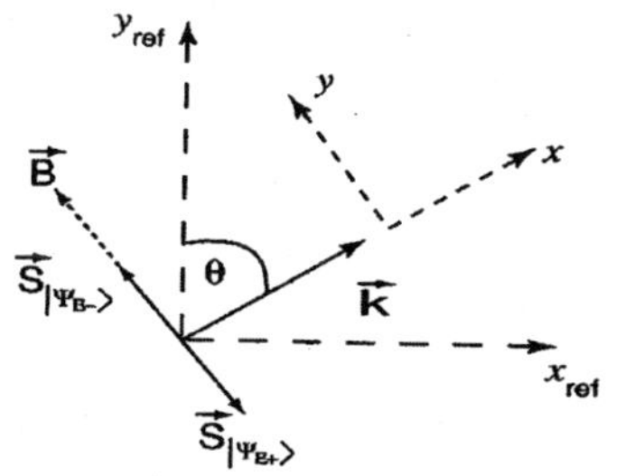

Fig. 6.2. The spin direction of the quantum state $|\psi_{E_\pm}\rangle$. Courtesy of Feve *et al.* (2002).

will be either parallel or antiparallel to the magnetic field and form the following two angles:

$$\angle(\boldsymbol{k}, \boldsymbol{S}_{|\psi_{E_-}\rangle}) = \frac{\pi}{2}; \quad \angle(\boldsymbol{k}, \boldsymbol{S}_{|\psi_{E_+}\rangle}) = \frac{\pi}{2}.$$

The magnitude of the effective magnetic field depends on the electron velocity. In general, the spin–orbit (S–O) splitting is smaller than the electron kinetic energy, e.g.

$$0.15\,\text{meV} \le \alpha k_{\text{F}} \le 1.5\,\text{meV} \quad (E_{\text{F}} = 14\,\text{meV}).$$

An electron with wave vector $\boldsymbol{k}$, is facing an effective magnetic field $\boldsymbol{B}(\boldsymbol{k}, \alpha)$. The axis x_γ is in parallel with $\boldsymbol{k}$. For 1D electrodes, γ indicates the electrode lead, then $\boldsymbol{k}$ is in the longitudinal direction.

6.3. The Rashba Effect and Dispersion (Feve *et al.*, 2002)

To further study the eigenstates and eigenvalues, assuming that $\boldsymbol{k}$ is along the electrode lead direction x_γ which has an angle θ_γ from the laboratory coordinate axis y_{ref}. The movement direction is labeled as ε: If the group velocity is positive, $v_{\text{g}} > 0$, then $\varepsilon \equiv a$; oppositely, $\varepsilon \equiv b$. $\sigma \equiv \pm$ represents the orientation of spin. Thus, the eigenstates and eigenvalues are written as

$$|\varepsilon, k, \sigma\rangle = \Phi(y_\gamma)\frac{e^{i\theta_\varepsilon k x_\gamma}}{\sqrt{2}}(e^{i\theta_\gamma/2}|\uparrow\rangle + \eta_\varepsilon\eta_\sigma e^{-i\theta_\gamma/2}|\downarrow\rangle)$$

$$E = \frac{\hbar^2 k^2}{2m} + \eta_\sigma \alpha k; \quad -\eta_\sigma k_0 \le k < \infty, \tag{6.102}$$

where $\Phi(y_\gamma)$ is the normalized horizontal wave function, $\eta_\varepsilon = +1(-1)$ corresponds to $\varepsilon = a(b)$, and $\eta_\sigma = \pm 1$ corresponds to $\sigma = \pm$. The dispersion relationship represented by Eq. (6.102) is shown in Fig. 6.3; the other part is mirrored of the figure, with the energy axis as the mirror axis.

It is noticed from Fig. 6.3 that for a certain energy, the corresponding states are not located at the same wave vector. In fact, they have a difference in wave vector as,

$$\Delta k \equiv |k(\varepsilon, E, +) - k(\varepsilon, E, -)| = 2k_0 = \frac{2m\alpha}{\hbar^2}. \tag{6.103}$$

The electrons in the state $|\varepsilon, k, +\rangle$ and $|\varepsilon, k, -\rangle$ have opposite spin orientations, but both of them are perpendicular to the movement direction $\boldsymbol{v}_{\text{g}}$.

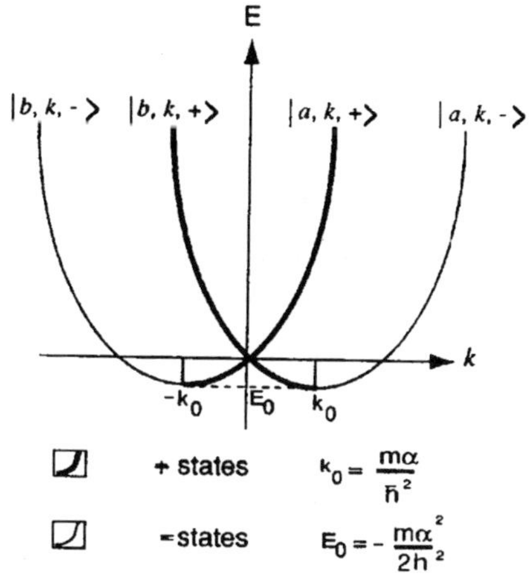

Fig. 6.3. The dispersion curve when spin–orbit are coupled. The range of $-\eta_\sigma k_0 \leq k < \infty$ is for state a, described by Eq. (6.89), while the range of $-\infty \leq k < \eta_\sigma k_0$ is for state b, the mirror of state a. Courtesy of Feve *et al.* (2002).

Note here that the movement direction is represented by $\boldsymbol{v}_\mathrm{g}$ rather than $\boldsymbol{k}$, although they may be consistent. It can be seen from Fig. 6.3 that at some energy positions, the positive (negative) value of k corresponds to negative (positive) value of $\boldsymbol{v}_\mathrm{g}$. Such an inconsistency can only occur if there exists SOI.

The density of states in the case of spin–orbit being coupled can be calculated from the following equation:

$$E = \frac{\hbar^2 k^2(E,\sigma)}{2m} + \eta_\sigma \alpha k(E,\sigma). \tag{6.104}$$

First the group velocity can be calculated as

$$v_{\mathrm{g}\sigma}(E) = \frac{1}{\hbar}\frac{\mathrm{d}E(k)}{\mathrm{d}k} = \frac{\hbar k(E,\sigma)}{m} + \eta_\sigma \frac{\alpha}{\hbar}. \tag{6.105}$$

Furthermore, from Eq. (6.103), we have

$$k(E,\sigma) = k(E,+) + (1-\eta_\sigma)\frac{m\alpha}{\hbar^2}.$$

Let

$$v_{\mathrm{g}\sigma}(E) = \frac{\hbar k(E,\pm)}{m} \pm \frac{\alpha}{\hbar} \equiv v_\mathrm{g}(E),$$

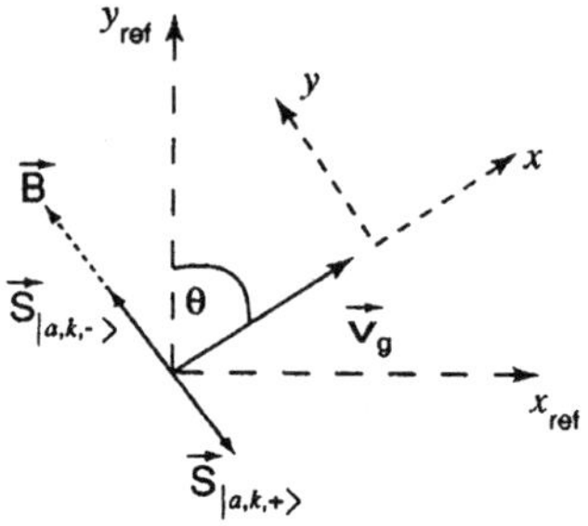

Fig. 6.4. The spin direction of states $|a, k, \pm\rangle$. Obviously, the density of states is not related to spin σ. Courtesy of Feve *et al.* (2002).

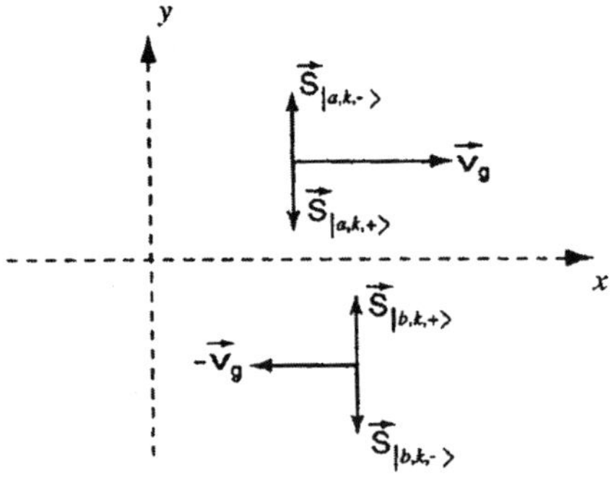

Fig. 6.5. Spin direction of x_σ^ε. Courtesy of Feve *et al.* (2002).

then

$$\rho(E) = \frac{\rho(k)}{\hbar v_g(E)} = \frac{L}{2\pi} \frac{1}{\hbar v_g(E)}. \tag{6.106}$$

Figure 6.4 shows the spin orientation of states $|a, k, \pm\rangle$, while Fig. 6.5 shows the spin of x_σ^ε, indicating in particular the relationship between spin and group velocity. These two figures correspond to the real setup of experiments.

6.4. Rashba Parameter α

At early times, the asymmetry of the confinement potential along the growth direction of the 2DEG plane is thought to be the origin of the average electric field along that axis, which leads to the Rashba SOI. However, this average field is actually zero for the bound state confined in a quantum well (QW). Further consideration based on the $\boldsymbol{k} \cdot \boldsymbol{p}$ model shows that the predominant mechanism of the SIA is related to the band discontinuation at the QW interfaces. In fact, the SOI energy is not necessarily proportional to the total potential drop $-\nabla V$ crossing over the quantum well.

6.4.1. *The $\boldsymbol{k} \cdot \boldsymbol{p}$ equation*

From the Bloch theorem,

$$\begin{aligned} \psi_{n\boldsymbol{k}}(\boldsymbol{r}) &= e^{i\boldsymbol{k}\cdot\boldsymbol{r}} u_{n\boldsymbol{k}}(\boldsymbol{r}) \\ u_{n\boldsymbol{k}}(\boldsymbol{r}+\boldsymbol{R}) &= u_{n\boldsymbol{k}}(\boldsymbol{r}), \end{aligned} \tag{6.107}$$

we have

$$\begin{aligned} &\left[\frac{p^2}{2m} + \frac{\hbar}{m}\boldsymbol{k}\cdot\boldsymbol{p} + V(\boldsymbol{r})\right] u_{n\boldsymbol{k}}(\boldsymbol{r}) \\ &\quad = \left[E_n(\boldsymbol{k}) - \frac{\hbar^2 k^2}{2m}\right] u_{n\boldsymbol{k}}(\boldsymbol{r}). \end{aligned} \tag{6.108}$$

Expanding in the vicinity of $\boldsymbol{k} = 0$, we get

$$\begin{aligned} &\left[H_0 + \frac{\hbar}{m}\boldsymbol{k}\cdot\boldsymbol{p}\right] u_{n\boldsymbol{k}}(\boldsymbol{r}) \\ &\quad = \left[E_n(\boldsymbol{k}) - \frac{\hbar^2 k^2}{2m}\right] u_{n\boldsymbol{k}}(\boldsymbol{r}), \end{aligned} \tag{6.109}$$

where

$$H_0 = \frac{p^2}{2m} + V(\boldsymbol{r}); \quad H_0 u_{n0}(\boldsymbol{r}) = E_n(0) u_{n0}(\boldsymbol{r}).$$

We know that the essentiality of the $\boldsymbol{k}\cdot\boldsymbol{p}$ method is to apply the precise solution at the $\boldsymbol{k} = 0$ point to approximate the general solutions in the entire $\boldsymbol{k}$ space, in which a second-order perturbation is used:

$$\begin{aligned} E_n(k) &= E_n(0) + \frac{\hbar}{m}\boldsymbol{k}\cdot\boldsymbol{p}_{nn} \\ &\quad + \frac{\hbar^2}{m^2}\sum_{n'\neq n}\frac{|\boldsymbol{k}\cdot\boldsymbol{p}_{nn'}|^2}{E_n(0) - E_{n'}(0)}, \end{aligned} \tag{6.110}$$

while the wave function under the first-order perturbation is

$$\begin{aligned}\boldsymbol{u}_{n\boldsymbol{k}}(\boldsymbol{r}) &= \boldsymbol{u}_{n0}(\boldsymbol{r}) + \frac{\hbar}{m}\sum_{n'\neq n}\frac{\boldsymbol{k}\cdot\boldsymbol{p}_{nn'}}{E_n(0)-E_{n'}(0)}\boldsymbol{u}_{n'0}(\boldsymbol{r}) \\ &= \sum_{n'\neq n} a_n \cdot u_{n'0}(\boldsymbol{r}). \end{aligned} \tag{6.111}$$

In the case of a single band, as

$$P_{nn} = 0 \quad \text{and} \quad \langle u_{n0}(\boldsymbol{r})|u_{n'0}(\boldsymbol{r})\rangle = \delta_{nn'},$$

we have

$$\begin{aligned} E_n(\boldsymbol{k}) - E_n(0) &= \sum_{\alpha,\beta} D^{\alpha\beta}k_\alpha k_\beta \\ &= \frac{\hbar^2}{2}\sum_{\alpha,\beta}\left(\frac{1}{m^*}\right)_{\alpha,\beta} k_\alpha k_\beta, \end{aligned} \tag{6.112}$$

where

$$\begin{aligned} D^{\alpha\beta} &= \frac{\hbar^2}{2m}\delta_{\alpha\beta} + \frac{\hbar^2}{2m^2} \\ &\cdot \sum_{n'\neq n}\frac{p^\alpha_{nn'}p^\beta_{n'n} + p^\beta_{nn'}p^\alpha_{n'n}}{E_n(0)-E_{n'}(0)} = \frac{\hbar^2}{2}\left(\frac{1}{m^*}\right)_{\alpha\beta}. \end{aligned} \tag{6.113}$$

Expanding the wave function $u_{n\boldsymbol{k}}(\boldsymbol{r})$ as

$$u_{n\boldsymbol{k}}(\boldsymbol{r}) = \sum_{n'} a_{n'}(\boldsymbol{k})u_{n'0}(\boldsymbol{r}) \tag{6.114}$$

and substituting Eq. (6.114) into Eq. (6.109), multiplying by $u^*_{n'0}(\boldsymbol{r})$, and integrating for the unit cell, we get,

$$\begin{aligned} &\sum_{n'}\left\{\left[E_n(0)+\frac{\hbar^2k^2}{2m}\right]\delta_{nn'} + \frac{\hbar}{m}\boldsymbol{k}\cdot\boldsymbol{p}_{nn'}\right\}a_{n'} \\ &\quad = E_n(\boldsymbol{k})a_n. \end{aligned} \tag{6.115}$$

In the case of double bands, we get

$$\begin{vmatrix} E_n(0)+\dfrac{\hbar^2k^2}{2m} - E & \dfrac{\hbar}{m}\boldsymbol{k}\cdot\boldsymbol{p}_{nn'} \\ \dfrac{\hbar}{m}\boldsymbol{k}\cdot\boldsymbol{p}_{nn'} & E_{n'}(0)+\dfrac{\hbar^2k^2}{2m} - E \end{vmatrix} = 0. \tag{6.116}$$

6.4.2. *The $\boldsymbol{k}\cdot\boldsymbol{p}$ treatment of SOI*

The Schrödinger equation including the SOI is written as

$$\left\{\frac{p^2}{2m}+V(\boldsymbol{r})+\frac{\hbar}{4m^2c^2}\boldsymbol{\sigma}\cdot(\nabla V\times\boldsymbol{p})\right\}\psi_{n\boldsymbol{k}}(\boldsymbol{r});$$
$$=E_n(\boldsymbol{k})\psi_{n\boldsymbol{k}}(\boldsymbol{r})$$

therefore,

$$\left\{\frac{p^2}{2m}+V(\boldsymbol{r})+\frac{\hbar}{m}\boldsymbol{k}\cdot\boldsymbol{p}+\frac{\hbar}{4m^2c^2}\boldsymbol{\sigma}\cdot(\nabla V\times\boldsymbol{p})\right.$$
$$\left.+\frac{\hbar}{4m^2c^2}\boldsymbol{\sigma}\cdot(\nabla V\times\boldsymbol{k})\right\}u_{n\boldsymbol{k}}(\boldsymbol{r})=E'u_{n\boldsymbol{k}}(\boldsymbol{r}). \tag{6.117}$$

The last term in the left is trivial, so that

$$\left\{\frac{p^2}{2m}+V(\boldsymbol{r})+\frac{\hbar}{m}\boldsymbol{k}\cdot\boldsymbol{p}+\frac{\hbar}{4m^2c^2}\boldsymbol{\sigma}\cdot(\nabla V\times\boldsymbol{p})\right\}u_{n\boldsymbol{k}}(\boldsymbol{r})=E'u_{n\boldsymbol{k}}(\boldsymbol{r}) \tag{6.118}$$

or

$$\sum_{n'}\left\{\left[E_n(0)+\frac{\hbar^2k^2}{2m}\right]\delta_{nn'}+\frac{\hbar}{m}\boldsymbol{k}\cdot\boldsymbol{p}_{nn'}\right.$$
$$\left.+\frac{\hbar}{4m^2c^2}\sigma\cdot(\nabla V\times\boldsymbol{p}_{nn'})\right\}a_{n'}=E_na_n. \tag{6.119}$$

6.4.3. *The eight bands model*

For the sake of convenience in the next chapter for discussing optical transitions, before introducing the theoretical solutions of the Rashba coefficient, we first provide the CB and valence band wave functions obtained by using the $\boldsymbol{k}\cdot\boldsymbol{p}$ method.

The CB wavefunctions are: $|S\uparrow\rangle, |S\downarrow\rangle$, with an energy of E_S, and the valence band wavefunctions are: $|X\uparrow\rangle, |Y\uparrow\rangle, |Z\uparrow\rangle, |X\downarrow\rangle, |Y\downarrow\rangle, |Z\downarrow\rangle$, with an energy of E_p. The generally adopted basis functions are:

$$|iS\downarrow\rangle,\quad \left|\frac{X-iY}{\sqrt{2}}\uparrow\right\rangle,\quad |Z\downarrow\rangle,\quad \left|-\frac{X+iY}{\sqrt{2}}\uparrow\right\rangle$$

and

$$|iS\uparrow\rangle,\quad \left|-\frac{X-iY}{\sqrt{2}}\downarrow\right\rangle,\quad |Z\uparrow\rangle,\quad \left|\frac{X+iY}{\sqrt{2}}\downarrow\right\rangle.$$

They form the eigenfunctions of a spherical harmonic function. The 8×8 interaction matrix is

$$\begin{bmatrix} \overline{\overline{\boldsymbol{H}}} & 0 \\ 0 & \overline{\overline{\boldsymbol{H}}} \end{bmatrix}.$$

At $k = k_z$,

$$\overline{\overline{\boldsymbol{H}}} = \begin{bmatrix} E_s & 0 & kP & 0 \\ 0 & E_p - \frac{\Delta}{3} & \sqrt{2}\Delta/3 & 0 \\ kP & \sqrt{2}\Delta/3 & E_p & 0 \\ 0 & 0 & 0 & E_p + \frac{\Delta}{3} \end{bmatrix}, \tag{6.120}$$

where the Kane parameter P and S–O splitting energy Δ are defined as

$$P = -i\frac{\hbar}{m}\langle S|p_z|Z\rangle \tag{6.121}$$

$$\Delta = \frac{3\hbar i}{4m^2c^2}\langle X|\frac{\partial V}{\partial x}p_y - \frac{\partial V}{\partial y}p_x|Z\rangle. \tag{6.122}$$

The real form of the matrix can be exampled by H_{11}:

$$\begin{aligned} H_{11} &= \langle iS\downarrow|H_0 + \frac{\hbar}{m}\boldsymbol{k}\cdot\boldsymbol{p} + \frac{\hbar}{4m^2c^2}\boldsymbol{\sigma}\cdot\nabla V\times\boldsymbol{p}|iS\downarrow\rangle \\ &= \langle S\downarrow|H_0|S\downarrow\rangle + \frac{\hbar}{m}\langle S\downarrow|\boldsymbol{k}\cdot\boldsymbol{p}|S\downarrow\rangle \\ &\quad + \frac{\hbar}{4m^2c^2}\langle S\downarrow|\sigma_z\left(\frac{\partial V}{\partial x}p_y - \frac{\partial V}{\partial y}p_x\right)|S\downarrow\rangle = E_S. \end{aligned} \tag{6.123}$$

The following results have been applied during the derivation:

$$\langle S\downarrow|\boldsymbol{p}|S\downarrow\rangle = \langle S|\boldsymbol{p}|S\rangle = 0 \tag{6.124}$$

$$\boldsymbol{\sigma}\cdot\nabla V\times\boldsymbol{p} = \sum_{i=x,y,z}\sigma_i(\nabla V\times\boldsymbol{p})_i \tag{6.125}$$

and

$$\langle\downarrow|\sigma_x|\downarrow\rangle = \langle\downarrow|\sigma_y|\downarrow\rangle = 0. \tag{6.126}$$

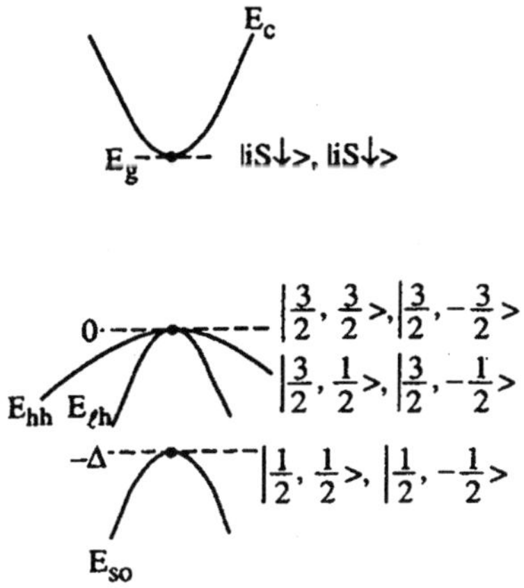

Fig. 6.6. The band edge energy and corresponding Bloch functions.

The obtained eigenvalues and the corresponding eigenfunctions are listed below, and shown in Fig. 6.6.

CB

$$E_{\rm c}(k) = E_{\rm g} + \frac{\hbar^2 k^2}{2m} + \frac{k^2 P^2 (E_{\rm g} + 2\Delta/3)}{E_{\rm g}(E_{\rm g} + \Delta)}$$

$$= E_{\rm g} + \frac{\hbar^2 k^2}{2m_{\rm e}^*}$$

$$\phi_{{\rm c},\alpha} = |iS\uparrow\rangle = \left|\frac{1}{2},\frac{1}{2}\right\rangle; \quad \phi_{{\rm c},\beta} = |iS\downarrow\rangle = \left|\frac{1}{2},\frac{1}{2}\right\rangle. \qquad (6.127)$$

Here in the quantum states expressed by the Dirac bracket at the right-hand side of the equation, the first number is the total angular momentum J and the second number is the magnetic component m_J of the total angular momentum,

Valence band:

Heavy hole:

$$E_{\rm hh}(k) = -\frac{\hbar^2 k^2}{2m_{\rm hh}^*}$$

$$\phi_{{\rm hh},\alpha} = \frac{-1}{\sqrt{2}}|(X+iY)\uparrow\rangle \equiv \left|\frac{3}{2},\frac{3}{2}\right\rangle$$

$$\phi_{{\rm hh},\beta} = \frac{-1}{\sqrt{2}}|(X+iY)\downarrow\rangle \equiv \left|\frac{3}{2},-\frac{3}{2}\right\rangle. \qquad (6.128)$$

Light hole:

$$E_{\rm lh}(k) = -\frac{\hbar^2k^2}{2m} - \frac{2k^2P^2}{3E_{\rm g}} = -\frac{\hbar^2k^2}{2m^*_{\rm lh}}$$

$$\phi_{{\rm lh},\alpha} = \frac{1}{\sqrt{6}}|(X-iY)\uparrow\rangle + \sqrt{\frac{2}{3}}|Z\downarrow\rangle \equiv \left|\frac{3}{2},-\frac{1}{2}\right\rangle$$

$$\phi_{{\rm lh},\beta} = \frac{1}{\sqrt{6}}|(X-iY)\downarrow\rangle + \sqrt{\frac{2}{3}}|Z\uparrow\rangle \equiv \left|\frac{3}{2},\frac{1}{2}\right\rangle. \tag{6.129}$$

Spin–orbit split band:

$$E_{\rm so}(k) = -\Delta - \frac{\hbar^2k^2}{2m} - \frac{k^2P^2}{3(E_{\rm g}+\Delta)} = -\Delta - \frac{\hbar^2k^2}{2m^*_{\rm so}}$$

$$\phi_{{\rm so},\alpha} = \frac{1}{\sqrt{3}}|(X-iY)\uparrow\rangle - \frac{1}{\sqrt{3}}|Z\downarrow\rangle \equiv \left|\frac{1}{2},-\frac{1}{2}\right\rangle$$

$$\phi_{{\rm so},\beta} = \frac{1}{\sqrt{3}}|(X-iY)\downarrow\rangle - \frac{1}{\sqrt{3}}|Z\uparrow\rangle \equiv \left|\frac{1}{2},\frac{1}{2}\right\rangle. \tag{6.130}$$

6.4.4. *The five bands model*

For the theoretical solution of the Rashba parameter α, here we only introduce a five bands model (Lassnig, 1985). As shown in Fig. 6.7, a five bands model is based on the band structure of one CB, two valence bands (a heavy hole one and a light hole one), plus two more CBs. Taking a quantum well

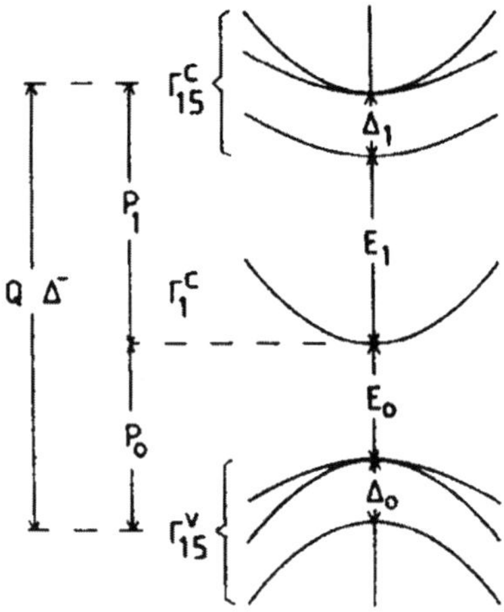

Fig. 6.7. The five bands model in the infinity of Γ point.

with a layer of material A sandwiched in two layers of B material as a simple example, with material A as the reference material. The Bloch function in B is expanded by the Bloch functions in A:

$$u_{l\boldsymbol{k}}^{(\mathrm{B})}(\boldsymbol{r}) = \sum_{m} c_{lm} u_{m\boldsymbol{k}}^{(\mathrm{A})}(\boldsymbol{r}), \tag{6.131}$$

where the expanding coefficients are

$$c_{lm} = \langle u_{m\boldsymbol{k}}^{(\mathrm{A})}(\boldsymbol{r}) | u_{l\boldsymbol{k}}^{(\mathrm{B})}(\boldsymbol{r}) \rangle. \tag{6.132}$$

The Hamiltonian is

$$H_{(\mathrm{A,B})} = -\frac{\hbar^2}{2m}\nabla^2 + V_{(\mathrm{A,B})} + \varphi(r), \tag{6.133}$$

where $V_{(\mathrm{A,B})}$ is the crystal potential and $\varphi(\boldsymbol{r})$ is the macroscopic electric potential of the heterojunction. If only two Γ_6 CBs with spin up and down, respectively, are considered, we have

$$H = \left[-\nabla\frac{\hbar^2}{2m^*(\boldsymbol{r})}\nabla + \Theta_{\mathrm{B}}(z)\Delta E_{\Gamma_6} + \varphi(\boldsymbol{r})\right]\cdot I + \frac{1}{\hbar}\alpha(\boldsymbol{r})\cdot\left(\sigma\times\frac{\hbar}{i}\nabla\right), \tag{6.134}$$

where $\Theta_{\mathrm{B}}(z)$ is taken as 1 in material B, and as 0 anywhere else.

To calculate the Rashba parameter α, we take the wavefunction in the following form:

$$\psi(\boldsymbol{r}) = \sum_{l} \psi_l(\boldsymbol{r}) u_{l0}^{(\mathrm{A})}(\boldsymbol{r}), \tag{6.135}$$

where $\psi_l(\boldsymbol{r})$ is the envelop function of the lth band; thus,

$$\left(-\frac{\hbar^2\nabla^2}{2m} + V_{(\mathrm{A})} + \varphi(\boldsymbol{r}) + \Theta_{\mathrm{B}}(z)[V_{(\mathrm{B})} - V_{(\mathrm{A})}]\right)\psi(\boldsymbol{r}) = E\psi(\boldsymbol{r}). \tag{6.136}$$

The matrix element is obtained by multiplying $u_{l0}^{*}(\boldsymbol{r})$ on both sides of the Schrödinger equation, and then integrating over r, with an assumption that the envelop function is homogeneous in the unit cell:

$$\begin{aligned} H_{il} = {} & \delta_{il}\left(-\frac{\hbar^2\nabla^2}{2m} + E_l^{(\mathrm{A})}(0) + \varphi(\boldsymbol{r})\right) + \frac{\hbar}{im}\nabla\cdot\boldsymbol{p}_{il}^{(\mathrm{A})} \\ & + \Theta_{\mathrm{B}}(z)\left[\sum_j C_{ji}C_{jl}^{*}E_j^{(\mathrm{B})}(0) - \delta_{il}E_l^{(\mathrm{A})}(0)\right], \end{aligned} \tag{6.137}$$

where

$$\left(-\frac{\hbar^2\nabla^2}{2m}+V_{(\mathrm{A,B})}\right)|u_{jk}^{(\mathrm{A,B})}(\boldsymbol{r})\rangle = E_j^{(A,B)}(0)|u_{jk}^{(\mathrm{A,B})}(\boldsymbol{r})\rangle$$

and

$$\boldsymbol{p}_{il}^{(\mathrm{A})} = \langle u_{i0}^{(\mathrm{A})}|\boldsymbol{p}|u_{j0}^{(\mathrm{A})}\rangle.$$

Finally, the Hamiltonian for the two states with spin up and spin down in the Γ_6 CB is

$$H_{il} = \frac{1}{2}\frac{\hbar}{i}\nabla(M_{il}^{-1})\frac{\hbar}{i}\nabla + \delta_{il}[E_l^{(\mathrm{A})}(0) + \Theta_B(z)\Delta E_l(0) + \varphi(\boldsymbol{r})], \tag{6.138}$$

where

$$M_{il}^{-1} = \delta_{il}\frac{1}{m}\boldsymbol{I} + \frac{2}{m^2} \times \sum_j \frac{\boldsymbol{P}_{ij}\otimes\boldsymbol{P}_{jl}}{E+(\hbar^2\nabla^2/2m) - E_j^{(\mathrm{A})}(0) - \Theta_B(z)\Delta E_j(0) - \varphi(\boldsymbol{r})}, \tag{6.139}$$

and meanwhile

$$\Delta E_l(0) = \sum_j |C_{jl}|^2 E_j^{(\mathrm{B})}(0) - E_l^{(\mathrm{A})}(0). \tag{6.140}$$

The subscripts i and l represent the spin up and down states, respectively, and $\Delta E_l(0)$ is the effective band offset at the interface.

The tensor in the first term of the Hamiltonian can be divided into a symmetric part and an asymmetric part:

$$-\sum_{\alpha,\beta}\hbar^2\nabla_\alpha M_{il\alpha\beta}^{-1}\nabla_\beta = -\hbar^2\sum_{\alpha,\beta}\{\nabla_\alpha M_{il\{\alpha\beta\}}^{-1}\nabla_\beta + \nabla_\alpha M_{il[\alpha\beta]}^{-1}\nabla_\beta\}, \tag{6.141}$$

where α and β represent the space directions x, y, z, etc. The first term is the kinetic energy, including the effective mass, and the second term is the Rashba interaction term. Then we have

$$H = \left(\frac{\hbar}{i}\frac{1}{2m^*(\boldsymbol{r})}\frac{\hbar}{i} + \nabla\Theta_B(z)\Delta E_{\Gamma_6} + \varphi(\boldsymbol{r})\right)\cdot\boldsymbol{I} - i\alpha(\boldsymbol{r})\cdot(\sigma\times\nabla). \tag{6.142}$$

When only two bands are considered in the valence band, the Rashba parameter can be explicitly expressed as

$$\alpha = \frac{\hbar^2}{6m}\nabla\left\{\frac{E_p}{[E - E_{\Gamma_7}^{(\mathrm{A})} - \Theta_B(z)\Delta E_{\Gamma_7} - \varphi(\boldsymbol{r})]} - \frac{E_p}{[E - E_{\Gamma_8}^{(\mathrm{A})} - \Theta_B(z)\Delta E_{\Gamma_8} - \varphi(\boldsymbol{r})]}\right\}, \tag{6.143}$$

where E_p is the $\boldsymbol{k}\cdot\boldsymbol{p}$ interaction parameter, describing the interaction intensity between the Γ_6 CB and the relevant Γ_7 and Γ_8 valence bands. In a quantum well, if only the first subband is considered, then the Hamiltonian is

$$H_{2\mathrm{D}} = \left(E_z^{\mathrm{sub}} - \frac{\hbar^2\nabla_\parallel^2}{2m^*} + \varphi(\boldsymbol{r}_\parallel)\right)\cdot \boldsymbol{I} + \frac{1}{\hbar}\alpha\cdot\sigma\times\frac{\hbar}{i}\nabla. \tag{6.144}$$

Finally, we find that for a 2DEG, the Rashba coefficient is (Schäpers *et al.*, 1998)

$$\begin{aligned}\alpha = {} & \frac{\hbar^2 E_p}{6m} e_z\langle\Psi_z|\left\{\left[\left(\frac{1}{[E - E_{\Gamma_7}^{(\mathrm{A})} - \varphi(z)]^2} - \frac{1}{[E - E_{\Gamma_8}^{(\mathrm{A})} - \varphi(z)]^2}\right)\right.\right.\\ & \times[1 - \Theta_B(z)]\varphi'(z) + \left(\frac{1}{[E - E_{\Gamma_7}^{(\mathrm{A})} - \Delta E_{\Gamma_7} - \varphi(z)]^2}\right.\\ & \left.\left.\left. - \frac{1}{[E - E_{\Gamma_8}^{(\mathrm{A})} - \Delta E_{\Gamma_8} - \varphi(z)]^2}\right)\Theta_B(z)\varphi'(z)\right]\right\}|\Psi_z\rangle\\ & + \frac{\hbar^2 E_p}{6m} e_z \sum_n\left(\frac{\Delta E_{\Gamma_7}}{2[E - E_{\Gamma_7}^{(\mathrm{A})} - \Delta E_{\Gamma_7} - \varphi(z_n)]^2} + \frac{\Delta E_{\Gamma_7}}{2[E - E_{\Gamma_8}^{(\mathrm{A})} - \varphi(z_n)]^2}\right.\\ & \left. - \frac{\Delta E_{\Gamma_8}}{2[E - E_{\Gamma_8}^{(\mathrm{A})} - \Delta E_{\Gamma_8} - \varphi(z_n)]^2} - \frac{\Delta E_{\Gamma_8}}{2[E - E_{\Gamma_8}^{(\mathrm{A})} - \varphi(z_n)]^2}\right) S_n|\Psi_z(Z_n)|^2.\end{aligned} \tag{6.145}$$

6.5. Deriving the Rashba Coefficient α from the SdH Oscillation

An electron moving in a 2DEG is generally described by a plane wave $\exp(i\boldsymbol{k}_\parallel\cdot\boldsymbol{r}_\parallel)$. Due to the lift of the spin degeneracy, the energy dispersion

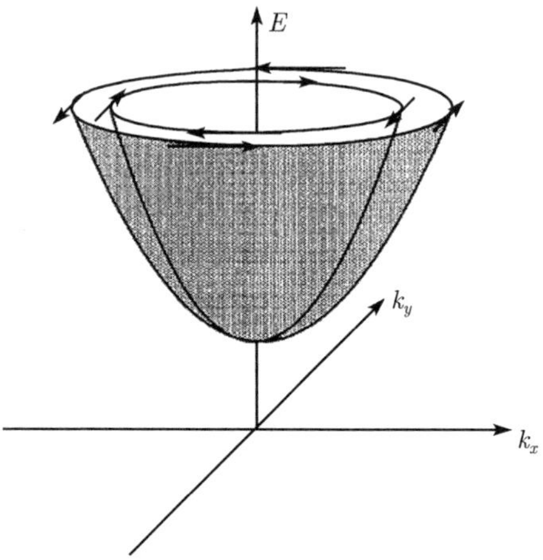

Fig. 6.8. The effect of the SOC to energy dispersion against $\boldsymbol{k}_{\|}$. The electron spin orientation is perpendicular to the 2DEG growth direction as well as to the direction of $\boldsymbol{k}_{\|}$. Courtesy of Schäpers *et al.* (1998, p. 4323).

curve split to two branches, as shown in Eq. (6.146) and Fig. 6.8.

$$E = E_z^{\text{sub}} + \frac{\hbar^2 k_{\|}^2}{2m^*} + \alpha|\boldsymbol{k}_{\|}|. \tag{6.146}$$

For a certain value of k, the energy difference between the two branches is $2\alpha k$ (refer to Fig. 6.8). More accurately, for the inner paraboloid (e.g. $k_x > 0$, spin down $\downarrow$, and $k_x < 0$, spin up $\uparrow$), the outer paraboloid (e.g. $k_x > 0$, spin up $\uparrow$, and $k_x < 0$, spin down $\downarrow$) represents a bit larger effective mass. It can be seen from the figure that for a k vector along the x-direction, $\uparrow$ represents the electron spin being along $+$ y-direction, while $\downarrow$ being along the opposite direction. More generally, if there is no external magnetic field, the electron spin orientation is perpendicular to the 2DEG growth direction as well as to the electron wave vector direction in the 2DEG. Starting from the dispersion relationship, one can calculate the density of states for the inner and outer paraboloids (labeled as $-$ and $+$):

$$Z_{\pm}(E) = \frac{1}{2}\frac{m^*}{\pi\hbar^2}\left(1 \mp \frac{1}{1 + [2(E - E_z^{\text{sub}})\hbar^2]/(\alpha^2 m^*)}\right). \tag{6.147}$$

This shows that the density of states for each paraboloid is half of the 2D density of states $m^*/\pi\hbar^2$, multiplied by a correction factor for the effective mass, which is dependent on energy. Integrating the density of states up to the Fermi level, one can get the real density of electrons for both of the outer and inner paraboloids, named as $n_{\pm}$.

For a 2D system with spin splitting, there will appear beating phenomenon in the SdH oscillation, exactly due to the spin splitting. Therefore, we can obtain two carreir densities, $n_{\pm}$, from the SdH measurement results. Taking the average value of $n = n_+ + n_-$ and the difference between the two $\Delta n = n_+ - n_-$, one can get the Rashba coefficient α from the following equation (Schäpers *et al.*, 1998)

$$\alpha = \frac{\Delta n \hbar^2}{m^*}\sqrt{\frac{\pi}{2(n - \Delta n)}}. \tag{6.148}$$

Upon application of a magnetic field, the total energy split due to the spin effect is approximately:

$$\delta = \Delta E_{\text{so}} + s g^* \mu_{\text{B}} B. \tag{6.149}$$

The spin splitting leads to a modulation of the SdH oscillation at the following frequency:

$$A \approx \cos \pi \frac{\delta}{\hbar \omega_{\text{c}}}. \tag{6.150}$$

Beating effect implies that there exist two oscillations with close frequency and similar amplitude. Two possible reasons may lead to beating as shown in Fig. 6.9 (for a sample used in Yang *et al.*, 2006):

(1) Electrons occupying two subbands with densities n_1 and n_2, respectively;
(2) Only one subband, but there are two densities $n_1^{\uparrow}$ and $n_1^{\downarrow}$ due to spin splitting.

In any case, a Hall measurement will get the total density of electrons n_{H}. Assuming there are two energy bands that are occupied, the average electron density in each band can be derived from the average value F_{av} of the SdH frequency:

$$n_{\text{SdH}} = 2\frac{e}{h}F_{\text{av}}. \tag{6.151}$$

But for a single spin-splitting band, the spin degeneracy factor 2 should not be included in the equation, i.e.

$$n_{\text{SdH}} = \frac{e}{h}F_{\text{av}}. \tag{6.152}$$

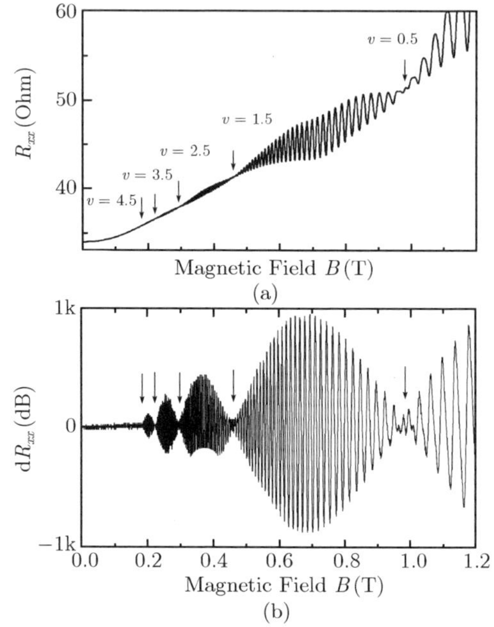

Fig. 6.9. The parallel resistivity ρ_{xx} as a function of the magnetic field. The nodes in the SdH oscillation are labeled by arrows. (a) Original data and (b) data with background deducted.

The total carrier density measured from the SdH experiment should be double of the above value and equal to the result obtained form the Hall measurement, i.e. $n_{\mathrm{H}} \approx 2n_{\mathrm{SdH}}$. This is readily tested from the experimental data.

In the experiment, the SOI parameter α can be determined by fast Fourier transform (FFT). We know that for an ideal 2D system, the period of the SdH oscillation is the reverse of the magnetic field. Taking the spin degenerate factor 2 into account, that period is expressed as

$$\Delta\left(\frac{1}{B}\right) = \left(\frac{e}{\pi\hbar}\right)\frac{1}{n_{\mathrm{s}}}. \tag{6.153}$$

From the result of FFT, we can see well distinguished two peaks, corresponding to two kinds (spin up and spin down) of carriers, as shown in Fig. 6.10.

A more precise method to determine the value of α is to derive that from the nodes in the beating spectrum of SdH oscillation. If the two peaks

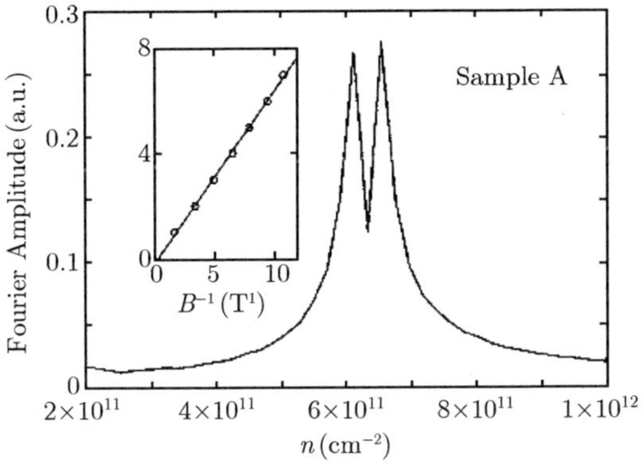

Fig. 6.10. FFT of the SdH oscillation in the range of 0.15–0.35 T. The inset shows the node index as a function of the inversion of the magnetic field, $1/B$. Courtesy of Schäpers *et al.* (1998, p. 4323).

in the FFT are indeed from a spin-split subband, then a classical model will provide a simple way to analyze the spin splitting. It is well known that a spin-split Landau level will have two very close frequencies with very similar amplitude, thus leading to beating. The nodes of the beating appear at $\nu = \delta/\hbar\omega_c$ taking half-integer numbers ($\pm 0.5, \pm 1.5$, etc.), i.e. $A \approx \cos \nu\pi = 0$, while the index of the nodes as a function of $(1/B)$ can be derived from Eq. (6.150):

$$i = \frac{2m^* \alpha k_F}{\hbar e} \left(\frac{1}{B} \right) + \left(\frac{g^* m^*}{2m_0} + \frac{1}{2} \right). \tag{6.154}$$

Plotting the node index as a function of $1/B$, one can get the SOI parameter α from the slope of the fitting line, while the value of δ, the spin-split energy, can be obtained from the intercept of the line on the vertical axis. Figure 6.10 provides an example of an SIA 2DEG (Yang *et al.*, 2006), with the sample structure and band structure shown in Figs. 6.11 and 6.12.

6.6. Spin-related Scattering and Spin Current

Spin current is one of the mostly focussed topics in spintronics studies. A pure spin current is equivalent to a flowing magnetization without electric current. The spin Hall effect is one of the mechanisms to create spin current, i.e. to drive a transverse spin current by a longitudinal electric

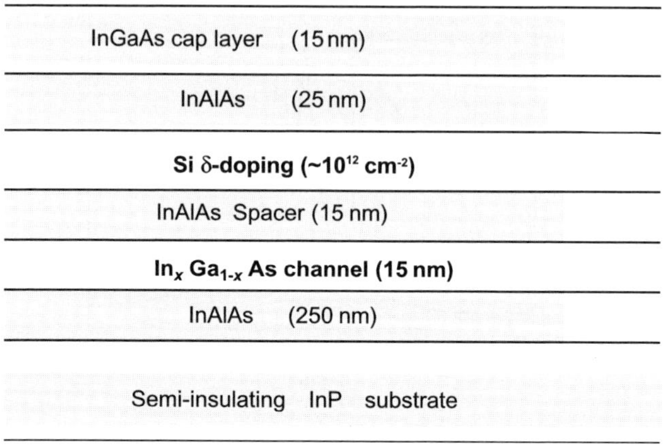

Fig. 6.11. The structure of the sample for the SdH oscillation measurement.

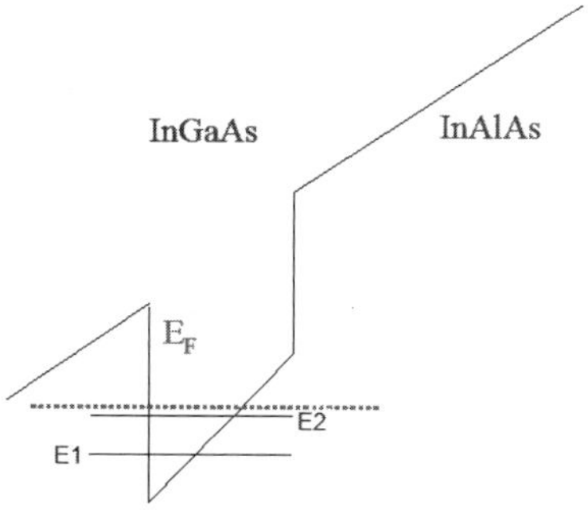

Fig. 6.12. Right-hand side: the band structure of the sample in Fig. 6.11, showing the formation of the SIA.

current, leading to a nonequilibrium accumulation of spin, and detectable magnetization effect at the boundaries (Koto *et al.*, 2004). Another spin current creating mechanism completely different from the spin Hall effect was introduced by Ganichev *et al.* (2006). In that mechanism, no electric current is required, rather the spin current is due to spin splitting caused by scatterings related to spin. To realize such a spin current, we need not apply an external electric field, the only condition is that the system should have an inversion asymmetry, i.e. to have a Rashba or a Dresselhaus effect. Simply speaking, as far as the electron gas in the system is heated up, the electron energy relaxation process which is related to spin will lead to spin splitting, and then create spin current induced by a magnetic field.

6.6.1. *Scattering related to spin*

Electron scattering refers to electron transition from a state with wave vector $\boldsymbol{k}$ to the other one with wave vector $\boldsymbol{k}'$. Generally, this process is treated as not being related to spin. But in a medium with gyroscopic symmetry, e.g. in GaAs quantum well or heterostructures, SOI adds a new term in the scattering probability, which is an asymmetrical and linear function of wave vector k and Pauli spin matrix σ, and related to spin. Obviously, this term is based on SIA and/or BIA, or in other words, on the Rashba and/or Dresselhaus effects. The electron asymmetrical scattering will lead to spin current, but not change the electron spin spectrum.

The real physical process that leads to the spin splitting is shown in Fig. 6.13(a) which shows the asymmetrical and spin-related excitation process due to the linear term of k and σ in the scattering matrix element. In the experiment of Ganichev *et al.* (2006), this excitation is realized by

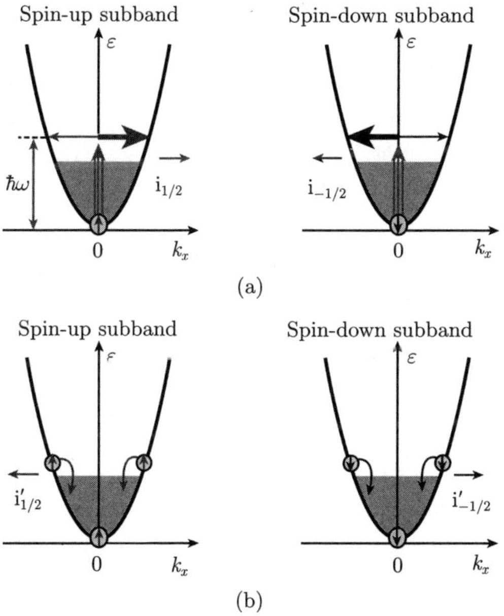

Fig. 6.13. The spin-related asymmetric excitation (a) and relaxation (b) processes due to the linear term about k and σ in the scattering matrix element. The spin-up and spin-down CBs are shown separately. It is assumed here that the scattering probability is larger for a state of positive k than that of negative k in the spin-up subband, while the situation is just the opposite in the spin-down subband. Courtesy of Ganichev *et al.* (2006).

terahertz indirect Drude absorption, but in fact, a similar effect can be realized by any technique for heating up the electron system.

As it is an indirect absorption, the momentum compensation involving phonons and/or impurities is necessary. In the figure, the left part is for excitation in the spin-up subband, while the right part is the excitation in the spin-down subband. The upward arrow represents the optical transition originating from the initial state $k_x = 0$, while the horizontal arrow represents the elastic scattering toward a final state with wave vector of positive or negative k'_x. Although for the sake of simplicity, we only discuss the case that the transition initial state is at $k_x = 0$, the principle applies to other transitions initiated form any state. As the scattering is related to spin, the scattering toward a final state with wave vector of positive or negative k'_x will have different probabilities. This character is indicated by different thicknesses of the arrows in the figure, toward left- and right-hand directions, respectively. Mathematically, as the asymmetrical term in the electron scattering is proportional to $[\boldsymbol{\sigma} \times \boldsymbol{k}']$, the scattering probability toward a final state with a positive or negative wave vector k'_x will just be opposite to each other for subbands of spin up and down. Note that the initial state is assumed to be $k_x = 0$, and we only consider the SIA case, i.e. a Rashba system. Similarly, the relaxation processes for the excited electrons are also asymmetrical, as shown in Fig. 6.13(b).

6.6.2. *Spin current and magnetocurrent*

The asymmetric scattering mentioned above leads to unbalanced distribution of the optically excited electrons; thus, in each spin subband, there should be current induced. However, as far as an electric current is concerned, the net result will be zero because the current induced in the two subbands with spin-up and spin-down, respectively, will have the same amplitude but opposite direction. Therefore, this current can only be taken as a spin current. This spin current can also accumulate spin at the sample boundaries, similar to the case of the spin Hall effect (Wunderlich *et al.*, 2005).

To create a measurable electric current, we have to apply a magnetic field to polarize the electron spin. The magnetic energy induced by the magnetic field splits the electron energy for subbands with different spin orientations, i.e. the Zeeman splitting. In Fig. 6.13(a), the two parabolic curves vertically shift $\pm g\mu_B B/2$, so that the electron density in the two subbands will be different. As the net current in each subband is proportional to its electron density, i.e. $\boldsymbol{i}_{\pm 1/2} \propto n_{\pm 1/2}$, when the electron density is different in the two subbands, the spin current created in the two subbands will no longer be cancelled by each other. In such a way, there will be not only the spin current, but also electric current in the system. The total current is

$$\boldsymbol{j} = e(\boldsymbol{i}_{+1/2} + \boldsymbol{i}_{-1/2}) = 4eS\boldsymbol{J}_{\text{spin}}, \tag{6.155}$$

where $S = (1/2)(n_{+1/2} - n_{-1/2})/(n_{+1/2} + n_{-1/2})$ is the average spin. Due to Zeeman effect, the external magnetic field makes a difference in the occupation for the two spin subbands. At equilibrium, the average spin is

$$\boldsymbol{S} = -\frac{g\mu_{\text{B}}\boldsymbol{B}}{4\bar{\varepsilon}}, \tag{6.156}$$

where g is the electron effective Landè factor, μ_{B} the Bohr magneton, and $\bar{\varepsilon}$ the characteristic energy of electron, which is just the Fermi energy E_{F} in degeneracy or the thermal energy $k_{\text{B}}T$ for the case of no degeneracy.

The samples in the experiment of Ganichev *et al.* (2006) are two structures grown by MBE in (001) direction n-type GaAs/AlGaAs and InAs/AlGaSb, respectively. The magnetic field is horizontal. The excitation laser is a linearly polarized teraHertz radiation, normally radiated on the surface of the sample. The photocurrent is measured along two directions, one being in parallel with and the other perpendicular to the magnetic field (i.e. along $x \| [1\bar{1}0]$ and $y \| [110]$ directions). The essential difference in this setup from the spin galvanic effect (SGE) described in the next chapter is that the effect of the photon spin (carried out by a circularly polarized (CP) light) is avoided. In the experiment, the angle α between the polarization plane and the x-axis can be adjusted. The most interesting results are shown in Fig. 6.14 which shows the photocurrent dependence on angle α under a fixed magnetic field.

The upper and lower parts in Fig. 6.14 show the photocurrent dependence on angle α when the current is perpendicular and parallel to the magnetic field B_y, respectively. The fitting equations are $j_x = j_1 \cos 2\alpha + j_2$ and $j_y = j \sin 2\alpha$, respectively. This dependence is not related to temperature and excitation wavelength, but the intensity of the photocurrent has a relationship with wavelength, as $j \propto \lambda^2$ which is just the spectroscopic characteristics of Drude absorption.

The constant term j_2 in the horizontal current is a background current, which can be interpreted by a phenomenological theory of magnetic field-induced photocurrent by Bel'kov *et al.* (2005). That theory phenomenologically analyzes the photocurrent dependence on the polarization direction

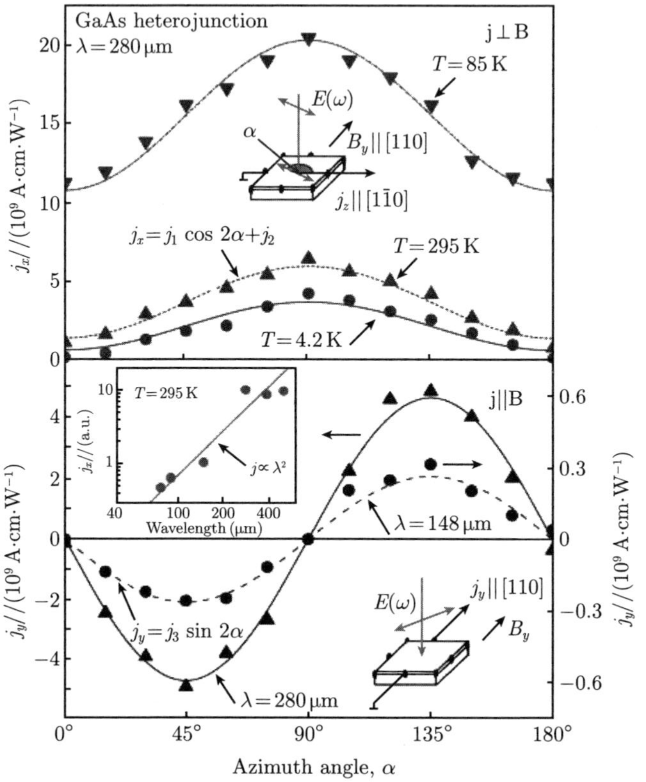

Fig. 6.14. The photocurrent of a GaAs heterostructure as a function of angle α. The experimental condition is $\boldsymbol{B} = 0.3\,\text{T}$. The inset is the geometric setup of the experiment. The other inset in the lower part is the wavelength dependence for horizontal geometric setup. *Upper part*: $\boldsymbol{j} \perp \boldsymbol{B} \| y, \lambda = 280\,\mu\text{m}$, $T = 4.2, 85, 295\,\text{K}$. The fitting equation is $j_x = j_1 \cos 2\alpha + j_2$. *Lower part*: $j \| \boldsymbol{B} \| y$, RT, $\lambda = 148, 280\,\mu\text{m}$. The fitting equation is $j_y = j \sin 2\alpha$. Courtesy of Ganichev *et al.* (2006, p. 609).

and magnetic field direction when a magnetic field is applied, without probing deeply into the microscopic origination. Under a linear approximation of the magnetic field intensity $\boldsymbol{B}$, the magneto-photogalvanic effect (MPGE) is given by:

$$j_\alpha = \sum_{\alpha\beta\gamma\delta} \phi_{\alpha\beta\gamma\delta} B_\beta \{E_\gamma E_\delta^*\} + \sum_{\beta\gamma} \mu_{\alpha\beta\gamma} B_\beta \widehat{e}_\gamma E_0^2 P_{\text{circ}}, \qquad (6.157)$$

where $I = E_0^2$, while the fourth-order tensor ϕ is symmetric to the last two subscripts, and E_i the component of the electric field $\boldsymbol{E}$ of the radiation light. In the following, the electric field will be represented by $\boldsymbol{E} = E_0\,\mathbf{e}$, with E_0 as the scalar value of the electric field: $E_0 = |\boldsymbol{E}|$ and $\mathbf{e}$ being the complex unit of the polarization state, $|\mathbf{e}| = 1$. Furthermore, $\{E_\gamma E_\delta^*\}$ is the symmetric product of the electric field and its complex conjugate:

$$\{E_\gamma E_\delta^*\} = \frac{1}{2}(E_\gamma E_\delta^* + E_\delta E_\gamma^*). \tag{6.158}$$

The right-hand side of Eq. (6.157) is a third-order tensor, induced by SOC, P_{circ} indicates the helicity of the incident light, and $\widehat{\mathbf{e}}$ the unit vector of the light propagation direction. It is readily seen that the second term of MPGE in Eq. (6.157) requires a CP light incidence, as will be discussed in the next chapter; while the first term will be nonzero even if the incident light is not polarized.

The above effects are closely correlated with the crystal orientation. We will focus on the situation in the work of Ganichev *et al.* (2006), i.e. quantum wells grown along (001) direction with III–V or II–VI zincblende structure of C_{2v} symmetry. In a system with C_{2v} symmetry, we choose a coordination system of $x \| [100]$ and $y \| [010]$, meanwhile it is more convenient for the MPGE current taking orientations as $x' \| [1\bar{1}0]$ and $y' \| [110]$. The reason is that the x' and y' axes are in the (110) and $(1\bar{1}0)$ planes of the crystal, while these two planes include the mirror reflection of the 2-fold axis C_2. In such a system of x', y', z, for a normal incident light and in-plane magnetic field, Eq. (6.157) can be simplified to:

$$\begin{aligned} j_{x'} &= S_1 B_{y'} I + S_2 B_{y'}(|e_{x'}|^2 - |e_{y'}|^2) I \\ &\quad + S_3 B_{x'}(e'_x e^*_{y'} + e_{y'} e^*_{x'}) I + S_4 B_{x'} I P_{\text{circ}}, \\ j_{y'} &= S'_1 B_{x'} I + S'_2 B_{x'}(|e_{x'}|^2 - |e_{y'}|^2) I \\ &\quad + S'_3 B_{y'}(e'_x e^*_{y'} + e_{y'} e^*_{x'}) I + S'_4 B_{y'} I P_{\text{circ}}. \end{aligned} \tag{6.159}$$

Parameters S_1 to S_4 and S'_1 to S'_4 are the nonzero elements in tensors ϕ and μ allowed by the C_{2v} symmetry. An in-plane current is obtained from the first term in the right-hand side of Eq. (6.159) (expressed in terms of S_1 and S'_1), which is not related to the light polarization. Whereas various terms in the above equation have different polarization dependences, they can be distinguished from the experimental data.

Linearly polarized light incidence: For a linearly incident light, various terms expressed by parameters S_2, S'_2 and S_3, S'_3 are proportional to $|e_{x'}|^2 - |e_{y'}|^2 = \cos 2\alpha$ and $e'_x e^*_{y'} + e_{y'} e^*_{x'} = \sin 2\alpha$, respectively, where α is the angle between the polarization plane and the x'-axis.

Therefore, the photocurrent expressed by the second term (S_2, S_2') in the right-hand side of Eq. (6.159) will reach a maximum when the light polarization plane is along x'- or y'-axis, while the photocurrent expressed by the third term (S_3, S_3') will reach a maximum when the light polarization plane is along bisector of x' and y'. The last term, being proportional to the helicity of the light, $P_{\rm circ}$, will not appear when the incident light is linearly polarized, which is one of the key effect (SGE) discussed in the next chapter.

Now we can write the MPGE photocurrent induced by normally incident linearly polarized light in the experiment of Ganichev *et al.* (2006) as

$$j_x = C_1 B_y (e_x^2 - e_y^2) I + C_2 B_y I, \tag{6.160}$$

$$j_y = C_3 B_y e_x e_y I, \tag{6.161}$$

where the coefficients C_i are evolved from the parameters S_i in Eq. (6.159). The second term in Eq. (6.160) is not related to polarization and only appears in the horizontal current, as the background in the upper part of Fig. 6.14. The other terms in Eqs. (6.160) and (6.161) are related to spin and are in good agreement with the experimental results.

The microscopic interpretation of Eq. (6.159) includes the nonequilibrium distribution of the photo-excited carriers, the cyclotron motion of the carrier distribution function, and the Hall deflection of the photocurrent. The interested readers may refer to Bel'kov *et al.* (2005), and the theoretical derivation of Ivchenko and Pikus (1983) and Ivchenko *et al.* (1988). Here we only provide a very simple introduction. Under the frame of spin density matrix, the impurity scattering-induced MPGE current can be written as

$$\begin{aligned} \boldsymbol{j} &= \sum_{s\boldsymbol{k}} e\boldsymbol{v}_{\boldsymbol{k}} \delta f_{s\boldsymbol{k}} \\ &= e\frac{2\pi}{\hbar} \sum_{s\boldsymbol{k}\boldsymbol{k}'} \tau_P (\boldsymbol{v}_{\boldsymbol{k}} - \boldsymbol{v}_{\boldsymbol{k}'}) |M_{s\boldsymbol{k},s\boldsymbol{k}'}|^2 (f_{s\boldsymbol{k}'} - f_{s\boldsymbol{k}}) \delta(\varepsilon_{\boldsymbol{k}} - \varepsilon_{\boldsymbol{k}'} - \hbar\omega), \end{aligned} \tag{6.162}$$

where $\boldsymbol{v}_{\boldsymbol{k}} = \hbar\boldsymbol{k}/m^*$ is the electron velocity, $\delta f_{s\boldsymbol{k}}$ the change of the electron distribution function in the subband s, $M_{s\boldsymbol{k},s\boldsymbol{k}'}$ the matrix element of the indirect optical transition and $\varepsilon_{\boldsymbol{k}} = \hbar^2 k^2/2m^*$ the electron in-plane movement kinetic energy. Under the first-order approximation of SOI, we get

$$M_{s\boldsymbol{k},s\boldsymbol{k}'} = \frac{eA}{c\omega m^*} \boldsymbol{e} \cdot (\boldsymbol{k} - \boldsymbol{k}') \mathrm{V}_{\boldsymbol{k}\boldsymbol{k}'} - 2\frac{eA}{c\hbar} \sum_{\alpha\beta} \mathrm{V}_{\alpha\beta} \sigma_\alpha e_\beta, \tag{6.163}$$

where $\boldsymbol{A} = A\boldsymbol{e}$ is the vector potential of the electromagnetic wave, c the light velocity, and $V_{\boldsymbol{k}\boldsymbol{k}'}$ the scattering matrix element (Ivchenko and Pikus, 1983):

$$V_{\boldsymbol{k}\boldsymbol{k}'} = V_0 + \sum_{\alpha\beta} \mathrm{V}_{\alpha\beta}\sigma_\alpha(k_\beta + k'_\beta), \tag{6.164}$$

where V_0 referes to scattering not related to spin, the second order pseudotensor $V_{\alpha\beta}$ makes an asymmetric contribution which is spin dependent and also linearly dependent on k. The first term on the right hand side of Eq. (6.163) corresponds to a transition in the CB via a virtual intermediate state, while the second one corresponds to a similar process in the valence band.

In a system with C_{2v} symmetry, there are only two nonzero elements in the matrix $V_{\alpha\beta}$, namely V_{xy} and V_{yx}. Using Eqs. (6.162)–(6.164), the expression of current $\boldsymbol{j}$ can be written. Assuming that the scattering is short ranged, and the impurities are static, then the matrix V is not relevant to wave vector, hence under the present geometric setup of the experiment, one can get:

$$j_x = -2(e_x^2 - e_y^2)V_{yx}S_y\frac{e\tau_\mathrm{P}}{\hbar \mathrm{V}_0}I\eta(\omega), \tag{6.165}$$

$$j_y = -4e_xe_yV_{xy}S_y\frac{e\tau_\mathrm{P}}{\hbar \mathrm{V}_0}I\eta(\omega). \tag{6.166}$$

Equation (6.165) does not include the spin irrelevant term in Eq. (6.160), as the relaxation process shown in Fig. 6.13(b) is not considered here, while that relaxation process is just the origin of the background current in the upper part of Fig. 6.14. Equations (6.165) and (6.166) include the polarization dependence for transverse and longitudinal cases, respectively, i.e.

$$e_x^2 - e_y^2 = \cos 2\alpha \quad \text{and} \quad 2e_xe_y = \sin 2\alpha. \tag{6.167}$$

Using these relationships to fit the experimental results shown in Fig. 6.14, the agreement is excellent.

Finally, the results from different samples are compared in Fig. 6.15. In the upper part of the figure, the GaAs transverse photocurrent dependence on the magnetic field for two polarization angles $\alpha = 0°$ and $\alpha = 90°$ is shown. Summing up or subtracting each other, the coefficients j_1 and j_2 can be obtained, as shown in the lower part of Fig. 6.15. Among the four samples, the InAs quantum well has a larger g value, and hence

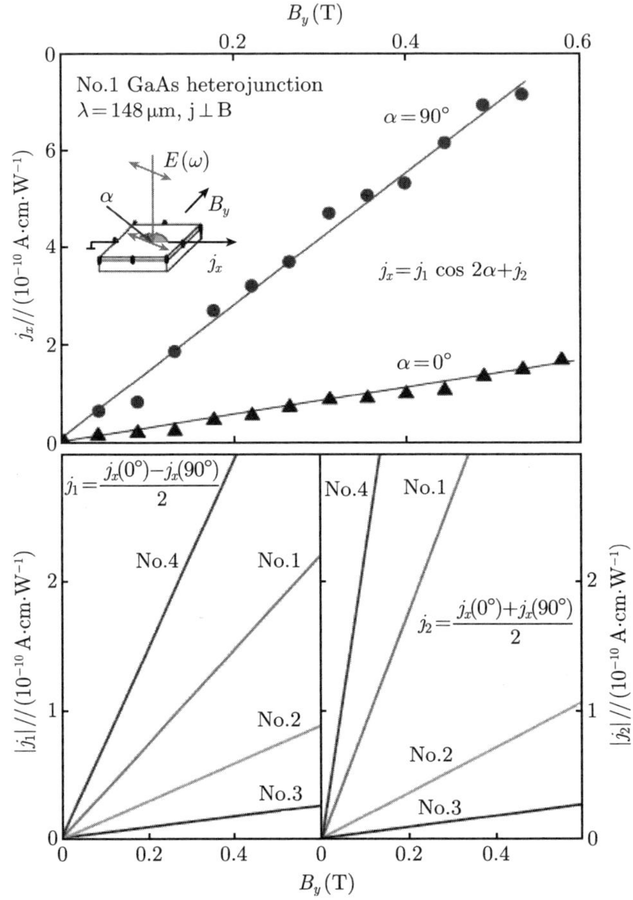

Fig. 6.15. Magnetic-field dependence of the transverse photocurrent. *Upper panel:* $\boldsymbol{j_x}(\boldsymbol{B})$ normalized by radiation intensity $\boldsymbol{I}$, measured for sample 1 at RT for $\alpha = 90°$ and $0°$. *Lower panels:* $\boldsymbol{j}_1(\boldsymbol{B})$ (left panel) and $\boldsymbol{j}_2(\boldsymbol{B})$ (right panel) obtained by subtracting and adding the currents for the two polarization directions for samples 1–4. Courtesy of Ganichev *et al.* (2006, p. 609).

greater average spin S, as well as greater MPGE photocurrent. Among the three GaAs samples, sample 1 is a heterojunction with triangle-shaped confinement potential; therefore, the SIA effect should be most remarked. Samples 2 and 3 have a structure similar to quantum well, but the modulation doping is asymmetric and symmetric, respectively, and therefore the SIA effect is larger for the asymmetric one. We can see that with decreasing strength of the SIA effect, the MPGE photocurrent is also reducing from samples 1 to 3. This is completely consistent with the physical picture of asymmetric scattering-induced current. In Eqs. (6.165) and (6.166), and other similar expressions of scattering mechanism, the SOI controls the intensity of the current via matrix $V_{\alpha\beta}$: the stronger the coupling, the more prominence the asymmetric scattering.

References

Bel'kov VV, Ganichev SD, Ivchenko EL, Tarasenko SA, Weber W, Giglberger S, Olteanu M, Tranitz H-P, Danilov SN, Petra, Schneider P, Wegschneider W, Weiss D, Prettl W. (2005) *J Phys Condens Matter* **17**: 3405.

Bir, Pikus GE. (1974) *Symmetry and Strain Induced Effects in Semiconductors.* Wiley, New York.

Bychkov YA, Rashba EI. (1984) *Pis Zh Eksp Teor Fiz* **39**: 66; *Sov Phys — JETP* **39**: 78.

Dressselhaus G. (1955) *Phys Rev* **100**: 580.

D'yakonov MI, Kachorovskiĭ V Yu. (1986) *Sov Phys Semicond* **20**: 110.

D'yakonov MI, Perel' VI. (1972) *Sov Phys Solid State* **13**: 3023.

Feve G, Oliver WD, Aranzana M, Yamamoto Y. (2002) *Phys Rev B* **66**: 155328.

Foreman BA. (2005) *Lecture Note of Graduate Study.* The Hong Kong University of Science and Technology.

Ganichev SD, Bel'kov VV, Tarasenko SA, Danilov SN, Giglberger S, Hofffmann C, Ivchenko EL, Weiss D, Wegscheider W, Gerl C, Schuh D, Stahl J, Boeck JD, Borghs G, Prettl W. (2006) *Nat Phys* **2**: 609.

Ivchenko EL, Pikus GE. (1983) *Bull Acad Sci USSR* **47**: 81.

Ivchenko EL, Lyanda-Geller Yu B, Pikus GE. (1988) *Ferroelectrics* **83**: 19.

Koto Y, Myers RC, Gosard AC, Awschalom DD. (2004) *Science* **306**: 1910.

Lassnig R. (1985) *Phys Rev B* **31**: 8076.

Schäpers Th, Engels G, Lange, Klocke Th, Hollfelder M, Lüth H. (1998) *J Appl Phys* **83**: 4324.

Wunderlich J, Kaestner B, Sinova J, Jungwirth T. (2005) *Phys Rev Lett* **94**: 047204.

Yang CL, He HT, Ding Lu, Cui LJ, Zeng YP, Wang JN, Ge WK. (2006) *Phys Rev Lett* **96**: 186605.

Chapter 7

OPTICAL RESPONSES OF ELECTRON SPINS IN SEMICONDUCTORS

In Chap. 6, we have introduced the lift-up effect of the spin degeneration due to SOI, and the consequent appearing of a linear term of vector $\boldsymbol{k}$ in the electron Hamiltonian. The splitting of the electron spin states has a fundamental significance in spintronics, as it provides a possibility of controlling the electron spin polarization by electric field and affects the rate of spin relaxation. In particular, as shown below, it might be applied to all-electric circuit for spin injection without an external magnetic field (Fabian and Das-Sarma, 2004).

Spin photogalvanic effect has been theoretically predicted some 30 years ago (D'yakonov and Petel', 1971) and was first observed in experiment more than 20 years ago (Balun *et al.*, 1984). The detailed research works are, however, only available in recent few years, and summarized by Ganichev and Prettl (2003). The essence of the spin photocurrent is the transfer between the photon spin and electron spin, and this transfer is a direct consequence of an effective magnetic field or effective momentum, both of which are induced by SOI.

Taking the Rashba effect as an example, in a 2DEG gas system which is perpendicular to z-direction, the linear term in the Hamiltonian is written as:

$$H = \alpha \boldsymbol{\sigma} \cdot (\widehat{\boldsymbol{z}} \times \boldsymbol{k}_{//}) = \alpha \boldsymbol{k}_{//} \cdot (\boldsymbol{\sigma} \times \widehat{\boldsymbol{z}}). \tag{7.1}$$

There are two equivalent expressions in Eq. (7.1), both have a clear physical meaning. In the equation, α is the Rashba coefficient, $\boldsymbol{\sigma}$ the spin vector, $\boldsymbol{k}_{//}$ the in-plane component of the electron momentum, and $\widehat{\boldsymbol{z}}$ the unit vector in z-direction. In the first expression, the bracket of a cross-product of two vectors corresponds to an effective magnetic field, whose dot product with the spin vector and multiplied by α will make a dimension of energy, so as becoming a term in the Hamiltonian. Obviously, this effective magnetic field is in the plane of the 2DEG and perpendicular to the electron momentum. In the second expression, the bracket of a cross-product of two vectors corresponds to an effective momentum which is also in the plane of the 2DEG and perpendicular to the direction of the electron spin. It is thus readily understood that spin current injection in the 2EDG plane will result in a pure momentum, hence inducing current. If the injection is induced optically, that is just the spin photogalvanic effect. On the other hand, current injection into such a 2DEG plane will induce spin polarization which can be detected by using optical techniques, such as Faraday or Kerr effects.

This chapter will introduce the optical responses of electron spin, mainly covering spin current induced by optical injection; spin polarization induced by electric field; the interference and experimental distinction between Rashba and Dresselhaus effects and their physical consequences; coherent-excitation-induced pure spin current and its detection; as well as some physical models of spintronic devices based on Rashba and Dresselhaus effects.

7.1. Spin of Photon or Polarization of Light

The wave functions of right-hand (RH) and left-hand (LH) CP light propagating in the $x-y$ plane are:

$$\begin{aligned} |R\rangle &= \frac{1}{\sqrt{2}}(|x\rangle + i|y\rangle), \\ |L\rangle &= \frac{1}{\sqrt{2}}(|x\rangle - i|y\rangle). \end{aligned} \tag{7.2}$$

In the $x-y$ plane of the electromagnetic field of the light, a rotation of angle φ produces new basis vectors such as:

$$\begin{aligned} |\bar{x}\rangle &= \cos\varphi|x\rangle + \sin\varphi|y\rangle \\ |\bar{y}\rangle &= -\sin\varphi|x\rangle + \cos\varphi|y\rangle, \end{aligned} \tag{7.3}$$

therefore

$$\begin{aligned} |\overline{R}\rangle &= e^{-i\varphi}|R\rangle \\ |\overline{L}\rangle &= e^{+i\varphi}|L\rangle, \end{aligned} \tag{7.4}$$

which can be used to represent RH and LH CP light.

It is well known that the general form of an angular momentum operator perpendicular to the electric field plane is written as

$$\widehat{j}_k = \widehat{l}_k + \widehat{s}_k = -i\hbar\frac{\partial}{\partial\varphi} \tag{7.5}$$

so that

$$-i\hbar\frac{\partial}{\partial\varphi}|\overline{L}\rangle = -i\hbar\frac{\partial}{\partial\varphi}e^{+i\varphi}|\overline{L}\rangle = +\hbar|\overline{L}\rangle, \tag{7.6a}$$

$$-i\hbar\frac{\partial}{\partial\varphi}|\overline{R}\rangle = -i\hbar\frac{\partial}{\partial\varphi}e^{-i\varphi}|R\rangle. \tag{7.6b}$$

Equation (7.6) implies that the eigenvalues of the LH and TH CP light are 1 and -1 respectively. Precisely, LH and RH CP light can be described as: when one is facing a light beam, if the electric field vector is evenly rotating anticlockwise, it is a LH CP light with a photon spin of $+1$, represented by σ_+; contrarily, if the electric field vector is rotating clockwise, it is a RH CP light with photon spin of -1, and represented by σ_-. A linearly polarized light can be easily decomposed into a LH CP plus a RH CP, but it is not an eigenstate of the angular momentum operator, and therefore there would be no spin of 0 for photons. It should thus be emphasized that only CP light has a well-defined photon spin, either $+1$, for LH CP, or -1, for RH CP light (Sakurai, 1994).

The matrix expressions of photon spin are:

$$\frac{\hat{j}^2}{\hbar^2} = \begin{pmatrix} 2 & 0 & 0 \\ 0 & 2 & 0 \\ 0 & 0 & 2 \end{pmatrix} \quad \frac{\hat{j}_z}{\hbar} = \begin{pmatrix} 1 & 0 & 0 \\ 0 & 0 & 0 \\ 0 & 0 & -1 \end{pmatrix}$$
$$\frac{\hat{j}_x}{\hbar} = \frac{1}{\sqrt{2}}\begin{pmatrix} 0 & 1 & 0 \\ 1 & 0 & 1 \\ 0 & 1 & 0 \end{pmatrix} \quad \frac{\hat{j}_y}{\hbar} = \frac{1}{\sqrt{2}}\begin{pmatrix} 0 & -i & 0 \\ i & 0 & -i \\ 0 & i & 0 \end{pmatrix} \tag{7.7}$$

7.2. Spin Conservation in Optical Transitions in Semiconductors

7.2.1. *Selection rules of optical transitions*

The optical transition probability is calculated by perturbation theory generally written in the form of a product of the optical transition matrix element and the density of states. For a band to band transition, the transition probability is given by the Fermi golden rule:

$$W = \frac{2\pi}{\hbar} \sum_{f,i} |\langle f|H_I|i\rangle|^2 (E_f - E_i + \hbar\omega), \tag{7.8}$$

where the subscripts i and j indicate the initial and final states, E_i and E_j are the energy levels, H_I is the interaction Hamiltonian, which is $-e\boldsymbol{r} \cdot \boldsymbol{E}$ in the case of electrical dipole approximation for the optical transitions. The summation for both initial and final states leads to reduced density of states. The transition probability is determined by the matrix element of the orbital part:

$$M \propto |\langle f|\boldsymbol{r} \cdot \boldsymbol{\eta}|i\rangle|, \tag{7.9}$$

where $\boldsymbol{\eta}$ is the polarization component of the light. The contribution from the spin part is simply:

$$\langle \uparrow \mid \uparrow \rangle = \langle \downarrow \mid \downarrow \rangle = 1 \quad \text{or} \quad \langle \uparrow \mid \downarrow \rangle = \langle \downarrow \mid \uparrow \rangle = 0. \tag{7.10}$$

Equation (7.10) shows that the electron spin is conserved in the transition process. For transitions between subbands, the situation is similar. The contribution of the photon spin is to change the magnetic component of the total angular momentum of the electron. Due to time-reversion symmetry, the magnetic component of the total angular momentum of the system is conserved. For example, one of the heavy hole states (β state) in Eq. (6.128) is

$$\phi_{\text{hh},\beta} = \frac{-1}{\sqrt{2}} |(X - iY) \downarrow\rangle \equiv \left|\frac{3}{2}, -\frac{3}{2}\right\rangle,$$

and the β state of electron is (Eq. (6.127))

$$\phi_{\text{c},\beta} = |iS \downarrow\rangle = \left|\frac{1}{2}, -\frac{1}{2}\right\rangle.$$

The two states share the same electron spin, then optical transitions between them are possible; furthermore, the difference between the magnetic components of the total angular momentum of the two states is 1 which can be compensated by photon spin in the optical transitions. In the case of absorption, the magnetic component of the total angular momentum increases 1; therefore, the absorbed photon spin should be 1, corresponding to a photon of σ_+, i.e. a LH CP light. On the contrary, in the case of light emission, the magnetic component of the total angular momentum reduces 1 which is "taken off" by the emitted photon, so that the photon's spin should also be 1, again corresponding to a photon of σ_+, i.e. a LH CP light.

On the other hand, the above heavy hole state has a different electron spin from the other electron state, α state:

$$\phi_{c,\alpha} = |iS\uparrow\rangle = \left|\frac{1}{2}, \frac{1}{2}\right\rangle,$$

so that the optical transitions are not allowed between them due to opposite directions of the electron spin.

Similarly, optical transitions can also occur between the electron β state and the light hole β state (Eq. (6.129)):

$$\phi_{\text{lh},\beta} = -\frac{1}{\sqrt{6}}|(X+iY)\downarrow\rangle + \sqrt{\frac{2}{3}}|Z\uparrow\rangle \equiv \left|\frac{3}{2}, \frac{1}{2}\right\rangle.$$

But one has to notice that only the first term in the above expression, $1/\sqrt{6}|(X+iY)\downarrow\rangle$, has the same electron spin direction of the electron β state, and thus can be involved in optical transitions with the latter. It is readily found that the square of the coefficient of this term, $1/\sqrt{6}$, is only one-third of the square of the coefficient of the heavy hole β state, $-1/\sqrt{2}$; hence, as the transition probability is proportional to the square of the matrix element, the transition probability between the electron β state and heavy hole β state is three times of that between the electron β state and the light hole β state ($1/2$ over $1/6$). We then reach the transition figure (Fig. 7.1) between various energy levels.

7.2.2. *Optical excitation for spin-split bands*

The spin degeneration lifting effect, or the splitting of the energy bands, has been analyzed theoretically in Chap. 6. In order to study the optical excitations in such a system in more detail, we will carry out some further phenomenological explanations and comparisons in this section.

The linear term of k in the Hamiltonian of the system is induced by SIA or BIA, named as the Rashba term or Dresselhaus term, respectively. In addition, the inversion asymmetry at the interface (IIA) can also lead to a linear term of k, which is similar to that induced by BIA, so that we will not distinguish them in the phenomenological analysis. We know that BIA is due to crystal symmetry; therefore, it only exists in zincblende or wurtzite structures, not in structures with inversion symmetry, such as the diamond structure. On the other hand, SIA is based on the intrinsic inversion asymmetry of heterostructures, thus not necessarily related to crystal structures, but may rather be induced by unequal interfaces, inversed interfaces, asymmetrical doping, asymmetrical quantum well structures,

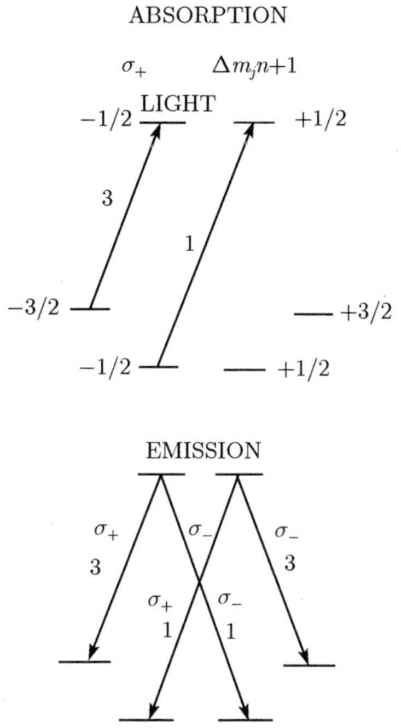

Fig. 7.1. Schematic figure of optical transitions. The numbers 3 and 1 are the weighting factors of the corresponding transitions.

and either external or built-in electric field. Heterostructures even made by inversion symmetrical materials, such as Si and Ge, may still have SIA. The most impressive advantage of SIA is that it can be controlled by an external electric field, which is significant for exploring spintronic devices.

In a symmetrical quantum well, the subbands is spin degenerate, but in an asymmetrical system, there will be a perturbation due to SOI:

$$H' = \sum_{lm} \beta_{lm} \sigma_l k_m, \tag{7.11}$$

where β_{lm} is a second-order pseudopotential, σ_l is the Pauli matrix which appears due to time inversion symmetry. Equation (7.11) can be separated into two terms according to the symmetry of $\sigma_l k_m$:

$$\sigma_l k_m = \{\sigma_l, k_m\} + [\sigma_l, k_m], \tag{7.12}$$

in which the symmetrical term

$$\{\sigma_l, k_m\} = \frac{1}{2}(\sigma_l k_m + \sigma_m k_l) \tag{7.13}$$

is induced by BIA and IIA, while the asymmetrical term:

$$[\sigma_l, k_m] = \frac{1}{2}(\sigma_l k_m - \sigma_m k_l) \tag{7.14}$$

is induced by SIA. Thus, the perturbation can be written as

$$H' = \sum_{lm}(\beta^s_{lm}\{\sigma_l k_m\} + \beta^a_{lm}[\sigma_l k_m]), \tag{7.15}$$

where β^s_{lm} and β^a_{lm} are the projections of the general pseudopotential on the symmetrical and asymmetrical aspects, respectively.

The concrete form of β_{lm} depends on the real representation of the quantum well point group. D_{2d}, C_{2v}, and C_s are the three point groups related to zincblende structures. Among them, D_{2d} corresponds to a symmetrically doped ideal quantum well grown along (001) direction. In a structure with such symmetry, only BIA and IIA exist. When additional asymmetry is introduced, e.g. nonequivalent interfaces, asymmetrical growth conditions or doping profiles, the symmetry of the system will be reduced to C_{2v}, implying SIA. In this case, the strength ratio between BIA (and/or IIA) and SIA depends on the real structure of the quantum well. The C_s group includes only two operations: identical and mirror reflection, it appears in quantum wells grown along (113) direction or off (001) direction.

The value of β_{lm} depends not only on the symmetry of the system, but also on the selection of the coordinate. For a system with D_{2d} or C_{2v} symmetry and grown along (001) direction, the pseudopotential tensor can be expressed in a coordinate of xyz with $x//[1\bar{1}0]$, $y//[110]$, and $z//[001]$. Here z-axis is the crystal growth direction, perpendicular to the plane of the quantum well. In these two point groups, x and y are both perpendicular to the reflection plane to which the major axis of the 2DEG is normal. For D_{2d} symmetry, β_{lm} is symmetrical, i.e. $\beta_{lm} = \beta^s_{lm}$. In the above coordination system, there are two nonzero elements, β_{xy} and β_{yx}, and $\beta_{yx} = \beta_{xy} = \beta^s_{yx}$; therefore, the Dresselhaus term in the Hamiltonian can be written as:

$$H'_{\mathrm{D}} = \beta^s_{yx}(\sigma_x k_y + \sigma_y k_x). \tag{7.16}$$

However, in C_{2v} symmetry, β_{lm} is asymmetrical, then a new term appears, i.e. the Rashba term:

$$H'_{\mathrm{R}} = \beta^a_{yx}(\sigma_x k_y - \sigma_y k_x). \tag{7.17}$$

In a zincblende compound semiconductor quantum well with BIA and SIA, the spin split, mainly contributed by the Rashba effect (Stein *et al.*,

1983; Stormer *et al.*, 1983), is about 10^{-10} to 10^{-9} eV/cm^{-1}, as determined by SdH oscillation experiment. In the case of C_s symmetry, usually taking a quantum well grown along (113) direction as an example, we use $x' = x//[1\bar{1}0]$, $y'//[33\bar{2}]$, $z'//[113]$, x is perpendicular to the plane of the only symmetrical operation, mirror reflection, and z' is the growth direction. The reduction of symmetry leads to a new term in the Hamiltonian for a crystal of C_s symmetry:

$$H'_A = \beta_{z'x}\sigma_{z'}k_x \tag{7.18}$$

that shows that in this system with low symmetry, the spin component perpendicular to the quantum well plane is also involved in SOI, or in other words, even in the normal injection situation, a CP light can still induce SGE in the quantum well plane. That is not possible for a system with higher symmetry, such as D_{2d} and C_{2v}.

In order to demonstrate the energy band structure with linear k terms more clearly, Fig. 7.2 depicts the electron energy as a function of k_x and k_y in a system with C_{2v} symmetry, where all of the pseudopotential coefficients are assumed to be positive. It can be seen that (from Fig. 7.2(a)), if only

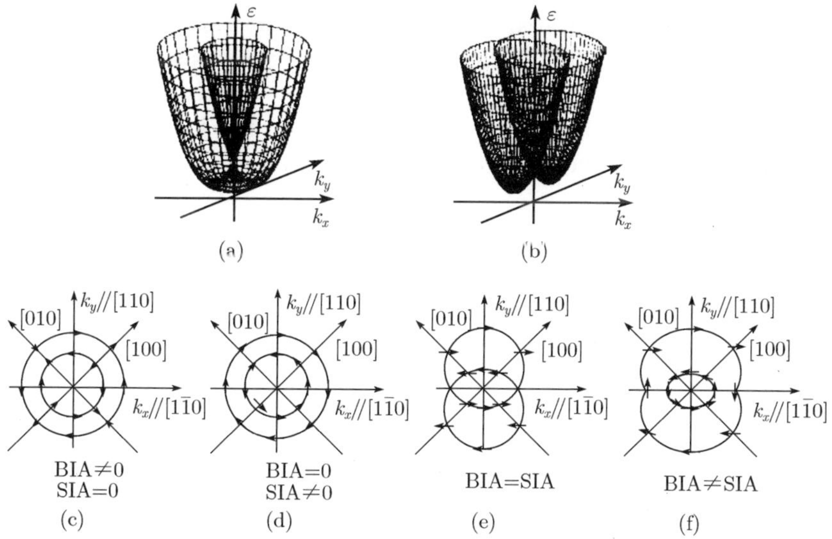

Fig. 7.2. Schematic figure of the 2D energy band structure for a system with C_{2v} symmetry and thus linear $\boldsymbol{k}$ terms in its Hamiltonian. (a) Only BIA or SIA exists. (b) BIA = SIA; (c–f) The distribution of the spin orientation at the Fermi level with various ratio of BIA/SIA. Courtesy of Ganichev and Prettl (2003).

BIA or SIA exists, the energy band structure consists of two paraboloids which are rotated about the energy axis and symmetrically displaced from $k = 0$ point. The equal-energy surfaces are two circles with the same center. But for the case with only BIA or only SIA, the orientation of the spins (as indicated by arrows in the figure) are different, as shown in Figs. 7.2(c) and 7.2(d). In particular, in the case of pure SIA, the electron spin will always be perpendicular to the wave vector $\boldsymbol{k}$. Another interesting result is that when BIA = SIA, the energy band structure consists of two rotated paraboloids, the rotation axis of each one displaces from $k = 0$ point (along k_y axis). The displacements are symmetrical but towards opposite directions (Fig. 7.2(b)). The electron spin at the equal-energy surface is always along $\pm k_x$ directions.

Figures. 6.8 and 7.2 demonstrate the 3D pictures of the energy band splitting induced by SOI. The interception of the picture in a 2D plane (say $y-z$ plane) is schematically shown in Fig. 7.3, in which the spin splitting in the heavy hole state is assumed small (Winkler, 2000). Based on the analysis in the last section and also in Fig. 7.1, if the excitation light is σ_+, and only the heavy holes are involved in the excitation process (for example, in the case of a quantum well where the energy levels of heavy and light holes are well separated) then the actual transitions are represented by the vertical dash arrow on the right hand (RH) side, as well as the vertical solid arrow on the left. In both processes, the electrons excited to the CB share the same spin, belonging to the same split subband. However, as that split band is not centrally symmetrical, the electrons excited by

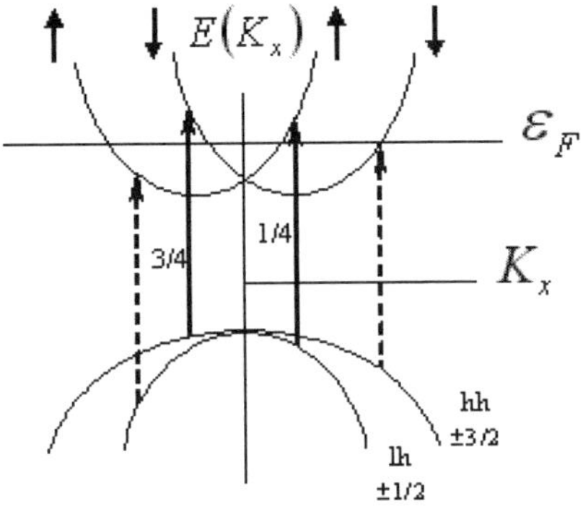

Fig. 7.3. The energy band splitting, and related optical transitions induced by SOI. Here the separation of heavy and light hole bands is neglected. The short arrows indicate the electron spin directions.

the two processes will have different momentum with opposite directions as well as different magnitudes. Therefore, as the transition probability is equivalent for the two processes, the result of the excitation will lead to a net electron momentum, producing current. That is the basic principle of the so-called circular photogalvanic effect (CPGE; see Balun *et al.*, 1984). The two solid vertical arrows in Fig. 7.2 show that nearby the Γ point, the heavy and light hole energy levels might be quasi-degenerate, and thus be excited at the same time. In this case, one may notice that the excited electrons have different spins but a similar magnitude of momentum in opposite directions. The momentum cannot be cancelled; however, as the transition probability from the heavy hole band is three times larger than that from the light hole band, the excitation still results in a current.

7.3. Spin Photocurrent Induced by Optically Injected Electron Spin in a Spin Splitting System

Microscopically, spin photocurrent is a result of spin polarization in gyrosymmetrical systems which have linear terms in its electron Hamiltonian. Generally speaking, the spin photocurrent is contributed by two mechanisms: optical excitation and scattering of photo-excited carriers. The first mechanism has been introduced in the last section, i.e. CPGE. The other one, SGE, first reported by Ganichev and Prettl (2002) in *Nature*, is based on nonsymmetrical scattering of photo-excited carriers, which is not necessarily induced by optical excitation. In the case of CPGE, the photocurrent is spin-polarized, which decays according to the free carrier relaxation time, while in the case of SGE, the current is not spin-polarized and decays according to spin relaxation time (SRT).

7.3.1. *Circular photogalvanic effect*

The excitation mechanism of CPGE is shown in Fig. 7.4(a) for interband excitation and in Fig. 7.4(b) for intraband excitation.

We will focus on interband excitations, taking the transitions from the heavy hole band to CB as examples. In a microscopic model, the ground state of the CB is split into

$$\varepsilon_{e1,\pm 1/2}(k) = \left[\left(\frac{\hbar^2 k_x^2}{2m_{e1}}\right) \pm \beta_{e1} k_x + \varepsilon_g\right], \tag{7.19}$$

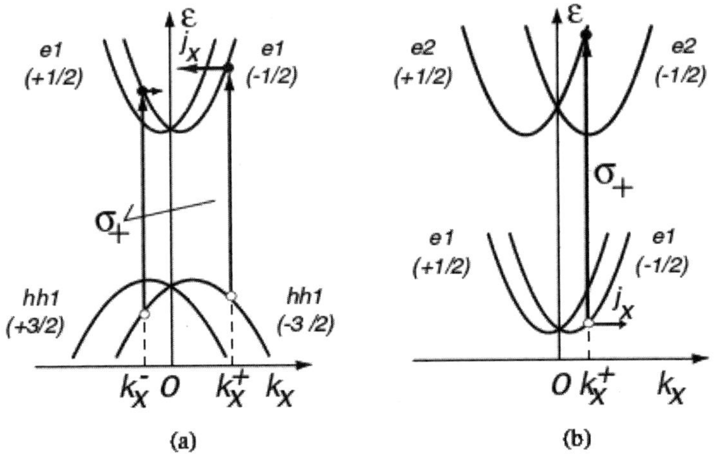

Fig. 7.4. Schematic depictions of optical excitations in CPGE induced by spin orientation: (a) for interband excitation, only heavy holes are shown; (b) for intra-CB excitations. Courtesy of Ganichev and Prettl (2003).

while the splitting of the heavy hole state is:

$$\varepsilon_{\mathrm{hh1},\pm 3/2}(k) = -\left[\left(\frac{\hbar^2 k_x^2}{2m_{\mathrm{hh1}}}\right) \pm \beta_{\mathrm{hh1}} k_x\right]. \tag{7.20}$$

If the excitation light is σ_+, then the transition from the $-3/2$ hole state to $-1/2$ electron state will occur. Following the conservation laws of energy as well as momentum, the k-value where the transition occurs should be

$$k_x^{\pm} = +\frac{\mu}{\hbar^2}(\beta_{\mathrm{e1}} + \beta_{\mathrm{hh1}}) \pm \sqrt{\frac{\mu^2}{\hbar^4}(\beta_{\mathrm{e1}} + \beta_{\mathrm{hh1}})^2 + \frac{2\mu}{\hbar^2}(\hbar\omega - \varepsilon_{\mathrm{g}})}, \tag{7.21}$$

where $\mu = (m_{\mathrm{e1}} m_{\mathrm{hh1}})/(m_{\mathrm{e1}} + m_{\mathrm{hh1}})$ is the reduced mass, $\hbar\omega$ the energy of the excitation photon, and ε_{g} the energy band gap. One can readily see that the average of k_x^+ and k_x^- is at $k_x = (\beta_{\mathrm{e1}} + \beta_{\mathrm{hh1}})\mu/\hbar^2$ rather than at $k_x = 0$; therefore, the velocity summation of the excited electrons is nonzero and can be expressed as

$$v_{\mathrm{e1}} = \frac{\hbar}{m_{\mathrm{e1}}}(k_x^- + k_x^+ - 2k_x^{\mathrm{min}}) = \frac{2(\beta_{\mathrm{hh1}} m_{\mathrm{hh1}} - \beta_{\mathrm{e1}} m_{\mathrm{e1}})}{\hbar(m_{\mathrm{e1}} + m_{\mathrm{hh1}})}, \tag{7.22}$$

where k_x^{min} is the minimum of the electron momentum in the CB. This net velocity implies that the contributions from the excited electrons at k_x^+ and k_x^- will not cancel each other, unless $\beta_{\mathrm{hh1}} m_{\mathrm{hh1}} = \beta_{\mathrm{e1}} m_{\mathrm{e1}}$, i.e.

the CB and valence band share the exact same splitting, which is quite unusual though. The calculated group velocities of the carriers at k_x^+ and k_x^- are of opposite sign as well as of different magnitudes, thus leading to a net current. Note that the electrons in the two opposite current will have opposite spins too, so that the current is also spin polarized. On the contrary, if the excitation light is σ_-, the corresponding transition is from the +3/2 state to +1/2 state, the whole process is just a reversed one of that with σ_+ light excitation in terms of k-dependence, and the resulting current direction and spin orientation are all reversed.

It is similar for the intraband excitations which can be readily seen from Fig. 7.4(b).

Phenomenological expression of CPGE can be written as follows:

$$j_\lambda = \sum_\mu \nu_{\lambda\mu} i(\boldsymbol{E} \times \boldsymbol{E}^*)_\mu, \tag{7.23}$$

$$i(\boldsymbol{E} \times \boldsymbol{E}^*)_\mu = \widehat{\boldsymbol{e}}_\mu E_0^2 P_{\text{circ}}, \tag{7.24}$$

where j is the current density, ν is a 2D pseudopotential tensor, $\boldsymbol{E}$ is the complex electric field intensity, and E_0, $\widehat{\boldsymbol{e}} = \boldsymbol{q}/q$, and $\boldsymbol{q}$ are the electric field magnitude, helicity of the CP light, the unit Poynting vector (indicating the propagation direction of the light), and the wave vector of the light in the medium, respectively. The induced photocurrent can be observed in experiment, which is proportional to the helicity of the injection light, P_{circ}:

$$P_{\text{circ}} = \frac{I_{\sigma_+} - I_{\sigma_-}}{I_{\sigma_+} + I_{\sigma_-}} = \sin 2\phi, \tag{7.25}$$

where ϕ is the phase difference between x and y components of the electric vector, and the variation of P_{circ} is from -1 (RH CP, σ_-) to $+1$ (LH CP, σ_+).

Tensors ν and β are essentially the same, sharing the same dependence on symmetry and coordination system. Equation (7.22) will take different forms under various point groups. In the case of D_{2d} symmetry, the nonzero elements in ν are $\nu_{xy} = \nu_{yx} = \nu^{(0)}$; thus, the photocurrents in the plane of the quantum well are

$$j_x = \nu^{(0)} \widehat{e}_y E_0^2 P_{\text{circ}} \quad \text{and} \quad j_y = \nu^{(0)} \widehat{e}_x E_0^2 P_{\text{circ}} \tag{7.26}$$

in which the square of the electric field magnitude is proportional to the power of the light radiation. According to the definition of the coordination, i.e. $x//[1\bar{1}0]$, $y//[110]$, $z//[001]$, if the light is propagating along one of the (110) directions, namely y- or x-direction, then either $\widehat{e}_x = 1$, $\widehat{e}_y = 0$ or $\widehat{e}_x = 0$, $\widehat{e}_y = 1$, being perpendicular to the resulting photocurrent. If the light is injected along (100) direction, i.e. $\widehat{e}_x = \widehat{e}_y = 1/\sqrt{2}$, then the photocurrent is longitudinal, along the same direction of the light propagation, because $j_x = j_y$. Therefore, one can observe an interesting result, i.e. when the wave vector of the light is rotating anticlockwise in the quantum well plane, the resulting photocurrent is rotating clockwise, and the two vectors may coincide with each other in some occasions.

When the symmetry is reduced to C_{2v}, there will be two independent elements in tensor ν, ν_{xy}, and $\nu_{yx} \neq \nu_{xy}$. We define that $\nu^{(1)} = \nu_{xy}$, $\nu^{(2)} = \nu_{yx}$, then

$$j_x = \nu^{(1)} \widehat{e}_y E_0^2 P_{\text{circ}} \quad \text{and} \quad j_y = \nu^{(2)} \widehat{e}_x E_0^2 P_{\text{circ}}. \tag{7.27}$$

Similar to the case of D_{2d}, when the wave vector of the light is along (110) direction, the photocurrent is also perpendicular to the wave vector of the light, but varies for different propagation directions. This is just the consequence of the unequal $[1\bar{1}0]$ and $[110]$ crystal axis, due to the 2D rotating axis in C_{2v} symmetry. If the light is propagating along (100) direction, i.e. $\widehat{e}_x = \widehat{e}_y = 1/\sqrt{2}$, then the photocurrent will be neither parallel nor perpendicular to the wave vector of the light, but rather makes an angle ψ ($\tan\psi = \nu_{xy}/\nu_{yx} = \beta_{xy}/\beta_{yx}$) between each other, as shown in Fig. 7.5.

It can be seen from Eqs. (7.26) and (7.27) that to achieve CPGE in a quantum-well system with high symmetry, such as D_{2d} and C_{2v}, the injected light must be oblique. Under a normal injection, $\widehat{e}$ is along [001] direction, hence $\widehat{e}_x = \widehat{e}_y = 0$, then naturally there will be no photocurrent. Differently, for a quantum well with C_s symmetry, due to the additional

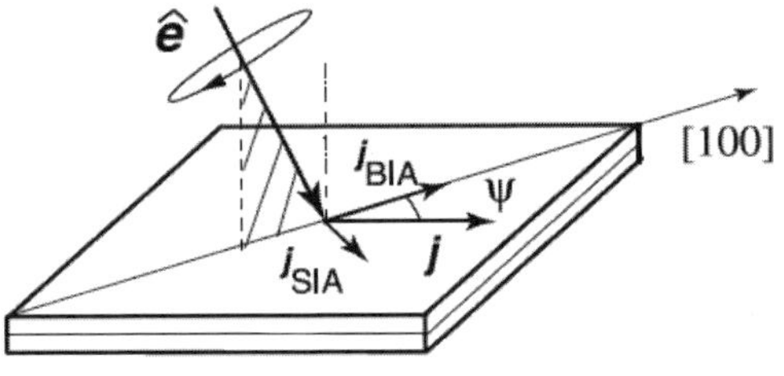

Fig. 7.5. CPGE induced by SIA and BIA in a C_{2v} system when CP light is oblique injected. The induced electron spin will have a component along the (100) direction. Courtesy of Ganichev and Prettl (2003).

ν_{xz} in the tensor, a normally injected optical excitation by using CP light may also induce spin photocurrent. Keeping the coordination as $x//[1\bar{1}0]$, for a quantum-well system of C_s symmetry, the general expressions of photocurrent are:

$$j_x = (\nu_{xy'}\widehat{e}_{y'} + \nu_{xz'}\widehat{e}_{z'})E_0^2 P_{\text{circ}} \quad \text{and} \quad j_{y'} = \nu_{y'x}\widehat{e}_x E_0^2 P_{\text{circ}} \tag{7.28}$$

so that for a normal injection, as $\widehat{e}_x = \widehat{e}_{y'} = 0$, $\widehat{e}_{z'} = 1$, a photocurrent is still achievable, with a direction perpendicular to the mirror reflection plane of C_s.

For systems with higher symmetry, the relationship between the oblique injection angle Θ_0 and photocurrent is certainly a question of concern. Assuming that the injection beam is in the (yz) plane, we have

$$\widehat{e}_x = t_p t_s \sin\Theta, \tag{7.29}$$

where Θ is the refraction angle, related to the injection angle Θ_0 by $\sin\Theta = \sin\Theta_0/\sqrt{\varepsilon^*}$, where ε^* is the dielectric constant of the quantum-well material, and t_p and t_s are the transmission coefficients of the two linearly polarized light, p and s. From the Fresnel equation, one can get:

$$t_p t_s = \frac{4\cos^2\Theta_0}{(\cos\Theta_0 + \sqrt{\varepsilon^* - \sin^2\Theta_0})(\varepsilon^*\cos\Theta_0 + \sqrt{\varepsilon^* - \sin^2\Theta_0})}. \tag{7.30}$$

The above results have been well verified in experiments. Whereas the pioneer work by Ganichev *et al.* used intraband excitation, Yang *et al.* (2006) used interband excitation to achieve CPGE in a C_{2v} system. It turned out that the interband excitation has an efficiency of two orders higher than that of intraband excitation. The results of Yang *et al.* are shown in Fig. 7.6. Two samples of asymmetrically doped InGaAs quantum-well systems, D and E, as shown in Fig. 7.6, were used. The only difference is that in sample E, the composition of indium in the well region has a gradient; hence, the SIA of that sample should be even stronger than that of sample D which has an even composed well region. The SdH measurement shows that the Rashba coefficient in samples D and E is 3.0×10^{-12} and 6.3×10^{-12} eVm, respectively. The results shown in Fig. 7.6 for the CPGE photocurrent dependence on the helicity and injection angle of the excitation light coincide well with the theoretical expectation. Quantitatively, we first notice that the CPGE current intensity ratio for samples E and D is similar to the ratio of their Rashba coefficient measured by SdH oscillation, which implies that the SOI in the two samples is mainly due to the Rashba effect.

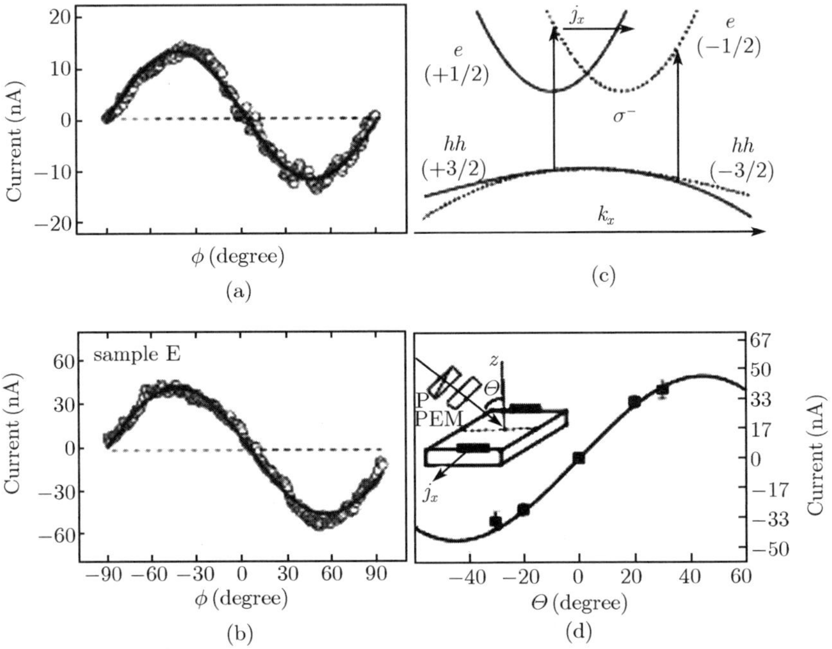

Fig. 7.6. (a) and (b) The CPGE photocurrent as a function of the helecity of the injection light for samples D and E. ϕ is the phase difference calculated by Eq. (7.25), which determines the helicity of the injection light. The injection angle is 30°, the laser power is 100 mW, $\lambda = 880$ nm, and $T = 10$ K. The solid line is the fitting by using $\sin 2\phi$. (c) The schematic of the optical excitation for a split band, taking σ_- excitation as an example. The horizontal arrow indicates the net current direction. (d) The photocurrent dependence on the injection angle Θ. The solid line is the fitting by using Eq. (7.30). Courtesy of Yang *et al.* (2006).

More important is that, as mentioned above, the CPGE current intensity is about two orders stronger than that using intraband excitation, which results from the higher efficiency of band to band excitations.

Yang *et al.* (2006) also reported the excitation wavelength dependence of the CPGE current, as shown in Fig. 7.7. It is noticed that when the excitation wavelength reaches some critical points, the sign of the CPGE current will reverse, indicating that the initial state of the transition has been transferred from heavy hole to light hole, and consequently the final state, as well as the current direction, will have different spin orientations. This result is consistent with the theoretical prediction by Golub (2003).

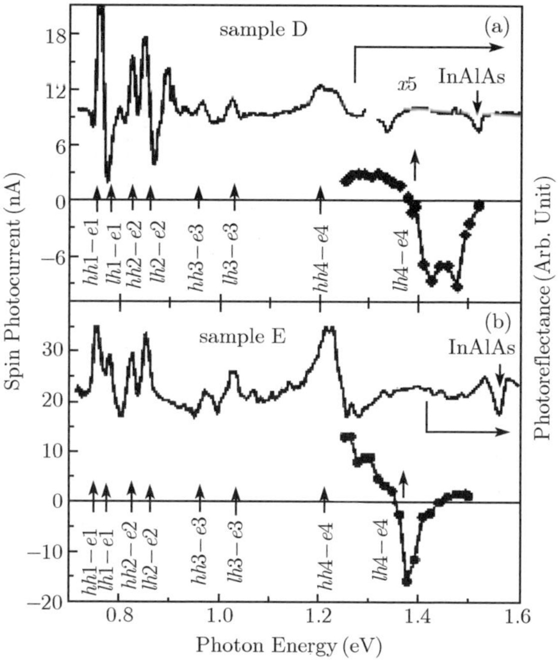

Fig. 7.7. The CPGE dependence on excitation wavelength for samples D and E. Modulated reflection spectrum is also given to identify the transition energy levels in the excitation. The arrows indicate the transitions initiated from the heavy hole states. Courtesy of Yang *et al.* (2006).

7.3.2. *Anomalous CPGE*

He *et al.* (2008) reported that under normal incidence of CP light at RT, a charge current with swirly distribution has been observed in the 2DEG in $Al_{0.25}Ga_{0.75}N/GaN$ heterostructures. They believe that anomalous charge current is produced by a radial spin current via the reciprocal spin Hall effect. It suggests a new way to research the reciprocal spin Hall effect and spin current on the macroscopic scale and at RT.

In their experiment, the intensity of the light spot has a Gauss profile; under normal incidence, an inhomogeneous spin density will be excited on the 2DEG plane shown in Fig. 7.8(a). Its gradient generates a diffused spin polarization current (SPC): $q_r^z = -D\nabla_r N_z$, where D is the electron diffusion coefficient, N_z the vector of spin density along z-direction, and r the radial direction on x−y plane. The SPC (spin polarized current) has a radial distribution and is proportional to both the polarization degree of the

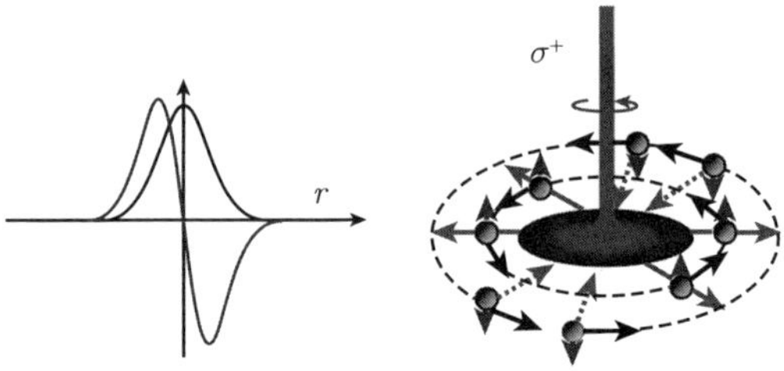

Fig. 7.8. (a) The spatial distribution of the inhomogeneous spin polarization density (dark curve) and its gradient (light curve). (b) Illustration of spin polarization electron's movement under normal incidence with σ^+ polarized light. The real (dashed) long arrows denote the flowing of electrons and the light short arrows show the spin direction. Dark short arrows denote the spin transverse force acting on electrons. Courtesy of He *et al.* (2008).

light and the z component of the light intensity (Zhou and Shen, 2007). Therefore, it can be further expressed as $q_r^z = -N_0 D\nabla G(r)\sin 2\varphi\cos\theta$, where θ is the incident angle, N_0 the average electron spin density with circular polarization ($\varphi = 45°$, 135) and normal incidence ($\theta = 0$), and $G(r)$ the spatial Gauss distribution of the light intensity. Reviewing the general theory of spin charge current coupling (Dyakonov, 2007), all possible charge currents related with a spin density N_z can be expressed as $j/e = \gamma\mu\boldsymbol{E} \times N_z + \gamma D\nabla \times N_z$, where e is the absolute value of electron charge, γ a dimensionless coupling constant proportional to the SOI, and μ and $\boldsymbol{E}$ the electron mobility and external electrical field, respectively. The first term describes a charge current generated by the anomalous Hall effect, where the N_z plays the role of a magnetic field. The second term describes that an SPC generates a charge current perpendicular to both N_z and its gradient which is referred as RSHE (reversed spin Hall effect; Hankiewicz *et al.*, 2005; Dyakonov, 2007). The absence of an external electrical field in the experiment excludes the anomalous Hall effect, thus the only reason for the swirly current is probably RSHE.

Due to the RSHE, the transverse charge current generated by SPC can be expressed as $\boldsymbol{j}/e = -N_0\gamma\boldsymbol{v}_y^z \times \widehat{z}$, where $\boldsymbol{v}_y^z = D\nabla_r G(r)$ is the effective velocity of electrons along y-direction with spin z polarization and $\widehat{z}$ is the unit vector perpendicular to the plane. In this process, spin-polarized electrons generate transverse displacements which can be described by a spin transverse force which is defined as $f_x(r) = (m^*\gamma/\tau_\mathrm{s})\boldsymbol{v}_y^z \times \widehat{z}$, where m^* and τ_s are the effective mass and SRT of electrons, respectively. The direction of $f(r)$ acting on the spin polarization electrons is perpendicular to both electrons' initial velocity and spin, so it leads to a transverse velocity $v_x(r) = f_x \tau_s / m^*$ for electrons in the relaxation time approximation. Therefore, if electrons with spin z polarization flowing with a radial distribution are subjected to a spin transverse force, it will generate a transverse current swirling over the spot center as shown in Fig. 7.8(b). Here one can find that the direction of the charge current [dashed loop shown in Fig. 7.8(b)] is consistent with the rotational direction of the electrical field component of the light. In fact, it implies the conversion of the angular momentum of the system. It is known that directed CPGE current is a transfer of the photon angular momentum into the linear momentum of the free charge carriers, which means a coupling of axial and polar vectors, and this coupling is only allowed by gyrotropic media (Ganichev and Prettl, 2003). Similarly, the swirly anomalous CPGE current can be considered as the transfer of the photon angular momentum into the angular momentum of the electron's orbit motion along the center of the light spot, which necessarily requires the uniformity of the direction of the swirly current and the electrical field component of the light. This transfer means a coupling of two axial vectors and should not be only limited to gyrotropic systems.

Figure 7.9 schematically shows the geometry of the light spot and two electrodes and the experimental data which indicate that the RSHE

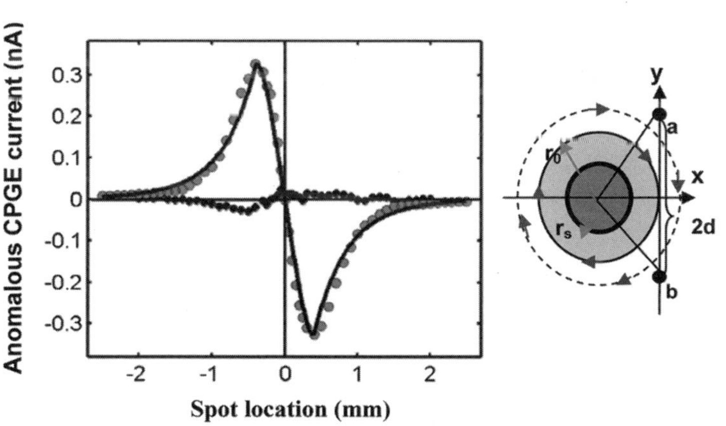

Fig. 7.9. The fitting (dark curve) of the experimental data (light dots). The dark dots are the experimental data obtained when the spot focussed with a diameter of about 0.1 mm. The inset is the geometry of the light spot and two electrodes in which r_o is the radius of the light spot, r_s the radius of saturated absorption, dark spots denote electrodes, and short arrows indicate the direction of the vortex electrical field. Courtesy of He *et al.* (2008).

converts a radial spin current into a swirly charge current via a spin transverse force.

7.3.3. *Spin galvanic effect (SGE)*

SGE can also be achieved by optical excitation, for example, when heavy and light holes are excited at the same time, as shown in the absorption part in Fig. 7.1. As the transition rate is different for transitions initiated from heavy and light hole states, there will be a difference in the numbers of spin-up and spin-down ($+1/2$ and $-1/2$) electrons, as shown in Fig. 7.10. That will lead to different occupation of states for different spins. A direct consequence is a fast relaxation from the occupied states to empty states and spin flip during the relaxation process. As indicated in Fig. 7.10, among the main relaxation processes, 1-2 and 3-4 are similar, and their effects will be cancelled, but 1-4 and 3-2 are very different in terms of relaxation rate, resulting in asymmetrical distribution in momentum for the same spin state, and hence leading to net momentum or spin current.

Figure 7.11 is the schematic of the SGE. CP light is normally injected onto the 2DEG plane, carrying photon spin oriented along the light propagation direction. The photon spin is transferred to electron spin via optical excitation, leading to electron spin polarization perpendicular to the 2DEG plane. But it is known that only the in-plane spin components appear in the SOI Hamiltonian. Electron spin precession induced by an in-plane magnetic field will, however, create electron spin components in the plane, as demonstrated by Ganichev and Prettl (2002) in their pioneer work. Assuming that

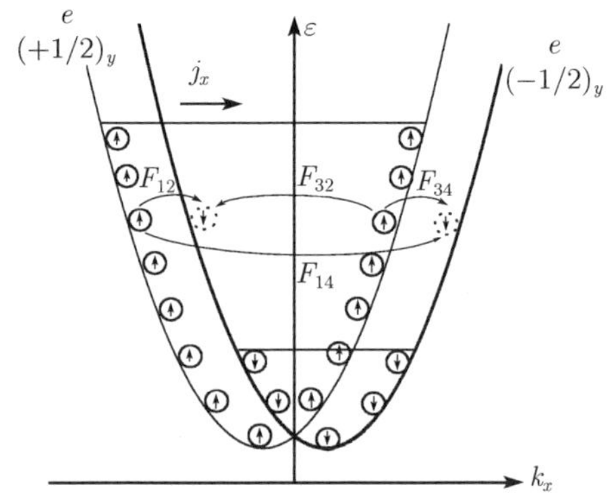

Fig. 7.10. Asymmetrical electron spin relaxation in a spin-split system. Courtesy of Ganichev and Prettl (2002).

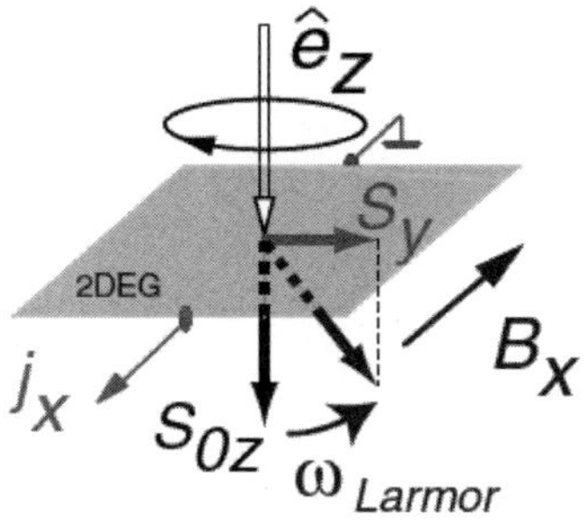

Fig. 7.11. The schematic of SGE experiment. Courtesy of Ganichev and Prettl (2002).

the optically injected electron spin along z-direction is S_{0z}, then, under the magnetic field B_x in x-direction, there will be a spin component along y-direction

$$S_y = -\frac{\omega_L \tau_{S\perp}}{1 + (\omega_L \tau_S)^2} S_{0z}, \tag{7.31}$$

where $\tau_S = \sqrt{\tau_{S//}\tau_{S\perp}}$, $\tau_{S//}$ and $\tau_{S\perp}$ are the longitudinal and transverse electron SRT, respectively, g the effective g-factor in plane, and $\omega_L = g\mu_B B_x/\hbar$ the Larmor frequency of the spin precession. Thus, we have

$$S_y = -\omega_l \tau_S S_{0Z} = \frac{g\mu_B \tau_S}{\hbar} B_x, \tag{7.32}$$

which implies that the spin polarization is proportional to the magnetic field intensity, and the current induced by the spin polarization can be expressed as

$$j_x = Q_{xy} S_y \approx e n_e \frac{\alpha}{\hbar} \frac{\tau_P}{\tau_S} S_y \propto \alpha B_x, \tag{7.33}$$

where Q_{xy} is a second-order pseudo-potential, proportional to B_{lm} in Eq. (7.10), reflecting the dependence of the x-direction current on the magnetic field. That current is an indication of SGE and a qualitative measure of the SOI strength.

7.3.4. *Linear photogalvanic effect*

One way to generate spin current is to use linearly polarized optical excitations in bulk III–V semiconductors and QWs with asymmetric band structures induced by strong SOC. A very recent example was reported by Dai *et al.* (2010). The left and right circular components of the linearly polarized light generate the same amount of carriers with opposite spins and velocities, leading to a spin photocurrent accompanied by no

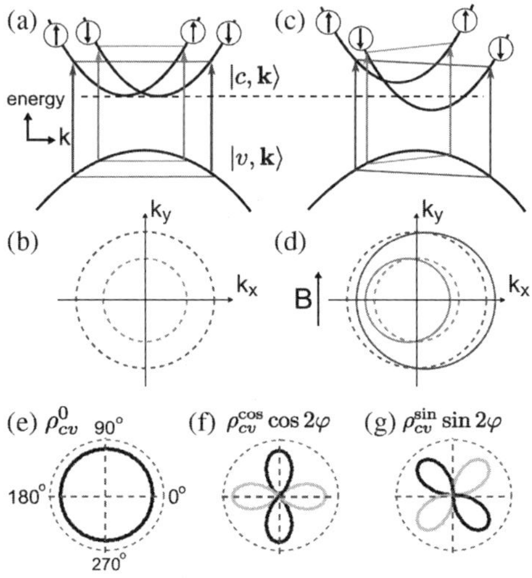

Fig. 7.12. Schematic description of direct optical excitations from a valence band $|\nu, \boldsymbol{k}\rangle$ to the spin-split CBs $|c = \pm, \boldsymbol{k}\rangle$ at magnetic field $\vec{B} = 0$ (a) and $\vec{B}$ along the y-axis (b). Dashed rings indicate the constant energy contours of photoexcited conduction electrons, which are symmetric at $\vec{B} = 0$ and shifted along $k_y(k_x)$ in the presence of a magnetic field $B_x(B_y)$. The circled arrows denote the relative spin orientations which are in-plane and perpendicular to the wave vector. (c) and (d) The top view of (a) and (b), respectively. (e)–(g) Wave vector angle (φ) dependence of photo-excited carrier density $\rho_{cv,k}$ in Eq. (7.36) at $\vec{B} = 0$. Only the results of one constant energy contour are shown. Dark (light) curves represent positive (negative) values. ρ^0_{cv} is always larger than $\rho^{\cos\,/\,\sin}_{cv}$ terms. Courtesy of Dai *et al.* (2010).

electric current. An in-plane magnetic field may induce an imbalance of photoexcited carriers with opposite spins in the spin-split bands, resulting in an electric photocurrent. Dai *et al.* (2010) report a measurement of magneto-electric photocurrent generated via direct interband transitions in an InGaAs/InAlAs 2DEG by a linearly polarized light.

Figure 7.12 presents a simple picture to qualitatively explain the magnetic-field-induced conversion from pure spin current to electric photocurrent. The picture [Figs. 7.12(a) and 7.12(b)] can be illustrated by a valence band $|\nu, \boldsymbol{k}\rangle$ and two spin-split conduction subbands $|c = \pm\boldsymbol{k}\rangle$ with band dispersions

$$E_\pm(\boldsymbol{k}) = \frac{\hbar^2}{2m^*}k^2 \pm \sqrt{(\alpha k_y + h_x)^2 + (\alpha k_x + h_y)^2}, \tag{7.34}$$

where $k_{x,y}$ are the in-plane wave vectors, α is the Rashba coefficient, and $h_{x;y}$ denote the Zeeman energies of the in-plane magnetic fields. In the k

space, the states involved in the interband transition form the constant energy contours which, in the absence of magnetic fields, are perfect concentric circles split by the SOI [Fig. 7.12(a)]. Because of the time-reversal symmetry, for each state on a constant energy contour, there always exists another state with opposite velocity and spin; therefore, they cancel one another to give no electric photocurrent, but a nonzero spin photocurrent. For the above $E_{\pm}(\boldsymbol{k})$, applying in-plane fields will lead to the Zeeman shifts of the bands and distort the constant energy contour to Fig. 7.12(b). This shift then breaks the balance of both velocities and photo-excited carrier density on the constant energy contours and leads to a finite electric photocurrent. Besides the Zeeman effect, a shift can also be induced by the diamagnetic mechanism, but the consequence should be qualitatively similar.

Dai *et al.* (2010) provided a theoretical model to explain the experimental observation. The electric photocurrent density j_x can be calculated by summing the velocities of the photo-excited carriers

$$j_x = -e \sum_{c,v,k} \rho_{cv,k} v^x_{c,k}, \tag{7.35}$$

where e is the electron charge, $\rho_{cv,k}$ is the carrier density, and $v^x_{c,k}$ denotes the x-axis velocity of $|c, \boldsymbol{k}\rangle$, with the definition of $v^x_{\pm\boldsymbol{k}} = (1/\hbar)\partial E_{\pm}(\boldsymbol{k})/\partial k_x$. The term of $\rho_{cv,k}$ turns out to be well described by the form

$$\rho_{cv,k} = \rho^0_{cv} + \rho^{\cos}_{cv} \cos 2\varphi \cos 2\theta + \rho^{\sin}_{cv} \sin 2\varphi \sin 2\theta, \tag{7.36}$$

where $\varphi \equiv k_y/k_x$ is the wave vector. As shown in Figs. 7.10(e)–(g), $\rho_{cv,\boldsymbol{k}}$ has an anisotropic φ dependence. Taking the above $\rho_{cv,\boldsymbol{k}}$ and $v^x_{\pm\boldsymbol{k}}$ into calculation, one immediately obtains a current density in the same form as

$$J_x(B_x, B_y, \theta) = C_0 B_y + C_y B_y \rho^{\cos}_{cv} \cos 2\theta + C_x B_x \cos 2\theta, \tag{7.37}$$

which well interprets the experimental results shown in Fig. 7.13.

7.4. Electric Field-induced Electron Spin Polarization in a Spin-Split System

Let us repeat Eq. (7.1):

$$H = \alpha\boldsymbol{\sigma} \cdot (\widehat{\boldsymbol{z}} \times \boldsymbol{k}_{//}) = \alpha\boldsymbol{k}_{//} \cdot (\boldsymbol{\sigma} \times \widehat{\boldsymbol{z}}). \tag{7.1}$$

The first expression of this equation implies that $(\widehat{\boldsymbol{z}} \times \boldsymbol{k}_{//})$ corresponds to an effective magnetic field which can polarize the electron spin; in other words, electron momentum (carried out by a current or provided by electric field) can induce electron spin polarization. Edelstein (1990) predicted

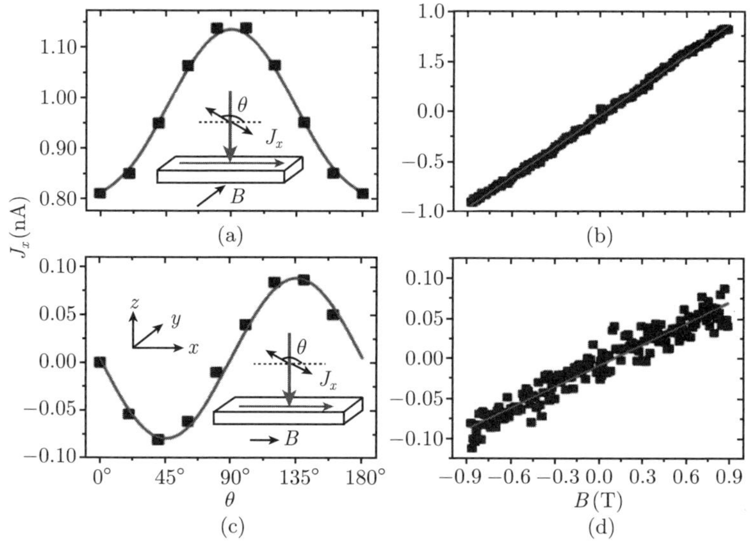

Fig. 7.13. Electric photocurrent $\boldsymbol{J_x}$ as a function of the polarization angle θ and in-plane magnetic fields $\boldsymbol{B}$, along the y-axis [(a)–(b)] and along the x-axis [(c)–(d)], respectively. $\boldsymbol{B} = 0.9\,\mathrm{T}$ in (a) and (c). $\theta = 20°$ in (b) and $\theta = 120°$ in (d). The light beam with power of 1 mW from a DFB laser diode at 1550 nm is normally incident on the sample and the focussed spot size is about 10 μm. Courtesy of Dai *et al.* (2010).

this phenomenon which may be viewed as a reverse process of SGE. Inoue *et al.* (2003) first derived the diffusion electric conductance tensor for a disorder 2DEG system with SOI. He proved that an external electric field can induce spin accumulation: $\langle S_y \rangle = 4\pi e\tau D\lambda E_x$, where $D = m/2\pi\hbar^2$ being the densities of states for each spin orientation, lifetime τ is the momentum relaxation time, and $\lambda = \alpha\langle E_z \rangle/\hbar$ represents the Rashba interaction. The results of Edelstein show that the electric current-induced spin polarization is also a manifest of the Rashba effect.

Experimentally, Yang *et al.* (2006) in their work injected electric current into the two samples (D and E) which have shown clear CPGE. To demonstrate the reversed effect of CPGE, they used the Kerr effect to measure the effect of spin polarization. It is well-known that the Kerr effect refers to the rotation of the polarization plane of a linearly polarized light when it is reflected from the surface of a magnetized medium. The experimental setup is shown in the inset of Fig. 7.14(b). The two polarizers are for polarizing and detecting the polarization, respectively, a PEM (photoelastic modulator) is in front of the detecting polarizer for measuring the electric current (in the x-direction) induced tiny tilt of the polarization plane of the reflected light, in comparison to that of the injected light which is polarized in the yz plane. To avoid the thermal effect, the KR (Kerr reflection) is taken as the difference between the results of current along x and $-x$ directions. The wavelength of the reflected light is adjusted to the wavelength corresponding to the transition from heavy hole to the third excite electron state, as determined by modulation reflection spectra.

The experimental results at $T = 10\,\text{K}$ for samples D and E are shown in Figs. 7.14(a) and 7.14(b), respectively. It can be seen that the KR is enlarged with increasing current, qualitatively in agreement with the theoretical prediction that $\langle S_y \rangle$ is proportional to the electric field E_x, but not being linearly dependent. There might be two reasons for the divergence. First, under low field conditions, the results are limited by the experimental errors. Secondly, the dependence of SRT on the electric field has to be

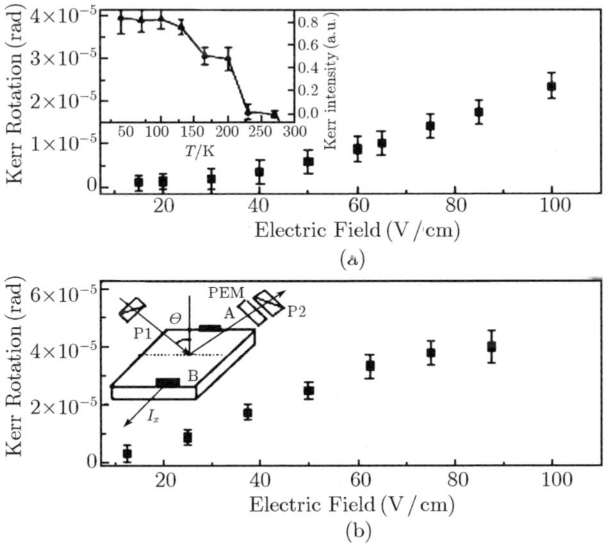

Fig. 7.14. The electric field dependence of the KR for samples D (a) and E (b), representing the current-induced spin polarization $\langle S_y \rangle$. $T = 10\,\text{K}$. The temperature dependence of the KR for sample D under the condition of $E_x = 50\,\text{V/cm}$ is shown in the inset of (a). The inset of (b) is the setup, with P_1 and P_2 being the polarizers. Courtesy of Yang *et al.* (2006).

taken into account. Silov *et al.* (2004) found that the SRT is reduced with increasing electric field intensity, leading to a reduction of spin polarization or spin density; therefore, the KR will be saturated at high field. It is worth noticing that comparing the experimental results from the two samples, the KR in sample E is again about three to four times of that in sample D, once more being consistent with the SOI intensity measured by SdH. In particular, the temperature dependence shows that for sample D, due to the simultaneous reduction in the SRT and momentum relaxation time, the experimental signal is quickly declined above 100 K, while for sample E, the signal is kept up to RT (the result is not shown here (Yang and Ge, 2006)). That result has a significance for practical applications, because the electric field-induced spin polarization might be a new mechanism for nonmagnetic spin sources, for being used in all-electric and nonmagnetic spin devices. Obviously, if the device can be operated at RT, it is quite significant.

7.5. The Experimental Distinction and Applications of the Rashba and Dresselhaus Effects

The Rashba and Dresselhaus effects, respectively, induced by SIA and BIA, may interfere and lead to anisotropy of the spin split. For some particular directions of momentum, the two effects may cancel each other (Ganichev and Prettl, 2003), and thus the so-called "anti-localization" (Knap *et al.*, 1996), spin relaxation along certain crystal axis (Averiev and Golub, 1999), and SdH beating frequency (Tarasenko and Arekiev, 2002), may disappear. Furthermore, some groups (Schliemann *et al.*, 2003) pointed out the importance of the co-existence of the Rashba and Dresselhaus effects, i.e. when the two coefficients are equal, namely $\alpha = \beta$, it is possible to build up nonballistic spin FET.

It is therefore interesting to distinguish the two effects in experiments. Ganichev *et al.* (2004) achieved that by measuring the angular distribution of the SPG effect, based on the fact that the two effects will make different contributions along some particular directions. From the angular distribution diagram of the SGE current in the quantum well plane, one can directly determine the ratio of the two effects in experiments.

To understand the principle of the above experiment, one has to know how the 2DEG electron dispersion and spin orientation of the electronic states would be affected by the Rashba and Dresselhaus effects. Take a zincblende quantum well grown along [001] direction as an example. The symmetry group is C_{2v}, and the SOI includes both the Rashba and Dresselhaus terms:

$$H_{SO} = \alpha(\sigma_x k_y - \sigma_y k_x) + \beta(\sigma_x k_x - \sigma_y k_y), \tag{7.38}$$

where x is the [100] direction, y is [010], and z is the growth direction, as shown in Fig. 7.15 which is the same as shown in Fig. 7.2, except that the crystal orientation is indicated, and it is assumed that $\alpha > \beta > 0$. Note that in Eq. (7.38), there is only a linear term, as the cubic term in the Dresselhaus effects may only affect the strength of β, but not the form of the Hamiltonian. In detail, when the cubic term is involved, the Dresselhaus coefficient $\beta = \gamma\langle k_z^2\rangle$ (please refer to Chapter 6) should be changed to $\beta = \gamma(\langle k_z^2\rangle - k^2/4)$, where γ is the bulk SOI constant.

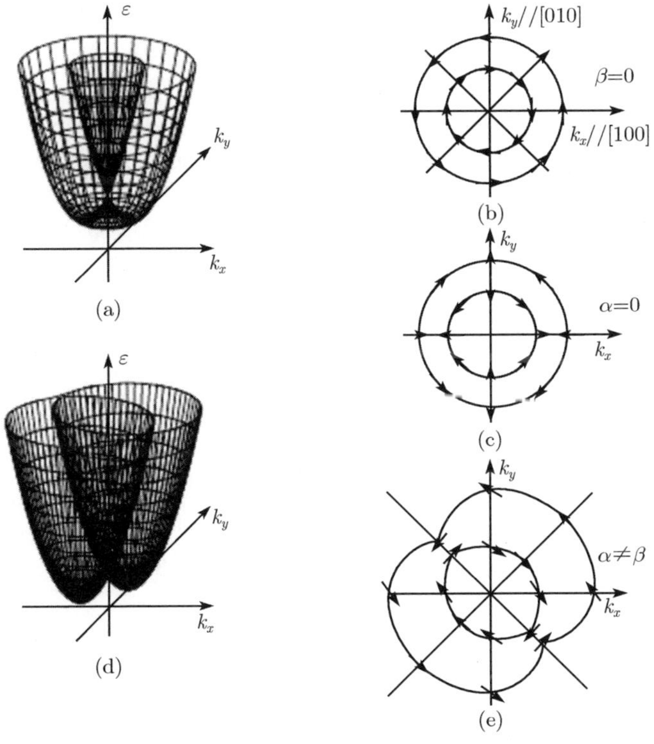

Fig. 7.15. The spin direction distribution for various strengths of SIA and BIA (only the linear term is taken into account) in a C_{2v} system: (a) only Dresselhaus effect; (d) both effects co-exist. Arrows show the spin orientation. Courtesy of Ganichev *et al.* (2004).

It can be seen from Fig. 7.6 that if only one of the two effects exists, the energy dispersion curves in the k_xk_y plane is similar, being two concentric circles. However, along some directions, the spin orientations of the eigenstates induced by the Rashba and Dresselhaus effects are just opposite to each other; and along some other directions, the spin orientations are perpendicular to each other. If the SOI is written as $H_{\text{SO}} = \boldsymbol{\sigma} \cdot \boldsymbol{B}_{\text{eff}}(k)$, with $\boldsymbol{B}_{\text{eff}}(k)$ as the effective magnetic field along the corresponding quantization axis, the distribution of the spin orientation is well classified. In concrete, one can see from Eq. (7.38) that for pure Rashba ($\beta = 0$) or pure Dresselhaus ($\alpha = 0$) cases, the effective magnetic field $\boldsymbol{B}_{\text{eff}}(k)$ will be either $\boldsymbol{B}_{\text{eff}}^{(\text{R})}(k) = \alpha(k_y, -k_x)$ or $\boldsymbol{B}_{\text{eff}}^{(\text{D})}(k) = \alpha(k_x, -k_y)$. Obviously, in the pure Rashba case, the spin orientation is always perpendicular to the momentum direction, while in the pure Dresselhaus case, the angle between the spin orientation and momentum direction depends on the momentum direction itself. In a C_{2v} system, when the Rashba and Dresselhaus effects co-exist, the momentum along two directions of $[1\bar{1}0]$ and $[110]$ leads to very different situations. If $k//[1\bar{1}0]$, the eigenvalue of the Hamiltonian is $\varepsilon = \hbar^2k^2/2m^* \pm (\alpha - \beta)|k|$, whereas when $\boldsymbol{k}//[110]$, $\varepsilon = \hbar^2k^2/2m^* \pm (\alpha + \beta)|\boldsymbol{k}|$, such an anisotropic dispersion is shown in Fig. 7.15(d), while the contour in the profile of the iso-energetic surface in the k_xk_y plane and the spin orientation corresponding to various momentum directions are shown in Fig. 7.15(e).

The SGE current is induced by an average spin in plane (Ganichev and Prettl, 2002, 2003), i.e.

$$\boldsymbol{j}_{\text{SGE}} \propto \begin{pmatrix} \beta & -\alpha \\ \alpha & -\beta \end{pmatrix} \boldsymbol{S}_{//}. \tag{7.39}$$

For any particular direction of $\boldsymbol{S}_{//}$, the SGE current $\boldsymbol{j}_{\text{SGE}}$ includes contributions from both Rashba and Dresselhaus effects, $\boldsymbol{j}_{\text{R}}$ and $\boldsymbol{j}_{\text{D}}$, respectively, and the amplitudes are $j_{\text{R}} \propto \alpha|\boldsymbol{S}_{//}|$ and $j_{\text{D}} \propto \beta|\boldsymbol{S}_{//}|$, respectively. Then their ratio is

$$\frac{j_{\text{R}}}{j_{\text{D}}} = \frac{\alpha}{\beta}. \tag{7.40}$$

When the direction of $\boldsymbol{S}_{//}$ coincides with a crystal axis (say along k_x or k_y), based on Eq. (7.39) and Fig. 7.6, the SGE current components along and perpendicular to $\boldsymbol{S}_{//}$ direction are just j_{D} and j_{R}, respectively; therefore, one can get the ratio of α/β from the experimental results.

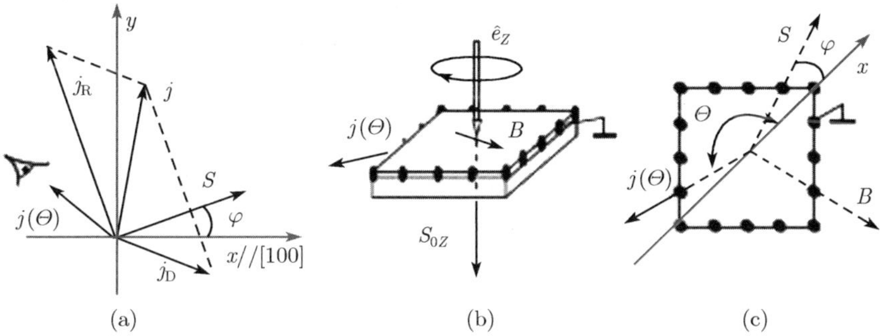

Fig. 7.16. (a) The angular dependence of the SGE current. (b) and (c) Geometry of the experiment. Courtesy of Ganichev *et al.* (2004).

The experiment conducted by Ganichev *et al.* (2004) is schematically shown in Fig. 7.16. The sample is an n-type single $InAs/Al_{0.3}Ga_{0.7}Sb$ quantum well grown along (001) direction by MBE. The free carrier concentration is $1.3 \times 10^{-12}\,cm^{-2}$, and the RT mobility is about $2 \times 10^4\,cm^2/Vs$. The edges of the sample are along [110] and [1$\bar{1}$0] directions, respectively. As shown in Fig. 7.16(b), eight electrodes are made to measure the current along various directions. The intraband optical excitation is normal to the sample, with a CP light. In order to avoid background signal influence, which is almost of the same amplitude of the signal, the authors took $j_{SGE} = (j_{\sigma_+} - j_{\sigma_-})/2$ as the SGE current. Similar to the principle in the work of Ganichev and Prettl (2002), the normally injected CP light will induce net electron spin along z-direction, then the in-plane components $S_{//} \propto \omega_L \tau_s$ will be created by Larmor precession under an in-plane magnetic field ($B = 1 - T$), here ω_L is the Larmor frequency and τ_s is the SRT. With the above experimental conditions, the SGE current increases linearly with the magnetic field, indicating that $\omega_L \tau_s < 1$; hence, the Hanle effect can be ignored. The angle between $\boldsymbol{S}_{//}$ and the magnetic field depends on the concrete condition of the spin relaxation. However, in an InAs quantum well, isotropic EY spin relaxation mechanism is predominant, so that the spin polarization of the photo-excited carrier, $\boldsymbol{S}_{//}$, is always perpendicular to the magnetic field and will change direction following the rotation of the magnetic field about the z-axis.

The contributions from the Rashba and Dresselhaus effects can be evaluated by measuring the SGE current along all possible directions Θ, for a fixed direction of $\boldsymbol{S}_{//}$. According to Eq. (7.39), j_R is always perpendicular to the direction of $\boldsymbol{S}_{//}$, while j_D has an angle 2φ from $\boldsymbol{S}_{//}$, where φ is the angle between $\boldsymbol{S}_{//}$ and the x-axis. The SGE current along any direction defined by Θ could then be a sum of the projections of j_{R} and j_{D} on that direction:

$$j_{\mathrm{SGE}} = j_{\mathrm{D}} \cos(\Theta + \varphi) + j_{\mathrm{R}} \cos(\Theta - \varphi). \tag{7.41}$$

In this measurement, the following three directions are particularly useful, as shown in Figs. 7.17(a)–7.17(c). In (a), the spin polarization $\boldsymbol{S}_{//}$ is

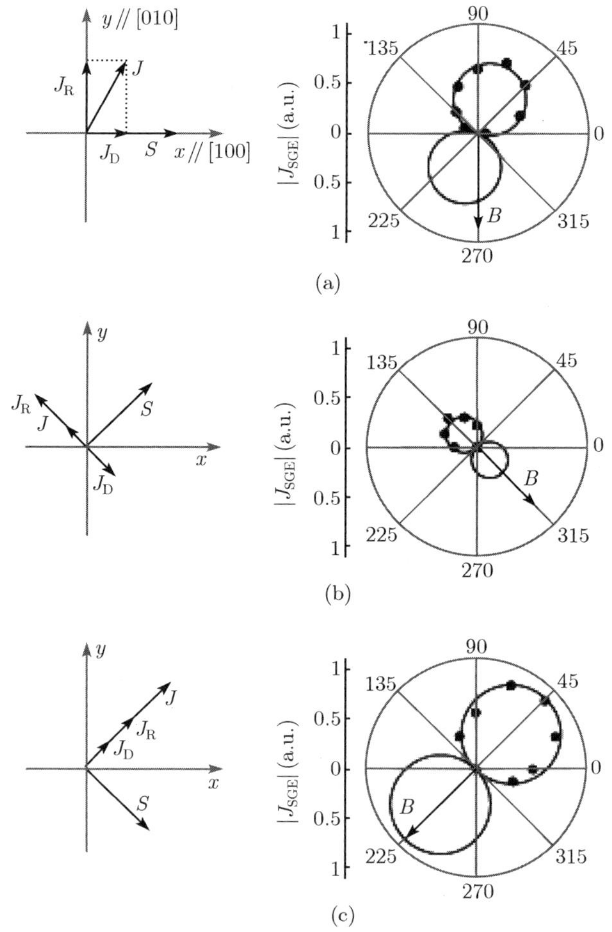

Fig. 7.17. SGE current in an n-type InAs. In the left part, the contribution to SGE current by three particular spin polarizations are shown. In the right part, the SGE current as a function of angle Θ is drawn in a polar coordinate. The current was measured under 10 kW radiation, and unified to the maximum current in (c) ($j_{\max} = 20\,\mu\mathrm{A}$). Courtesy of Ganichev *et al.* (2004).

along [100] direction ($\varphi = 0$); then according to Eq. (7.41), the current along [100] direction ($\Theta = 0$) and [010] direction ($\Theta = \pi/2$) are just j_R and j_D, respectively, as shown in the left part of Fig. 7.17(a). In Fig. 7.17(b), $\boldsymbol{S}_{//}$ is along [110] direction ($\varphi = \pi/4$); then according to Eq. (7.41), the SCE current should have a maximum along $\Theta = 3\pi/4$ direction: $j = j_R - j_D$. If $\boldsymbol{S}_{//}$ is along $[1\bar{1}0]$ direction (Fig. 7.17(c), $\varphi = -\pi/4$), then the maximum current appears at $\Theta = \pi/4$ direction: $j = j_R + j_D$. With the obtained data of $j_R - j_D$ and $j_R + j_D$, one can readily get $j_R/j_D = \alpha/\beta$.

In the right part of Fig. 7.17, the experimental results of Ganichev *et al.* (2004) are shown in polar coordinates, and the current amplitude is unified by taking the maximum value of $j = j_R + j_D$ as being one. The ratio of the Rashba and Dresselhaus current can be directly obtained from the ratio of $j_R/j_D = j(\pi/2)/j(0)$, shown in Fig. 7.17(a) or calculated from the maximum current in Figs. 7.17(b) and 7.17(c). The two approaches achieve the same value, i.e. $j_R/j_D = 2.15 \pm 0.25$. Using the value of $\alpha/\beta = 2.15$ to calculate Eq. (7.41), one can precisely fit the right part of Fig. 7.13, and there is no need for any additional parameters. The value of 2.15 is also consistent with theoretical prediction by Lommer *et al.* (1988) and other experimental results (Knap *et al.*, 1996).

7.6. Spin Electronic and Opto-electronic Devices

We will focus on two device models relevant to Rashba and Dresselhaus effects and a mechanism which transfers spin current to measurable electric current in possible spintronic devices.

7.6.1. *Spin FET based on combined Rashba and Dresselhaus effects*

The most well-known spin device model was the spin FET proposed by Datta and Das (1990). The basic principle of that spin FET is applying the Rashba effect in the semiconductor channel region to modulate spin of the injected electrons. The magnitude of the modulation can be controlled by the gate voltage. In the Datta and Das model, both the source and drain are magnetic materials; hence, in both regions, the electron spin is polarized. Due to the mismatching of the impedance between magnetic metal and semiconductor, and interface scattering, there is a great difficulty in injecting as well as in flowing out for the spin-polarized electrons. DMSs are thought to be a good candidate for the source and drain material, but the Curie's temperature is still too low for a real device. As such, the realization of the Datta-Das device is still yet to achieve.

There is a very smart idea to combine the Rashba and Dresselhaus effects in a device, proposed by Hall *et al.* (2003), to achieve an entire semiconductor, no magnetic materials FET. The basic working principle of that model is shown in Fig. 7.18. The material system used is an AlSb/GaSb/InAs heterostructure with a lattice constant of 6.1 Å. The unique advantage of that system is a huge spin splitting in the valence band of GaSb and large electron mobility in InAs. Especially in such a structure along (110) direction, the effective magnetic fields induced by the Rashba and Dresselhaus effects are perpendicular to each other (see Fig. 7.10), and the strength of the Dresselhaus effect is stronger than that of the Rashba effect (Cartoixa *et al.*, 2001). The channel structure is similar to that of the Datta–Das model. The key improvement is that both the spin creation

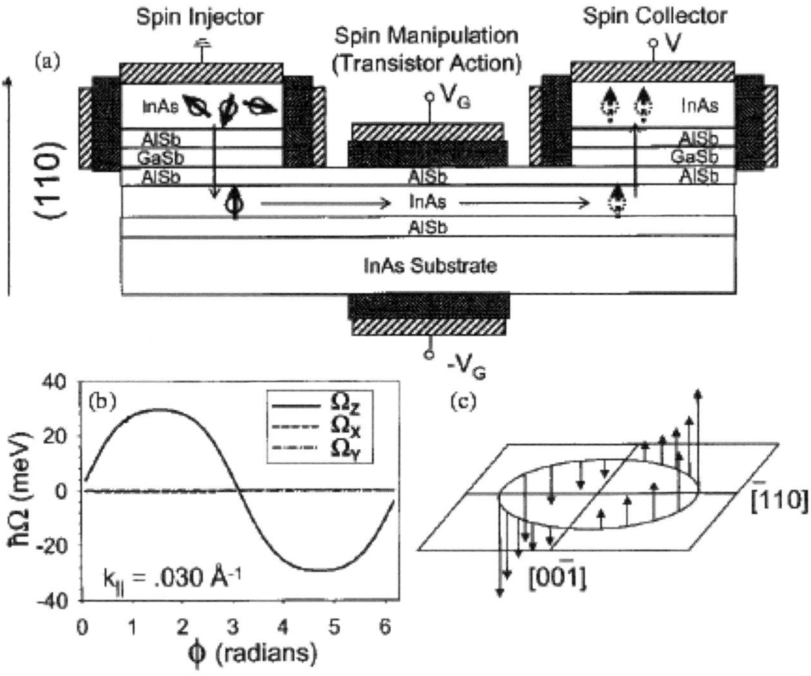

Fig. 7.18. (a) Schematic of a nonmagnetic FET. (b) The calculated crystal magnetic field dependence on angle ϕ ($\phi = 0$ corresponds to the $[00\bar{1}]$ direction) at the HH1 energy level and $k_{//} = 0.03\,\text{Å}^{-1}$ in a GaSb/AlSb quantum well. (c) Schematic of the calculated magnetic field in (b). Courtesy of Hall *et al.* (2003).

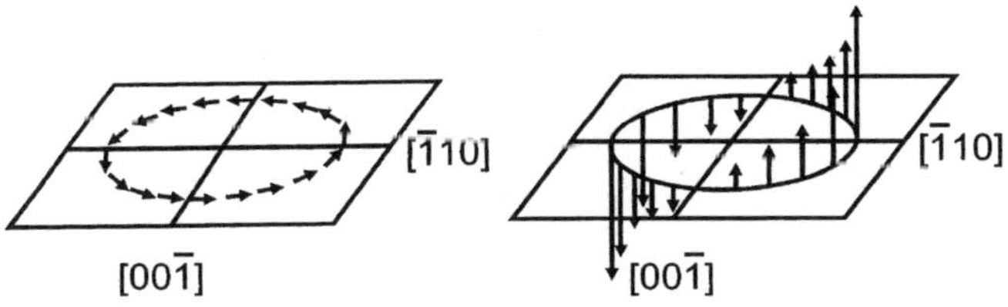

Fig. 7.19. The effective magnetic field induced by Rashba (a) and Dresselhaus (b) effects in a GaSb quantum well grown along the [110] direction. Courtesy of Hall *et al.* (2003).

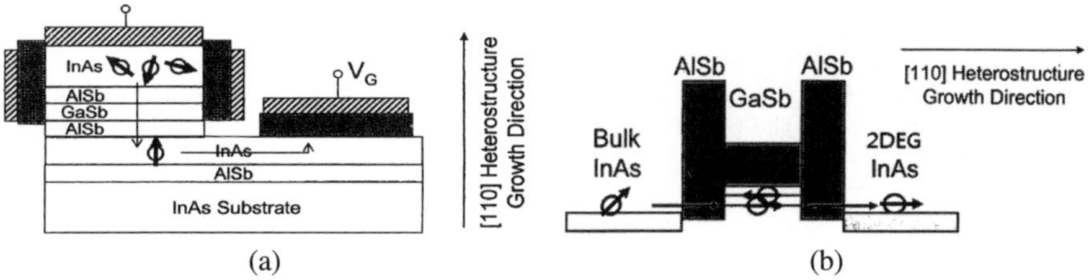

Fig. 7.20. Schematic of spin interband resonant tuneling: (a) profile; (b) side view. Courtesy of Hall *et al.* (2003).

and detection are achieved by using spin resonant interband tunneling. The principle of which is shown in Fig. 7.20. It can be seen from Figs. 7.18(b) and 7.19(b) that (i) the effective magnetic field induced by the Dresselhaus effect is along the growth direction of the quantum well; (ii) for each spin subband, on the two sides of the [00$\bar{1}$] axis in k space, the spin orientations of the carriers are opposite to each other. In other words, the two opposite momentum directions, along [$\bar{1}$10] and [1$\bar{1}$0], respectively, lead to opposite spin orientations as well, and the spin is perpendicular to momentum. As the spin quantization in a quantum well normal to (110) direction would naturally choose to be along the growth direction, the magnetic field along the same growth direction induced by the Dresselhaus effect will not make precession movement of the spin; therefore, the spin lifetime T_1 is largely enhanced.

The principle of the spin resonant interband tunneling effect is shown in Fig. 7.20, with (a) as the profile of the spin FET device, showing the hole spin resonant tunneling in the source region. A horizontal electric field is applied along [$\bar{1}$10] direction, providing additional net momentum to

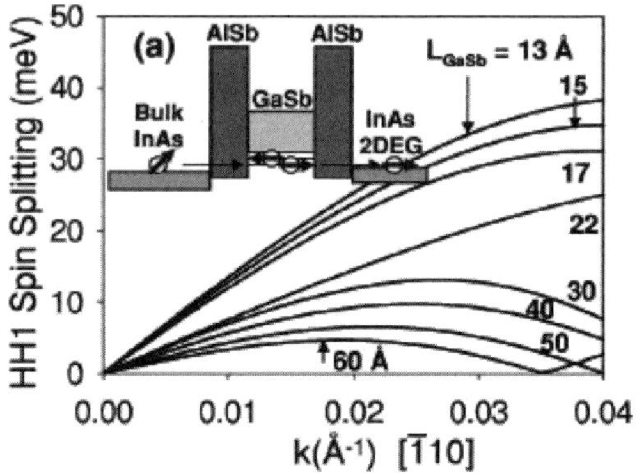

Fig. 7.21. The parallel momentum dependence of the calculated spin splitting of HH1 in a GaSb/AlSb ($L_{\mathrm{AlSb}} = 60\,\mathrm{Å}$) quantum well. The inset is the same as Fig. 7.20(b), indicating the position of the spin splitting in real space. Courtesy of Hall *et al.* (2003).

carriers, then the effective magnetic field in the GaSb source region will be along one of the vertical directions, [110] or $[\bar{1}\bar{1}0]$, and hence spin splitting is along z-direction, as in the valence band shown in Fig. 7.16(b). Calculation result shows that the spin splitting of the HH1 level in GaSb may be over 30 meV, as shown in Fig. 7.21. Such a big gap will certainly make the resonant tunneling to easily occur, leading to a natural spin polarization for the carriers injected from the InAs source to the InAs 2DEG channel, thus achieving a nonmagnetic spin injection.

The electrons in the channel are moving horizontally with spin in vertical direction. In the channel, the Rashba effect is predominant, the magnetic field induced by which is in plane, leading to spin precession. The drain and source are symmetrical, being spin selective too. In such a structure, the spin is modulated by gate voltage and detected by resonant tunneling. An FET is thus produced. In the drain region, a straightforward way is to apply a horizontal electric field, as that in the source region. But there is another way which consumes less energy: no external lateral bias being applied for the drain, there is, however, an intrinsic lateral voltage due to horizontal ballistic current in the 2DEG resulted from spin polarization along a certain direction. The sign of that voltage directly indicates the spin orientation of the electrons flowing into the drain. Reducing energy consumption is a great advantage for spin devices, so that the latter technique has important applications.

7.6.2. *Spin light resources — LED and laser*

Due to spin splitting in the energy bands of semiconductors, it should be possible to control the photon polarization in the light emission process from semiconductors, to produce spin light sources, i.e. spin LED and spin lasers. The light emitted from those devices will be purely CP. Holub and Bhattacharya (2007) had made a summary of progress on this topic.

(1) *Working principle*

In a spin light source, the spin-polarized electrons (holes) are injected into the active area where they will radiatively recombine with nonpolarized holes (electrons) and emit light. The emitted light would be predominated by one of the CP (LH or RH) photons. Such a connection between electron spin and photon polarization has great potential in practical applications. A fundamental significance of this connection is that a polarized light source can be used to directly detect electron spin in a semiconductor heterostructure, and hence motivate the development of spintronics, toward the ultimate target of RT spin devices. In addition, the capability of modulating light polarization in spintronics will greatly enhance the bandwidth in optical communication and easily ensure the security of communication while the laser has super characters, such as low threshold but high intensity and polarization stability. It thus has potential in cryptology, re-locatable optical inner connection, and advanced optical switching and optical modulators.

Electron spin injection is usually using a ferromagnetic material as the spin source. In a ferromagnetic material, the electron spin subbands with spin up and spin down are split related to energy and momentum, due to exchange interaction, hence with different occupations. The state with higher occupation rate is named spin majority state and the other is the spin minority state. Obviously, the spin orientation of the spin majority (minority) state is along (anti) the magnetization direction of the ferromagnetic material. The difference in the occupation rate leads to the net spin polarization in ferromagnetic materials, which is generally expressed as

$$\Pi_{S,X} = \frac{X_{\uparrow} - X_{\downarrow}}{X_{\uparrow} + X_{\downarrow}}, \tag{7.42}$$

where $X_{\uparrow}$ and $X_{\downarrow}$ represent the corresponding electron numbers of spin up and spin down. The real injection process is complex, and which one being the majority or minority of spin depends on the band structure of the ferromagnetic material and the tunneling mechanism; these are beyond the scope of this chapter. The injection efficiencyη is defined as

$$\eta = \frac{\Pi_{\text{inj}}}{\Pi_{S,X}}, \tag{7.43}$$

where $\Pi_{\text{inj}} = (n_\uparrow - n_\downarrow)/(n_\uparrow + n_\downarrow)$, $n_\uparrow$ and $n_\downarrow$ are the carrier concentrations of spin up and down. A number of factors affect the injection efficiency, including interface quality, defect and impurity concentration, band structure, etc. It is a great challenge to enhance the injection efficiency for material growth and band structure engineering.

The most serious problem in spin injection is still the mismatching of impedance between ferromagnetic and semiconductor materials. Rashba (2000) has suggested to adopt tunneling contact as a solution, and DMS is another approach, which is discussed in relevant chapters in this book.

After the spin injection, coherent transport is the next problem. Only if the spin polarization can be well kept from the moment of injection till the radiative recombination for emitting CP light, the spin light source can be realized. Three spin relaxation mechanisms affect the spin coherent transport, namely the EY, DP, and BAP processes. These mechanisms lead an asymmetric spin distribution to equilibrium. In concrete, the EY effect results from SOI which is due to the fact that the ions in the crystal lattice will induce involvement of opposite spin in the electron wave function. As far as an electron changes its orbit momentum by scattering, there will be a possibility of spin flip, no matter the scattering mechanism is spin related or not. Impurities, interfaces, defects, and phonons may all create spin flip between up and down state via SOI, and then lead to degeneration of the spin polarization. For an electron in the CB of energy E, the EY spin relaxation rate $[\tau_{\text{S}}^{EY}(E)]^{-1}$ is proportional to the electron momentum scattering rate $[\tau_{\text{P}}(E)]^{-1}$:

$$\frac{1}{\tau_{\text{S}}^{EY}(E)} = A\left(\frac{E}{E_{\text{g}}}\right)^2\left(\frac{\Delta}{E_{\text{g}}+\Delta}\right)^2\frac{1}{\tau_{\text{P}}(E)}, \tag{7.44}$$

where A is a dimensionless constant, Δ is the spin–orbit split of the valence band. The DP spin relaxation process is based on the lack of reflection symmetry in a semiconductor, which leads to degeneration lift, as the Rashba and Dresselhaus effects. As discussed above, the effective magnetic fields induced by the Rashba and Dresselhaus effects depend on electron momentum. The electron spin may have the Larmor precession around those effective magnetic fields. The DP spin relaxation rate is thus proportional to the electron momentum scattering rate, $[\tau_{\text{P}}(E)]^{-1}$, and in contradiction to the EY case:

$$\frac{1}{\tau_{\mathrm{S}}^{\mathrm{DP}}(E)} = Q\alpha^2 \frac{E^3}{\hbar^2 E_{\mathrm{g}}} \tau_{\mathrm{P}}(E), \tag{7.45}$$

where Q is another dimensionless constant, indicating the strength of the SOI. Finally, BAP results from electron–hole exchange interaction which is particularly important in p-type semiconductors. The electron–hole exchange interaction is due to the spin of electron and hole, corresponding to an effective magnetic field acting upon electrons. Electron spin precesses around such a field, and then relaxes. The BAP rate increases with the spacial overlap of the wave functions of electron and hole, and therefore becomes more important at low temperatures and under strong confining potential, such as in quantum wells and QDs.

The relative importance of the above three mechanisms depends on the structure design and experimental conditions, but in general, EY is predominant in semiconductors with narrow band gap and strong SOI; the spin relaxation of holes also basically results from the EY mechanism. The DP is predominant in semiconductors with middle-sized band gap, at higher temperatures, and with low hole concentration. In a p^+-semiconductor with heavy doping, the BAP is predominant at low temperatures, but DP becomes predominant at higher temperatures. In a system with quantum confinement, the spin dynamics is quite different from that in bulk materials, as all of the above three mechanisms would be affected by mobility, charge localization, and band structure. Recent studies show that the spacial confinement in low-dimensional systems will restrict the movement of electrons and thus the DP spin relaxation, leading to long spin lifetime. In QDs, the enhanced spin lifetime may be longer than the radiative recombination lifetime, so that the spin polarization is no longer related to temperature. That leads to a huge difference between QD and QW (quantum well) spin LEDs. But it is also true that in QDs, the superfine interaction between electron and nuclear may induce decoherence for localized electrons confined in the QDs and bound to donors, leading to reduction of spin lifetime.

(2) *Polarization effect in light emission*

The CP light emitted from a spin light source results from radiative recombination of spin-polarized carriers, hence being a detection technique for spin polarization. Based on the spin reservation rule in optical transitions, as discussed in Sec. 7.2, the helicity of the emitted light directly

Table 7.1. Electron and hole states at Γ point.

Energy band	$\|J, m_j\rangle$	Wave function
CB	$\|\frac{1}{2}, +\frac{1}{2}\rangle$	$\|S\rangle \uparrow$
	$\|\frac{1}{2}, -\frac{1}{2}\rangle$	$\|S\rangle \downarrow$
Heavy hole	$\|\frac{3}{2}, +\frac{3}{2}\rangle$	$-\frac{1}{\sqrt{2}}(\|P_x\rangle + i\|P_y\rangle) \uparrow$
	$\|\frac{3}{2}, -\frac{3}{2}\rangle$	$+\frac{1}{\sqrt{2}}(\|P_x\rangle - i\|P_y\rangle) \downarrow$
Light hole	$\|\frac{3}{2}, +\frac{1}{2}\rangle$	$-\frac{1}{\sqrt{6}}(\|P_x\rangle + i\|P_y\rangle) \downarrow -2\|P_z\rangle \uparrow$
	$\|\frac{3}{2}, -\frac{1}{2}\rangle$	$+\frac{1}{\sqrt{6}}(\|P_x\rangle - i\|P_y\rangle) \uparrow +2\|P_z\rangle \downarrow$

Table 7.2. Transition probability for allowed transitions and the light polarization emitted from the surface perpendicular to the z-direction.

Transition	Matrix element M_{if}	Δm_j	Light polarization	$\|M_{ij}\|^2$
CB↑→ HH↑	$\langle\frac{3}{2}, +\frac{3}{2}\|\,\mu_x\,\|\frac{1}{2}, +\frac{1}{2}\rangle$	−1	σ_+	$\frac{1}{2}\,\|\langle P_x\|\,\mu_x\,\|S\rangle\|^2$
CB↓→ HH↓	$\langle\frac{3}{2}, -\frac{3}{2}\|\,\mu_x\,\|\frac{1}{2}, -\frac{1}{2}\rangle$	+1	σ_-	$\frac{1}{2}\,\|\langle P_x\|\,\mu_x\,\|S\rangle\|^2$
CB↑→ LH↑	$\langle\frac{3}{2}, -\frac{1}{2}\|\,\mu_x\,\|\frac{1}{2}, +\frac{1}{2}\rangle$	−1	σ_-	$\frac{1}{6}\|\langle P_x\|\mu_x\|S\rangle\|^2$
CB↓→ LH↓	$\langle\frac{3}{2}, +\frac{1}{2}\|\,\mu_x\,\|\frac{1}{2}, -\frac{1}{2}\rangle$	+1	σ_+	$\frac{1}{6}\|\langle P_x\|\mu_x\|S\rangle\|^2$

reveals the situation of the spin polarization and its variation under external perturbation or due to relaxation. The above two tables show the electron and hole states (Table 7.1) and the transition processes corresponding to emission of CP light (Table 7.2). The Hamiltonian of optical dipole transition is:

$$\boldsymbol{H}_1 = \boldsymbol{\mu} \cdot \boldsymbol{\varepsilon}, \tag{7.46}$$

where $\boldsymbol{\mu}$ is the dipole moment and $\boldsymbol{\varepsilon}$ the electric field of the light. The electric field is almost constant in the scale of an atom, hence the components of $\boldsymbol{\varepsilon}$ can be taken as constants. In order to calculate the restriction of J and m_j for optical transitions, it is more convenient to write the vector operator $\boldsymbol{\mu}$ as a circular tensor $\mu^1_{\pm 1}$ for optical transitions σ_+ and σ_-. In this case, the Wigner–Eckart theorem points out that only if $\Delta m_j = \pm 1$, the matrix element will not be zero. As the total angular momentum is conserved, the surface emitted light corresponding to $\Delta m_j = \pm 1$ would be of $\sigma_\mp$ polarization, namely, when $\Delta m_j = -1$, the light emitted along $+z$-direction is of σ_+ polarization, while the light emitted along the $-z$-direction would be of σ_- polarization.

It should be noted that the above selection rule is precisely valid only at the $k = 0$ (Γ) point. Away from Γ point, the light and heavy holes will mix with each other to a certain extent; therefore, the optical transitions will not be so ideal. However, the above selection rule is still a good approximation. It can be seen from Table 7.2 that the transition with a heavy hole involved has a probability three times larger than that with a light hole involved. In a bulk material, the heavy and light holes are degenerate at the Γ point where if the transition probability ratio is taken into account, then the following relationship between the electron spin and light polarization can be obtained:

$$\Pi_{\text{CP}} = \frac{I(\sigma_+) - I(\sigma_-)}{I(\sigma_+) + I(\sigma_-)} = \frac{(3n_\downarrow + n_\uparrow) - (3n_\uparrow + n_\downarrow)}{(3n_\downarrow + n_\uparrow) + (3n_\uparrow + n_\downarrow)}$$
$$= \frac{n_\downarrow - n_\uparrow}{2(n_\downarrow + n_\uparrow)} = \frac{\Pi_{\text{inj}}}{2}, \tag{7.47}$$

where $I(\sigma_+)$ and $I(\sigma_-)$ are the intensities of the light of σ_+ or σ_- polarization, respectively, and $n_\uparrow$ and $n_\downarrow$ are the concentration of electrons with up and down spins, respectively. Because of the ratio of 3 between the heavy hole and light hole involved transition probabilities, the light polarization and electron spin will have an opposite sign, i.e. σ_- (σ_+) light corresponding to 1/2 ($-1/2$) electron state. The relationship described by Eq. (7.43) is especially suitable to the case of quantum wells, in which the central wave packet function describing the electron and hole states is slightly affected by the confinement potential. However, the degeneracy at Γ point is likely lifted due to confinement and epitaxial strain. For example, in an $In_{1-x}Ga_xAs/GaAs$ quantum well with suppression strain, the heavy hole level is higher than that of the light hole and the latter is then not involved in optical transitions. In this case, the electron spin will be equal to the light polarization in magnitude, but the sign is still opposite to each other:

$$\Pi_{\text{CP}} = \frac{I(\sigma_+) - I(\sigma_-)}{I(\sigma_+) + I(\sigma_-)} \cong \frac{3n_\downarrow - 3n_\uparrow}{3n_\downarrow + 3n_\uparrow} = \frac{n_\downarrow - n_\uparrow}{n_\downarrow + n_\uparrow} = \Pi_{\text{inj}}. \tag{7.48}$$

The optical selection rule in semiconductors is schematically shown in Fig. 7.22.

The source and drain areas in a spin light source is mainly related to magnetic or DMS materials, which is beyond the scope of this chapter. Here

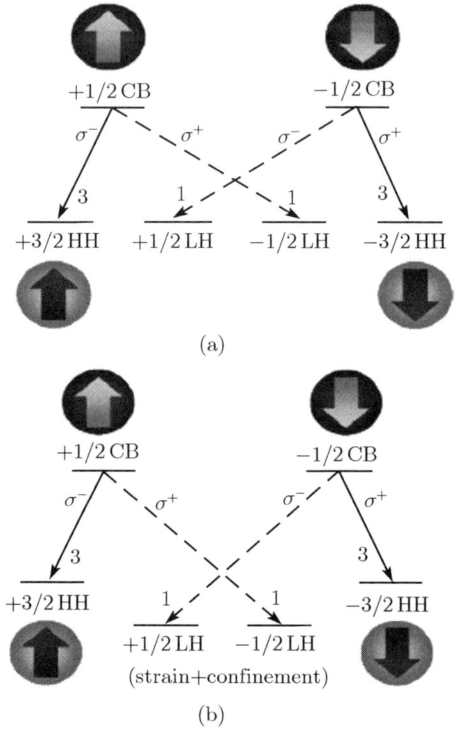

Fig. 7.22. The allowed interband dipole transitions and corresponding light polarization. (a) Bulk material, heavy and light holes degenerate. (b) Quantum well with quantum confinement and epitaxy strain. The heavy and light holes degeneracy is lifted. Courtesy of Holub and Bhattacharya (2007).

we will focus on the effect of the structure and composition of the semiconductor active region on spin light source. An ideal active area should have a high efficiency of light emission and long spin lifetime. Compared with bulk materials, quantum wells have outstanding advantages regarding the above requirement. First, the carriers are strongly confined in the well, which provides a longitudinal resolution for detecting spin in the heterostructure, as one can reach the location of the spin precisely. By varying the thickness of the intrinsic doped isolating layer of a spin LED, one can get information of the spin transport length. Secondly, the width of a quantum well is much smaller than that of other areas in the heterostructure, and the emitted light due to recombination of spin carriers can be easily distinguished from light emission from other areas, and not likely to be absorbed in other areas either. Finally, the SRT in a well-designed quantum well can be much longer than that in a bulk material. For example, in a GaAs quantum well grown along (110) direction, due to suppression of the DP effect, the SRT along the growth direction can reach milliseconds at RT.

There are three detecting modes to study the relationship between light polarization emitted from the device and the spin of the injected carriers: the Faraday mode, quasi-Voigt mode, and inclined Hannel mode, as shown in Fig. 7.23. The Faraday mode is the most popular one in which the carrier spin and the light polarization are fairly directly transferred, and hence easy to understand the optical selection rule behind the transition. It can be seen from Fig. 7.23(a) that in the Faraday mode the external magnetic field is parallel to the direction of the light propagation, which is the growth direction of the material, and hence perpendicular to the device surface. The difficulty of that mode is that in general the easy magnetic orientation is in the device plane; therefore, it is hard to make magnetization along the growth direction. The quasi-Voigt mode is particularly convenient for the laterally emitting device in which the magnetic field is still parallel to the light propagation direction, but perpendicular to the growth direction. However, in this case, those originally applied selection rules in the Faraday mode may no longer be valid, making the relationship between the carrier spin and light polarization more complex. As the easy magnetization axis is in the same plane, the residual spin and optical polarization should be observable, and it is expected to use a residual state to

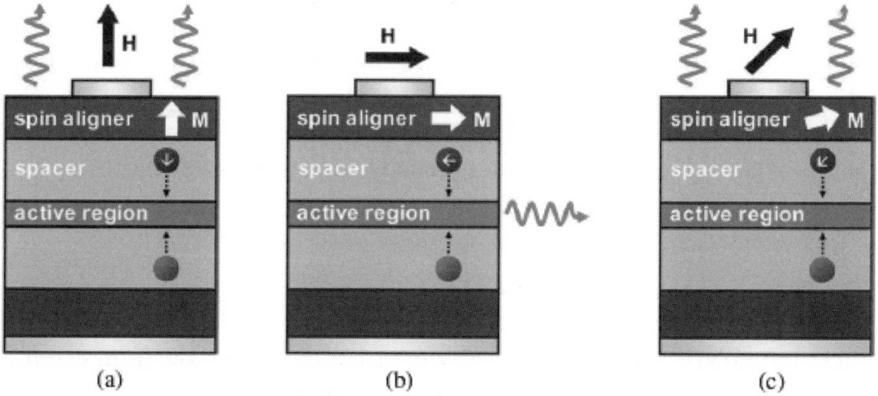

Fig. 7.23. The schematic of the three modes of spin LED: (a) Faraday mode; (b) quasi-Voigt mode; (c) inclined Hannel mode. Courtesy of Holub and Bhattacharya (2007).

control the spin-polarized light source. The emitted light is collected from a cleaved surface; therefore, there is no need to pass through a ferromagnetic layer, so that the anonymity due to the magnetic field can be ignored, and the parasitic contribution induced by dichroist re-absorption can also be reduced to a lowest level. In the declined Hannel mode, the ferromagnetic layer in the source region is magnetized first along the easy magnetization direction, i.e. the magnetization is in plane, so that the injected electrons are spin-polarized also in the plane. Then a magnetic field B is applied at an angle of 45° from the normal direction of the plane, making the electron spin in the active layer precess with a Larmor frequency Ω, and thus producing a vertical component of the spin. As this spin component transfers to light polarization, one can still observe CP light emitting from the normal direction of the device surface, which can be used to evaluate the injected spin in the plane. In this mode, the relationship between the optical polarization Π_{CP} and the spin polarization Π_{inj} is (Van Dorpe *et al.*, 2003),

$$\Pi_{\mathrm{CP}} = \frac{I(\sigma_+) - I(\sigma_-)}{I(\sigma_+) + I(\sigma_-)} = \frac{\Pi_{\mathrm{inj}} \cos\theta}{4} \frac{T_{\mathrm{S}}}{\tau} \frac{(\Omega \cdot T_S)^2}{1 + (\Omega \cdot T_S)^2} + \frac{\Pi_{\mathrm{inj}} \sin\theta}{4} \frac{T_{\mathrm{S}}}{\tau} \frac{1 + (\Omega \cdot T_{\mathrm{S}})^2/2}{1 + (\Omega \cdot T_{\mathrm{S}})^2}. \tag{7.49}$$

The ratio of T_{S}/τ represents the spin detection efficiency in the active layer, T_{S} is the spin lifetime ($T_{\mathrm{S}}^{-1} = \tau^{-1} + \tau_{\mathrm{S}}^{-1}$, τ is the electron–hole recombination time constant, τ_{S} is the SRT). Declined application of a small magnetic field would be enough to produce a spin component along z-direction, which can be used to detect the spin injection without changing the magnetization of the active area. Such a technique can also be used to distinguish the injected spin from some side effects, such as spin polarization induced by magneto-optical effect and Zeeman effect.

(3) *Spin-polarized vertical-cavity surface emitter layers (VCSELs)*

The polarization characteristics of a semiconductor laser are determined by the polarization properties of the gain medium and optical cavity. As such, the polarization characteristics of lasers emitting light from the top surface or side surface are fundamentally different. The side emission laser emits light with linearly polarized light along two perpendicular directions, one is parallel (TE polarization) and the other perpendicular (TM) to the plane of the active layer, and the output light polarization will be finally determined by the optical mode selection of the waveguide. In contrast, a VCSEL is a quasisymmetrical device, there is no selection to the polarization, and therefore hard to design, stabilize, and control its polarization characteristics. In general, the direction of the polarized light emission in the transverse mode from VCSELs is randomly distributed and varied with injection current and temperature, as well as with individual devices. The stabilization and control of the polarization can be achieved by introducing geometric asymmetry or anisotropy to optimize a certain polarization. Considering the complex relationship between the electron spin and optical polarization, controlling the electron spin is one of the approaches to select the optical polarization.

In comparison with conventional VCSELs, the spin VCSELs have a number of advantages, such as the stability of intensity and polarization of the emitted light, controlling of the polarization, and low threshold. We have already seen that the radiative combination of spin-up electron and spin-up heavy hole will produce LH CP (σ_+) light, while the recombination of the spin-down electron and spin-down heavy hole will emit RH CP (σ_-) light; as such, the spin-polarized electrons will selectively couple with one of the laser modes, emitting either LH or RH CP light. Therefore, the spin polarization of the injected electrons determines the polarization of the emitted light, i.e. the manipulation of the electron spin may control the polarization of the emitted light. With a bias, even incomplete electron spin polarization may lead to completely CP light. Figure 7.24 is a gain

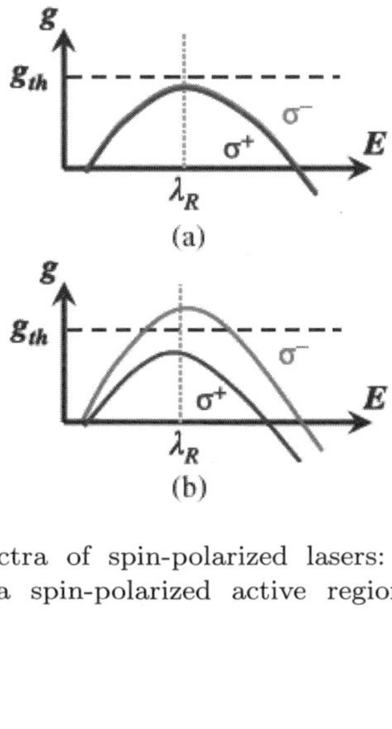

Fig. 7.24. The gain spectra of spin-polarized lasers: (a) nonpolarized injection; (b) carrier occupation in a spin-polarized active region. Courtesy of Holub and Bhattacharya (2007).

spectrum of a polarization-selected laser with a bias lower than the threshold for the injected carriers being polarized or nonpolarized. When the pumping current is not spin-polarized, the gain spectra for the LH and RH CP light are coincident; while if the pumping current is spin-polarized, the gain is inhomogeneous, and the selected CP light mode will reach threshold first. Around the threshold, the gain of that mode will be over the threshold, while the other mode's gain will be reduced. Therefore, a small spin polarization may induce a huge difference in the output power for LH and RH CP light.

In spin VCSELs, the carrier–photon dynamics can be analyzed by spin-related rate equations, in which the spin-up and sign-down electron concentration in the barrier is taken as $n_{\mathrm{b}}^{\pm}$ while that in the active layer as $n^{\pm}$, and $S_{\pm}$ represent the concentration of photons of LH and RH CP. The six parameters are connected by the following rate equations (Rudolph *et al.*, 2005):

$$\frac{\partial n_{\mathrm{b}}^{\pm}}{\partial t} = \frac{n_{\mathrm{b}}^{\pm}}{\tau_{\mathrm{cap}}} + \frac{1 \pm P_{\mathrm{spin}}}{2} \frac{P_{\mathrm{pump}}}{qV_{\mathrm{b}}} \mp \frac{n_{\mathrm{b}}^{+} - n_{\mathrm{b}}^{-}}{\tau_{\mathrm{s,b}}}, \tag{7.50a}$$

$$\frac{\partial n^{\pm}}{\partial t} = \frac{V_{\mathrm{b}}}{V_{\mathrm{MQW}}} \frac{n_{\mathrm{b}}^{\pm}}{\tau_{\mathrm{cap}}} - v_{\mathrm{g}} g(n^{\pm}, S_{\mp}) S_{\mp} - B_{sp} \frac{n^{\pm}(n^{+} + n^{-})}{2}$$
$$\mp \frac{n^{\pm}(n^{+} - n^{-})}{\tau_{\mathrm{s}}} - C \frac{n^{\pm}(n^{+} + n^{-})^{2}}{2}, \tag{7.50b}$$

$$\frac{\partial S_{\mp}}{\partial t} = \Gamma v_{\mathrm{g}} g(n^{\mp}, S_{\pm}) S_{\pm} + \Gamma \beta B_{\mathrm{sp}} \frac{n^{\mp}(n^{+} + n^{-})}{2} - \frac{S_{\pm}}{\tau_{\mathrm{ph}}}, \tag{7.50c}$$

where the carrier capture time τ_{cap} includes two steps: the carrier relaxation in the way of transport from the barrier to the well and that within the well. P_{pump} is the polarization of the pumping current I_{pump}, and if the pumping current is purely spin up (down), $P_{\mathrm{pump}} = 1(-1)$. The spin flipping time in the barrier and well is $\tau_{\mathrm{s,b}}$ and τ_{s} respectively. The volume ratio of the barrier layer and well layer, $V_{\mathrm{b}}/V_{\mathrm{MQW}}$ is also included in Eq. (7.50b), in order to normalize the concentrations $n_{\mathrm{b}}^{\pm}$ and $n^{\pm}$. It is known that the optical gain depends on the density of photons and will be saturated at a high photon density. Such a nonlinear dependence of the optical gain on the photon density is introduced into the rate equations by a phenomenological gain suppression factor ε, in the way that the gain of the laser cavity is written as $g(n, S) = (\mathrm{d}g/\mathrm{d}n)(n - n_{\mathrm{tr}})(1 + \varepsilon S)^{-1}$, where $\mathrm{d}g/\mathrm{d}n$

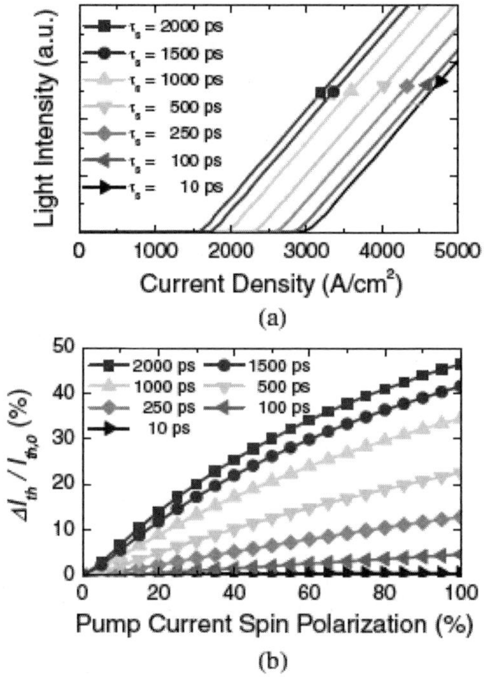

Fig. 7.25. (a) The relationship between the light intensity and pumping current as a function of SRT. (b) The relationship between the threshold current and the spin polarization of the pumping current as a function of electron SRT. The data in the figures are calculated by using the rate equation (7.46), for a GaAs VCSEL working at RT. The device parameters are $dg/dn = 7.5 \times 10^{-16}\,\text{cm}^2$, $\tau_{\text{cap}} = 20\,\text{ps}$, $n_{\text{tr}} = 2 \times 10^{17}\,\text{cm}^{-3}$, $\varepsilon = 1.5 \times 10^{-17}\text{cm}^3$, $\beta = 1.7 \times 10^{-4}$, $\Gamma = 003$, $B_{\text{sp}} = 1.0 \times 10^{-10}\,\text{cm}^3\text{s}^{-1}$, $C = 3.5 \times 10^{-30}\,\text{cm}^6\text{s}^{-1}$, $\tau_{\text{ph}} = 1\,\text{ps}$, and $v_{\text{g}} = 0.86 \times 10^{10}\,\text{cms}^{-1}$. Courtesy of Holub and Bhattacharya (2007).

is the differential gain and n_{tr} is the penetrated carrier concentration. B_{sp} and C are the bimolecular radiative recombination coefficient and Auger recombination coefficient, respectively. Finally, v_{g} is the group velocity of light, β is the spontaneous radiation factor and τ_{ph} is the cavity lifetime of photons. Figure 7.25 shows an example for a GaAs/AlGaAs quantum well VCSEL and calculated using Eq. (7.50), indicating the relationship between light intensity and current ($L - I$: Fig. 7.25(a)) under a pumping of 100% spin polarization and various SRTs (from 10 ps to 2 ns), and the corresponding relationship between the threshold and spin polarization of the pumping current (b). It can be seen that if the SRT is long enough, i.e.

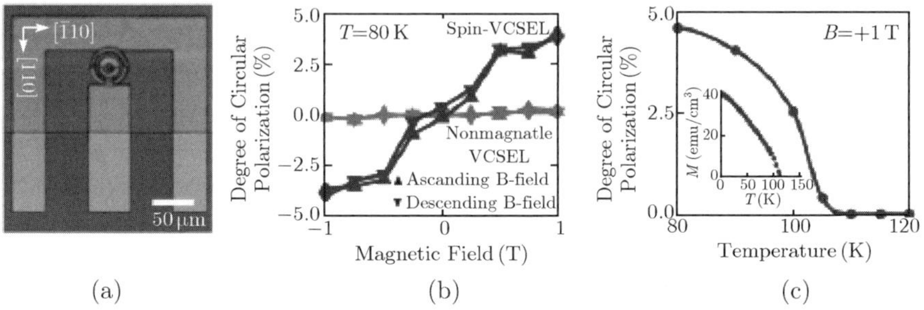

Fig. 7.26. (a) Topology of spin VCSEL. (b) The helicity dependence on external magnetic field for spin and nonmagnetic VCSELs. (c) The helicity dependence on the temperature for spin VCSELs; the inset is the magnetization of the (Ga,Mn)As layer on the temperature, showing that the Curie temperature is 113 K. Courtesy of Holub and Bhattacharya (2007).

>1 ns, then the threshold current can be reduced to 50% at RT with 100% spin-polarized pumping current. This is no doubt a very exciting result.

The electrically injected spin VCSELs were first reported by Holub *et al.* (2005). They use a (Ga,Mn)As layer to achieve spin arrangement, injecting spin-polarized holes into the (In,Ga)As multiquantum well active layer. The Curie temperature of the (Ga,Mn)As layer is 113 K, so that one has to carefully design the low-temperature working condition, including the difference between the resonant wavelength of the gain cavity of the laser and the maximum wavelength of the gain for the material. The topology of the device is shown in Fig. 7.26(a); the growth is carried out by using MBE. Meanwhile, a nonmagnetic device is also grown using a heavily Be doped GaAs layer to replace the (Ga,Mn)As layer for no spin polarization hole injection. Polarization-resolved electrically pumped luminescence measurement result is shown in Fig. 7.26(b), which shows that the spin VCSEL has a spin lagging effect, while there is no lagging for the nonmagnetic device. The best optical polarization Π_{CP} achieved at 80 K and under 1 Tesla magnetic field was 4.6%, and the optical polarization can be kept till 105 K, slightly below the Curie temperature, as shown in Fig. 7.26(c).

7.6.3. *Detecting spin current by electric current*

One of the key problems in spin semiconductor devices is how to detect the spin polarization, after the spin has been injected and manipulated. Using a spin light source is one way to detect the spin polarization, but in that case, the electron spin polarization has to be converted to optical polarization.

How to detect spin polarization by a convenient electric technique is still a challenge. Here we will mainly introduce an approach reported by Cui *et al.* (2007). In that work, the spin injection is achieved optically. It has been proved both theoretically and experimentally that a normally incident linearly polarized light into a 2DEG system will produce spin current (Bhat *et al.*, 2005; Li *et al.*, 2006), i.e. optically excited electrons move toward the opposite direction with reversed spin polarization, which is not photocurrent and cannot be detected electrically. The experimental setup is shown in Fig. 7.27(a). The sample is in the shape of crossed bars carved on a sample of 2DEG which is formed at one interface between the only doped barrier and a 14 nm thick InGaAs/InAlAs quantum well grown on a semi-insulating [001] InP substrate by MBE. The SIA is achieved by δ doping on top of the well in the vertical direction. The spin decoherent time at 77 K measured by time-resolved Kerr effect is $\tau \approx 106$ ps (Fig. 7.23(b)).

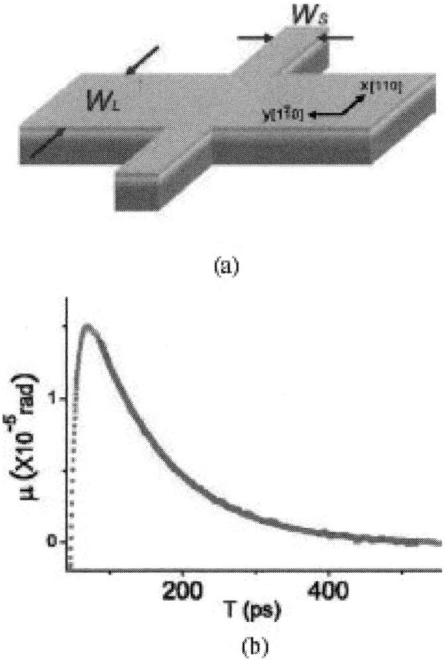

Fig. 7.27. (a) Schematic view of the 2DEG sample and the experimental setup. The 20 and 200 μm wide planar channels are carved by standard photolithography and wet etching. (b) TRKR measurement finds a spin decoherent time of 106 ps. Dark line, experimental data; grey line, fitting with a single exponential function $\exp(-t/\tau_s)$. Courtesy of Cui *et al.* (2007).

As the Fermi velocity is estimated to be $U_F \approx 5.0 \times 10^5$ m/s, the coherent length is therefore about $l_S \approx 53\,\mu$m, between the half-width of 200 and 20 μm for the two perpendicular bars, namely $W_S/2 < l_S < W_L/2$. The interband excitation source is normally incident linearly polarized light focusing on a 10 μm diameter spot. The sample is immersed in a liquid nitrogen optical cryostat. The electric currents passing through the terminals are monitored by voltage drops at a symmetrically loaded resistor of 633 Ω. The voltages are read out with a lock-in amplifier, which is triggered by the PEM which modulated the incident light polarization to be oscillating between two orthogonal directions with a frequency of 50 kHz. Possible nonpolarization-related phenomena, e.g. Demper effect and thermoelectric effects, were thus clearly eliminated.

The experimental results are shown in Fig. 7.28. No measurable current along the wide channels (y-axis, between A and B electrodes) was observed

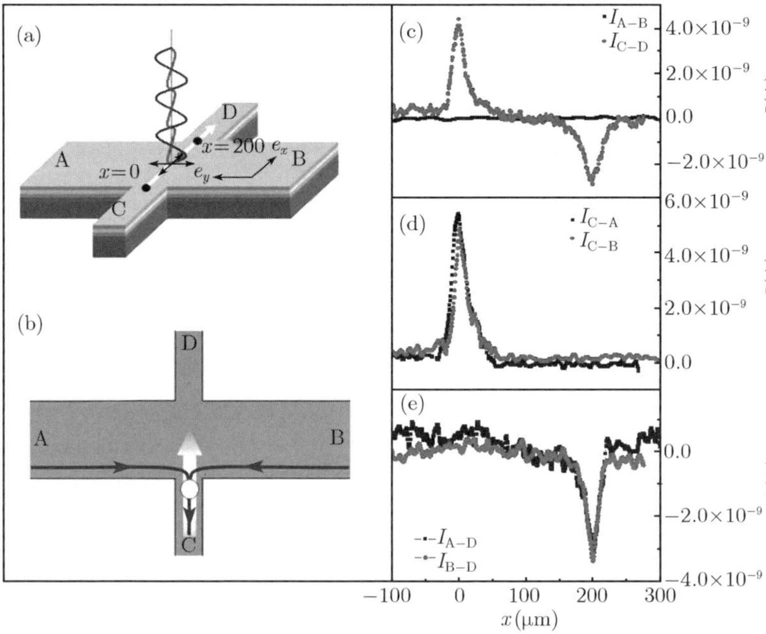

Fig. 7.28. (a) Linearly polarized light beam, polarized along the x-axis as indicated by grey color or along the y-axis as indicated by dark color, scans along the narrow channel (x-axis) at normal incidence. (b) Flow pattern of measured electric currents for the light spot along the narrow channel. (c)–(e) Typical charge currents through all the terminals as a function of the light spot position at x-axis. Courtesy of Cui *et al.* (2007).

on the background noise of tens of picoamperes, under normal incidence, though a significant photocurrent was observed if the CP light is shed at an oblique angle on the same sample. However, when linearly polarized light along the x- or y-axis is normally incident onto the plane near the electric channel junction, as illustrated in Fig. 7.28(a), reproducible currents along the narrow channels (between relevant electrodes, as indicated in the figure) under unbiased conditions can be observed. Figure 7.28(b)–(e) shows the currents passing through the terminals as functions of the spot position when the incident light is along the 20 μm channel (x-axis). The current peaks at $x = 0$ and 200 μm correspond to the positions of the two channel junctions and vanish at a distance of 50 μm away from the peak position. Note that the current curves of $I_{\mathrm{C-A}}$ and $I_{\mathrm{C-B}}$, and of $I_{\mathrm{A-D}}$ and $I_{\mathrm{D-B}}$, almost overlap, indicating that the current flows inward or outward simultaneously through the channel junction. The spatial distribution of the observed current is substantially different from the current due to Hall effect or reciprocal spin Hall effect, whereas the Hall current is unidirectional. The flow pattern of the currents and the absence of a net electric current along the wide channel are the signature of the observed effect, as summarized in Fig. 7.28(b). The vanishing of all currents at the middle point of the C–D channel, $x = 100$ μm, evidently indicates that the phenomena are not due to conventional photocurrents.

To further examine the relationship between the observed currents and polarization of the incident light, the polarization plane of the incident light is rotated while the light spot is fixed at the peak position of $I_{\mathrm{C-A}}$ and $I_{\mathrm{C-B}}$. The current varies with angle ϕ between the light polarization plane and y-axis, and a good fit is obtained with the form of $\cos 2\phi$, as shown in Fig. 7.29. Obviously, the polarization dependence of the current is in excellent agreement with the calculated angle dependence of the spin current (Li *et al.*, 2006; Zhou and Shen, 2007). The result of the laser scanning along the edge of channel A–B at $x = 10$ μm is shown in Fig. 7.30, which shows the importance of the edge effect and the excitation position. The results of $I_{\mathrm{A-B}}$, $I_{\mathrm{A-D}}$, and $I_{\mathrm{B-D}}$ are not shown, as they are all negligible. All of the results clearly show that when the laser spot is beyond a distance of the coherent length from the cross points ($y = 0$, $x = 0$, and $y = 0$, $x = 200$ μm), there will be no electric current signal. One can conclude that the observed current is definitely not the general spin current (as introduced in Sec. 7.3), and that $I_{\mathrm{A-C}} = I_{\mathrm{B-C}}$ certainly excludes the possibility of a diffusion mechanism for generating the photocurrent.

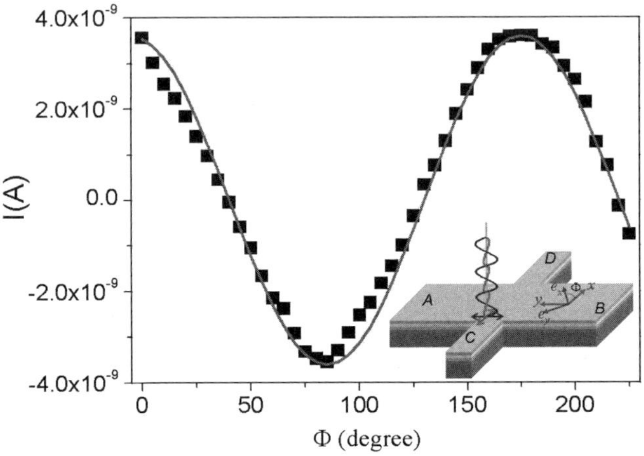

Fig. 7.29. The charge current (back square) at $x = 0$ varies with the polarization angle on ϕ of light relative to the lab. Frame, as shown in the inset, i.e. the $x-z$ plane corresponds to $\phi = 0$. The grey curve is a fit of cos 2ϕ. Courtesy of Cui *et al.* (2007).

Cui *et al.* (2007) interpreted the observed electric current as a result of scattering of the spin current near the junction of the cross-bars. First, the optical excitation serves as the source of a spin current with its spin polarization lying in the plane and perpendicular to the direction of the electron motion. The effective distance of this spin current is estimated to be within l_S, which has the relation of $W_S/2 < l_S < W_I/2$. Keeping this picture in mind, the observed effect can then be well understood as follows. In the case that the light spot scans along the central line of the narrow channel C–D in Fig. 7.30, the geometric boundaries of the sample set a limit to the motion of the excited electrons and lead them to two opposite directions with opposite in-plane spin polarizations. Two beams of the excited electrons compose the same spin current, which is then scattered by the edge of the crossing region to produce a transverse electric current. On the other hand, when the light spot scans along the central line of the wide channel A–B (in Fig. 7.30), the spin current decays to zero before reaching the edges of the crossing region. Therefore, no electric current circulates. While in the case that the light spot moves along the edge, also shown in Fig. 7.30, when the distance between the spot and edge is within l_S, the edge may again impose a constraint to the movement of the electrons and lead to an outward (inward) electric current perpendicular to the edge, and thus a measurable current between C–A/C–B or D–A/D–B. Also as the spin coherent length at liquid nitrogen temperature is longer

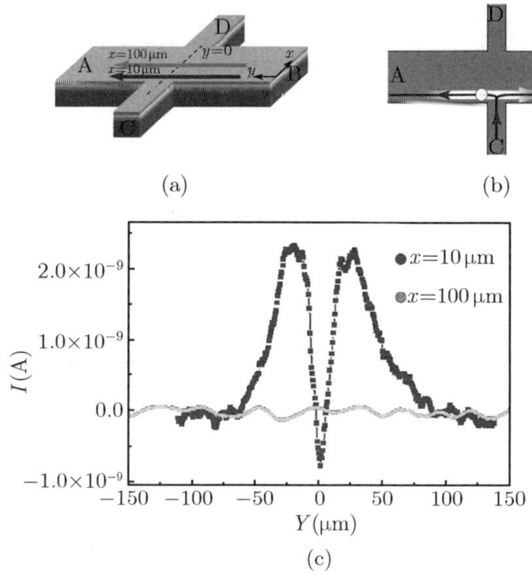

Fig. 7.30. (a) Two scanning routes at $x = 10\,\mu\text{m}$ (dark line) and $x = 100\,\mu\text{m}$ (grey line). (b) Schematic of the directions of the current flow. (c) The charge current $I_{\text{A-C}}$ ($I_{\text{B-C}}$ is the same) at $x = 10\,\mu\text{m}$ (in dark) and $x = 100\,\mu\text{m}$ (in grey). Courtesy of Cui *et al.* (2007).

than the width $W_s = 20\,\mu\text{m}$ of the narrow channel, the impurity effect or other interactions would not be crucial in this experiment.

The conversion of spin current to electric current has been discussed extensively in the context of the reciprocal spin Hall effect where the separated spins are out of plane (Bhat *et al.*, 2005). In the present case, the spin current is polarized within the plane, as the linearly polarized light does not generate out-of-plane spin polarization due to conservation of the total angular momentum. As checked by Li *et al.* (2006), an electron of spin $-y$ polarization will be equally scattered along x and $-x$ axes, and hence the resulting scattered electric current has such a pattern: either both being outward or both inward, in the transverse directions of the spin current. The experimental observation of the electric current in Cui *et al.* (2007) is in excellent qualitative agreement with this picture. For the case of in-plane polarized spin current as in the present experiment, the conversion rate for the spin current to electric current is estimated to be $0.3 \sim 1\%$. Although there are some unknown parameters of the sample, the amplitude of the injected spin current J_{xy} by the linearly polarized light can be roughly estimated by using the measurable photocurrent J_x injected by the CP light at a small oblique angle Θ away from normal. A theoretical estimation gives

$$\frac{J_x(\Theta)}{J_{xy}} \approx 0.17\Theta\left(\frac{2e}{\hbar}\right) \tag{7.51}$$

by solving the semiconductor Bloch optical equation near the Γ point of the InGaAs 2DEG (Zhou and Shen, 2007). Taking the light spot size, geometry of the setup, and laser power into account, it is estimated from the experimental data that $J_x(\Theta) \approx 94\Theta\,\text{nA}$ (Yang *et al.*, 2006); hence, the injected spin current is estimated to be approximately $550\,\text{nA}(\hbar/2e)$. The observed electric current is several nanoamperes, say 4 nA for the peaks in Fig. 7.24(c). The conversion rate from the spin current to charge current is roughly about 0.8%, close to the numerically simulated values.

References

Averiev NS, Golub LE. (1999) *Phys Rev B* **60**: 15583.

Balun AA, Zakharchenya BP, Roganchev BP, Tkachunk MN, Fleisher VG. (1984) *Sov Phys — JETP* **40**: 1293.

Bhat RDR, Nastos F, Najmaie Ali, Sipe J. (2005) *Phys Rev Lett* **94**: 096603.

Cartoixa X, Ting DZ-Y, Daniel ES, McGill TC. (2001) *Supperlattices Microstruct* **30**: 309.

Cui X-D, Shen S-Q, Li J, Yang C, Ge W, Zhang F-C. (2007) *Appl Phys Lett* **90**: 242115.

Dai JF, Lu HZ, Yang CL, Shen SQ, Zhang FC, Cui XD. (2010) *Phys Rev Lett* **104**: 246601.

Datta S, Das B. (1990) *Appl Phys Lett* **56**: 665.

Dyakonov MI. (2007) *Phys Rev Lett* **99**: 126601.

D'yakonov MI, Petel' VI. (1971) *Sov Phys — JETP* **13**: 144.

Edelstein VM. (1990) *Solid State Commun* **73**: 233.

Fabian Zutic J, Das-Sarma S. (2004) *Rev Mod Phys* **76**: 323, 2004.

Ganichev SD, Prettl W. (2002) *Nature (London)* **417**: 153.

Ganichev SD, Prettl W. (2003) *J Phys: Condens Matter* **15**: R935.

Ganichev SD, Bel'kov VV, Golub LE, Ivchenko EL, Schneider Petra, Giglberger S, Eromos J, De Boeck J, Borghs G, Wegschneider W, Weiss D, Prettl W. (2004) *Phys Rev Lett* **92**: 256601.

Golub LE. (2003) *Phys Rev B* **67**: 235320.

Hall KC, Wayne HL, Gündoğdu K, Flatté ME. (2003) *Appl Phys Lett* **83**: 2937.

Hankiewicz EM, Li J, Jungwirth Y, Niu Q, Shen SQ, Sinova J. (2005) *Phys Rev B* **72**: 155305.

He XW, Shen B, Chen YH, Zhang Q, Han K, Yin CM, Tang N, Xu FJ, Tang CG, Yang ZJ, Qin ZX, Zhang GY, Wang ZG. (2008) *Phys Rev Lett* **101**: 147402.

Holub M, Bhattacharya P. (2007) *J Phys D: Appl Phys* **40**: R179–R203.

Holub M, Shin J, Chakrabarti S, Bhattacharya P. (2005) *Appl Phys Lett* **87**: 091108.

Inoue JI, Bauer GEW, Molenkamp LW. (2003) *Phys Rev B* **67**: 033104.

Knap W, Skierbiszewski C, Zduniak A, Litwin-Staszewska E, Bertho D, Kobbi F, Robert JL, Pikus GE, Pikus FG, Iordanskii SV, Mosser V, Zekentes K, Lyanda-Geller YuB. (1996) *Phys Rev B* **53**: 3912.

Li J, Dai X, Shen SQ, Zhang FC. (2006) *Appl Phys Lett* **88**: 162105.

Lommer G, Malcher F, Rossler U. (1988) *Phys Rev Lett* **60**: 728.

Rashba EI. (2000) *Phys Rev B* 62: R16267.

Rudolph J, Dohrmann S, Hägele D, Oestreich M, Stolz W. (2005) *Appl Phys Lett* **87**: 241117.

Sakurai JJ. (1994) *Modern Quantum Mechanics.* Addison Wesley Longman.

Schliemann J, Egues JC, Daniel L. (2003) *Phys Rev Lett* **90**: 146801.

Silov A Yu, Blajnov PA, Wolter JH, Hey R, Ploog KH, Averkiev NS. (2004) *Appl Phys Lett* **85**: 5929.

Stein D, von Klitzing K, Weimann G. (1983) *Phys Rev Lett* **51**: 130.

Stormer HL, Schlesinger Z, Chang A, Tsui DC, Gossard AC, Weigmann W. (1983) *Phys Rev Lett* **51**: 126.

Tarasenko SA, Arekiev NS. (2002) *JETP Lett* **75**: 552.

Van Dorpe P, Motsnyi VF, Nijboer M, Gooverts E, Safarov VI, Das J, Van Roy W, Borghs G, De Boeck J. (2003) *J Appl Phys* **42**: L502.

Winkler R. (2000) *Phys Rev B* **62**: 4245.

Yang CL, He HT, Ding Lu, Cui LJ, Zeng YP, Wang JN, Ge WK. (2006) *Phys Rev Lett* **96**: 186605.

Yang CL, Ge WK. (2006) Private communications.

Zhou B, Shen SQ. (2007) *Phys Rev B* **75**: 045339.

Chapter 8

MANIPULATION OF SPIN COHERENT ELECTRONS

8.1. Experimental Methods

It is found experimentally that the spin coherent times T_2 in n-type semiconductors are very long. In n-ZnSe, the T_2 at RT reaches several nanoseconds (Kikkawa *et al.*, 1997), whereas in n-GaAs the T_2 at low temperatures can reach several hundred nanoseconds. This long spin coherent time can be used to derive a spin coherent polarization current through a macroscopic distance in a uniform semiconductor crystal (Kikkawa and Awschalom, 1999). Especially, it was found that the long spin coherent time is insensitive to the defect density in the sample (Beschoten *et al.*, 2001), which expand greatly the material range to be used for spin coherence.

Physically, the coherence and decoherence processes of electron spin deals with an electron spin assemble composed of electrons ($\sim 10^{15}\,\mathrm{cm}^{-3}$) excited by a femtosecond laser pulse. At the initial stage, all light-excited spins are added to form a pure magnetization, which rotates with the Larmor frequency. If all electrons have no interaction with each other, the decay of the magnetization reflects the intrinsic spin decoherence of a single electron. Generally, the decay is exponential with time, characteristic by a transverse SRT T_2. Except the uniform decoherence, there is also the nonuniform effect, for example, the variation of the g-factor, or local magnetic field, which leads to the extra decay (or dephasing) of spin polarization. Even if all electrons move coherently, the nonuniform effect results in the broadening of the orientation distribution of the electron spins. Thus, we call the measured experimentally decay time as T_2^*, which is only the lower limit of T_2, unless the nonuniform effect is eliminated. In the case of existence of an electron–electron spin interaction, there occurs a complicated process. Because the spin–spin interaction between the electrons $\mathbf{s}_i \cdot \mathbf{s}_j$ commutes with the total electron spin and its movement equation, it can induce a hidden decoherence of a single electron without change in the pure magnetization, which cannot be detected in the measurement of the Larmor precession. Therefore, the measurement of the Larmor precession only reveals the loss rate of electron coherence to the outside system, which is called the decoherence of electron spin by environment.

The main experimental methods to measure the electron spin decoherence are shown in Fig. 8.1. Figure 8.1(a) is the Hanle measurement, whose theorem is the Hanle effect, explained by Eq. (5.74) and its discussion. When a CP light beam is illuminated perpendicularly on the sample and excites electrons with spins in the definite direction. The applied magnetic field in the sample plane (perpendicular to the light beam) makes the electron spins to precess around the magnetic field direction, resulting in a decrease in the circular polarization of penetrated light with increasing magnetic field. The circular polarization $P(B)$ as a function of the magnetic

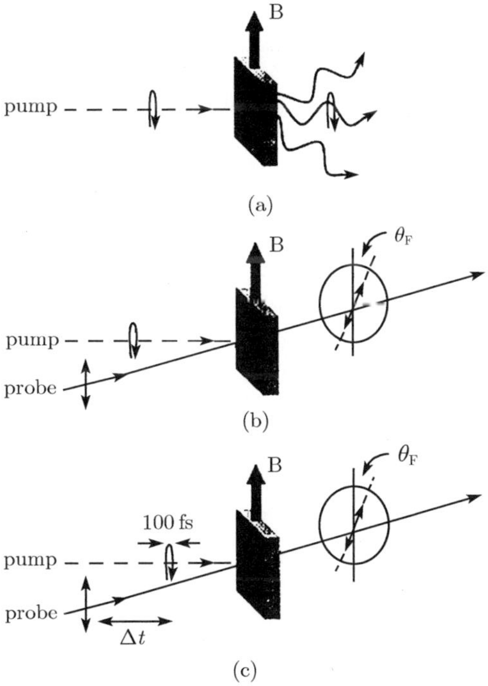

Fig. 8.1. Schematic figures of (a) Hanle measurement, (b) static-state Faraday rotation measurement, and (c) TRFR measurement.

field is Lorentz type,

$$P(B) = \frac{P_0}{\tau} \int_0^{\infty} \mathrm{d}t \left[\exp\left(-\frac{t}{T_s} \right) \cos(\Omega_L t) \right] = \frac{P(0)}{1 + \Omega_{\mathrm{L}}^2 T_{\mathrm{s}}^2}, \tag{8.1}$$

where Ω_{L} is the Larmor precession frequency and T_{s} is the spin lifetime. The spin lifetime can be obtained from the $P(B)$–B curve.

Figure 8.1(b) shows the Faraday rotation measurement. The pure spin polarization induces different refractive indices for left and right polarized lights. When a linearly polarized light (probe beam) penetrates the sample, the polar angles of two CP lights shift relatively, resulting in the rotation of linearly polarized angle. The rotation angle θ_{F} is proportional to the component of the spin polarization angle on the light transmission direction. The same effect is also valid for the reflected light, called as the Kerr effect.

Figure 8.1(c) shows the TRFR measurement. A mode-locked Ti:sapphire laser provides 100 fs pump and probe pulses whose relative delay is adjusted by a conventional delay line, and the repetition frequency is 76 GHz (Kikkawa and Awschalom, 1998). The pump beam is used to excite electrons from the valence band to CB, and produce the spin polarized electrons. The probe beam is linearly polarized, delays by Δt after the pump beam to reach the sample.

At the time that the probe beam reaches the sample, the Faraday (or Kerr) rotation angle, i.e. the instantaneous measurement of the spin polarization is given by

$$\theta_{\mathrm{F}} = A \exp\left(-\frac{\Delta t}{T_2^*} \right) \cos\left(\frac{g\mu_{\mathrm{B}} B \Delta t}{\hbar} \right). \tag{8.2}$$

If the total spin is the coherent addition of the spin eigenstates, its projection on the probe beam direction is a cosine function of time due to the Larmor precession. Figure 8.2 shows the TRFR for n-doped GaN epilayers at $T = 5$ and 300 K (Beschoten *et al.*, 2001). Using Eq. (8.2) to fit the experimental curves, one can obtain the spin lifetime T_2^* and the electron g-factor. Fast Fourier transforms (FFTs) of the TRFR spectra in Fig. 8.2(a) give $g = 1.94$, independent of doping, which is consistent with values from electron spin resonance (ESR) and reflects the small SOC in GaN. The spin coherence persists till RT (Fig. 8.2(b)) and yields a spin lifetime of ~35 ps.

If the spin decoherence time exceeds several nanoseconds, it needs much long pump-probe delay time Δt in the TRFR measurement. It is difficult in technique to increase Δt by elongating the light path. It is found that when the spin precession is commensurate with the pulse repetition,

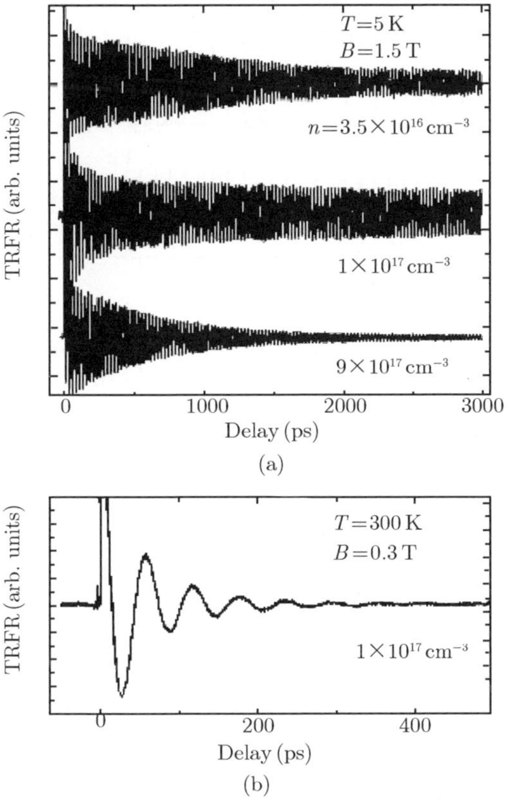

Fig. 8.2. (a) TRFR for n-doped GaN epilayers at $T = 5\,\mathrm{K}$ and $B = 1.5\,\mathrm{T}$ for various electron densities. (b) TRFR of $n = 1 \times 10^{17}\,\mathrm{cm}^{-3}$ sample at $T = 300\,\mathrm{K}$ and $B = 0.3\,\mathrm{T}$.

the spin polarization and the TRFR signal increase. Using this "resonant spin amplification" (RSA) Kikkawa and Awschalom (1998) obtained precise measurements of spin lifetimes well in excess of t_{rep}. When $g\mu_{\mathrm{B}}Bt_{\mathrm{rep}}/\hbar$ is integer times of 2π, the measured spin polarization will be strengthened resonantly, otherwise it will be suppressed. Thus, for a fixed delay time Δt, one measures the Faraday rotation angle θ_{F} as the magnetic field changes, one finds that the θ_{F} is a periodic function of magnetic field. Figure 8.3 shows the RSA spectra for $n = 10^{16}\,\mathrm{cm}^{-3}$ GaAs sample at small magnetic field with $\Delta t = 10\,\mathrm{ps}$ (a) and $1\,\mathrm{ns}$ (b) (Kikkawa and Awschalom, 1998). The peaks correspond to $B = 0$ and $g\mu_{\mathrm{B}}Bt_{\mathrm{rep}}/\hbar = \pm n\pi$, respectively. The electron g-factor can be determined by the spacing distance of resonance peaks and the spin lifetime T_2^* from the width of the peak. $T_2^* \sim 130\,\mathrm{ns}$ is

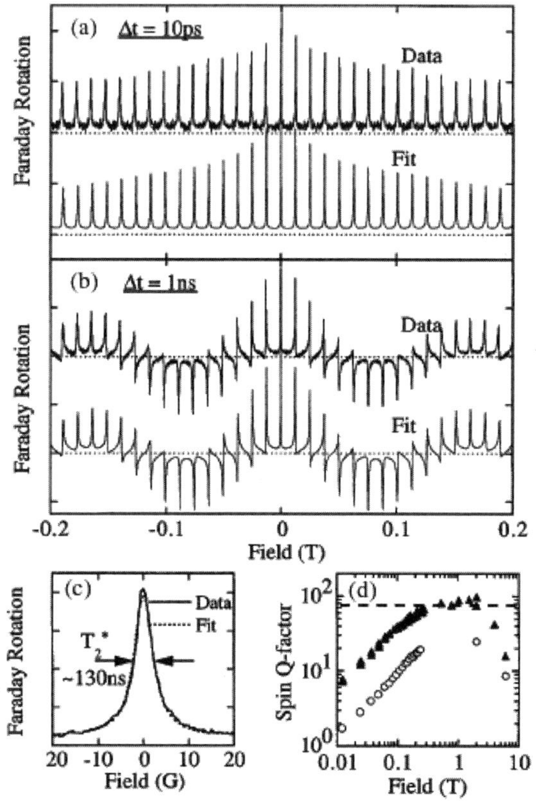

Fig. 8.3. RSA spectra for $n = 10^{16}\,\text{cm}^{-3}$ GaAs sample at small magnetic field with $\Delta t = 10\,\text{ps}$ (a) and $1\,\text{ns}$ (b). The upper and lower curves are experimental and fitting results. (c) An expanded peak of the central resonance in (a). The fit given by the dotted line corresponds to a spin lifetime of 130 ns. (d) Spin Q factor versus field on a log–log plot at $T = 5\,\text{K}$, $N_{\text{ex}} = 2\times10^{14}\,\text{cm}^{-3}$ (solid triangles) and $3\times10^{15}\,\text{cm}^{-3}$ (open circles).

given in Fig. 8.3(c). The shape of resonance peaks can be fitted by addition of the exponential delays of spin polarization,

$$M_{\text{s}}(\Delta t, B) = \sum_n \Theta(\Delta t + nt_{\text{rep}}) A \exp\left(-\frac{\Delta t + nt_{\text{rep}}}{T_2^*}\right) \times \cos\left[\frac{g\mu_{\text{B}} B(\Delta t + nt_{\text{rep}})}{\hbar}\right]. \tag{8.3}$$

From Figs. 8.3(a) and 8.3(b), it is seen that the fitting curves are in good agreement with the experimental curves.

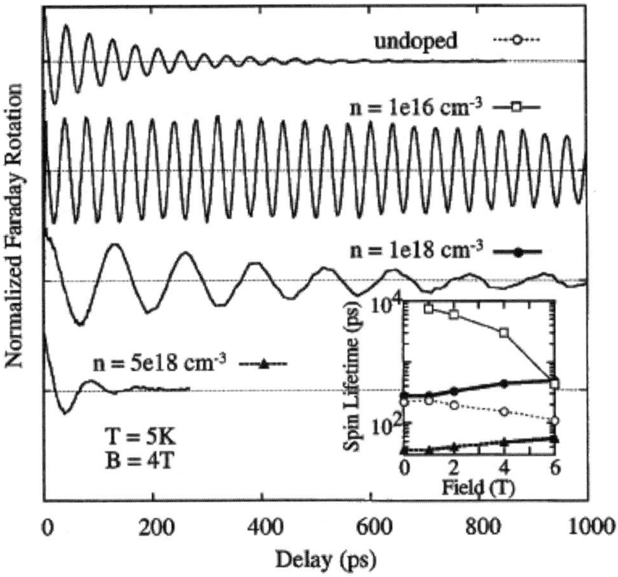

Fig. 8.4. TRFR for undoped and n-type GaAs at $T = 5\,\text{K}$ and $B = 4\,\text{T}$. The inset shows T_2^* versus magnetic field.

8.2. Electron Spin Coherence in Semiconductor Bulk Materials

The samples are Si-doped GaAs, the electron densities at RT are $n = 10^{16}$, 10^{18}, and $5 \times 10^{18}\,\text{cm}^{-3}$, the mobilities are $\mu = 5400$, 2300, and $1340\,\text{cm}^2/\text{V s}$, respectively, and one semi-isolated GaAs $n \approx 0$. Figure 8.4 shows the TRFR for undoped and n-type GaAs at $T = 5\,\text{K}$ and $B = 4\,\text{T}$ (Kikkawa and Awschalom, 1998). A change in the Larmor frequency with doping reflects an energy dispersion in the GaAs CB g-factor. The free carrier spin lifetime T_2^* is a nonmonotonic function of carrier density, abruptly increasing between the $n \approx 0$ and $n = 10^{16}\,\text{cm}^{-3}$ samples and steadily decreasing thereafter. The inset shows T_2^* versus field, spanning over two orders of magnitude and ranging from nearly 10 ns for $n = 10^{16}\,\text{cm}^{-3}$ to less than 40 ps for $n = 5 \times 10^{18}\,\text{cm}^{-3}$. Qualitatively, T_2^* increases with the field for the two highest doses and decreases for the two lowest doses.

For $n = 10^{16}\,\text{cm}^{-3}$ GaAs sample, T_2^* is longer than the pulse repetition time t_{rep}; thus, it needs the RSA technique to precisely measure T_2^*, and the results are shown in Fig. 8.5 (Kikkawa and Awschalom, 1998). Figure 8.5(b)

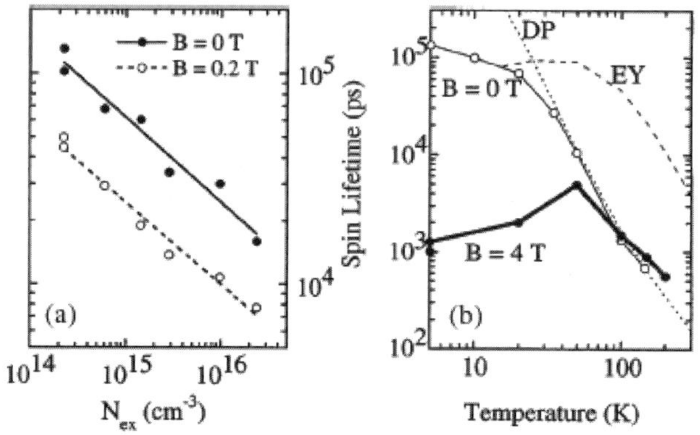

Fig. 8.5. (a) T_2^* versus N_{ex} for n-GaAs samples at $T = 5\,K$, $B = 0$, and $0.2\,T$. Solid lines are power-law fits to $T_2^* \sim N_{ex}^{-0.4}$. (b) T_2^* versus T for $n = 10^{16}\,cm^{-3}$, $N_{ex} = 2 \times 10^{14}\,cm^{-3}$ sample at $B = 0$ and $4\,T$. Dotted and dashed lines are DP and EY predicted results, respectively.

shows the T_2^* versus T for $n = 10^{16}\,cm^{-3}$, $N_{ex} = 2 \times 10^{14}\,cm^{-3}$ sample at $B = 0$ and $4\,T$. Dotted and dashed lines are DP and EY predicted results, respectively. Below about 50 K, the spin lifetime becomes strongly field-dependent, splitting into high- and low-field regimes. The low-field behavior shows a good agreement with the DP mechanism above $T = 30\,K$, below which a weaker temperature dependence $T_2^* \sim (kT)^{-1/2}$ is suggestive of electron–electron scattering. Assuming the classical statistics, the theory gives $T_{e-e} \sim N^{-1}(k_B T)^{-1/2}$, where N is the total electron density $N = n + N_{ex}$. Only when $N_{ex} \sim n$, the electron–electron interaction starts to act on.

GaN has wide band gap (~3.54 eV), high defect density (~$10^8\,cm^{-3}$); on the other hand, it has a sharp absorption edge and strong luminescence. The TRFR measurement found that the spin lifetime of GaN is very long and persists till RT (Fig. 8.2). From Fig. 8.2(a), it is seen that $n = 10^{17}\,cm^{-3}$ sample has the longest lifetime. Figure 8.6 shows the spin lifetime T_2^* versus temperature and magnetic field for $n = 3.5 \times 10^{16}\,cm^{-3}$ sample (Beschoten *et al.*, 2001). The temperature dependences of T_2^* are similar to that of GaAs (Fig. 8.5). The magnetic field dependence of T_2^* (Fig. 8.6(c)) can be fitted as $T_2^* \propto B^{-\alpha}$, where $\alpha = 0.52$, 0.93, and 0.13. The existence of long lifetime in GaN, and surprising similarity with GaAs, indicates that these two materials have same spin scattering mechanisms.

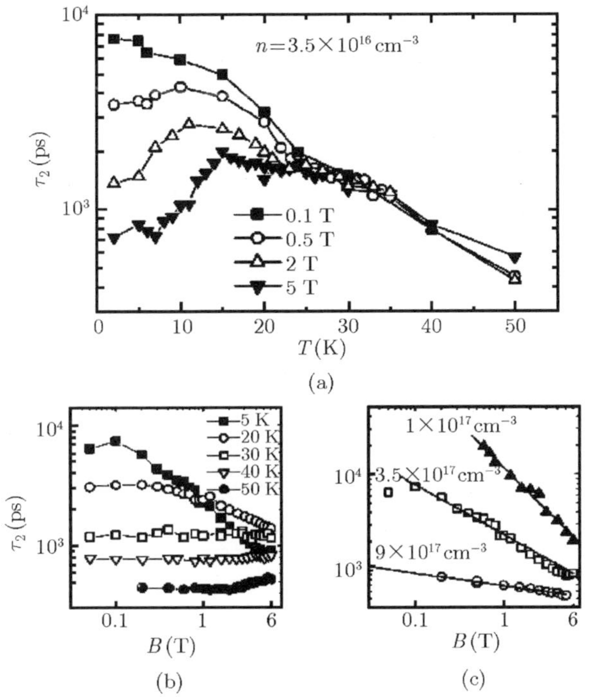

Fig. 8.6. (a) Spin lifetime T_2^* versus temperature at various field for $n = 3.5 \times 10^{16}\,\text{cm}^{-3}$ sample. (b) T_2^* versus magnetic field at various temperatures for the same sample. (c) T_2^* versus magnetic field on a log–log plot at $T = 5\,\text{K}$ for three samples. Solid lines are power-law fits to $T_2^* \propto B^{-\alpha}$, yielding $\alpha = 0.52$ (open squares), 0.93 (filled triangles), and 0.13 (open circles).

8.3. Electron Spin Coherence in Semiconductor QDs

In order to study the spin relaxation mechanisms in semiconductors, Gupta *et al.* (1999) and Gupta and Awschalom (2001) used TRFR to probe spin dynamics in chemically synthesized CdSe QDs. Figure 8.7 shows the TRFR of core and core/shell (~3 ML CdS) QDs at $B = 0$ and $0.25\,\text{T}$, and $T = 6$ and $282\,\text{K}$ (Gupta *et al.*, 1999). In Fig. 8.7(a), the TRFR can be fitted with a biexponential decay, yielding lifetimes of order 100 ps and 3 ns. The decay lifetimes are generally ~20% shorter for the core/shell samples than the same core samples with the same diameter. In Fig. 8.7(b), the TRFR of the two samples at $B = 0.25\,\text{T}$ is qualitatively similar, although small differences in oscillation frequency and decay exist. Figure 8.7(c) demonstrates the persistence of spin precession at $T = 282\,\text{K}$ for the 5.7 nm core

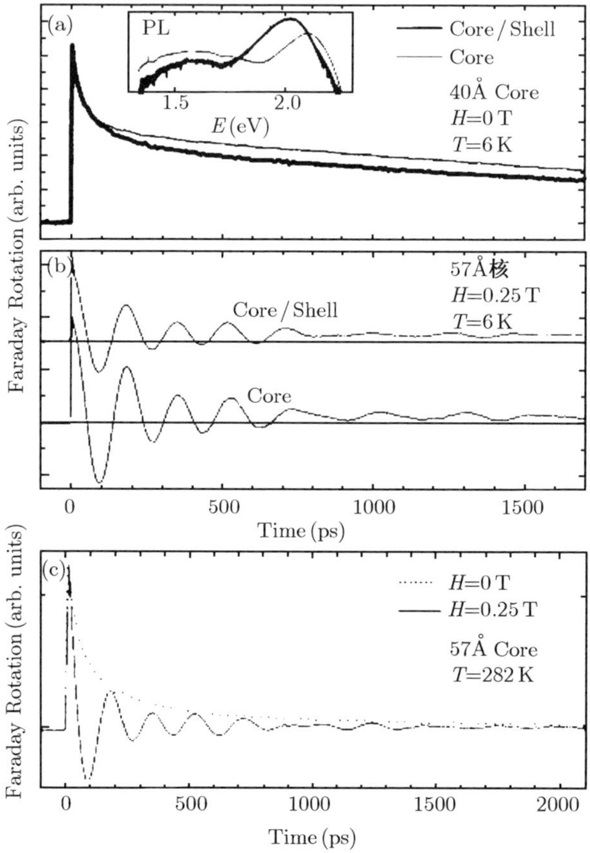

Fig. 8.7. (a) TRFR of core and core/shell (~3 ML CdS) QDs with 4 nm diameter cores at $B = 0\,\mathrm{T}$ and $T = 6\,\mathrm{K}$. (b) Spin precession at $B = 0.25\,\mathrm{T}$ and $T = 6\,\mathrm{K}$ in core and core/shell (~3 ML CdS) QDs with 5.7 nm diameter cores. (c) TRFR at $B = 0\,\mathrm{T}$, 0.25 T and $T = 282\,\mathrm{K}$ of the 5.7 nm QDs.

QDs. The zero-field decay yields spin relaxation lifetime of 50 ps and 0.4 ns. Measurements from $T = 6$–$280\,\mathrm{K}$ indicate little change in signal magnitude or decay up to 100 K, and no change in the oscillation frequency over the temperature range studied. The long spin lifetime at RT can be attributed to the quantum confined levels in QDs which decreases the spin relaxation probabilities and increases the stability of spin coherence.

The experiments found that the spin dynamics in QDs is insensitive to the energy relaxation, temperature, excitation intensity, etc. The spin coherence is observed not only in the high-quality sample,

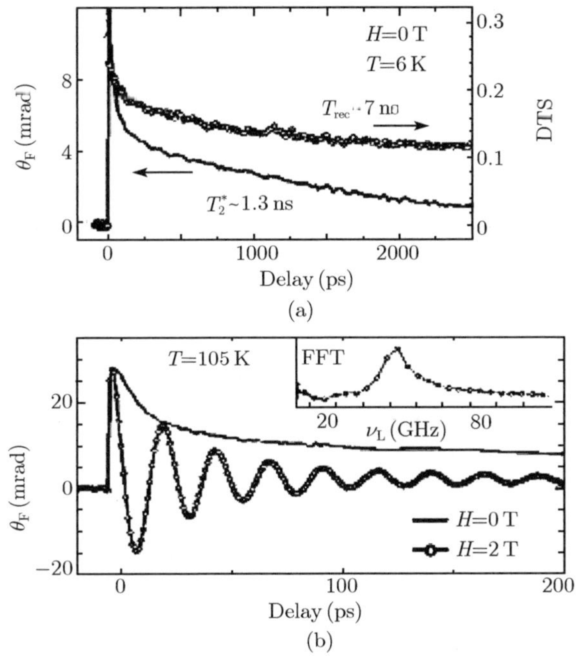

Fig. 8.8. (a) TRFR and DTS of CdS_xSe_{1-x} ($x \sim 0.5$) QDs at $B = 0\,T$ and $T = 6\,K$. (b) TRFR at $B = 0\,T$, $2\,T$ and $T = 105\,K$. The inset shows the FFT of 2 T data.

but also in samples with maximum size and composition distributions. Figure 8.8 shows the TRFR in CdS_xSe_{1-x} QDs grown in glass (Gupta and Awschalom, 2001). This material is one of a series of commercially available semiconductor-doped glasses (Corning 2-73) consisting of ~6 nm diameter CdS_xSe_{1-x} ($x \sim 0.5$) QDs, and the size distribution is ~25%. The differential transmission spectra (DTS) is also a method to measure the spin polarization, defined as $DTS(E_d) = [T(E_d) - T_0(E_d)]/T_0(E_d)$, where $T(E_d)$ and $T_0(E_d)$ are the probe transmissions detected at an energy E_d with and without the pump, respectively. In Fig. 8.8(a), the decay of DTS at fixed E_d generally reflects both energy relaxation and recombination processes. Energy relaxation in this sample is relatively fast (<1 ps), indicating that slower dynamics reflects a recombination lifetime t_{rec}. Both decays are typically biexponential, with components <100 ps and >1 ns. At $T = 6\,K$, the zero-field spin lifetime T_2^* reaches 1.3 ns, and at $T = 105\,K$, $B = 0$ and $2\,T$, the T_2^* decreases to 1 ns and 60 ps, respectively. This is similar to previous results in CdSe QDs (Fig. 8.7), where the field-dependent spin lifetime were attributed to inhomogeneous dephasing effects arising from a distribution of g-factors in the sample. The inset of Fig. 8.8(b) shows a FFT of the 2 T data that indicate a Larmor frequency of 42.7 GHz, corresponding to a spin g-factor of 1.52. This value is close to one of the g-factors observed in the CdSe QDs. They also observed the spin coherence in the samples at $T = 290\,\text{K}$, where the zero-field lifetime is ~360 ps, representing a 60% decrease versus the value at $T = 6\,\text{K}$. There is no change in the g-factor is observed over the entire temperature range studied ($T = 6$–$290\,\text{K}$).

The spin dynamics in CdS/CdSe/CdS QD quantum well (QDQW) is studied using TRFR (Berezovsky *et al.*, 2005). Figure 8.9(a) shows the

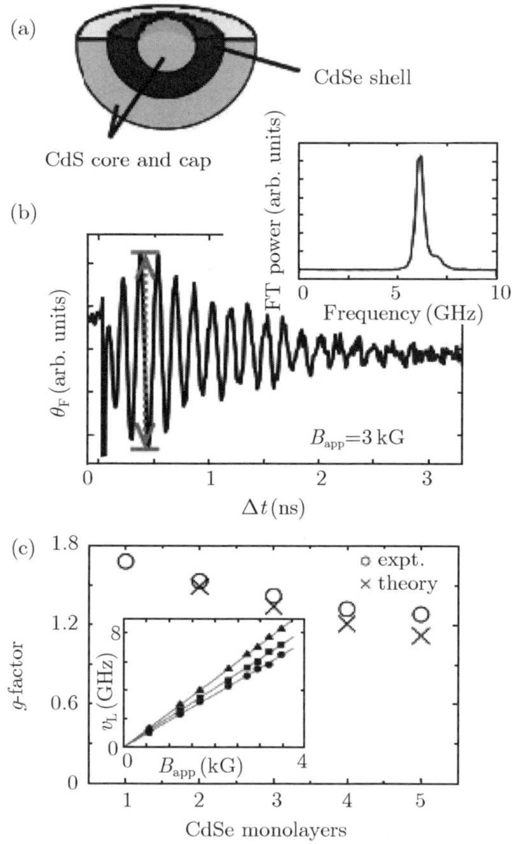

Fig. 8.9. (a) Schematic figure of QDQW. (b) TRFR from a QDQW with $n_{\text{CdSe}} = 3$ at $T = 294\,\text{K}$ and $B = 0.3\,\text{T}$. The inset shows the FT spectrum of the data. (c) g-factor as a function of CdSe quantum well width n_{CdSe}. The inset shows the precession frequencies versus magnetic field B for $n_{\text{CdSe}} = 1$ (triangles), 3 (squares), 5 (circles) samples.

schematic figure of QDQW, where the central core and the outside cap are CdS, the layer between them is the CdSe layer, whose width can be changed $n_{\mathrm{CdSe}} = 1$–5 monolayers, and the width of each monolayer is 0.43 nm. Figure 8.9(b) shows the typical TRFR data from a QDQW with $n_{\mathrm{CdSe}} = 3$ at $T = 294\,\mathrm{K}$ and $B = 0.3\,\mathrm{T}$. The inset shows the Fourier transform (FT) of the data. A second precession frequency was observed, as indicated both by the small shoulder in the FT spectrum and the beating in the TRFR data. The similar behavior has been observed in CdSe nanocrystals (Gupta *et al.*, 2002). The spin lifetime T_2^* is of order 2 or 3 ns for all samples measured. The spin lifetime is essentially temperature-independent between RT and 5 K.

The inset of Fig. 8.9(c) shows the main precession frequency versus magnetic field B for $n_{\mathrm{CdSe}} = 1$ (triangles), 3 (squares), 5 (circles) samples. They increase linearly with B, but with different slopes, indicating that the g-factors in three samples are different. The measured g-factor for each sample as a function of n_{CdSe} (circles), in comparison with the theoretical values (crosses) are shown in Fig. 8.9(c). The g-factor decreases with increasing quantum well width n_{CdSe}, it is because the width of the quantum well CdSe layer is smaller, the wave function in the well diffuses into neighbor CdS barrier layers, and the measured g-factor is a weighted average of the CdSe and CdS g-factors.

The experiments found that when the probe wavelength is near the absorption edge of QDQW, the amplitude of TRFR oscillates with the probe wavelength. Figure 8.10 shows the amplitude of the Faraday rotation angle, θ_{F}, as a function of probe wavelength for $n_{\mathrm{CdSe}} = 3, 4, 5$ samples (Berezovsky *et al.*, 2005). While the absorption signal only shows a featureless staircase-like behavior with no distinct resonances, the amplitude of the TRFR signal exhibits several pronounced resonances close to the absorption edge. The TRFR amplitudes provide information not only on the spin dynamics, but also on the individual exciton transitions in QDQWs. The CB and valence-band levels of spherical QDQWs are calculated with $k \cdot p$ theory. The energies of the lowest CB states $1\mathrm{S_e}$, $1\mathrm{P_e}$, $2\mathrm{S_e}$ and valence-band states $1\mathrm{P}_{3/2}$, $1\mathrm{S}_{3/2}$, $1\mathrm{D}_{5/2}, \ldots$ are obtained, which decrease with increasing n_{CdSe}. The electron g-factor can be calculated from the electron ground wave functions,

$$g = g_{\mathrm{CdSe}} \int_{r_1}^{r_2} \mathrm{d}r |\psi_{1S_e}(r)|^2 + g_{\mathrm{CdS}} \left(\int_0^{r_1} + \int_{r_2}^{r_3} \mathrm{d}r |\psi_{1S_e}(r)|^2 \right), \tag{8.4}$$

where r_1 and r_2 are the inner and outer radii of the CdSe quantum well and r_3 is the radius of the QD. The calculation results are shown in Fig. 8.9(c), the agreement is good for narrow QDQWs.

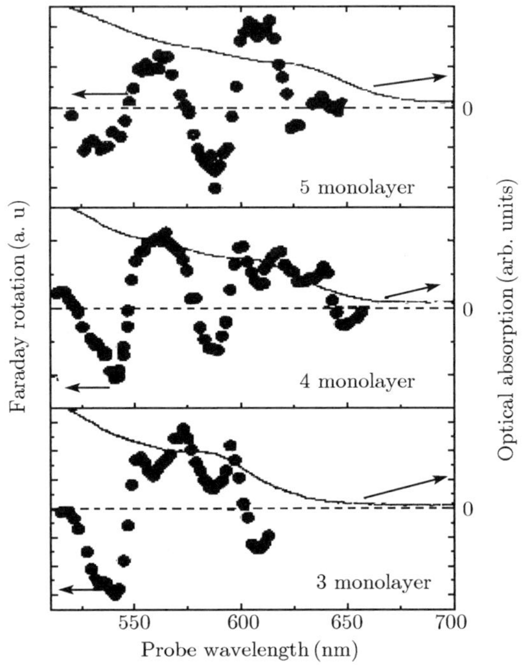

Fig. 8.10. Amplitude of the Faraday rotation angle, θ_F, as a function of probe wavelength for $n_{CdSe} = 3, 4, 5$ samples. The numerical value for θ_F was defined as the difference between the local maximum and minimum of the oscillations in the TRFR data nearest to $\Delta t = 500\,\text{ps}$ (as shown by the dotted line in Fig. 8.9(b)). The solid line is the absorption spectra.

The amplitude of the TRFR signal as a function of probe energy, $\theta_F(E)$, can be calculated from the calculated single-particle spectrum,

$$\theta_F(E) = CE \sum_{\sigma=\pm 1, \Phi_v} \sigma |\langle 1S_e, \downarrow | \hat{p}_x + \sigma i \hat{p}_y | \Phi_v \rangle|^2 \times \frac{E - E_{X,v}}{(E - E_{X,v})^2 + \gamma_v^2}, \quad (8.5)$$

where C is a constant, $E_{X,v}$ is the transition energy of the CB state to valence-band state, γ_v is the line width, and the summation is over all valence-band states contributing to the transitions to the CB ground state. The valence-band states are $|\Phi_v\rangle = 1S_{3/2}$, $2S_{3/2}$, $3S_{3/2}$, and the resonance peaks in Fig. 8.10 correspond to the transitions $1S_{3/2}$–$1S_e$, $2S_{3/2}$–$1S_e$, and $3S_{3/2}$–$1S_e$. The transition energies are in good agreement with the experimental resonance energies. Therefore, the exciton transitions in QDQWs can be assigned by the TRFR spectrum, not by the absorption spectrum. In contrast to the resonance energies, the spectral weight of the different resonances is not well reproduced by the theory. A possible explanation is that the line width is a function of energy.

8.4. Spatial Movement of Spin Coherent Electrons in Semiconductors

The above sections discussed the time decay of the electron coherent spins, i.e. spin relaxation. In the following sections, we will study the special movement of the spin-polarized electrons. Only after knowing the rule of the spatial movement of spin-polarized electrons, one can design various spintronics devices. For that purpose, TRFR was developed to the time and spatially resolved Faraday rotation (TSRFR), where the laser beams are normally incident on the sample and focussed to spots with FWHM of 14 μm, whose position can be controlled with a stepper-motor-driven mirror, enabling spatial scanning in one dimension.

8.4.1. *Spatial movement of coherent spin in compound semiconductors*

Early in 1999, Kikkawa and Awschalom (1999) reported time-resolved optical studies of nonlocal Faraday rotation in n-type bulk GaAs, which show macroscopic lateral transport of coherently precessing electronic spins over distances exceeding 100 μm.

The compound semiconductor (zincblende or wurtzite) crystals has the BIA, resulting in Dresselhaus spin splitting (Dresselhaus, 1955), which has a quantized axis perpendicular to the electron wave vector $\boldsymbol{k}$. Besides, the SIA in heterostructures results in RSS (Bychkov and Rashba, 1984), which induces an effective magnetic field perpendicular to $\boldsymbol{k}$ and the symmetry-broken axis $\boldsymbol{n}$.

In the bulk material, the similar spin splitting can be induced by strain (Kato *et al.*, 2004a). There are at least two possible mechanisms for strain-induced zero-magnetic-field spin splitting. Uniform strain can modify the magnitude of the BIA term, whereas a strain gradient can break the inverse symmetry, thus introducing SIA. In the latter case, the axis of asymmetry needs to be in the z-direction in order to produce B_{int} in the plane of the sample for in-plane $\boldsymbol{k}$.

The coherent electron spin precession angles exceeding 3π in zero magnetic field has been observed as the spin drift distances is greater than 60 μm in ~13 ns. Samples grown by MBE consists of a 2 μm n-GaAs layer

($n = 3 \times 10^{16}\,\mathrm{cm}^{-3}$) acting as a spin probe layer at the surface, and a $2\,\mu\mathrm{m}$ $Al_{0.4}Ga_{0.6}As$ layer underneath serving as a stressor/etch-stop layer. The semi-insulting GaAs (001) substrate is removed by chemical etching in order to form a rectangular film of $\sim 100\,\mu\mathrm{m} \times 300\,\mu\mathrm{m}$. The strain is produced in GaAs layer due to the mismatch of the lattice constants of GaAs and $Al_{0.4}Ga_{0.6}As$. The Ni/GeAu ohmic contacts are formed on the surface in order to apply an in-plane electric field $\boldsymbol{E}$ along the $[1\bar{1}0]$ direction.

Figures 8.11(a)–(d) show the spatiotemporal evolution of a spin packet FR at zero magnetic field at 5 K for $E = 0$, 33, 67, 100 V/cm (Kato *et al.*, 2004a). The spin packet drifts and coherently precesses as time evolves, and its velocity increases with increasing E. Line sections along the motion of the packet (Fig. 8.11(e)) show that the precession frequency itself is increasing with E.

The observed internal effective magnetic field B_{int} is characterized by analyzing the B_{ext} dependence of the FR signal at a fixed delay. In the

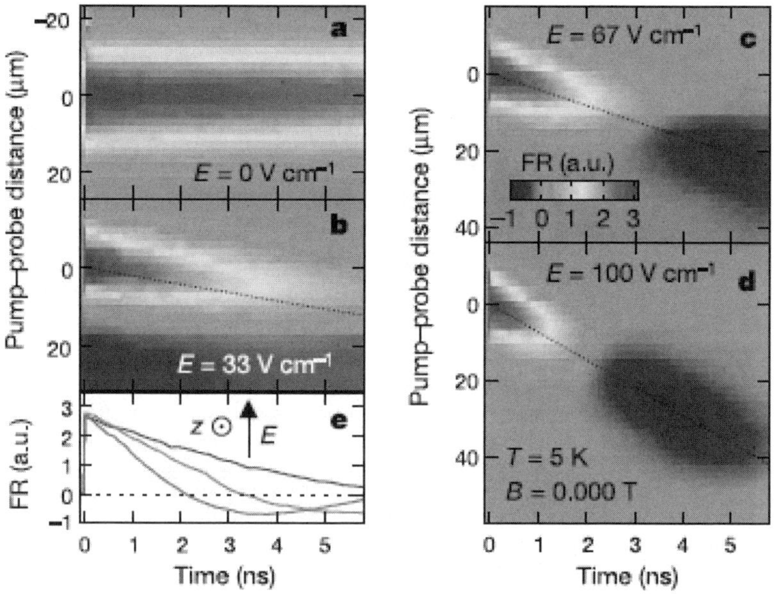

Fig. 8.11. Spatiotemporal evolution of a spin packet at zero magnetic field at 5 K. (a)–(d) FR as a function of temporal and spatial separation between pump and probe pulses for $E = 0$, 33, 67, 100 V/cm. (e) FR signals along dotted line in (b) (up), (c) (middle), and (d) (down).

Voigt geometry with $E = 0\,\text{V/cm}$, the FR signal can be described by

$$\theta_F \sim S_0 \exp\left(-\frac{\Delta t}{T_2^*}\right)\cos\left(\frac{g\mu_B B_{ext}\Delta t}{\hbar}\right). \tag{8.6}$$

The signal is periodic as a function of B_{ext} and symmetric around $B_{ext} = 0\,\text{T}$ at a fixed $\Delta t = 13.1\,\text{ns}$ (top curve of Fig. 8.12(a)) (Kato *et al.*, 2004a). The sharp peaks are due to a summation of contributions from successive pulses at laser repetition rate.

When the electric field E is applied perpendicular to B_{ext}, the curve is shifted in the B_{ext} axis (Fig. 8.12(a), middle curve), indicating that B_{int} is along B_{ext}, i.e. perpendicular to both z and E. The sign of B_{int}

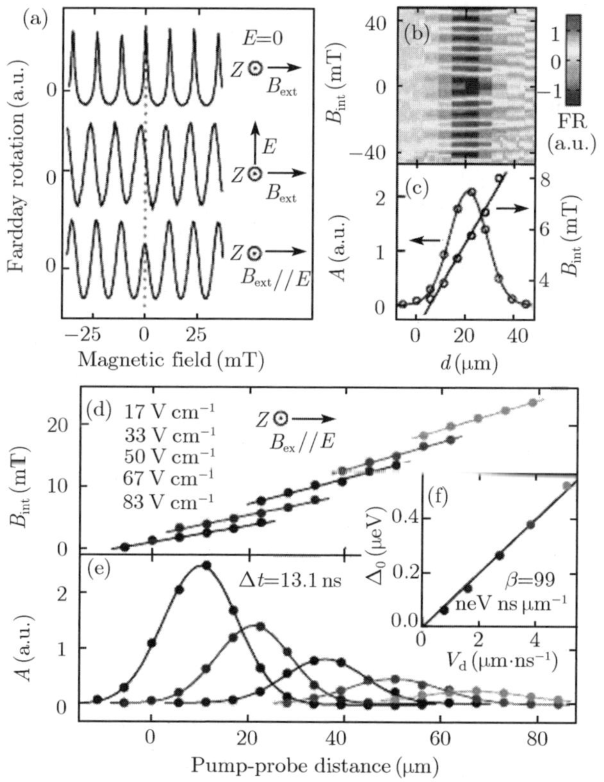

Fig. 8.12. (a) FR signal versus magnetic field B_{ext} in different geometries of E and B_{ext}. (b) FR at $\Delta t = 13.1\,\text{ns}$ as a function of d and B_{ext} with $E = 33\,\text{V/cm}$. (c) Fitting parameter A and B_{int} as a function of d. (d) and (e) B_{int} and A, respectively, as a function of d for different E. (f) Spin splitting Δ_0 as a function of wave packet drift velocity v_d.

can be determined from the direction of the shift, and it changes when the direction of E is reserved. For the case of E and B_{ext} being parallel (Fig. 8.12(a), bottom curve), assuming that B_{int} is independent of B_{ext}, the data can be fitted to

$$\theta_{\text{F}} \sim A \cos\left(\frac{g\mu_{\text{B}}\sqrt{B_{\text{ext}}^2 + B_{\text{int}}^2}}{\hbar}\right). \tag{8.7}$$

The measured FR at $E = 33\,\text{V/cm}$ for different pump-probe distances d are shown in Fig. 8.12(b), and the parameters A and B_{int} are plotted in Fig. 8.12(c) as a function of d, where A represents the spatial distribution of the spin wave packet. In order to evaluate B_{int}, one fit A with a Gaussian function to determine the center position of the spin packet. B_{int} is not a constant, linearly varies with d. The procedure is repeated for different E (Figs. 8.12(d) and 8.12(e)), and the spin splitting $\Delta_0 = g\mu_{\text{B}}B_{\text{int}}$ at the center of the spin packet is plotted in Fig. 8.12(f) as a function of v_{d}. From the figure, $\Delta_0 = \beta v_{\text{d}}$, so that the linear variation of B_{int} with d is caused by the spin drift velocity v_{d}.

The strain plays a critical role in generating B_{int}. Similar measurement was performed for the sample without the substrate removal process to avoid strain. The β is suppressed by an order of magnitude for both E along the [110] and [1–10] directions. There are at least two possible mechanisms for strain-induced zero-magnetic-field spin splitting: Dresselhaus splitting due to BIA and Rashba splitting due to SIA. The Kerr rotation spectra indicate that the β_{BIA} increases with stain, in contrast, β_{SIA} does not vary systematically with strain; thus, BIA mechanism caused by strain is the main spin-splitting mechanism. By adding functionality using strain without magnetic field, simple epilayers, rather than sophistical quantum structures, could be used for spintronics applications.

8.4.2. *Current-induced spin polarization*

The polarization of conduction electron spin due to an electrical current is observed in strained nonmagnetic semiconductors, which is called "current-induced spin polarization" (CISP). The schematics of device and experimental setup are shown in Figs. 8.13(a) and 8.13(b) (Kato *et al.*, 2004b). The sample is a 500 nm strained n-$In_{0.07}Ga_{0.93}As$ epilayer (Si-doped, $n = 3 \times 10^{16}\,\text{cm}^{-3}$), capped with 100 nm of undoped GaAs. The Ni/GeAu Ohmic contacts are added to both sides of the sample. The electric and magnetic fields are along the same x-direction, the z-direction is

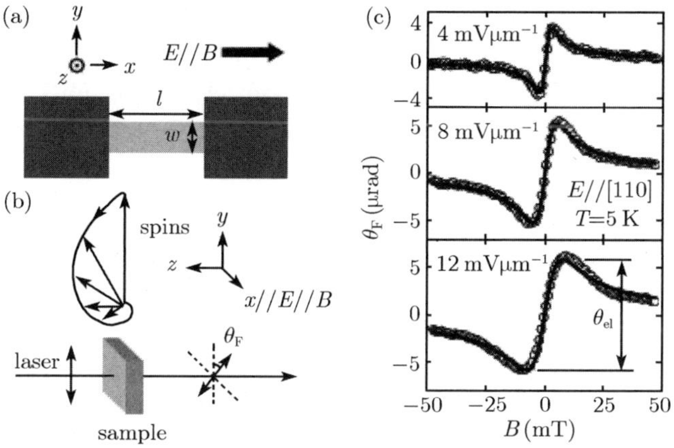

Fig. 8.13. (a) Device schematic and sample geometry. Dark areas are contacts and light area is the InGaAs channel. (b) Schematic of experimental setup and the Larmor precession of spins excited by electric current. (c) Voltage-induced θ_F as a function of B for $E = 4$, 8, and 12 mV/μm at $T = 5$ K, from a device with $w = 60$ μm and $l = 200$ μm.

perpendicular to the sample. A linearly polarized light along the z-direction penetrates the sample, the Faraday rotation angle θ_F is measured. The strain-induced asymmetry is in the z-direction, the electron wave vector $\boldsymbol{k}$ is in the x-direction, so the spin of the spin-splitting state due to the Dresselhaus or Rashba term is in the y-direction. This spin precesses in the magnetic field B around the x-direction, as shown in Fig. 8.13(b). Figure 8.13(c) shows the voltage-induced θ_F as a function of B for different E.

The θ_F–B curve can be explained by assuming a constant orientation rate for spins polarized along the y-axis. Under static illumination, the z component of spin per unit volume ρ_z is written as

$$\rho_z = \int_0^\infty dt \left[\gamma \exp\left(-\frac{t}{\tau}\right) \sin(\omega_L t)\right] = \frac{\rho_{el}\omega_L \tau}{(\omega_L \tau)^2 + 1}, \tag{8.8}$$

where γ is the number of spins oriented along the y-axis, τ is the inhomogeneous transverse spin lifetime, $\omega_L = g\mu_B B/\hbar$ is the electron Larmor frequency, and $\rho_{el} = \gamma\tau$ is the steady-state spin density due to electric excitation. Notice the difference in Eqs. (8.1) and (8.8): the integrated part in Eq. (8.1) is a cosine function and the integration result is a even function of ω_L, while that in Eq. (8.8) is a sine function; thus, the result is an odd function of ω_L, as shown in Fig. 8.13(c). By fitting the curves in Fig. 8.13(c) with Eq. (8.8), one obtains that the spin lifetime is about 5–10 ns and decreases with increasing electric field E.

In order to get the magnitude of the spin polarization γ induced by electric field, the authors measured the pump-probe TRFR in the absence of electric fields with the same probe power. Assuming 50% polarization from CP light, the total number of optical injected spins per pulse is

$$\rho_{\mathrm{op}} = \frac{P\lambda}{8\pi^2 \sigma_x \sigma_y d\hbar c f_{\mathrm{rep}}}, \tag{8.9}$$

where P is the absorbed power of the pump beam, σ_x, σ_y are the standard deviations of the laser intensity in the x- and y-directions, respectively, λ is the wavelength, d is the sample thickness, and f_{rep} is the repetition frequency of pulse. Comparing θ_{el} and θ_{op} obtained from two experiments, one obtains

$$\rho_{\mathrm{el}} = \frac{\rho_{\mathrm{op}}\theta_{\mathrm{el}}}{(2\theta_{\mathrm{op}})}, \quad \gamma = \frac{\rho_{\mathrm{el}}}{\tau}. \tag{8.10}$$

Figure 8.14 shows the temperature and electric field dependences of ρ_{el}, τ, γ, and η (Kato *et al.*, 2004b). ρ_{el} does not change significantly up to

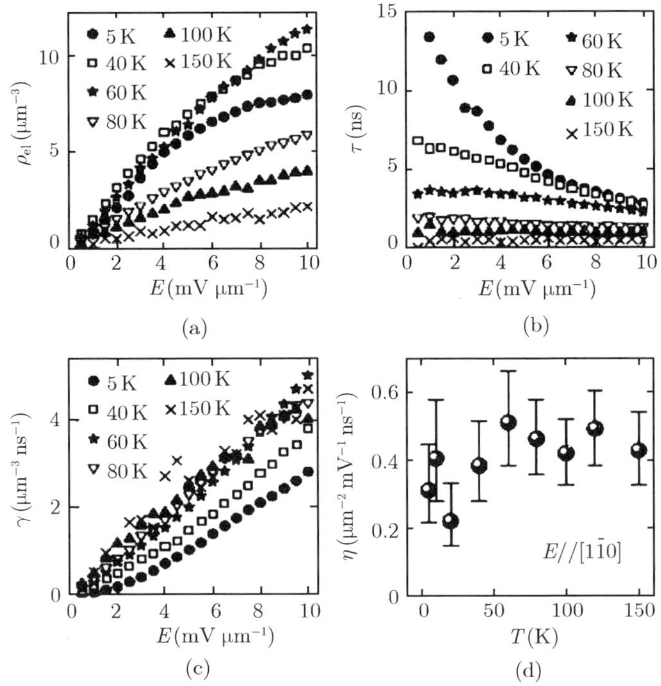

Fig. 8.14. Temperature and electric field dependences of ρ_{el} (a), τ (b), γ (c), and η (d). $E\|[1\bar{1}0]$, and data taken from a device with $w = 60\,\mu$m and $l = 200\,\mu$m.

$T = 60\,\text{K}$, at higher temperatures, ρ_{el} becomes smaller due to the decline of τ (Fig. 8.14(b)). The τ decreases obviously with increasing E. It is seen from Figs. 8.14(c) and 8.14(d) that γ and η do not considerably change up to $T = 150\,\text{K}$, where η is the slope of a linear fit to γ versus E. The signal is below the noise level for $T > 150\,\text{K}$ due to shorter τ and lower sensitivity of Faraday rotation.

The microscope origin of the CISP is not clear now. The strain strengthens inverse asymmetry and induces inner magnetic field B_{int}, resulting in spin polarization. In general n-GaAs samples, the B_{int} has a strong correlation with the strain magnitude, but the CISP has little correlation with B_{int}. In n-ZnSe samples, the magnitude of CISP is comparable with that in GaAs, but B_{int} has not been measured in n-ZnSe samples.

8.5. Spin Hall Effect

In analogy to the conventional Hall effect, the spin Hall effect has been proposed to occur in paramagnetic systems as a result of SOI and refers to the generation of a pure spin current transverse to an applied electric field even in the absence of applied magnetic field. A pure spin current can be thought of as a combination of a current of spin-up electrons in one direction and a current of spin-down electrons in the opposite direction, resulting in a flow of spin angular momentum with no net charge current. Similar to charge accumulation at the sample edges, which causes a Hall voltage in the conventional Hall effect, spin accumulation is expected at the sample edges in the spin Hall effect. Early theoretical studies predicted a spin Hall effect originating from asymmetries in scattering for up and down spins, which is referred to as the extrinsic spin Hall effect. Recently, it has been pointed out that there may exist an intrinsic spin Hall effect that arises as a result of the band structure, even in the absence of scattering (Murakami *et al.*, 2003; Sinova *et al.*, 2004).

8.5.1. *Optical observation of the spin Hall effect*

The unstrained GaAs sample and the experimental geometry are shown in Fig. 8.15(a) (Kato *et al.*, 2004c). The sample with a channel parallel to [110] with width $w = 77\,\mu\text{m}$, length $l = 300\,\mu\text{m}$, and mesh height $h = 2.3\,\mu\text{m}$. The applied magnetic field B_{ext} is along the x-direction, the electric field along the y-direction, and the z-direction is perpendicular to the sample. Static KR probes the electron spin polarization in the sample.

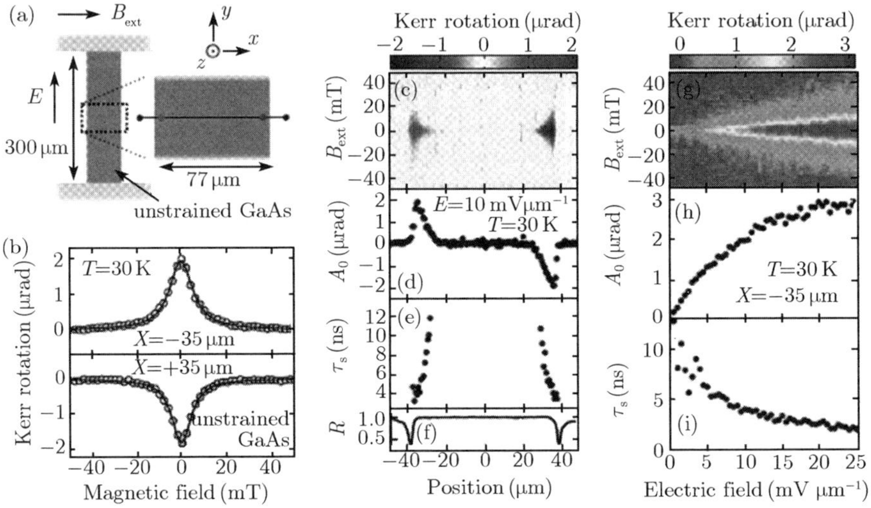

Fig. 8.15. (a) Schematic of the unstrained GaAs sample and the experimental geometry. (b) KR angle θ_K versus magnetic field B_{ext} at two sample edges $x = -35\,\mu m$ and $x = +35\,\mu m$ and $T = 30\,K$. (c) θ_K as a function of x and B_{ext} for $E = 10\,mV/\mu m$. (d)–(f) A_0, τ_s, and reflectivity R as a function of x, respectively. (g) θ_K as a function of E and B_{ext} at $x = -35\,\mu m$. (h) and (i) A_0, τ_s, versus E, respectively.

A mode-locked Ti:sapphire laser excites spin-polarized electrons in GaAs, a linearly polarized light beam is directed along the z-axis and is incident normal to the sample. The polarization axis of the reflected beam rotates by an amount proportional to the net magnetization of the electrons along the laser propagation direction. The rotation angle is called KR angle θ_K. Figure 8.15(b) shows the KR angle θ_K versus magnetic field B_{ext} at two sample edges $x = -35\,\mu m$ and $x = +35\,\mu m$. The signal changes the sign for the two edges of the sample, indicating an accumulation of electron spins polarized in the $+z$-direction at $x = -35\,\mu m$ and in the $-z$-direction at $x = +35\,\mu m$.

Physically, the electron SOC Hamiltonian is

$$H_{so} = -\frac{\hbar}{4m_0^2c^2}(\nabla V \times \boldsymbol{p}) \cdot \boldsymbol{\sigma}, \tag{8.11}$$

where ∇V is the gradient of the crystal field potential, also including the potential of applied electric field E; thus, ∇V is along the y-direction. Suppose that there is a charge current along the x-direction, the spin $\sigma_z = \pm 1$ state splits. The electrons are at the ground state, producing spin polarization. Similarly, the electron moving in the $-x$-direction also produces spin polarization, but the spin direction is opposite to those moving in the $+x$-direction. Therefore, the electrons of opposite spin accumulate at the two edges of the sample, and the net charge current equals zero.

The θ_K–B curve in Fig. 8.15(b) corresponds to the measurement of Hanle effect (Eq. (8.1)). The electron spins along the z-direction precess around the magnetic field in the x-direction. The spin lifetime τ_s is long enough, so that the spins have precessed many circles in the time of τ_s. In the static-state condition, the averaged KR angle equals

$$\theta_K = \frac{A_0}{(\omega_L \tau_s)^2 + 1}, \tag{8.12}$$

where ω_L is the precession frequency, and the sign of A_0 is decided by the initial spin. The opposite signs of the θ_K–B curves at $x = \pm 35\,\mu m$ indicate that the electron spins polarized in the $+z$-direction at $x = -35\,\mu m$ and in the $-z$-direction at $x = +35\,\mu m$. This is the direct evidence of the spin Hall effect.

Figure 8.15(c) shows the θ_K as a function of x and B_{ext} for $E = 10 mV/\mu m$, and Figs. 8.15(d)–8.15(f) show the A_0, τ_s, and reflectivity R as a function of x, respectively. It is seen that A_0, which is a measure of the spin density, has a maximum at the two edges and falls off rapidly with distance from the edge, disappearing at the center of the channel. The width of the Lorentzian of the KR curve becomes narrower as the distance from the edge increases, corresponding to an apparent increase in the spin lifetime (Fig. 8.15(e)).

The effect of the applied electric field E is also considered. Figure 8.15(g) shows the θ_K as a function of E and B_{ext} at $x = -35\,\mu m$, and Fig. 8.15(h) and 8.15(i) show the A_0, τ_s, versus E, respectively. Increasing E leads to larger spin accumulation A_0, evidencing that the SOC effect produces the spin Hall effect (Eq. (8.11)). The A_0 saturates at large E because of shorter τ_s with increasing E (Fig. 8.15(i)).

The homogeneity of the effect over the entire sample is studied by taking a 2D image of KR. Here, instead of taking a full B_{ext} scan, B_{ext} is sinusoidally modulated. The measured signal is then proportional to the second derivative of a B_{ext} scan around $B_{ext} = 0\,mT$ and can be regarded as a measure of the spin density n_s. Figures 8.16(a) and 8.16(b) show the 2D images of (a) spin density n_s and (b) reflectivity R, respectively. The polarization at the edges is uniform over a length of $150\,\mu m$ but decreases near the contacts. It is expected as unpolarized electrons are injected at the contacts.

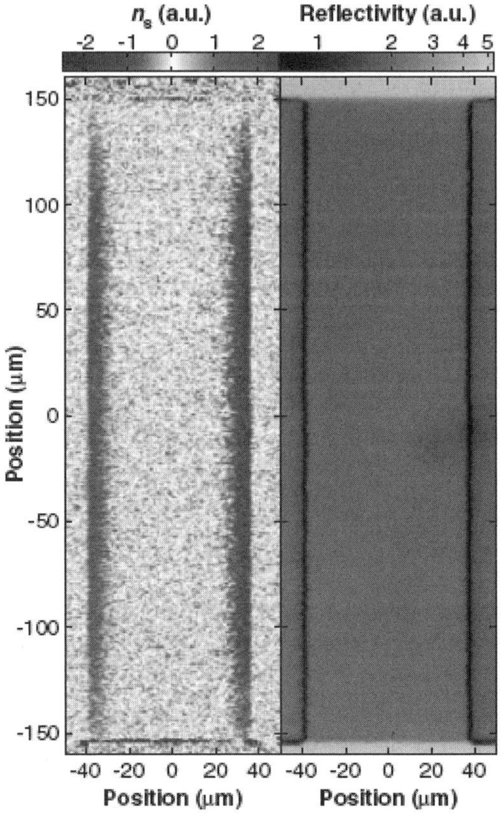

Fig. 8.16. 2D images of (a) spin density n_{s} and (b) reflectivity R, for the unstrained GaAs sample measured at $T = 30\,\mathrm{K}$ and $E = 10\,\mathrm{mV}/\mu\mathrm{m}$.

Kato *et al.* (2004c) indicated that the origin of the observed spin Hall effect is likely to be extrinsic, as the intrinsic effect in only expected in systems with spin splitting that depends on the electron wave vector $\boldsymbol{k}$. Although k^3 spin splitting in bulk GaAs may give rise to an intrinsic spin Hall effect, this is unlikely because negligible spin splitting has been observed in unstrained n-GaAs.

The strained InGaAs sample offers the possibility of displaying the intrinsic spin Hall effect in addition to the extrinsic effect. Using reciprocal space mapping with XRD it is determined that the in-plane strain along [110] and $[1\bar{1}0]$ to be -0.24% and -0.60%, respectively, and the strain along [001] to be $+0.13\%$. The strain produces an effective internal magnetic field B_{int}, which is parallel to the applied magnetic field B_{ext}.

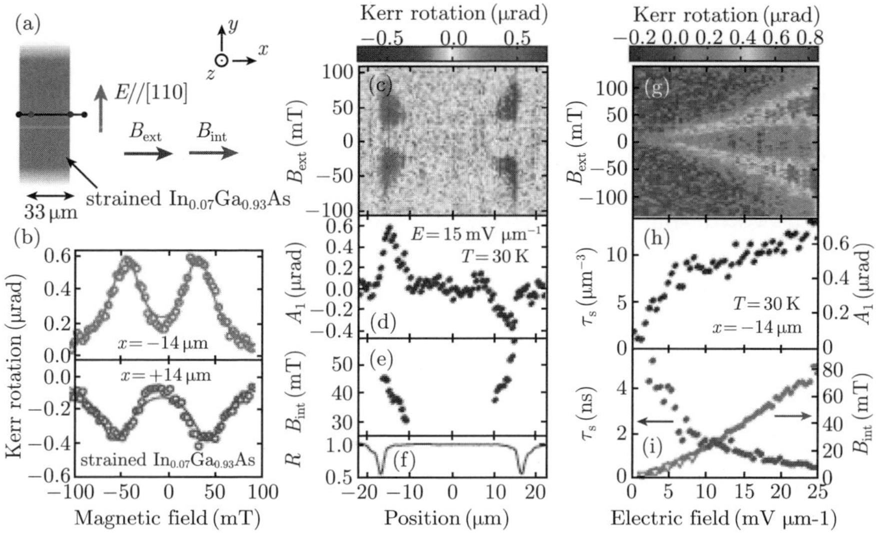

Fig. 8.17. The spin Hall effect in strained $In_{0.07}Ga_{0.93}As$ at $T = 30\,K$. (a) Schematic of the sample and the geometry of electric and magnetic fields. (b) θ_K as a function of B_{ext} at $x = -14\,\mu m$ and $+14\,\mu m$ for $E = 15\,mV/\mu m$. (c) θ_K as a function of x and B_{ext} for $E = 15\,mV/\mu m$. (d)–(f) A_0, τ_s, and reflectivity R as a function of x, respectively. (g) θ_K as a function of E and B_{ext} at $x = -14\,\mu m$. (h) and (i) A_0, $.\tau_s$, and B_{int} versus E, respectively.

Figure 8.17(b) shows the θ_K as a function of B_{ext} at $x = -14\,\mu m$ and $+14\,\mu m$ for $E = 15\,mV/\mu m$ (Kato *et al.*, 2004c). The spin polarization is suppressed at $B_{ext} = 0\,mT$ and two peaks offset from $B_{ext} = 0\,mT$ appear, which is attributed to the presence of B_{int}. Because electron spins respond to the sum of B_{ext} and B_{int}, the two peaks can be fitted by the following formula:

$$\theta_K = \frac{A_1}{(\omega_+\tau_s)^2 + 1} + \frac{A_1}{(\omega_-\tau_s)^2 + 1}, \tag{8.13}$$

where

$$\omega_+ = \frac{g\mu_B}{\hbar}(B_{ext} + B_{int}), \quad \omega_- = \frac{g\mu_B}{\hbar}(B_{ext} - B_{int}) \tag{8.14}$$

are precession frequencies for the plus and minus cases of B_{int} and B_{ext}.

Spatial dependence in this sample (Fig. 8.17(c)) reproduces the characteristic signatures of the spin Hall effect. The polarization is localized at the channel edges and has an opposite sign for the two edges, disappearing at the center (Fig. 8.17(d)). The splitting of the peaks ($\sim B_{int}$) decreases with increasing distance from the channel edge (Fig. 8.17(e)). The electric field dependence (Fig. 8.17(g)) exhibits similarity to the unstrained GaAs sample, showing polarization saturation (Fig. 8.17(h)) as the spin lifetime diminishes (Fig. 8.17(i)).

The spin-polarized electrons accumulated at the edges will diffuse from the edges to center of the sample. Ignoring the position dependence of τ_s, the static-state solution of the spin drift–diffusion equation is

$$n_s(x) = n_0 \frac{\sinh(x/L_s)}{\cosh(w/2L_s)}, \tag{8.15}$$

where $L_s = (D_s \tau_s)^{1/2}$ is the spin diffuse length, D_s is the spin diffuse coefficient, $n_0 = i_s \tau_s / L_s$ is the spin density at the edge, $i_s = n\langle \sigma_z v \rangle$ is the spin current density, v is the velocity operator, and w is the channel width. By fitting Eq. (8.15) with the experimental $n_s(x)$ for different E, one can obtain the electric field dependence of the L_s, $j_s = ei_s$ and the spin Hall resistivity $\rho_s = E/j_s$. The L_s changes a little with E and j_s increases with increasing E. The spin Hall resistivity ρ_s decreases with increasing E.

The spin Hall effect has been also observed in wide-gap semiconductor ZnSe. The experiments found that only in samples of high electron density $n = 9 \times 10^{18}\,\text{cm}^{-3}$, the spin Hall effect can be observed. Figure 8.18(a) shows the schematic of the ZnSe sample and experimental geometry, and Fig. 8.18(b) shows the θ_K versus B for sample edges $y = \pm 48\,\mu\text{m}$ at $T = 20\,\text{K}$, $E = 3\,\text{mV}/\mu\text{m}$ (Stern *et al.*, 2006). The $\theta_K \sim B$ curve can be fitted by Eq. (8.12); from this, the amplitude of the spin accumulation A_0 (θ_{el}) is obtained as a function of E, as shown in Fig. 8.18(c). The A_0 increases linearly with increasing E. The A_0, τ, and R as a function of y are shown in Fig. 8.18(d), which are similar to those of GaAs (Figs. 8.15(d–f)).

The sign and magnitude of the accumulated spins are found by direct comparison with CISP (Fig. 8.13 and related discussion) in a geometry with $E \| B$. At 20 K, the peak spin density near the edges is approximated $n_0 \approx 16\,\text{spins}/\mu\text{m}^3$. The fits gives the diffuse length $L_s = 1.9 \pm 0.2\,\mu\text{m}$. The spin current density along the y-direction $|j_y^s| = L_s n_0/\tau$, and the spin Hall conductivity is calculated equal to $\sigma_{SH} = -j_y^s/E_x = 3 \pm 1.5\,\Omega^{-1}\text{m}^{-1}/|e|$.

Figure 8.19(a) shows the θ_K of CISP versus B at $T = 295\,\text{K}$ and $E = 8.9\,\text{mV}/\mu\text{m}$, and Fig. 8.19(b) shows θ_K of the spin Hall effect versus B at sample edges $y = \pm 48\,\mu\text{m}$, where the circular points are experimental results, and the lines are fitting results (Stern *et al.*, 2006). The spin accumulation at the sample edges caused by the spin Hall effect persist until RT. Figure 8.19(c) shows the temperature dependence of n_0, τ, L_s,

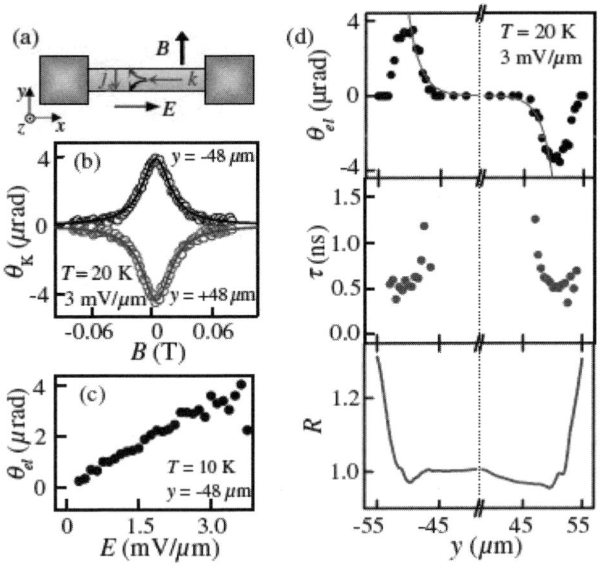

Fig. 8.18. (a) Schematic of the ZnSe sample and experimental geometry. (b) $\theta_{\rm K}$ versus B for sample edges $y = \pm 48\,\mu\text{m}$ at $T = 20\,\text{K}$, $E = 3\,\text{mV}/\mu\text{m}$. (c) Electric field dependence of the spin accumulation amplitude $\theta_{\rm el}$. (d) A_0, τ, and R as a function of y.

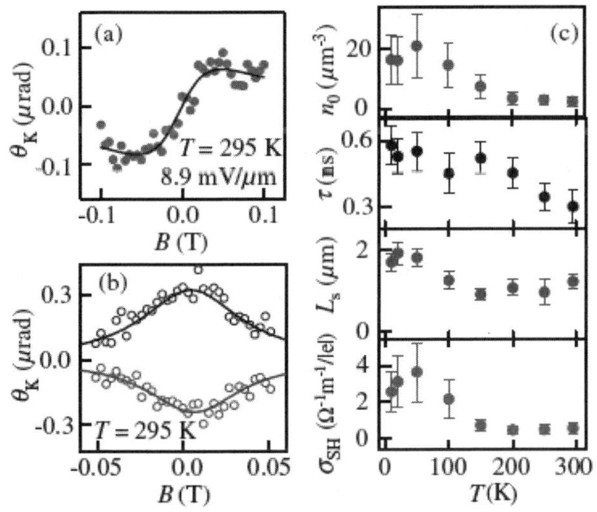

Fig. 8.19. (a) $\theta_{\rm K}$ of CISP versus B at $T = 295\,\text{K}$ and $E = 8.9\,\text{mV}/\mu\text{m}$. (b) $\theta_{\rm K}$ of the spin Hall effect versus B at sample edges $y = \pm 48\,\mu\text{m}$. (c) Temperature dependence of n_0, τ, $L_{\rm s}$, and $\sigma_{\rm SH}$.

and σ_{SH}. They all decrease with increasing temperature. The temperature increases from $T = 20$ to $295\,K$, the spin polarization density n_0 decreases by one order of magnitude, L_s decreases from 1.9 to $1.2\,\mu m$, and the spin Hall conductivity decreases to $0.5\,\Omega^{-1}m^{-1}/|e|$.

8.5.2. *Spin Hall effect in 2DEGs*

The spin Hall effect observed in unstrained GaAs (Kato *et al.*, 2004c) is considered as due to the extrinsic effect, because in GaAs the k linear term-related effective magnetic field is negligible. However, theoretical work argued that the cubic Dresselhaus term of k in GaAs could produce a non-negligible intrinsic spin Hall effect. Sih *et al.* (2005) designed a (110)-oriented quantum well (2DEG), where the Dresselhaus field is along the growth direction, i.e. perpendicular to the 2DEG plane. While in conventional (001) heterostructures, the Dresselhaus field is in the 2DEG plane. The spin polarization in the 2DEG is spatially resolved using low-temperature scanning KR microscopy in the Voigt geometry. In contrast to previous measurements, the spin Hall profile shows a complex structure and the current-induced spin polarization is out-of-plane.

Figure 8.20(a) shows the device schematic and measurement geometry, the electric field $E\|[001]$ (Sih *et al.*, 2005). Figure 8.20(b) shows the Kerr angle θ_K versus B_{ext} for $x = \pm 56.4\,\mu m$ at $V_p = 2\,V$, which has no difference with that of bulk GaAs, indicating that the electrons of opposite spins accumulate at the sample edges. Figure 8.20(c) shows the scan of θ_K in the B_{ext}–x plane, except at the sample edge $x = \pm 58.6\,\mu m$ the Kerr angle peak amplitude A_0 has a maximum, it also has a second maximum at $x = \pm 56.4\,\mu m$. This structure was not observed in measurements on bulk epilayers. The A_0, τ_s, and R as a function of x for $V_p = 2\,V$ (circular dots) and $V_p = 1.5\,V$ (square dots) shown in Fig. 8.20(c) are asymmetric for the right and left edges. At $V_p = 1.5\,V$, the accumulation of spin-polarized electron at the edges is not observed.

In the [001]-oriented device, electrically induced spin polarization is observed only at the edges of the channel. In contrast, devices fabricated along the $[1\bar{1}0]$, $[1\bar{1}1]$, and $[\bar{1}12]$ directions also show spin polarization at the centre of the channel. Figure 8.21(b) shows the Kerr angle .θ_K at $x = 0\,\mu m$ versus B_{ext} at $T = 30\,K$, $V_p = 2\,V$, for E parallel to three crystal directions (Sih *et al.*, 2005). As the polarization is along the growth direction and depends on the direction of $\boldsymbol{E}$ relative to the crystal axis, Sih *et al.* (2005) attributed this effect to the Dresselhaus field. The application of

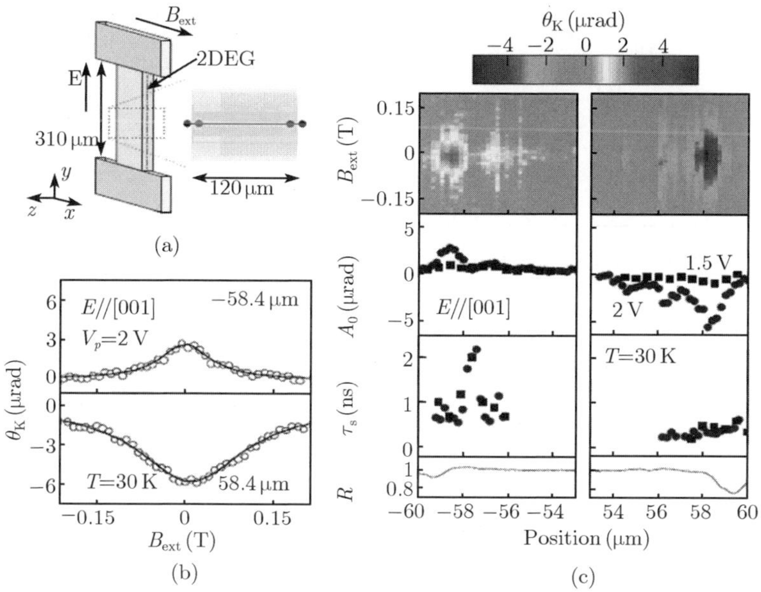

Fig. 8.20. (a) Device schematic and measurement geometry. (b) Kerr angle θ_K versus B_{ext} for $x = \pm 58.4\,\mu m$ at $V_p = 2\,V$. (c) At $V_p = 2\,V$, scan of θ_K in the B_{ext}–x plane; A_0, τ_s, and R as a function of x for $V_p = 2\,V$ (circular dots) and $V_p = 1.5\,V$ (square dots).

an electric field results in a nonzero average drift velocity of the electrons, which produces a nonzero effective magnetic field that orients spins. The results in Fig. 8.21(b) are consistent with the calculated B_{int} due to the cubic Dresselhaus field in a (110) QW. The theory also predicted that B_{int} should be zero for $\boldsymbol{E}\|[001]$, as observed.

Figure 8.21(c) shows the scan of θ_K in the B_{ext}–x plane near the edges for $\boldsymbol{E}\|[1\bar{1}0]$. A_0 is negative across the entire channel, and $|A_0|$ increases with increasing voltage. In the range of $-52\,\mu m < x < 52\,\mu m$, $|A_0|$ is nearly constant across the channel, and a negative peak in A_0 is evident near the right edge. The data for $V_p = 3\,V$ suggest that there may be two spin Hall peaks at $x = 55.5\,\mu m$ and $x = 57.5\,\mu m$, similar to the [001] device (Fig. 8.20(c)). τ_s changes little over the entire channel regime, is about 2 ns. For the case of $E\|[1\bar{1}1]$, A_0 is positive across the entire channel, and $|A_0|$ is nearly constant in the range of $-26\,\mu m < x < 26\,\mu m$ of the channel with width 68 μm. However, A_0 has a positive peak at $x = -31\,\mu m$ and diminishes near the right edge of the channel.

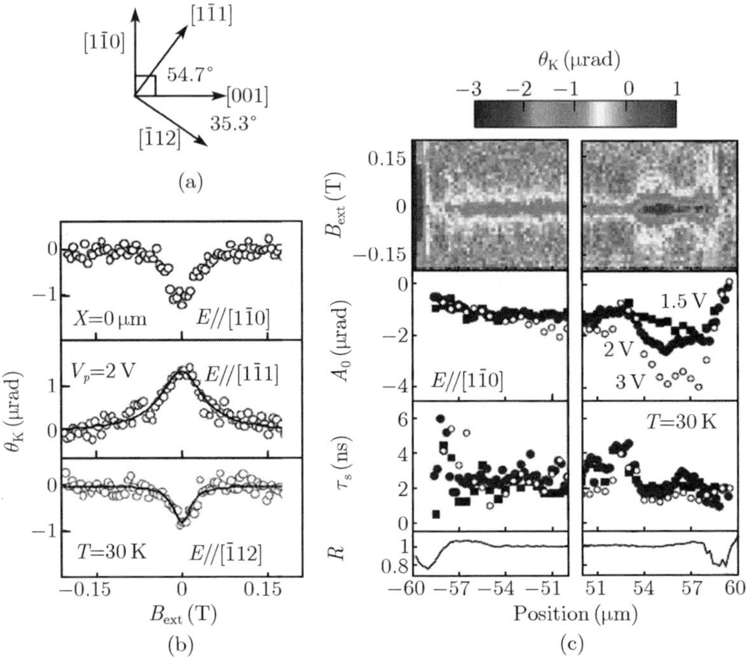

Fig. 8.21. (a) Relative orientations of crystal directions in the (110) plane. (b) Kerr angle θ_K at $x = 0\,\mu$m versus B_{ext} at $T = 30$ K, $V_p = 2$ V, for E parallel to three crystal directions. (c) Scan of θ_K in the B_{ext}–x plane for $E\|[1\text{-}10]$; A_0, τ_s, and R as a function of x for $V_p = 1.5$ V (solid square dots), $V_p = 2$ V (solid circular dots), and $V_p = 3$ V (open circular dots).

Spin–orbit engineering in 2D systems allows for the manipulation of the magnitude and direction of the internal fields for sourcing spin polarization in nonmagnetic semiconductors. Moreover, these interactions can be used to operate on electron spins by changing the direction of current, thereby enabling new degrees of control for quantum confined spintronic devices.

8.6. Generation of Spin Current

The spin Hall effect refers to the generation of a pure spin current transverse to an applied electric field even in the absence of applied magnetic field. A pure spin current can be thought of as a current of spin-up electrons in one direction and a current of spin-down electrons in the opposite direction, resulting in a flow of spin angular momentum with no net charge current. The device using the pure spin current will be very useful, because its power consumption is very small. The accumulation of spin electrons at the sample edges has been determined certainly, but the direct measurement of the spin current is difficult.

8.6.1. *Spin current generated by the spin Hall effect*

There arises a problem from the spin Hall effect: whether the spin polarization observed at the sample edges is due to the edge effect, such as depletion near the edge resulting in a local spin splitting from the Rashba effect, or due to a bulk spin current. In order to solve the problem, Sih *et al.* (2006) designed a device in which the effects of the boundary of the electric field are separated from edge effects. The device has transverse channels as shown in Fig. 8.22(a) to allow spins to drift into regions in which there is minimal electric current. The main channels have length $l = 316\,\mu$m, width $w = 60\,\mu$m, and the width of the transverse arm is $40\,\mu$m. The transverse arms extend out 10–40 μm from the side of the main channel. The coordinate origin is taken at the centre of the channel.

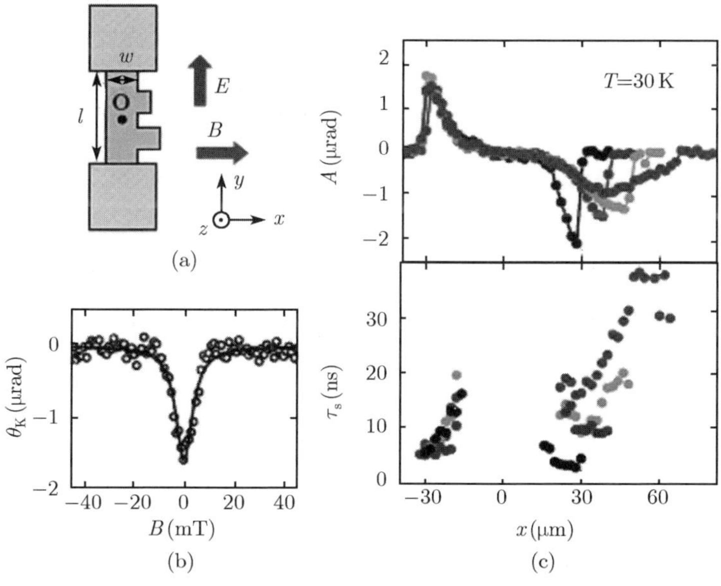

Fig. 8.22. (a) Device schematic and experimental geometry. (b) Kerr angle θ_K at $(x, y) = (26\,\mu\text{m}, 0\,\mu\text{m})$ versus B at $T = 30$ K, $V = 6$ V. (c) θ_K peak amplitude A and lifetime τ_s as a function of x at the arm length $a = 0\,\mu$m, $10\,\mu$m, $20\,\mu$m, and $40\,\mu$m. A have non-zero values at arm ends, respectively.

Figure 8.22(b) shows the Kerr angle θ_K at $(x, y) = (26\,\mu\text{m}, 0\,\mu\text{m})$ versus B (Sih *et al.*, 2006), where it is just the edge of the channel. The data can be fit to a Lorentz function to determine the amplitude A and the transverse spin lifetime τ_s. Figure 8.22(c) shows θ_K peak amplitude A and lifetime τ_s as a function of x at the arm length $a = 0\,\mu\text{m}$, 10, 20, and $40\,\mu\text{m}$. In the presence of the side arms, the spin accumulation has a different spatial profile that is dependent on the arm length. The spin accumulation is not always largest near the channel edge, and the maximum A is at larger x for longer arm length. This means that the spins drift from the main channel and toward the end of the side arm, and the spin polarization is not a local effect caused by the sample boundary.

The spin-polarized electrons cannot be generated in the side arm, where there should be minimal electric current. The authors proposed the following model: the spins are generated in the main channel and then they drift into the side arm. Assuming that the spin coherent time τ_s is a constant independent of x and averaging the spin orientations of the precessing electrons over the Gaussian distribution of their arrival times, one obtains the spin polarization at x_2,

$$S_z(x_1, x_2, B) = \int_0^\infty \frac{S_0}{\sqrt{4\pi Dt}} \exp\left[-\frac{(x_2 - x_1 - v_{\text{sd}}t)^2}{4Dt} - \frac{t}{\tau_s}\right] \cos(\omega_L t)\text{d}t, \tag{8.16}$$

where D is the spin-diffuse coefficient, S_0 is the initial spin polarization at x_1, and v_{sd} is the draft velocity. $S_z(x_2, B)$ is obtained by integrating Eq. (8.16) over x_1 from -30 to $+30\,\mu\text{m}$ (width of the main channel). Figure 8.23(a) shows the main channel and the arm with length and width of $40\,\mu\text{m}$, and the positions of P_1, P_2, and P_3, and Fig. 8.23(b) shows the θ_K versus B measured at P_1, P_2, and P_3 (Sih *et al.*, 2006). The fitting results with $S_z(x_2, B)$ give $\tau_s = 11.4\,\text{ns}$, $D = 10\,\text{cm}^2/\text{s}$, and $v_{\text{sd}} = 1.6 \times 10^5\,\text{cm/s}$.

To check the uniformity of the spin polarization along the longitudinal (y) direction, authors performed spatial scans of θ_K along the paths shown by dashed lines in Fig. 8.23(a). Figure 8.23(c) shows the θ_K peak amplitude A as a function of y from $y = -156$ to $25\,\mu\text{m}$ for paths across different x. For $x = -26\,\mu\text{m}$, i.e. the left edge of the channel, A increases from zero at the contact to a maximum value over $50\,\mu\text{m}$ and remains unchanged. For $x = 26\,\mu\text{m}$, the amplitude drops near the side arm due to spin drift into the side arm. Measurements taken across the side arm, at $x = 36$ and $56\,\mu\text{m}$, show that the amplitude is largest near the edges of the side arm, which may also be due to spin drift.

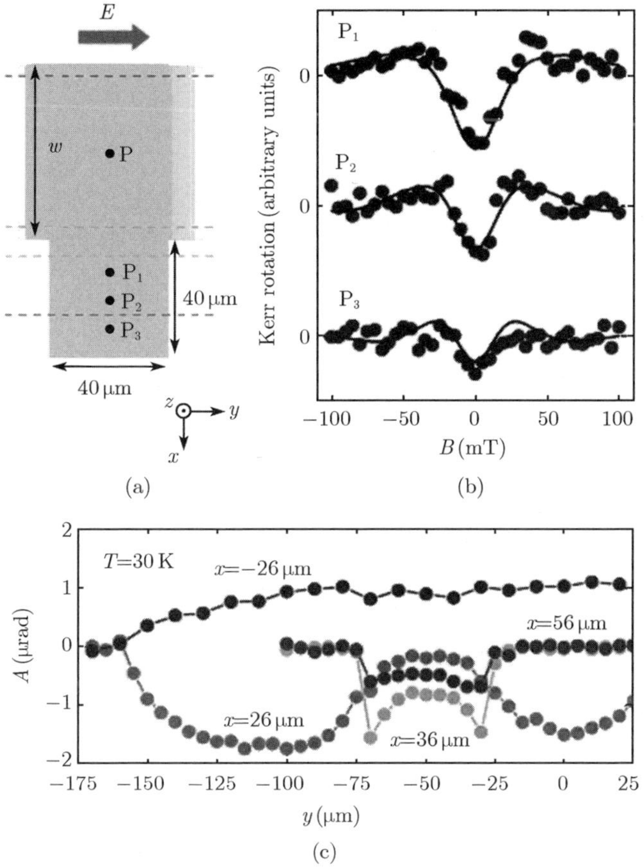

Fig. 8.23. (a) Main channel and the arm with length and width of 40 μm, and the positions of P_1, P_2, and P_3. (b) θ_K versus B measured at P_1, P_2, and P_3. The lines are fitting results with $S_z(x_2, B)$. (c) θ_K peak amplitude A as a function of y from $y = -156$ to 25 μm for $x = -26$, 26, 36, and 56 μm.

8.6.2. *Spin current generated by a two-color light field*

In nonmagnetic semiconductors, Bhat and Sipe (2000) proved that by the quantum of one- and two-photon absorption processes in a two-color light field, an electric current with polar asymmetry can be injected. By appropriate choice of beam polarization, the current is spin-polarized. It is possible to generate spin currents with no associated electrical current. Here the quantum interference sorts the carriers as they are injected, sending those of one spin in one direction and those of the opposite spin in the opposite

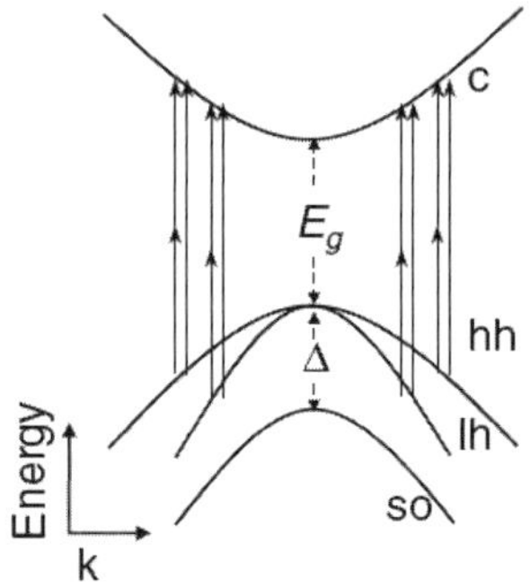

Fig. 8.24. One- and two-photon transition processes of electron excitations. The c, hh, lh and so are conduction, heavy hole, light hole, and spin–orbit splitting bands, respectively.

direction. In the independent particle approximation, the two-color light field

$$\boldsymbol{E}(t) = \boldsymbol{E}_\omega \mathrm{e}^{-\mathrm{i}\omega t} + \boldsymbol{E}_{2\omega}\mathrm{e}^{-\mathrm{i}2\omega t} + \text{c.c. (complex conjugate)}, \tag{8.17}$$

causes transitions from the semiconductor ground state $|0\rangle$ to states $|csvp\boldsymbol{k}\rangle$ by exciting an electron from the valence band v with spin p to CB c with spin s, while leaving its wave vector $\boldsymbol{k}$ unchanged as shown in Fig. 8.24.

The spin current is the average value of the pseudotensor,

$$\dot{K}^{ab} = \langle \hat{v}^a \hat{S}^b \rangle, \tag{8.18}$$

where $\hat{\boldsymbol{v}}$ is the velocity operator, $\hat{\boldsymbol{S}}$ is the spin operator, and a, b represent x, y, z Cartesian components. No spin current will be generated by one- or two-photon absorption alone. Only when the two absorptions exist simultaneously, the interference between them will generate spin current. The interference term,

$$\dot{K}^{ab}_{I,e} = \frac{2\pi}{L^3} \sum_{cvss'p\boldsymbol{k}} \langle cs\boldsymbol{k}|\hat{v}^a \hat{S}^b|\rangle \times \Omega^{(1)*}_{csvp\boldsymbol{k}} \Omega^{(2)}_{cs'vp\boldsymbol{k}} \delta[2\omega - \omega_{cv}(\boldsymbol{k})] + \text{c.c.}, \tag{8.19}$$

$$\dot{K}^{ab}_{I,h} = -\frac{2\pi}{L^3} \sum_{cvspp'\boldsymbol{k}} \langle vp'\boldsymbol{k}|\hat{v}^a \hat{S}^b|\rangle \times \Omega^{(1)*}_{csvp\boldsymbol{k}} \Omega^{(2)}_{csvp'\boldsymbol{k}} \delta[2\omega - \omega_{cv}(\boldsymbol{k})] + \text{c.c.}, \tag{8.20}$$

where $|ns\boldsymbol{k}\rangle$ is a Bloch state, $\Omega^{(1)}$ and $\Omega^{(2)}$ are one-photon and two-photon amplitudes, respectively,

$$\begin{aligned} \Omega^{(1)}_{csvp\boldsymbol{k}} &= i\frac{e}{2\hbar\omega} \boldsymbol{E}_{2\omega} \cdot \boldsymbol{v}_{csvp}(\boldsymbol{k}), \\ \Omega^{(2)}_{cs'vp\boldsymbol{k}} &= -\left(\frac{e}{\hbar\omega}\right)^2 \times \sum_{ns''} \frac{[\boldsymbol{E}_\omega \cdot \boldsymbol{v}_{cs'ns''}(\boldsymbol{k})][\boldsymbol{E}_\omega \cdot \boldsymbol{v}_{ns''vp}(\boldsymbol{k})]}{\omega_{cv}(\boldsymbol{k})/2 + \omega_{vn}(\boldsymbol{k})}, \end{aligned} \tag{8.21}$$

where $\boldsymbol{v}_{nsms'}(\boldsymbol{k})$ is the velocity matrix element between Bloch states. For cubic semiconductors with the point group $T_{\rm d}$, O, or O_h, $\dot{K}_I^{ab}$ can be divided into four terms with independent parameters A_i ($i = 1$–4), which can be represented by three parameters: band gap $E_{\rm g}$, spin-split energy Δ, and Kane energy $E_{\rm p}$.

The same model gives a current injection,

$$\dot{J}_I^a = -iD\frac{e}{\hbar}[B_1 E_\omega^a(\boldsymbol{E}_\omega \cdot \boldsymbol{E}_{2\omega}^*) + B_2 E_{2\omega}^{*a}(\boldsymbol{E}_\omega \cdot \boldsymbol{E}_\omega)] + \text{c.c.}, \tag{8.22}$$

where the coefficients B_1 and B_2 can also be represented by three parameters.

Two cases of light fields are considered:

Case 1: Same circular polarizations along the z-direction,

$$\boldsymbol{E}_{\omega/2\omega} = E_{\omega/2\omega} e^{i\phi_{\omega/2\omega}} \frac{(\hat{\boldsymbol{x}} \pm i\hat{\boldsymbol{y}})}{\sqrt{2}}. \tag{8.23}$$

The electrical current injection is

$$\dot{\boldsymbol{J}}_I = \frac{\sqrt{2}\hat{\boldsymbol{m}} D B_1 E_\omega^2 E_{2\omega} e}{\hbar}, \tag{8.24}$$

where D is a constant,

$$\hat{\boldsymbol{m}} = \hat{\boldsymbol{x}} \sin(2\phi_\omega - \phi_{2\omega}) \pm \hat{\boldsymbol{y}} \cos(2\phi_\omega - \phi_{2\omega}). \tag{8.25}$$

The spin current injection is

$$\dot{K}_I^{ab} = \mp\sqrt{2} E_\omega^2 E_{2\omega} D[(A_1 - A_4)\hat{m}^a \hat{z}^b + (A_2 + A_4)\hat{z}^a \hat{m}^b]. \tag{8.26}$$

Recalling from Eq. (8.18) that the first index of K^{ab} is associated with the carrier velocity, one can see that the first term in Eq. (8.26) shows that the electrical current is partially spin-polarized. The spin polarization rate of the current is defined by $\dot{K}^{az}/(\dot{J}^a/e) = \mp f\hbar/2$. In this case, $f = 2(A_1 - A_4)/B_1$. Using parameter values appreciate to GaAs, $f \approx 0.57$. The second term in Eq. (8.26) represents spins pointing along the $\hat{\boldsymbol{m}}$ that move along the z-direction. Since there is no net electrical current in the z-direction, this is a pure spin current. It arises because the electrons with velocity in the positive z-direction have an average spin opposite to those with velocity in the negative z-direction. In GaAs, the second term is a factor of 7.7 smaller than the first term. The situation is schematically shown in Fig. 8.25(a).

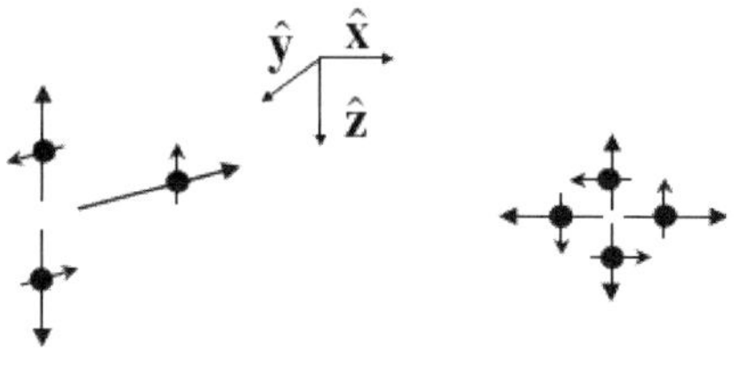

Fig. 8.25. Schematic illustration of the net electron motion combining the information of K^{ab} and J^a for (a) Case 1 with CP beams along the z-direction. (b) Case 2 with two cross-polarized beams.

Case 2: Two linear polarizations perpendicular to each other,

$$\boldsymbol{E}_\omega = E_\omega e^{i\phi_\omega}\hat{\boldsymbol{x}}, \quad \boldsymbol{E}_{2\omega} = E_{2\omega} e^{i\phi_{2\omega}}\hat{\boldsymbol{y}}. \tag{8.27}$$

In this case, since the beams are linearly polarized, there is no spin injection, $\dot{\boldsymbol{S}}_e = \dot{\boldsymbol{S}}_h = 0$.

The electrical current depends sinusoidally on the relative phase of the two beams,

$$\dot{\boldsymbol{J}}_I = 2\left(\frac{e}{\hbar}\right) B_2 D E_\omega^2 E_{2\omega} \sin\left(2\phi_\omega - \phi_{2\omega}\right)\hat{\boldsymbol{y}}. \tag{8.28}$$

According to the Kane model, $B_2 \ll B_1$, so that the charge current is much smaller than that in the case 1. The spin current is

$$\dot{K}_I^{ab} = -2E_\omega^2 E_{2\omega} D\cos(2\phi_\omega - \phi_{2\omega}) \times [(A_1 + A_3)\hat{x}^a\hat{z}^b + (A_2 - A_3)\hat{z}^a\hat{x}^b], \tag{8.29}$$

where the first term arises from carrier motion along $\boldsymbol{E}_\omega$ with spin aligned along the beam propagation direction (z), and the second term arises from carrier motion along the beam propagation direction with spins aligned along $\boldsymbol{E}_\omega$, as shown in Fig. 8.25(b). Both of them are pure spin currents, since there is no electrical current in either direction. For the other cases, for example, the beams have opposite circular polarizations or the beams are collinearly polarized, and there is no or small spin current injection.

Stevens *et al.* (2003) demonstrated all-optical quantum interference injection and control of a ballistic pure spin current in GaAs/AlGaAs quantum wells.y The sample is an MQW consisting of 10 periods of 14 nm

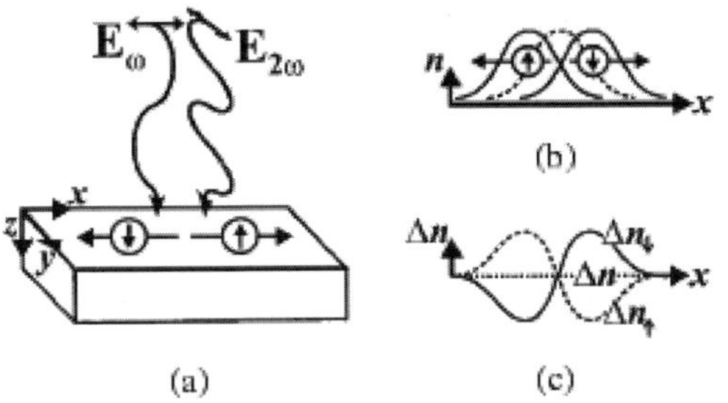

Fig. 8.26. (a) Experimental geometry. (b) Schematic representation of the initial electron distribution (dashed Gaussian profile) and the subsequent motion of the spin-down (up) electrons in the $+x(-x)$ direction, yielding no net charge current. (c) The corresponding change in spin-up, spin-down, and total electron density resulting from this motion.

wide GaAs wells and 17 nm thick $Al_{0.3}Ga_{0.7}$ As barriers grown on a (001)-oriented GaAs substrate. Figures 8.26(a) shows the experimental geometry (Stevens *et al.*, 2003), same as case 2 discussed above. Two laser pulses propagate in the $+z$ directions and have orthogonal linear polarizations, with the ω beam polarization in the x-axis and the 2ω beam polarization along the y-axis. The two beams are incident normal to the sample, causing electron excitation as shown in Fig. 8.24. The quantum interference effect can be produced as shown by Eqs. (8.19) and (8.20).

According to the theoretical prediction of Bhat and Sipe *et al.* (2000), the two injected spin currents traveling along the x-direction are given by (Eq. (8.29)),

$$\boldsymbol{J}_{\pm x}^{\mp \sigma} = \pm \dot{J}_1 \cos(\Delta\phi)\ddot{\boldsymbol{x}}, \tag{8.30}$$

where the upper index represents the spin direction, the subscript represents the current direction, $\Delta\phi = 2\phi_\omega - \phi_2\omega$ and $\phi_\omega(\phi_{2\omega})$ is the phase of the beam. Initially [dashed line in Fig. 8.26(b)], there are an equal number of spin-up and spin-down electrons produced, and for $\Delta\phi = 0$, the spin-down (spin-up) electrons should have a preferred ballistic momentum in the $+x(-x)$ direction, and shift toward $+x(-x)$. As a result, the density distributions of spin-up, spin-down, and total electrons are shown in Fig. 8.26(c).

The spatially resolved pump-probe technique was used to monitor the spin separation resulting from these oppositely directed spin-polarized currents. The pulses with wavelengths of 1.42 and 0.71 μm, which have orthogonal linear polarizations, are recombined collinearly, overlapped in time with definite relative phase, and focussed onto the sample at normal incidence. The spin-up or spin-down carrier population at each position is determined by varying the probe polarization and using the circular polarization optical selection rules for excitonic transitions. Since the probe is tuned to the heavy-hole excitonic resonance, the change in transmission of a right CP probe is a measure of spin-down electron density $n_\downarrow$ (or change in that density, $\Delta n_\downarrow$). Conversely, a left CP probe is sensitive to the spin-up electron density $n_\uparrow$ (or $\Delta n_\uparrow$). While a linearly polarized probe is sensitive to the total electron density n (or Δn).

Figure 8.27(a) shows the differential transmission ΔT of left-polarized probe versus phase difference $\Delta\phi$ at $x = -7.5\,\mu\text{m}$, $x = 0\,\mu\text{m}$, and $x = +7.5\,\mu\text{m}$, the solid lines are sinusoidal fits to the data (Stevens *et al.*, 2003). The spin-up populations $\Delta n_\uparrow$ at $x = -7.5\,\mu\text{m}$ and $+7.5\,\mu\text{m}$ both vary sinusoidally with $\Delta\phi$, clearly demonstrating quantum interference control

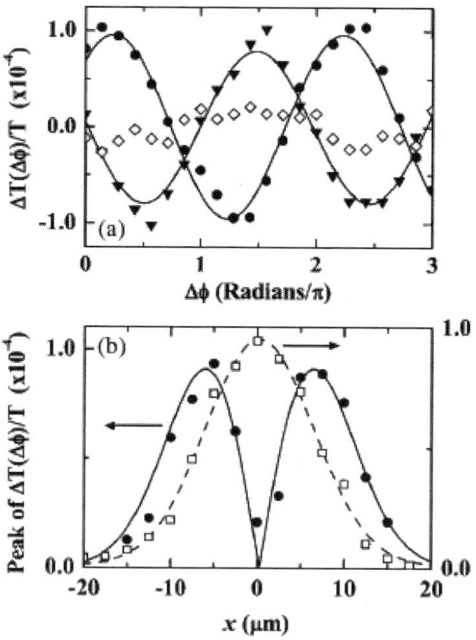

Fig. 8.27. (a) Differential transmission ΔT of left-polarized probe versus phase difference $\Delta\phi$ at $x = -7.5\,\mu\text{m}$ (solid triangles), $x = 0\,\mu\text{m}$ (open diamonds), and $x = +7.5\,\mu\text{m}$ (solid circles). Solid lines are sinusoidal fits to the data. (b) Differential transmission as a function of probe position x. The open squares show $\Delta T(n)/T$, and the solid circles show the peak of $\Delta T(\Delta\phi)/T$. The dashed line is a Gaussian fit to $\Delta T(n)/T$, while the solid line is the absolute value of the derivative of that Gaussian.

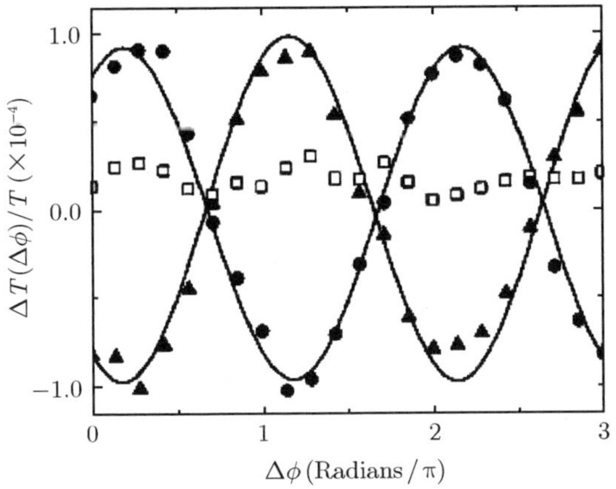

Fig. 8.28. Differential transmission $\Delta T/T$ versus $\Delta\phi$ for three different probe polarizations: left circular $\hat{\sigma}^-$ (solid circles), right circular $\hat{\sigma}^+$ (solid triangles), and linear (open squares). Solid lines are sinusolidal fits to the data.

(QUIC) of the spin-up electrons. The phases of the oscillations at these two points differ by π. $\Delta n_\uparrow$ at $x = 0$ shows almost no phase dependence, indicating that for any $\Delta\phi$, an equal number of spin-up electrons move into and out of the probed region.

Figure 8.27(b) shows the differential transmission ΔT peak as a function of probe position x. The peak value of $\Delta n_\uparrow(x, \Delta\phi)$ at each position follows the initial $|\partial n(x)/\partial x|/2$, verifying that the measured is a QUIC current, rather than QUIC population change. From the relative amplitudes of the solid and dashed curves in Fig. 8.27(b), it is estimated that each of the profiles moves $\sim 10\,\mu$m, resulting in a separation of the two spin profiles of $\sim 20\,\mu$m.

For a given $\Delta\phi$, the spin-up and spin-down electrons move in opposite directions, as predicted by Eq. (8.30). Figure 8.28 shows $\Delta T/T$ versus $\Delta\phi$ for three different probe polarizations measured at $x = +7.5\,\mu$m (Stevens *et al.*, 2003). The spin-up and spin-down densities are almost exactly π out of phase at one position, and the total density has no change, which verifies furthermore that there is no net charge current.

8.7. Spin Dynamics in Semiconductors

There has been a perfect transport theory of charge current in semiconductors; this section will introduce the transport theory of spin current. First we recall the transport theory of charge current; the largest simplification of the charge current model is the introduction of local chemical potential. Suppose that in a special spatial region the carriers are in equilibrium with other carriers, but can be nonequilibrium with carriers in other regions. When the average free path of the carrier is smaller than the characteristic length of the region, the assumption of local chemical potential is tenable. In this case, the space-related properties of carriers in one simple band can be described by two parameters: local chemical potential $\varepsilon_\mu(\boldsymbol{r})$ and the electrical potential $\phi(\boldsymbol{r})$.

The current is related directly to the variation of the total chemical potential $\varepsilon_\mu - \phi$,

$$\boldsymbol{J} = \sigma \nabla (\varepsilon_\mu - \phi), \tag{8.31}$$

where σ is the conductivity. The σ of semiconductors can be written as the product of the carrier density n and the mobility μ, and only n depends on ε_μ. The current conservation demands,

$$\nabla \cdot \boldsymbol{J} = 0. \tag{8.32}$$

The third equation is the Poisson equation of the electric field,

$$\nabla \cdot \boldsymbol{E} = -\frac{e}{\varepsilon}(n - n_0), \tag{8.33}$$

where n_0 is the equilibrium carrier density, and ε is the dielectric constant. In semiconductors, a small change of carrier density will influence apparently conductivity; thus, the σ in Eq. (8.31) should be determined by solving self-consistently Eqs. (8.31)–(8.33).

In low-doped semiconductors the drift–diffusion equation (Eq. (8.34)) is used more frequently. For nondegenerate carriers, $\varepsilon_\mu = k_\mathrm{B} T \ln(n/n_0)$; thus, Eq. (8.31) can be written as,

$$\boldsymbol{J} = \sigma \boldsymbol{E} + \frac{\sigma k_\mathrm{B} T}{n} \nabla n = \sigma \boldsymbol{E} + eD\nabla n, \tag{8.34}$$

where D is the diffusion coefficient. From Eq. (8.34), the current can be written as a gradient of a scalar quantity $\boldsymbol{J} = \nabla \xi$. In general case, the current can be calculated with chemical potential expression or the drift–diffusion equation.

8.7.1. *Mobility and diffusion coefficient of spin current*

It is found in experiments (Kikkawa and Awschalom, 1999) that in semiconductors the spin diffusion length is much longer than the diffusion length of charge carriers. Flatté and Byers (2000) analyzed the origins. The origins of the differences in spin diffusion between semiconductors and metals are (1) the much greater spin relaxation lifetime in semiconductors, (2) the ineffectiveness of screening in semiconductors relative to metals, and (3) the possibility of controlling whether carriers in a degenerate band or not by small perturbations.

The second origin is that in metal the local charge density variations are screened out on length scales of angstroms by free carriers. In semiconductors, the local variation of the electron density at $\boldsymbol{x}$ $\Delta n(\boldsymbol{x})$ is balanced by a local change in the valence hole density $\Delta p(\boldsymbol{x})$ due to the electric neutrality requirement. The $\Delta n(\boldsymbol{x})$–$\Delta p(\boldsymbol{x})$ constraint in semiconductors has key implications for the motion of packets of changed carrier density. The motion of holes tends to be much slower (due to their lower mobility) than that of electrons, so hole mobility and diffusion tend to dominate the properties of a packet consisting of both electron and hole density variations.

Figure 8.29 shows several packets, where (a) is the charge packets in a n-type semiconductor (Flattè and Byers, 2000). The $n_\uparrow$ and $n_\downarrow$ have same density variations $\Delta n_\uparrow$ and $\Delta n_\downarrow$, there is no spin polarization, which is called the charge packet. Meanwhile, the hole density $p_\uparrow$ and $p_\downarrow$ have same variations. Figure 8.29(b) shows the spin packet, where the spin-polarized electrons $\Delta n_\uparrow$ are injected into an undoped semiconductor by, for example, optical excitation. Meanwhile, the hole densities have corresponding variations $\Delta p_\uparrow$ and $\Delta p_\downarrow$. Figure 8.29(c) shows the spin packet in an n-type semiconductor, where the $-\Delta n_\downarrow$ is produced corresponding to the $\Delta n_\uparrow$ without variation of hole density due to the background electrons. This single-band spin packet does not drag a local inhomogeneous hole density

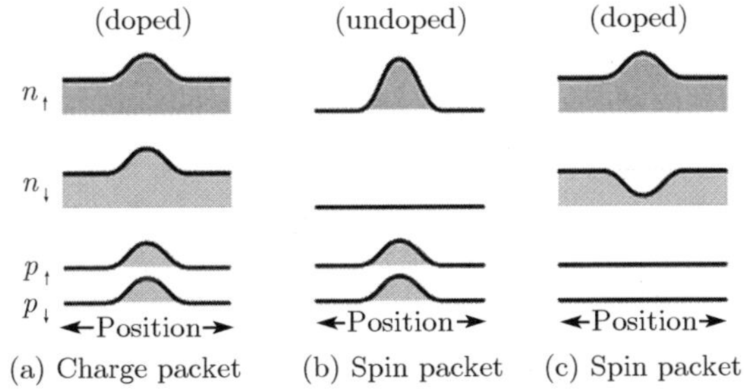

Fig. 8.29. (a) Charge packets in an n-type semiconductor, (b) spin-polarized packets in an undoped semiconductor, and (c) spin-polarized packets in an n-type semiconductor.

with it, and thus its mobility and diffusion properties are very different from those of a spin packet in the undoped semiconductor.

Let $\tau_{\rm r}$, $\tau_{\rm s,e}$, and $\tau_{\rm s,h}$ be carriers recombination lifetime, electron spin relaxation lifetime, and hole spin relaxation lifetime, respectively. In semiconductors, $\tau_{\rm s,h} \ll \tau_{\rm r} \ll \tau_{\rm s,e}$. The spin packet in Fig. 8.29(c) can be generated by excitation of CP light. During the recombination time, the unpolarized holes annihilate an equal number of spin-up and spin-down electrons, leaving behind excess spin polarization in the CB. The motion of a charge packet (Fig. 8.29(a)) involves both a conduction and valence disturbance and is described by an ambipolar mobility and diffusion constants,

$$\mu_a = \frac{(n-p)\mu_{\rm e}\mu_{\rm h}}{n\mu_{\rm e} + p\mu_{\rm h}}, \quad D_a = \frac{n\mu_{\rm e}D_{\rm h} + p\mu_{\rm h}D_{\rm e}}{n\mu_{\rm e} + p\mu_{\rm h}}, \tag{8.35}$$

where $D_{\rm e}$, $\mu_{\rm e}$ and $D_{\rm h}$, $\mu_{\rm h}$ are diffusion coefficients and mobility of electrons and holes, respectively. For n doping ($n \gg p$), $D_{\rm a} \sim D_{\rm h}$ and $\mu_{\rm a} \sim \mu_{\rm h}$, so the diffusion and the mobility of the charge packet are dominated by the holes. The mobility and the diffusion coefficient of the spin packet in Fig. 8.29(c) are given in analogy with the two-species case of Eq. (8.35),

$$\begin{aligned} \mu_{\rm s} &= \frac{(n_\downarrow + n_\uparrow)\mu_{\rm e\downarrow}\mu_{\rm e\uparrow}}{n_\downarrow\mu_{\rm e\downarrow} + n_\uparrow\mu_{\rm e\uparrow}}, \\ D_{\rm s} &= \frac{n_\downarrow\mu_{\rm e\downarrow}D_{\rm e\uparrow} + n_\uparrow\mu_{\rm e\uparrow}D_{\rm e\downarrow}}{n_\downarrow\mu_{\rm e\downarrow} + n_\uparrow\mu_{\rm e\uparrow}}. \end{aligned} \tag{8.36}$$

For nonmagnetic semiconductors, $n_\uparrow = n_\downarrow$, $\mu_{\rm e\uparrow} = \mu_{\rm e\downarrow}$, $D_{\rm e\uparrow} = D_{\rm e\downarrow}$, so Eq. (8.36) predicts $\mu_{\rm s} = \mu_{\rm e}$, $D_{\rm s} = D_{\rm e}$. Thus, the mobility and diffusion coefficient of the spin packet are the same as those of the background electrons.

For a single species,

$$eD = -\mu \frac{\int_0^\infty N(E) f_0(E) \widehat{d}E}{\int_0^\infty N(E)(f_0(E)/\partial E)\widehat{d}E} = 2\mu k_{\rm B}T \frac{F_{1/2}[E_{\rm F}/k_{\rm B}T]}{F_{-1/2}[E_{\rm F}/k_{\rm B}T]}, \tag{8.37}$$

where $N(E)$ is the density of states, and $f_0(E)$ is the Fermi distribution function. For a bulk parabolic band, $N(E) = (2E)^{1/2}m^{*3/2}/(\pi^2\hbar^3)$, $F_n(\xi) = \int_0^\infty x^n[\exp(x-\xi)+1]^{-1}$. In the low density limit, $[\partial f_0(E)/\partial E] = -f_0(E)/k_{\rm B}T$; thus, $eD = \mu k_{\rm B}T$, which is the Einstein relation. For degenerate systems of high density, $eD/\mu k_{\rm B}T > 1$. Therefore, $eD/\mu k_{\rm B}T$ is adjustable by doping in semiconductors.

Figure 8.30 shows the $eD/\mu k_{\rm B}T$ of spin packet and charge packet in n-GaAs at $T = 1.6$ and 300 K as a function of background electron density

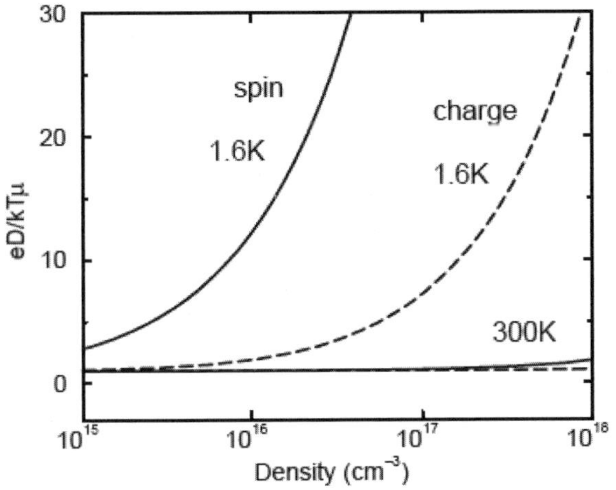

Fig. 8.30. $eD/\mu k_B T$ of spin packet (solid lines) and charge packet (dashed lines) in n-GaAs at $T = 1.6\,K$ and 300 K as a function of background electron density n.

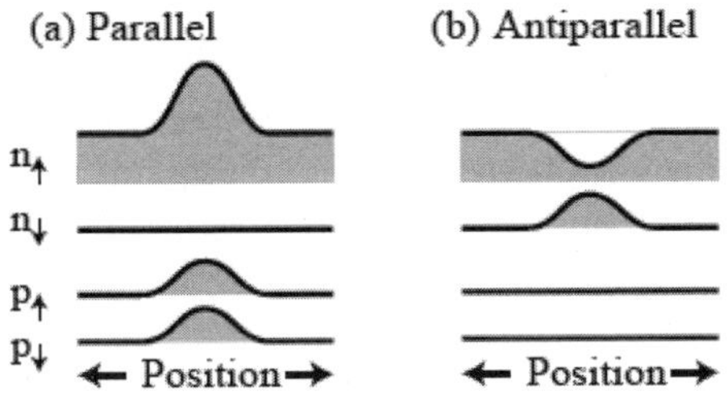

Fig. 8.31. (a) Parallel spin packet in an n-type DMS and (b) antiparallel spin packet.

n (Flattè and Byers, 2000). The $eD/\mu k_B T$s of spin packet are much larger than those of charge packet. At $n = 10^{16}\,cm^{-3}$, the $eD/\mu k_B T$ of spin packet ~12, larger than that of charge packet by one order of magnitude, in agreement with the experimental result that the spin diffusion length is larger than the charge diffusion length by one order of magnitude.

In n-type DMSs, for example BeMnZnSe, the CB has large splitting in a small magnetic field due to the sp–d exchange interaction, i.e. the giant Zeeman splitting effect. In the magnetic field ~1 T and at low temperatures, 100% spin polarization can be reached, as shown in Fig. 8.31(a). For this semiconductor in equilibrium $\Delta n_\uparrow > 0$, here ↑ is defined as the low-energy spin direction in the magnetic field. Meanwhile, the hole densities have variations $\Delta p_\uparrow$ and $\Delta p_\downarrow$, constituting parallel spin packet with $\Delta n_\uparrow$.

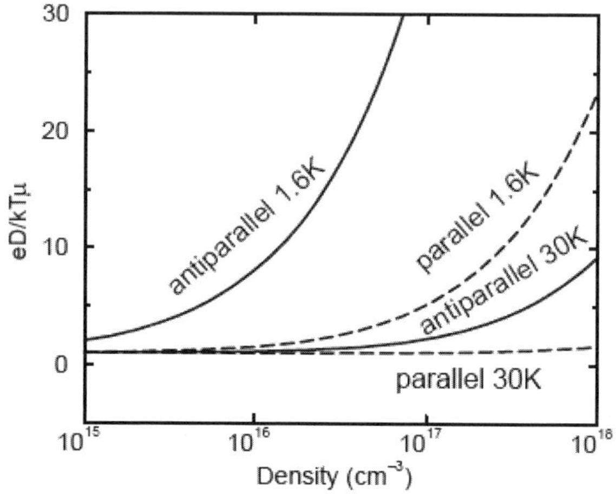

Fig. 8.32. $eD/\mu k_B T$ of antiparallel and parallel spin packets in n-$Be_xMn_yZn_{1-x-y}Se$ at $T = 1.6$ and 30 K as a function of electron density.

Another case is shown in Fig. 8.31(b), where the spin-polarized antiparallel spin packet is produced by optical excitation, $\Delta n_\downarrow > 0$, and $\Delta n_\uparrow < 0$. Meanwhile, there is no variation in the hole density. Such a single-band spin pocket would have diffusion and mobility properties entirely determined by those of the majority electrons and thus would move over an order of magnitude faster.

Figure 8.32 shows the $eD/\mu k_B T$ of antiparallel and parallel spin packets in n-$Be_xMn_yZn_{1-x-y}Se$ at $T = 1.6$ K and 30 K as a function of electron density (Flattè and Byers, 2000). The $eD/\mu k_B T$s of the antiparallel spin packet are much larger than those of the parallel spin packet. The behavior of the spin packet in a spin-polarized p-type semiconductor is completely opposite. Here a parallel spin packet of hole would require a conduction electron component ($n \ll p$). The minority carriers (electrons) would determine the mobility and diffusion coefficient (larger) of such a packet. An antiparallel spin packet of hole would consist entirely of holes and would have a much smaller mobility and diffusion coefficient.

8.7.2. *Effect of electric field on spin-polarized current*

The gradient of the spin-polarized electron density can produce space separation of opposite spin electrons; now we consider the drift–diffusion equation of spin-polarized electrons in an applied electric field (Yu and Flattè, 2002a,b). Assume that there is only spin pocket of electron in n-type semiconductor (see Fig. 8.29(c)), then the electron currents of spin up and spin down can be written as

$$\begin{aligned} \boldsymbol{j}_\uparrow &= \sigma_\uparrow \boldsymbol{E} + eD_\uparrow \nabla n_\uparrow, \\ \boldsymbol{j}_\downarrow &= \sigma_\downarrow \boldsymbol{E} + eD_\downarrow \nabla n_\downarrow, \end{aligned} \tag{8.38}$$

which consist of the drift current and the diffusion one. Here $n_{\uparrow(\downarrow)}$ is the up-spin (down-spin) electron density deviation from its equilibrium value $n^0_{\uparrow(\downarrow)}$, and $\sigma_{\uparrow(\downarrow)}$ and $D_{\uparrow(\downarrow)}$ are conductivity and diffusion constant of two kinds of electrons, respectively.

The continuity equations for up-spin and down-spin electrons in systems including spin-flip process are

$$\begin{aligned} \frac{\partial n_\uparrow}{\partial t} &= -\frac{n_\uparrow}{\tau_{\uparrow\downarrow}} + \frac{n_\downarrow}{\tau_{\downarrow\uparrow}} + \frac{1}{e}\nabla \cdot \boldsymbol{j}_\uparrow, \\ \frac{\partial n_\downarrow}{\partial t} &= -\frac{n_\downarrow}{\tau_{\downarrow\uparrow}} + \frac{n_\uparrow}{\tau_{\uparrow\downarrow}} + \frac{1}{e}\nabla \cdot \boldsymbol{j}_\downarrow, \end{aligned} \tag{8.39}$$

where $\tau_{\uparrow\downarrow}(\tau_{\downarrow\uparrow})$ is the spin-flip scattering time. In a steady state, $\partial n_{\uparrow(\downarrow)} = 0$, we have

$$\begin{aligned} \nabla\sigma_\uparrow \cdot \boldsymbol{E} + \sigma_\uparrow \nabla \cdot \boldsymbol{E} + eD_\uparrow \nabla^2 n_\uparrow &= \left(\frac{n_\uparrow}{\tau_{\uparrow\downarrow}} - \frac{n_\downarrow}{\tau_{\downarrow\uparrow}}\right) e, \\ \nabla\sigma_\downarrow \cdot \boldsymbol{E} + \sigma_\downarrow \nabla \cdot \boldsymbol{E} + eD_\downarrow \nabla^2 n_\downarrow &= \left(\frac{n_\downarrow}{\tau_{\downarrow\uparrow}} - \frac{n_\uparrow}{\tau_{\uparrow\downarrow}}\right) e, \end{aligned} \tag{8.40}$$

where $\nabla \cdot \boldsymbol{E} = -e(n_\uparrow + n_\downarrow)/\varepsilon$ and ε is the dielectric constant.

In the n-type semiconductor, because the spin polarization can be created without changing the electron density,

$$n_\uparrow + n_\downarrow = 0. \tag{8.41}$$

Multiplying two equations in Eq. (8.40) by $\sigma_\downarrow$ and $\sigma_\uparrow$, respectively, subtracting one from the other, eliminating the terms containing $\nabla \cdot \boldsymbol{E}$, and using Eq. (8.41), we obtain

$$\nabla^2(n_\uparrow - n_\downarrow) + \frac{\mu}{eD} e\boldsymbol{E} \cdot \nabla(n_\uparrow - n_\downarrow) - \frac{n_\uparrow - n_\downarrow}{L^2}, \tag{8.42}$$

where μ and D are the effective mobility and diffusion constant, respectively,

$$\begin{aligned} \mu &= \frac{\sigma_\uparrow \mu_\downarrow + \sigma_\downarrow \mu_\uparrow}{\sigma_\uparrow + \sigma_\downarrow}, \\ D &= \frac{\sigma_\uparrow D_\downarrow + \sigma_\downarrow D_\uparrow}{\sigma_\uparrow + \sigma_\downarrow}. \end{aligned} \tag{8.43}$$

The spin diffusion length

$$L = \sqrt{D\tau_{\rm s}}, \quad \frac{1}{\tau_{\rm s}} = \frac{1}{\tau_{\uparrow\downarrow}} + \frac{1}{\tau_{\downarrow\uparrow}}. \tag{8.44}$$

Equation (8.43) indicates that the behavior of spin transport is controlled by the minority spin species (σ small), analogous to ambipolar charge transport, where minority charge carriers determine the behavior.

For nonmagnetic systems, such as semiconductors, $\mu_\uparrow = \mu_\downarrow = \mu$, $D_\uparrow = D_\downarrow = D$. For ferromagnetic systems, μ and D equal approximately those of lower-conductivity spin species. Here we assign the down-spin electrons; thus, $\mu = \mu_\downarrow$, $D = D_\downarrow$. The coefficient of the second term in Eq. (8.42) μ/eD has been given in Eq. (8.37), which is simplified to the Einstein relation in the nondegenerate case.

Assume that there is continuous imbalance spin current injected at $x = 0$ $(n_\uparrow - n_\downarrow)|_0$, the imbalance spin electron density will drift and diffuse along the $\pm x$ directions. Let

$$n_\uparrow - n_\downarrow = A\exp\left(-\frac{x}{L}\right). \tag{8.45}$$

Inserting Eq. (8.45) into Eq. (8.42) and taking $\lambda = 1/L$, we obtain the equation of λ,

$$\lambda^2 - \lambda\frac{\mu E}{D} - \frac{1}{L^2} = 0. \tag{8.46}$$

Solving the second-order algebraic equation (8.46), and considering the boundary conditions at $x = 0$ and $\pm\infty$, we obtain

$$\begin{aligned} n_\uparrow - n_\downarrow &= (n_\uparrow - n_\downarrow)|_0 \exp\left(\frac{-x}{L_{\rm d}}\right), \quad x > 0 \\ n_\uparrow - n_\downarrow &= (n_\uparrow - n_\downarrow)|_0 \exp\left(\frac{x}{L_u}\right), \quad x < 0 \end{aligned} \tag{8.47}$$

It is assumed in Eq. (8.42) that the electric field is along the $-x$. direction, so that the $L_{\rm d}$ and $L_{\rm u}$ are the downstream and upstream spin diffusion lengths, respectively.

$$\begin{aligned} L_{\rm d} &= \left[-\frac{eE}{2}\frac{\mu}{eD} + \sqrt{\left(\frac{eE}{2}\frac{\mu}{eD}\right)^2 + \frac{1}{L^2}}\right]^{-1}, \\ L_{\rm u} &= \left[\frac{eE}{2}\frac{\mu}{eD} + \sqrt{\left(\frac{eE}{2}\frac{\mu}{eD}\right)^2 + \frac{1}{L^2}}\right]^{-1}. \end{aligned} \tag{8.48}$$

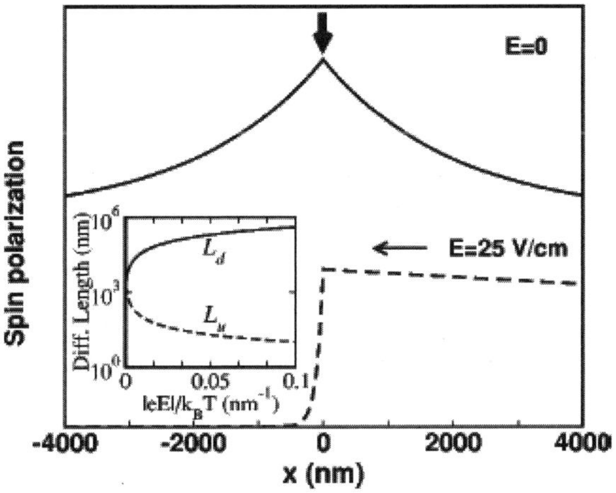

Fig. 8.33. $T = 3\,\mathrm{K}$, in electric fields $E = 0$ and $25\,\mathrm{V/cm}$, the spin polarization $n_\uparrow - n_\downarrow$ as a function of position for a spin imbalance injected at $x = 0$. The inset shows the upstream (dashed line) and downstream (solid line) diffusion lengths as a function of the electric field.

Figure 8.33 shows at $T = 3\,\mathrm{K}$, in electric fields $E = 0$ and $25\,\mathrm{V/cm}$, the spin-polarization $n_\uparrow - n_\downarrow$ as a function of position for a spin imbalance injected at $x = 0$. When $E = 0$, the spin polarization distribution along the $\pm x$ directions is symmetric, and the downstream and upstream diffusion lengths equal the intrinsic diffusion length L. When $E \neq 0$, obviously the downstream spin diffusion length $L_\mathrm{d} > L$, and the upstream diffusion length $L_\mathrm{u} < L$, which is caused by the electric field drift movement. The inset shows the upstream (dashed line) and downstream (solid line) diffusion lengths as a function of electric field; they increase and decrease with increasing electric field, respectively. Define a critical electric field,

$$\frac{eE_\mathrm{c}}{2k_\mathrm{B}T} = \frac{1}{L}. \tag{8.49}$$

When $E > E_\mathrm{c}$, L_d and L_u deviate obviously from L. If take $L = 2\,\mu\mathrm{m}$, at $T = 300\,\mathrm{K}$, $E_\mathrm{c} = 125\,\mathrm{V/cm}$; $T = 3\,\mathrm{K}$, and $E_\mathrm{c} = 1.25\,\mathrm{V/cm}$.

8.8. Coherent Manipulation of a Single Spin in Semiconductors

Single-spin systems in semiconductors broadly fall into two categories: atomic impurities and QDs. One of prime example of atomic impurities is the nitrogen–vacancy (N–V) colour center in diamond, which consists of a substitutional nitrogen atom next to a missing carbon vacancy. This N–V centre has a paramagnetic electron spin and a strong optical transition at a visible wavelength, which allows optical imaging of single spins.

QDs, on the other hand, behave like atoms in many cases. But QDs allow flexible control over the confinement potential and tend to be easier to excite optically. Spins in neighboring coupling QDs overlap strongly and will form two-particle wavefunctions such as spin singlet and triplet states.

QDs with a tunable number of electrons are routinely fabricated from a 2DEG that confines the carriers to a plane. Confinement in the remaining two dimensions is achieved by electric fields through metallic surface gates above the 2DEG. GaAs has been the material of choice for many years for QDs. More recently, motivated by the detrimental effect of lattice nuclear spins on the coherence times of electron spins, QDs have been also studied in materials such as silicon and carbon that can be isotropically purified to obtain a lattice that is free of nuclear spins. The second type of QDs is the self-assembled growth QDs by MBE. For instance, small islands of InGaAs can be created within a matrix of a semiconductor with a larger bandgap, such as GaAs. Another example of growth-defined dots is nanocrystal QDs, which is isolated and confines charge carriers in its small size.

8.8.1. *Single-spin rotations*

Koppens *et al.* (2006) reported the controllable coherent oscillations of a single electron spin in a QD. They isolated with certainty a single electron in each of two coupled QDs, as shown in Fig. 8.34(a). The spin of this electron can be reliably initialized to the ground state, spin-up, via optical pumping or by thermal equilibration at sufficiently low temperatures and strong static magnetic field ($T = 100\,\text{mK}$ and $B_{\text{ext}} = 1\,\text{T}$), as shown in Fig. 8.34(c) the left first one. The spin states are also very long-lived, with relaxation times of the order of milliseconds. The state of the coupling dots can be described via the occupations (m, n) of the left and right dots, the initial state is (0,1). When an electron enters the left dot (with rate Γ_1) starting from (0,1), the two-electron system that is formed can be either a singlet S(1,1) or a triplet T(1,1). When the system is in T(1,1), current is blocked as shown in Fig. 8.34(c), the left second one, because the energy of the T(1,1) state is much higher than that of the S(1,1) state. From S(1,1), further current is possible via a transition to S(0.2) (with rate Γ_m), as shown in Fig. 8.34(c), the left fourth one. How be the T(1,1) state changed

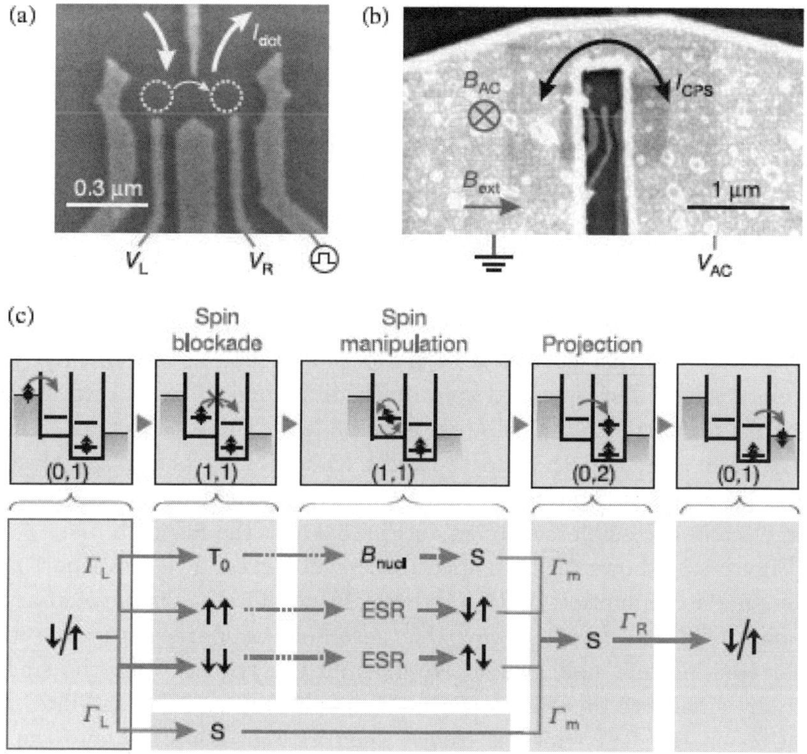

Fig. 8.34. (a) Scanning electron microscope (SEM) image of the two coupling QDs system in 2DEG. White arrows indicate the current flow through the two coupled dots (dotted circles). (b) SEM image of a device similar to the one used in the experiment, and the directions of the magnetic fields. (c) Diagrams illustrating the transport cycle in the spin blockade regime.

into the S(1,1) state? This means that the spin of the second electron flips from up to down. The most commonly used technique for inducing spin flips is ESR. ESR is the physical process whereby electron spins are rotated by an oscillating magnetic field B_{ac} with frequency f_{ac}, which is resonant with the spin precession frequency in an external magnetic field B_{ext}, oriented perpendicularly to B_{ac} (Fig. 8.34(b)). $hf_{ac} = g\mu_{\text{B}}B_{\text{ext}}$, where μ_{B} is the Bohr magneton and g the electron spin g-factor. The spin of the second electron flips via ESR, and the state changes from T(1,1) to S(1,1), as shown in Fig. 8.34(c).

In a double-dot system, spin flips can be detected through the transition of an electron from one dot to the other rather than between a dot and a reservoir. Thus, there is no need for the electron spin Zeeman splitting to exceed the temperature of the electron reservoirs ($\sim$100 mK). The experiment can be performed at a smaller static magnetic field, and consequently with lower excitation frequencies.

To detect the ESR-induced spin rotations, the authors used electric transport measurements through the two dots in series in the spin blockade regime where current flow depends on the relative spin state of the electrons in the two dots. When the state of two electron spins is in the T(1,1) state, the left electron cannot move to the right dot because the right dot's triplet is much higher in energy. The electron also cannot move back to the lead and therefore further current flow is blocked as soon as any of the triplet state is formed. This is called spin blockade regime. If two electrons form a double-dot singlet S $=\uparrow\downarrow - \downarrow\uparrow$, it is possible for the left electron to move to the right dot, and then to the right lead, leaving behind an electron in the right dot with spin $\uparrow$ or spin $\downarrow$, since the right dot singlet state is energetically accessible, as shown in Fig. 8.34(c), the left fifth one.

Figure 8.35 shows the ESR spectroscopy observed in the transport measurements as a function of $B_{\rm ext}$, where the satellite peaks develop at the resonant field $B_{\rm ext} = \pm hf_{\rm ac}/g\mu_{\rm B}$. The zero-field peak arises from the inhomogeneous nuclear field, which admixes all the triplets with the singlet. The key signature of ESR is the linear dependence of the satellite peak location on the RF, which is clearly seen in Fig. 8.35(c), where the RF frequency is varied from 10 to 730 MHz. From a linear fit through the top of the peaks, a g-factor of 0.35 ± 0.01 was obtained, which is in agreement with that of confined electrons in GaAs QD.

The authors also rotated the spin by applying RF bursts with a variable length. When one of the electrons is rotated over $(2n+1)\pi$, the two-electron state evolves to $\uparrow\downarrow$ or $\downarrow\uparrow$, giving a maximum contribution to the current. However, no electron flow is observed after rotations of $2n\pi$, where two parallel spins in the two dots are found after the RF burst. In this way, about eight oscillations of the spin state (the so-called Rabi oscillations) are observed during a microsecond burst (Koppens *et al.*, 2006).

8.8.2. *Quantum control of a single QD spin using ultrafast optical pulses*

Although the rotation of a single electron spin has been successfully demonstrated using ESR (Koppens *et al.*, 2006), the time required to achieve rotation with this technique is rather long, typically longer than a few

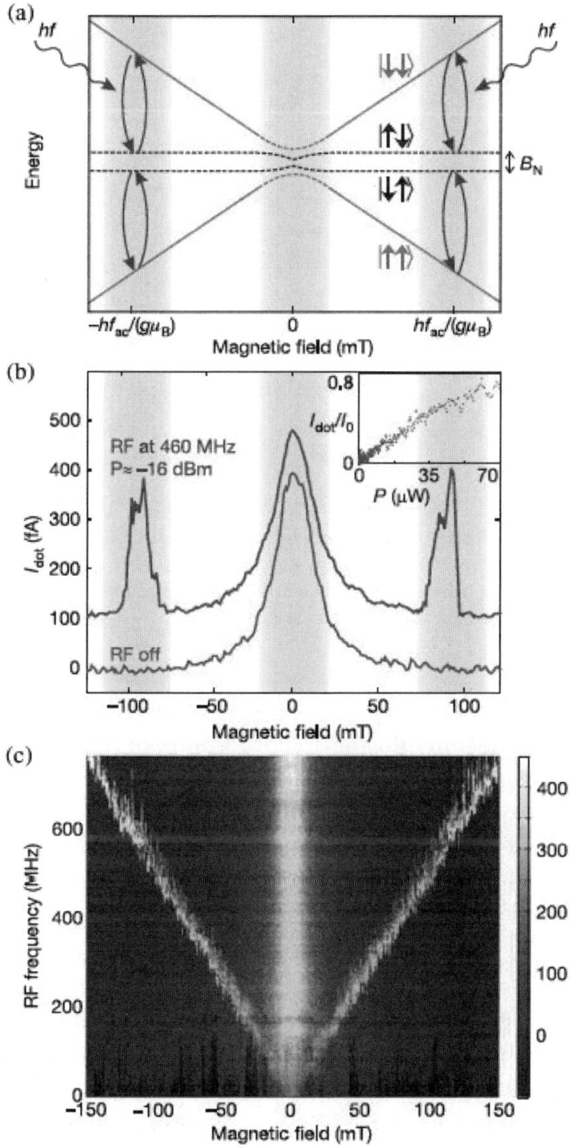

Fig. 8.35. (a) Energy diagram showing the relevant eigenstates of two electrons in a double-dot, subject to an external magnetic field and nuclear fields. (b) Current measured through the double-dot as a function of B_{ext} in the spin-blockade regime, with (offset by 100 fA) and without a RF magnetic field. Satellite peaks appear as the external magnetic is swept through the spin resonance condition. (c) Currents through the dots as functions of B_{ext} and RF.

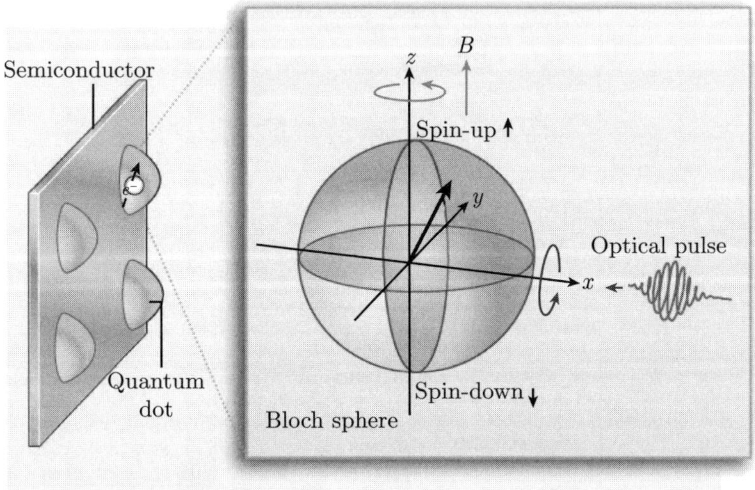

Fig. 8.36. Quantum state of the electron's spin represented by a vector from the origin of the Bloch sphere to a point on its surface: the spin-up and spin-down states are at the north and south poles, respectively. The spin rotation about the x-axis is produced by a CP optical pulse injected along the x-axis.

nanoseconds. For quantum information processing to remain effective, rotation must be performed within a timescale that is much shorter than the spin's decoherence time. Press *et al.* (2008) proposed that the short rotation time may be achieved using ultrafast optical pulses. The scheme to rotate a single electron spin using a picosecond pulse is schematically shown in Fig. 8.36 (Edamatsu, 2008).

In place of the microwave field used in ESR, a CP laser pulse is applied along the x-axis, which produces the spin rotation about the x-axis. A single electron is confined in the QD. The electron spin states $|\downarrow\rangle$ and $|\uparrow\rangle$ are split by an externally applied field $B_{\text{ext}} = 7\,\text{T}$, aligned parallel to the z-axis (Voigt geometry) as shown in Fig. 8.36. A narrowband, continuous-wave laser optically drives the $|\downarrow\rangle \rightarrow |\uparrow\downarrow, \Downarrow\rangle$ transition with rate Ω_{p}, where $|\Downarrow\rangle$ denotes heavy hole state with spin down. Spontaneous decay quickly initializes the electron into the $|\uparrow\rangle$ state. After spin rotation, the population in the $|\downarrow\rangle$ state is measured using the same optical pumping process. If the spin is rotated to $|\downarrow\rangle$, the QD will emit a single photon from the $|\uparrow\downarrow, \Downarrow\rangle \rightarrow |\uparrow\rangle$ transition, which can be detected using a single-photon counter. The count rate is proportional to the instantaneous population in $|\downarrow\rangle$.

The optical initiation is calibrated by measuring the single-photon count rate as a function of optical pumping power, P_{OP}, following a fixed rotation through angle $\Theta = \pi$. The signal saturates at around $P_{\mathrm{OP}} \approx 15\,\mu\mathrm{W}$ as the population in the $|\downarrow\rangle$ state becomes almost completely initialized to $|\uparrow\rangle$. By comparing the count rates immediately before and after the rotation pulse, the spin initialization fidelity is estimated to be $F_0 = 92 \pm 7\%$.

Figure 8.37(a) shows the single photon count rate as a function of rotation pulse power P_{RP}, where the Rabi oscillations between the spin states are evident (Press *et al.*, 2008). Figure 8.37(b) shows the rotation angle as a function of rotation pulse power, giving an empirical fit $\Theta \propto P_{\mathrm{RP}}^{0.68}$ in the range $\pi \leq \Theta \leq 13\pi$. The amplitude of the Rabi oscillations shrinks owing to incoherent processes such as trion dephasing. This decreasing amplitude is well fit by an empirical exponential decay proportional to $\exp(-\Theta/8.6\pi)$, as shown in Fig. 8.37(c).

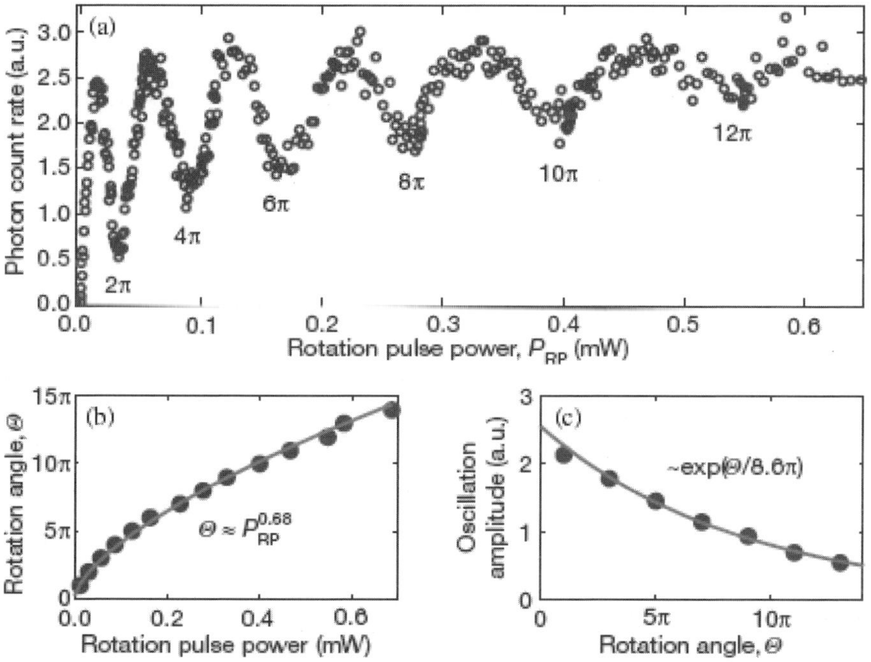

Fig. 8.37. (a) Single-photon count rate as a function of rotation pulse power P_{RP}. (b) Rotation angle as a function of rotation pulse power. (c) Amplitude of measured Rabi oscillations as a function of the rotation angle.

8.9. Spin Polarization and Transport in Silicon

So far, spin transport has been measured in direct-bandgap semiconductors, excluding a wide range of semiconductors with indirect bandgaps. Especially, Si has long been predicted a superior semiconductor for spintronics with enhanced lifetime and transport length due to low spin–orbit scattering and lattice inversion symmetry. More recently, motivated by the detrimental effect of lattice nuclear spins on the coherent times of electron spins, QDs have been studied in materials such as silicon and carbon that can be isotopically purified to obtain a lattice that is free of nuclear spins.

8.9.1. *Measurement and control of spin transport in silicon*

Appelbaum *et al.* (2007) demonstrated CB spin transport across $10\,\mu m$ in an undoped Si device that operates by spin-dependent ballistic hot-electron filtering through ferromagnetic thin films for both spin injection and spin detection. As it is not based on magnetoresistance, the hot-electron spin injection and spin detection avoids impedance mismatch issues and prevents interference from parasitic effects. The clean collector current shows independent magnetic and electrical control of spin precession, thus confirming the spin coherent drift in the CB of silicon.

Figure 8.38 shows the operating principle and schematic band diagram of the Si spin device. At a constant emitter voltage V_e, the first collector current I_{c1} at the NiFe contact and the second collector current I_{c2} at the In contact with the n-Si substrate, under optional voltage bias V_{c1} across the single-crystal undoped float-zone Si (FZ-Si) drift region, are measured. A tunnel junction injects unpolarized hot electrons from the Al emitter into the ferromagnetic $Co_{84}Fe_{16}$ base, forming the emitter current I_e. Spin-polarized hot electrons transport over the Schottky barrier and into the FZ-Si CB, forming the current I_{c1}, with their spin parallel to the magnetization of the $Co_{84}Fe_{16}$. The $Ni_{80}Fe_{20}$ base is used as the ballistic hot-electron spin filter, forming the current I_{c2} into the n-Si substrate CB, which is dependent on the relative magnetizations of both ferromagnetic layers. When they are parallel, I_{c2} is higher than when they are antiparallel, but only if electron spin polarization is maintained through the undoped Si layer.

Figure 8.39(a) shows the injector tunnel junction current–voltage characteristics, I_c–V_e relationship, which is nonlinear as expected.

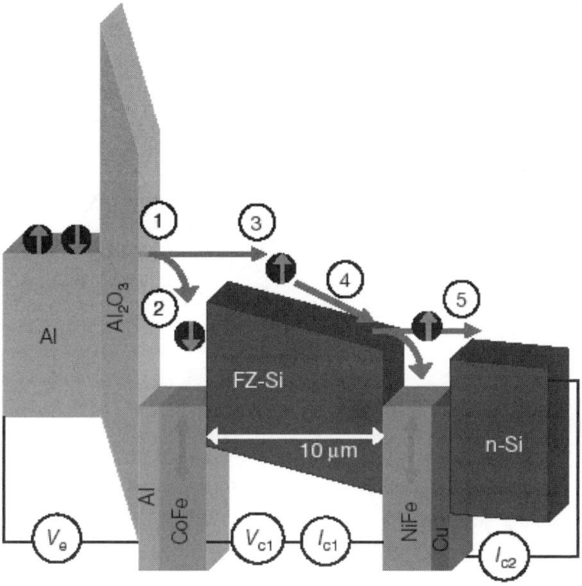

Fig. 8.38. Schematic band diagram of the Si spin device.

Figure 8.39(b) shows the simultaneous measurement of inject current I_{c1}–V_e relationship, demonstrating a threshold in I_{c1} at $V_e = -0.8\,V$. This represents that the 0.8 eV potential energy is needed for the electrons to exceed the $Co_{84}Fe_{16}$/FZ-Si Schottky barrier height. After a vertical transport through 10 μm FZ-Si, some of these electrons travel ballistically through the $Ni_{80}Fe_{20}$/Cu film and into the n Si collector, resulting in a current I_{c2} detected at the In contact with the substrate. This signal rises above the detection limit at an emitter voltage of $-1.2\,V$, as shown in Fig. 8.39(c).

Figure 8.40(a) shows the dependence of I_{c2} on the magnetic field perpendicular to the film plane with $V_{c1} = 0\,V$. The measurement begins with the field at -5 kOe and decreases toward a smaller field value (light line). there is a peak near -700 Oe and a dip near -350 Oe. Reversing the field scan direction (dark line) results in similar features reflected on the magnetic field axis. In the perpendicular magnetic field, the electron spins will precess around the magnetic field, the magnetic field of the extrema seen in Fig. 8.40(a) can be identified approximately as the conditions for π and 2π precession. The precessing spin direction, projected on the measurement axis defined by $Ni_{80}Fe_{20}$ magnetization, causes oscillation in I_{c2}.

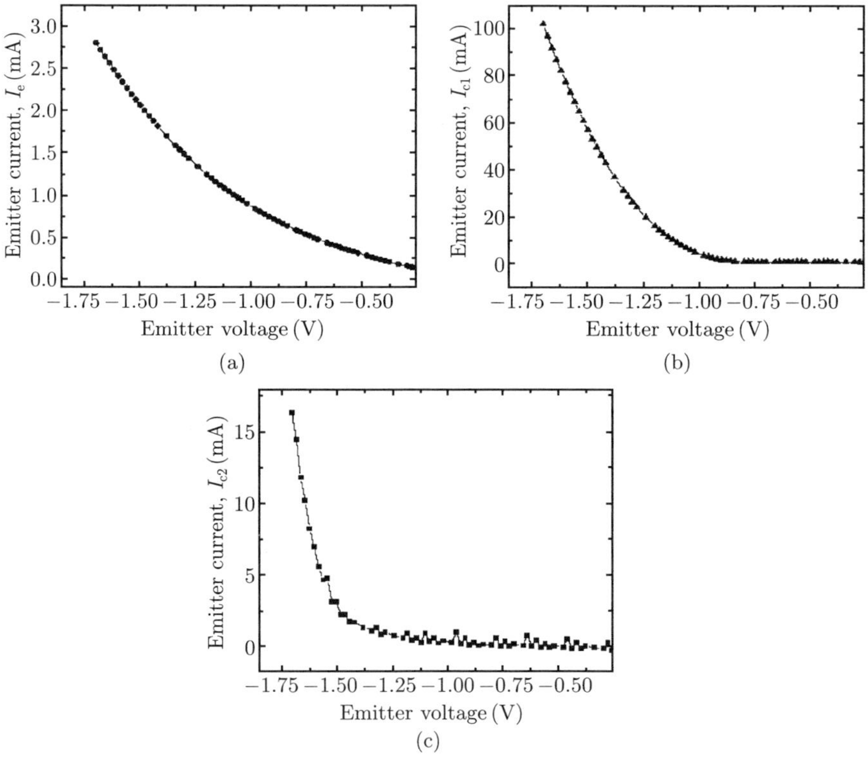

Fig. 8.39. Simultaneously measured current dependence on tunnel-junction emitter voltage V_e at 85 K. (a) I_e, (b) I_{c1} for $V_{c1} = 0$ V, and (c) I_{c2}.

Because of the action of random diffusion, spin dephasing occurs with higher precession angles, resulting in damped higher-order oscillations (Hanle effect). Therefore, just a few multiples of π precession angle are seen. Figures 8.40(b) and 8.40(c) show the results under voltage bias of $V_{c1} = 0.5$ and 1.0 V, respectively. When the electric drift increases with an applied voltage bias, the transit time is reduced, and the precession angle at any fixed magnetic field is consequently also reduced. This pushes the extrema to higher values of perpendicular magnetic fields, and precession angles up to 4π are seen, as shown in Figs 8.40(b) and 8.40(c). In this way, maxima correspond to even multiples of π and minima to odd multiples of π. The large number of precession extrema and their expected electric file dependence are conclusive evidence of strong coherent spin transport in the FZ-Si.

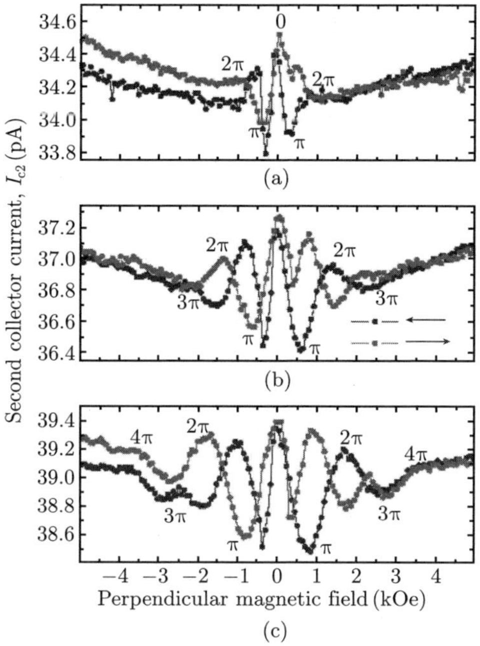

Fig. 8.40. Spin precession and dephasing in a perpendicular magnetic field at constant emitter voltage $V_e = -1.8\,\text{V}$ and 85 K. I_{c2} as a function of magnetic field under voltage bias of (a) $V_{c1} = 0\,\text{V}$, (b) 0.5 V, (c) 1.0 V.

Using the spin precession frequency $g\mu_B/h = 2.8\,\text{MHz/Oe}$, the transit time can be calculated from the magnetic field value of the first extremum, where the average precession angle is approximately π. Using $\pm 350\,\text{Oe}$ as the position of the extrema in Fig. 8.40(a), the $\tau_{\text{transit}} \approx 0.5\,\text{ns}$ is obtained, which gives an approximately $2 \times 10^6\,\text{cm}\,\text{s}^{-1}$ drift velocity through the 10 μm-thick FZ-Si. The positions of the precession extrema in Figs. 8.40(b) and 8.40(c) indicate transit time of ~0.3 and ~0.2 ns, respectively.

As the electron transit time is controlled by applied voltage V_{c1}, the normalized magnetocurrent change in $I_{c2}(\Delta I_{c2})$ can be used to deduce the spin lifetime in FZ-Si. Assuming a single exponential decay law, and using the measured transit times from Figs. 8.40(a)–(c), the spin lifetime τ_{sf} (at 85 K) can be calculated using the ratios of

$$\frac{\Delta I_{c2}(V_{c1})}{I_{c1}(V_{c1})} \propto \exp\left[-\frac{\tau_{\text{transit}}(V_{c1})}{\tau_{\text{sf}}}\right]$$

for each pair of data. Using simultaneously measured $I_{c1}(0\,\mathrm{V}) = 118\,\mathrm{nA}$, $I_{c1}(0.5\,\mathrm{V}) = 132\,\mathrm{nA}$, and $I_{c1}(1.0\,\mathrm{V}) = 139\,\mathrm{nA}$, the spin lifetime $\tau_{sf} \approx 1\,\mathrm{ns}$ is obtained. This value is consistent with spin lifetimes in the direct-bandgap semiconductor GaAs measured at similar temperatures using optical techniques.

8.9.2. *Electrical injection of spin polarization in silicon at RT*

Dash *et al.* (2009) demonstrated RT electrical injection of spin polarization into n-type and p-type silicon from a ferromagnetic tunnel contact, spin manipulation using the Hanle effect, and the electrical detection of the induced spin accumulation. The extracted spin lifetime is greater than 140 ps for conduction electrons in heavily doped n-type silicon and greater than 270 ps for holes in heavily doped p-type silicon at 300 K. The spin diffusion length is greater than 230 nm for electrons and 310 nm for holes.

Figure 8.41 shows the device geometry and the energy diagram of magnetic contact with n-Si. In Fig. 8.41(a), the left contact consists of an oxide insulator and a ferromagnetic-metal electrode (FM). The right contact is

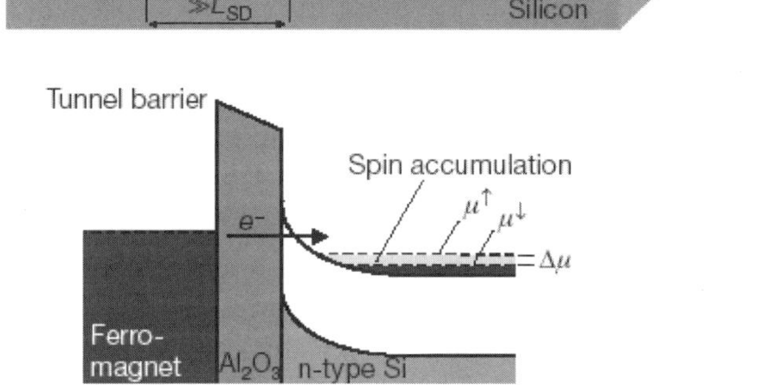

Fig. 8.41. Device geometry and energy diagram of magnetic contact with n-Si. (a) Three-terminal device for injection and detection of spin polarization in Si. (b) Energy band profile of the junction.

used to derive the source current and the middle contact to detect the voltage, which are placed away from the active interface by more than several spin diffusion length L_{SD}. A spin-polarized tunnel current is electrically injected from the ferromagnet into the Si, producing an imbalance in the Si CB or in the valence band, as shown in Fig. 8.41(b) for n-type Si. This is described by different electrochemical potentials, $\mu^{\uparrow}$ and $\mu^{\downarrow}$, for the up and down spin directions, respectively, and a spin accumulation, $\Delta\mu = \mu^{\uparrow} - \mu^{\downarrow}$. The orientation of the spin polarization is determined by the magnetization direction of the ferromagnet, which is parallel to the interface, i.e. in-plane.

When a perpendicular magnetic field is applied, the spin accumulation is reduced by means of the Hanle effect, as shown in Fig. 8.42(a).

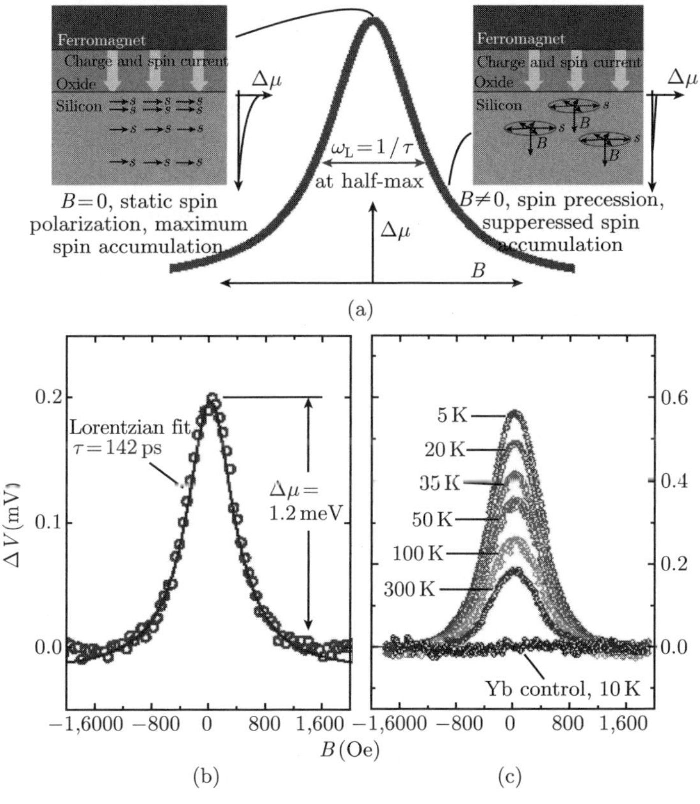

Fig. 8.42. (a) Hanle effect, producing a decay of the net spin accumulation $\Delta\mu$, due to spin precession in the magnetic field. (b) Detected ΔV across the n-Si–FM contact tunnel junction at $T = 300\,\text{K}$, as a function of magnetic field. (c) Detected ΔV as a function of magnetic field for various temperatures.

The magnetic field causes the precession of the spins at the Larmor frequency. As a result, the spin accumulation decays as a function of B with an approximately Lorentzian line shape given by $\Delta\mu(B) = \Delta\mu(0)/[1+(\omega_L\tau)^2]$, where ω_L is the Larmor frequency and τ is the spin lifetime. Similar data (Fig. 8.42(c)) were obtained over the full range of temperatures, with the detected ΔV being larger at low temperatures. Hence, the RT (300 K) data shown in Figs. 8.42(b) and 8.42(c) demonstrate the electrical injection of spin polarization into n-type silicon from a ferromagnetic tunnel contact, the Hanle precession of the electron spins, and the electrical detection of the spin accumulation.

The spin accumulation at the tunnel interface is obtained from $\Delta V = \text{TSP} \times \Delta\mu/2$, where TSP is the tunnel spin polarization of the ferromagnet–insulator interface. Using the known TSP value of 0.3 for Al_2O_3–$Ni_{80}Fe_{20}$ at 300 K, $\Delta\mu = 1.2$ meV at 300 K is obtained. From the half-width of the Hanle curve (for which $\omega_L = 1/\tau$), the spin lifetime $\tau = 142$ ps for the heavily doped n-Si with an electron density of $1.8. \times 10^{19}$ cm^{-3} at 300 K is obtained. Impurity scattering by the high density of donors is expected to reduce the spin lifetime, compared with low-doped n-Si of 10 ns, in which the EY mechanism due to photon scattering is dominant.

The spin diffusion length $L_{SD} = (D\tau)^{1/2}$ is obtained in the Si, where D is the diffusion constant $D = 3.7$ cm^2/s at 300 K as derived from the measured electron mobility. With $\tau = 142$ ps, the $L_{SD} = 230$ nm at RT for the heavily doped n-type Si. Such values are sufficient to transfer spin information over the typical length ($L < 100$ nm) of the channels of modern silicon transistors.

Figure 8.43(a) shows the detected ΔV across the p-Si–FM contact tunnel junction as a function of the applied magnetic field. The sample is boron-doped p-type Si with a measured hole density of $4.8. \times 10^{18}$ cm^{-3} at 300 K. A clear Hanle signal is observed, demonstrating electrically induced spin polarization of holes in the valence band of p-type Si. From the width of the Hanle curve, a value of $\tau = 270$ ps for the hole spin lifetime at 300 K is extracted, which is larger than that for electrons in n-type Si. It may be caused because the density of acceptor impurities in p-type Si is less than that of the donor impurities in n-type Si. For $\tau = 270$ ps and the measured hole mobility of 117 cm^2/V s ($D = 3.6$ cm^2/s), a hole spin diffusion length of $L_{SD} = 310$ nm at RT is obtained. Figure 8.43(b) shows the variation of the spin–RA product (the product of spin resistance $\Delta V/I$, and area) with V for p-type device. Just as for n-type Si, this product is nearly constant as a function of bias voltage for the polarity in which (hole) carriers are injected

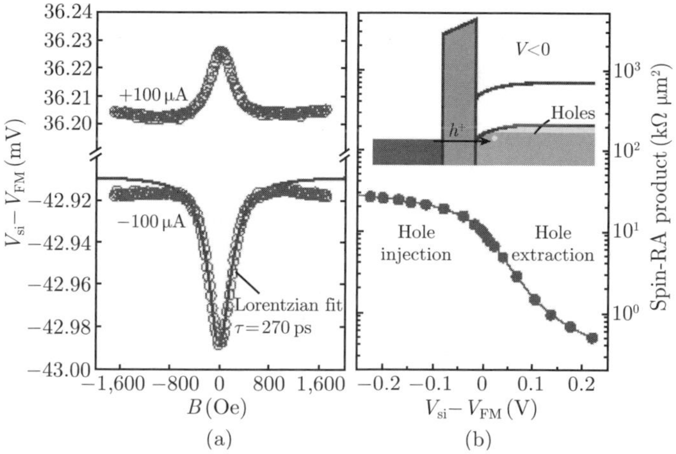

Fig. 8.43. Spin accumulation of holes in p-type Si at 300 K. (a) Detected ΔV across the p-Si-FM contact tunnel junction as a function of the applied magnetic field. Two curves correspond to constant currents −100 and +100 μA, respectively. (b) Energy band diagram of hole tunneling and spin–RA product versus applied bias voltage at 300 K.

into the Si ($V < 0$ in this case) and exhibit a faster decay when the spin accumulation is created by extrating (hole) carriers from the Si ($V > 0$).

References

Appelbaum I, Huang B, Monsma DJ. (2007) *Nature* **447**: 295.

Berezovsky J, Ouyang M, Meier F, Awschalom DD. (2005) *Phys Rev B* **71**: R081309.

Beschoten B, Johnston-Halperin E, Young DK, *et al.* (2001) *Phys Rev B* **63**: R121202.

Bhat RDR, Sipe JE. (2000) *Phys Rev Lett* **85**: 5432.

Bychkov YA, Rashba EI. (1984) *J Phys C* **17**: 6039.

Dash SP, Sharma S, Patel RS, de Jong MP, Jansen R. (2009) *Nature* **462**: 491.

Dresselhaus G. (1955) *Phys Rev B* **100**: 580.

Edamatsu K. (2008) *Nature*, **456**: 182.

Flatté ME, Byers JM. (2000) *Phys Rev Lett* **84**: 4220.

Gupta JA, Awschalom DD. (2001) *Phys Rev B* **63**: 085303.

Gupta JA, Peng X, Alivisatos AP, Awschalom DD. (1999) *Phys Rev B* **59**: R10421.

Gupta JA, Awschalom DD, Efros Al L, Rodina AV. (2002) *Phys Rev B* **66**: 125307.

Kato Y, Myers RC, Gossard AC, Awschalom DD. (2004a) *Nature* **427**: 50.

Kato YK, Myers RC, Gossard AC, Awschalom DD. (2004b) *Phys Rev Lett* **93**: 176601.
Kato YK, Myers RC, Gossard AC, Awschalom DD. (2004c) *Science* **306**: 1910.
Kikkawa JM, Awschalom DD. (1998) *Phys Rev Lett* **80**: 4313.
Kikkawa JM, Awschalom DD. (1999) *Nature (London)* **397**: 139.
Kikkawa JM, Smorchkova IP, Samarth N, Awschalom DD. (1997) *Science* **277**: 1284.
Koppens FHL, Buizert C, Tielrooij KJ, *et al.* (2006) *Nature* **442**: 766.
Murakami S, Nagaosa N, Zhang SC. (2003) *Science* **301**: 1348.
Press D, Ladd TD, Zhang BY, Yamamoto Y. (2008) *Nature* **456**: 218.
Sih V, Myers RC, Kato YK, Lau WH, Gassard AC, Awschalom DD. (2005) *Nature Physics* **1**: 31.
Sih V, Lau WH, Myers RC, Horowitz VR, Gossard AC, and Awschalom DD. (2006) *Phys Rev Lett* **97**: 096605.
Sinova J, Culcer D, Niu Q, Sinitsyn NA, Jungwirth T, MacDonald AH. (2004) *Phys Rev Lett* **92**: 126603.
Stern NP, Ghosh S, Xiang G, Zhu M, Samarth N, Awschalom DD. (2006) *Phys Rev Lett* **97**: 126603.
Stevens MJ, Smirl AL, Bhat RDR, Najmaie A, Sipe JE, van Driel HM. (2003) *Phys Rev Lett* **90**: 136603.
Yu ZG, Flatté ME. (2002a) *Phys Rev B* **66**: 201202R.
Yu ZG, Flatté ME. (2002b) *Phys Rev B* **66**: 235302.

Chapter 9

SPIN-POLARIZED ELECTRON AND DOMAIN WALL TRANSPORT

9.1. Spin Transport in Magnetic Semiconductor 2DEG

Spin transport in magnetic semiconductor 2DEG has attracted intensive attention in the past few years. Magnetic semiconductor 2DEG consists of two subsystems: free electrons and local magnetic moments coupled by the s–d exchange interaction. The quantum transport property is heavily influenced by the s–d exchange interaction. Therefore, one can obtain strongly spin-polarized current by engineering the electron band structure and distribution of magnetic impurities (Smorchkova and Samarth, 1996; Smorchkova *et al.*, 1997)

Figure 9.1 is the schematic of the magnetic semiconductor 2DEG samples. The magnetic ions Mn^{2+} are localized in several single layers, showing antiferromagnetic correlation between them, while the free electrons distribute in the ZnCdSe quantum well. The transport property is measured in a Hall bar at $T = 300$ mK, the magnetic field can be tuned to 17 T With increasing Mn concentration, the mobility decreases significantly due to the spin scattering processes. Figure 9.2 shows the measurement of the integer quantum Hall effect in the samples A and B at 4.2 K.

There are several character energy scales in the magnetic semiconductor 2DEG: the Landau level, Zeeman splitting. $\Delta E_s = g\mu_B B$, the Fermi energy, the Coulomb interaction, and temperature. In a conventional nonmagnetic semiconductor 2DEG, the spacing between the Landau levels is much larger than the Zeeman splitting. Due to the Giant Zeeman effect, the Zeeman splitting is comparable with the spacing between the Landau levels even at small magnetic fields. In the sample A without magnetic ions, the Zeeman splitting can be resolved when the magnetic field $B = 6$ T at $T = 4.2$ K. But in the sample B, the electron spins will be fully polarized at $B = 2$ T. The spin splitting is $\Delta E_s = (\Delta E)_{\max} B_{5/2}(5\mu_B B/k_B T_{\text{eff}})$, $B_{5/2}$ is the

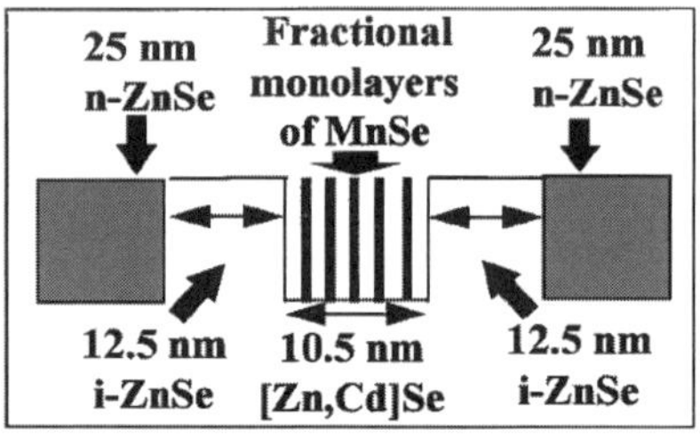

Fig. 9.1. Schematic of the magnetic semiconductor 2DEG sample. The sample consists of a modulation-doped quantum well placed in between two n-type ZnSe doping layers. The quantum well is digitally doped with MnSe layers $(Zn_{0.80}Cd_{0.20}Se)_{m-f}(MnSe)_f$, $m = 5$, $f = 0, 0.125, 0.25, 0.5$ (corresponds to the samples A, B, C, and D, respectively) (Smorchkova, Samarth, Kikkawa, Awschalom, 1997).

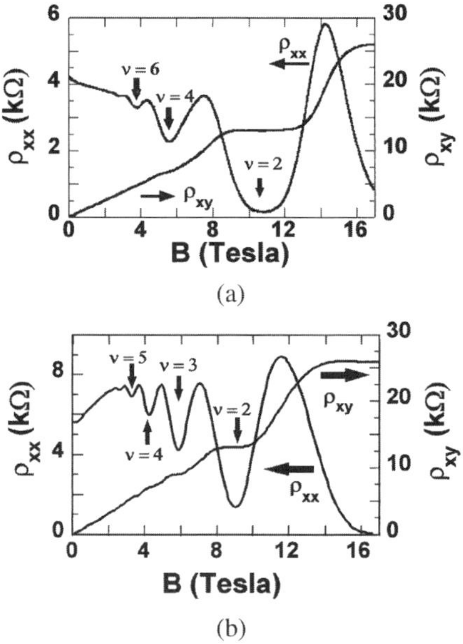

Fig. 9.2. Integer quantum Hall effect in the samples A (the upper panel) and B (the lower panel).

Brillioun function, and $T_{\text{eff}} = T + T_0$ is the pseudotemperature including the antiferromagnetic interaction between magnetic ions.

The important difference between magnetic and nonmagnetic semiconductor 2DEGs is the low-field longitudinal MR. This can be found in

Fig. 9.2 and attributed to the quantum correction of the disorder scattering to the conductivity. Magnetic fields destroy the weak localization and lead to the decrease in the MR, while the electron–electron interaction in the spin-splitted Landau levels increases the MR. The low-field MR in sample A can be understood from the standard theory of the weak localization, but the behavior of sample B is more complicated. Large spin splitting is caused by a drastic reduction in the Hartree contributions to the particle–hole diffusion channel (Smorchkova, *et al.*, 1997; Lee and Ramakrishnan, 1981, 1985), In addition, the correction caused by the magnetic ion scattering is also important (Hikami *et al.*, 1980).

Diluted 2DEGs ($n \approx 10^{11}\,\mathrm{cm}^{-2}$) display a surprising positive MR (can approach 700%) at small perpendicular and/or in-plane magnetic fields ($B = 0.1\,\mathrm{T}$) (Smorchkova *et al.*, 1998); as shown in Fig. 9.3. This feature arises from the suppression of the spin-related hopping process between local states. One can see the delocalization of electron states at strong magnetic fields. The perpendicular magnetic fields can induce a quantum transition between the spin-polarized insulating phase and $v = 1$ quantum Hall liquid.

The s–d exchange interaction leads to the beating pattern in the longitudinal MR (Teran *et al.*, 2003). The experimental measurements can be explained well based on the quantum Boltzmann transport theory (Yang *et al.*, 2005). Considering the impurity scattering, the conductivity can calculated (Vasilopolous *et al.*, 1989) as

$$\sigma_{xx} = \frac{e^2}{2K_{\mathrm{B}}T}\frac{1}{A}\sum_{i,j} f_i(1-f_i)W_{ij}(\langle i|x|i\rangle - \langle j|x|j\rangle)^2, \qquad (9.1)$$

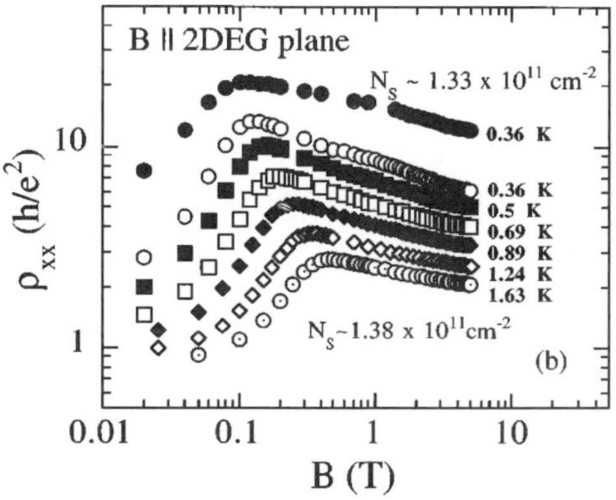

Fig. 9.3. Magnetoresistance versus in-plane magnetic field (Smorchkova *et al.*, 1998).

where $|i\rangle$ is the Landau level, W_{ij} is the transition rate induced by the impurity from the initial state $|i\rangle$ to final state $|j\rangle$, and $f(E)$ is the Fermi distribution function. The impurity potential is adopted as the screened Coulomb potential. The spin-up and spin-down electron conductivities including the Landau-level broadening are

$$\begin{aligned}\sigma_{xx}^{\uparrow} &= \frac{e^2}{2K_{\mathrm{B}}T}\frac{1}{A}\sum_{nk_y.n'k'_y} f_{nk_y\uparrow}(1-f_{n'k'_y\uparrow})\\ &\quad\times W_{nky\uparrow.n'k'_y\uparrow}(\langle nk_y|x|nk_y\rangle - \langle n'k'_y|x|n'k'_y\rangle)^2,\\ \sigma_{xx}^{\downarrow} &= \frac{e^2}{2K_{\mathrm{B}}T}\frac{1}{A}\sum_{nk_y.n'k'_y} f_{nk_y\downarrow}(1-f_{n'k'_y\downarrow})\\ &\quad\times W_{nky\downarrow.n'k'_y\downarrow}(\langle nk_y|x|nk_y\rangle - \langle n'k'_y|x|n'k'_y\rangle)^2.\end{aligned} \tag{9.2}$$

The current and electron spin polarizations are

$$\begin{aligned}P_J &= \frac{\sigma_{xx}^{\uparrow}-\sigma_{xx}^{\downarrow}}{\sigma_{xx}^{\uparrow}+\sigma_{xx}^{\downarrow}},\\ P_c &= \frac{n_{\mathrm{e}}^{\uparrow}-n_{\mathrm{e}}^{\downarrow}}{n_{\mathrm{e}}^{\uparrow}+n_{\mathrm{e}}^{\downarrow}}.\end{aligned} \tag{9.3}$$

In Fig. 9.4, the experiment measurement is compared with the theoretical results. The beating pattern in the longitudinal MR arises from the interplay between the magnetic field and the s–d exchange interaction. The spin splitting of the Landau levels induced by the s–d exchange interaction increases linearly at small magnetic fields, but saturates at strong magnetic fields. This feature results in the crossing between different Landau levels (see Fig. 9.5(b)), and consequently leads to the beating pattern in the longitudinal MR. Interestingly, the electron spin polarization is quite small, but the current polarization can be very strong, which is caused by the strong spin polarization near the Fermi energy (see the solid and dashed curves in Fig. 9.5(a)).

These results reveal that the strong spin-polarized current can be obtained in magnetic semiconductor 2DEG by designing some microstructures properly even without external magnetic fields (Yang *et al.*, 2006a).

We consider a magnetic semiconductor quantum well with ferromagnetic metal stripes deposited on it, which generate a periodic magnetic field in the quantum well. Spin-up and spin-down electrons feel different

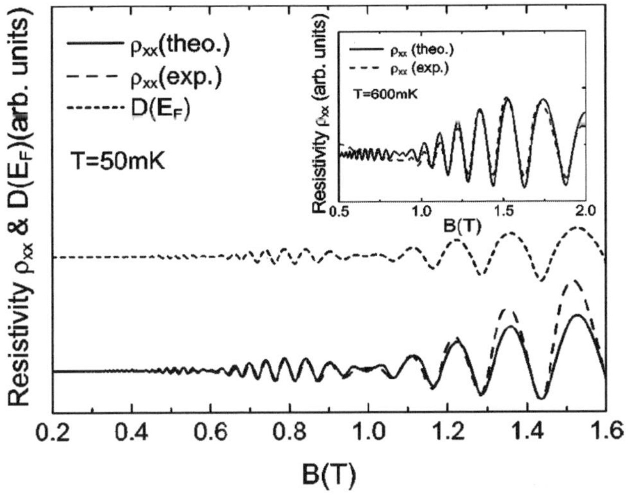

Fig. 9.4. The longitudinal MR of a magnetic semiconductor 2DEG at different temperatures $T = 50$ and 600 mK (see the inset). The solid (dashed) curves are the experimental (theoretical) results. The dotted curves describe the density of states at the Fermi energy (Yang *et al.*, 2005).

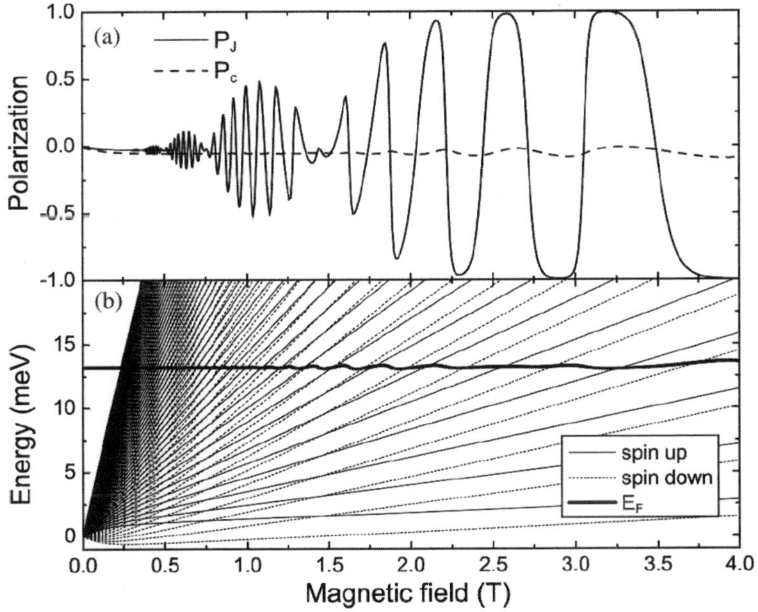

Fig. 9.5. (a) The electron (the dashed curves) and current spin polarization (the solid curves). (b) The Landau levels versus the magnetic fields (Yang *et al.*, 2005).

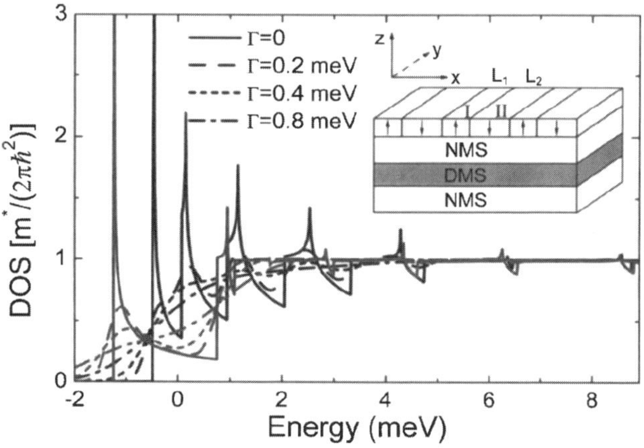

Fig. 9.6. Schematic of the magnetic semiconductor 2DEG under a periodic modulation of magnetic fields. The black (grey) solid lines denote the spin-up (spin-down) density of states of electrons. Γ indicates the different level broadenings (Yang *et al.*, 2006a).

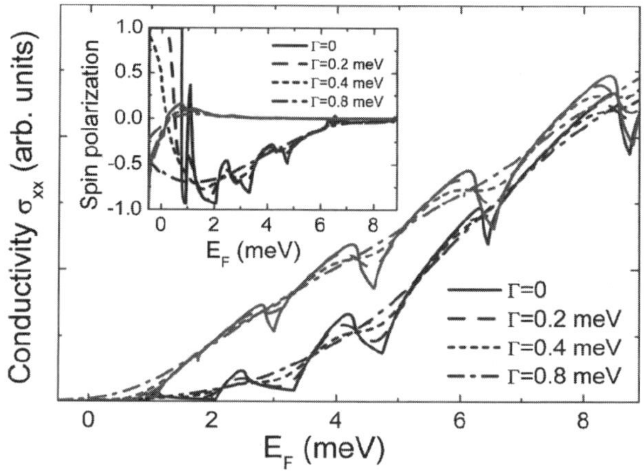

Fig. 9.7. The conductivity parallel to the ferromagnetic metal stripes against the Fermi energy; The inset shows the spin polarization as a function of the Fermi energy; the line shapes are the same as in Fig. 9.6 (Yang *et al.*, 2006a).

periodic potentials due to the giant Zeeman effect. This feature leads to the formation of the spin-dependent minibands, as shown in Fig. 9.6. In Fig. 9.7, the longitudinal conductivity and the current spin polarization are calculated from Eqs. (9.2) and (9.3) as a function of the Fermi energy.

From this figure, one can see clearly that the spin-polarized transport can be realized even without external magnetic fields.

SOI and the s–d exchange interaction can exist simultaneously in magnetic semiconductor 2DEG. The interplay between them can affect the magnetotransport property significantly, which has been studied extensively in the past few years (Gui *et al.*, 2004). Very recent experiment demonstrates that one can extract the spin splitting caused by the Rashba SOI and the s–d exchange interaction from the SdH oscillation of the longitudinal MR. In n-type $Hg_{0.98}Mn_{0.02}Te/Hg_{0.3}Cd_{0.7}Te$ quantum wells, the spin splitting caused by the Rashba SOI is comparable with that induced by the s–d exchange interaction. In Fig. 9.8, the SdH oscillations vary significantly with increasing the temperature and the gate voltage. The temperature mainly affects the spin splitting induced by the s–d exchange interaction, the electric field, i.e. the gate voltage, changes the spin splitting caused by the Rashba SOI. The competition between these spin splitting induces the variation of the Landau levels near the Fermi energy, and consequently influences the SdH oscillation.

The theoretical calculation predicts the resonant spin polarization by tuning the SOI and s–d exchange interaction (Yang *et al.*, 2006b). The Landau levels in 2DEGs are coupled by the SOIs, leading to the mixing of the Landau levels. Within the single band effective-mass theory, the $n = 0$ spin-down Landau level is decoupled with the other Landau levels in the presence of the Rashba SOI alone. The longitudinal MR can be calculated based on the Landau levels and its wavefunctions in Eqs. (9.4) and (9.5) (see Fig. 9.8). From this figure, one can clearly see that the temperature and electric fields can change the beating pattern of the longitudinal MR significantly.

$$
\begin{aligned}
E_n^{(\pm)} &= (n \pm \delta)\hbar\omega_c, \\
E_0 &= \frac{h\omega_c - \Delta_Z}{2}, \\
\delta &= \sqrt{\frac{[\Delta_Z/(\hbar\omega_c) - 1]^2}{4} + 2n\left[\frac{\alpha}{l_B \hbar\omega_c}\right]^2}, \\
\psi_{n,k_y}^{(\pm)} &= \frac{e^{ik_y y}}{\sqrt{L_y}[C_n^{(\pm)}\phi_{n-1}|\uparrow\rangle + D_n^{(\pm)}\phi_n|\downarrow\rangle]}, \\
\psi &= \frac{e^{ik_y y}}{\sqrt{L_y}\phi_0|\downarrow\rangle},
\end{aligned}
\tag{9.4}
$$

where $\Delta_Z = g_e\mu_B B - N_0\alpha_{ex}x\langle S_z\rangle$, L_y is the length of the sample along the y-axis, $\{\varphi_n\}$ is the wavefunction of the 1D harmonic oscillator located

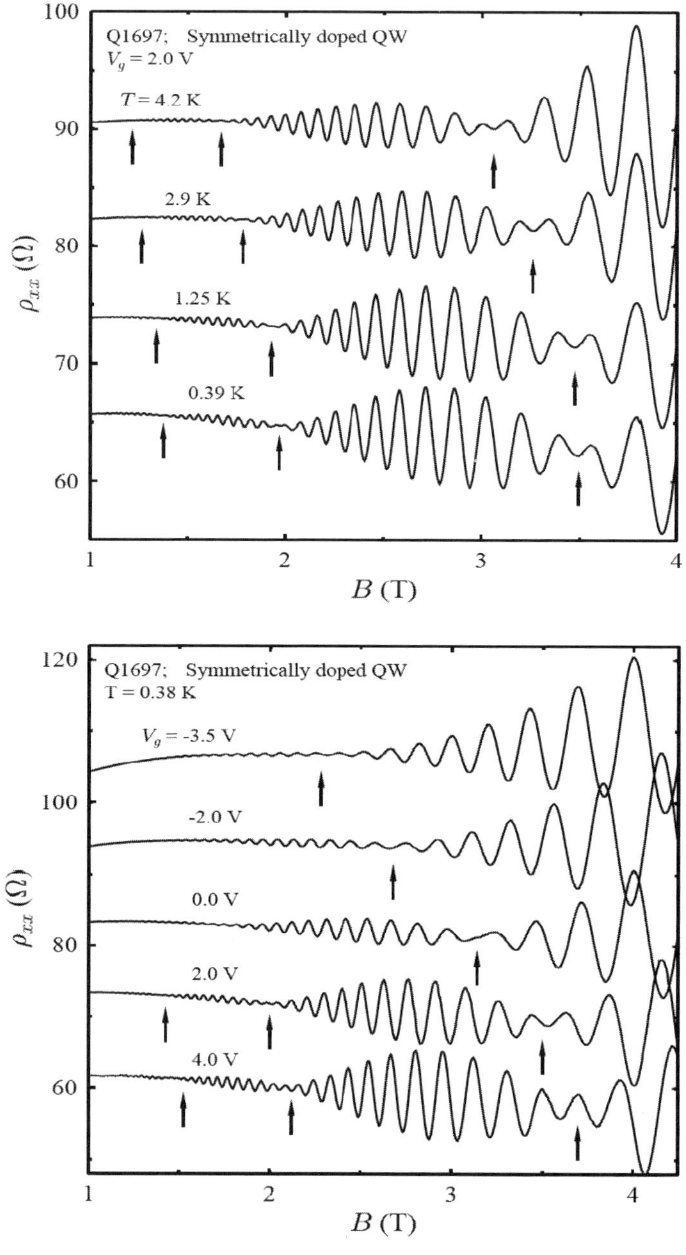

Fig. 9.8. The SdH oscillation in n-type magnetic semiconductor $Hg_{0.98}Mn_{0.02}Te$/ $Hg_{0.3}Cd_{0.7}Te$ quantum wells at different temperatures and electric biases (Yang *et al.*, 2006b).

at $x = x_0$ (Yang *et al.*, 2006b),

$$\begin{aligned}
x_0 &= -k_y l_B^2, \quad l_B = \sqrt{\frac{\hbar}{eB}}, \\
C_n^{(+)} &= -D_n^{(-)} = \frac{P_n}{A_n}, \quad C_n^{(-)} = D_n^{(+)} = \frac{1}{A_n}, \quad \text{for } \Delta_Z < \hbar\omega_c, \\
C_n^{(+)} &= D_n^{(-)} = \frac{1}{A_n}, \quad C_n^{(-)} = -D_n^{(+)} = -\frac{P_n}{A_n}, \quad \text{for } \Delta_Z > \hbar\omega_c, \\
P_n &= \frac{\sqrt{2n}\, l_B^{-1}\alpha}{|\Delta_Z - \hbar\omega_c|/2 + \hbar\omega_c\delta}, \quad A_n = \sqrt{1 + P_n^2}.
\end{aligned} \tag{9.5}$$

The calculation also shows that one can obtain strongly spin-polarized current by tuning the electric field properly. Figure 9.9 plots the current spin polarization in 2DEGs against the Rashba coefficient α at $B = 0.5\,\text{T}$ and $T = 1\,\text{K}$, the electron concentration $n_e = 4 \times 10^{11}\,\text{cm}^{-2}$, and the effective magnetic ion concentration $x_{\text{eff}} = 0.02$. In this figure, a strongly spin-polarized current can be generated at a specific value of the strength of the Rashba SOI α. The underlying physics can be understood from the inset. When the electric field increases, i.e., the strength of the Rashba SOI α, the Fermi energy decreases. The Fermi energy crosses with the lowest

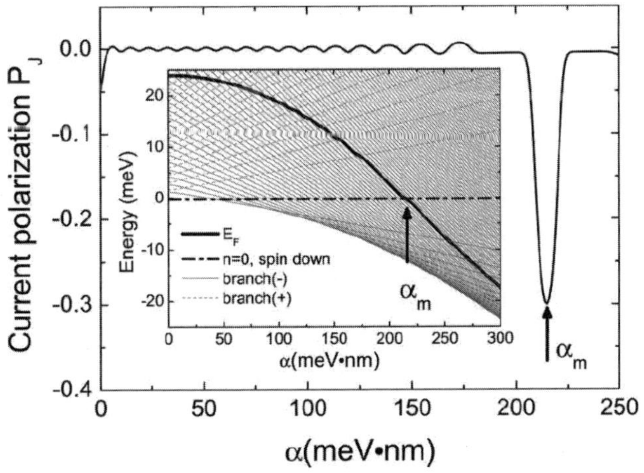

Fig. 9.9. The current spin polarization in 2DEG is a function of the Rashba coefficient α at $B = 0.5\,\text{T}$ and $T = 1\,\text{K}$, when $n_e = 4 \times 10^{11}\text{cm}^{-2}$, $x_{\text{eff}} = 0.02$, the inset shows the variation of the Femi energy and Landau level against the Rashba coefficient α (Yang *et al.*, 2006b).

spin-down Landau level at $\alpha = 220\,\text{meV}\,\text{nm}$, resulting in a strong current spin polarization.

In the presence of both the Rashba and Dresselhaus SOIs (the latter usually exists in the conventional zincblende semiconductor), the situation becomes more completely since all the Landau levels are coupled together (Yang and Chang, 2006). The beating pattern in the SdH oscillation disappears with increasing the strengths of the Rashba and Dresselhaus SOIs.

9.2. Spin Transport Through QDs

One of the central issues in semiconductor spintronic devices is to develop a technique to manipulate electron spin without external magnetic fields (Ohno *et al.*, 2000; Kato *et al.*, 2004). Spin injection into a single QD has attracted intensive attentions. So far, there are many theoretical works studying spin transport through nonmagnetic semiconductor QDs with ferromagnetic metal electrodes. These studies predict that spin accumulation will appear inside the dots (Brataas *et al.*, 1999; Imamura *et al.*, 1999; Martinek *et al.*, 1999, 2002; Rudzinski and Barnaś, 2001; Wetzels *et al.*, 2005; Weymann *et al.*, 2005). Recently, spin injection in a nonmagnetic metal nanoparticle was realized experimentally, which implies that the spin accumulation occurs inside the particle. MTJs consisting of single InAs QD with ferromagnetic Co leads were fabricated and its MR demonstrates the spin-dependent transport through the QD. One can see clearly the hyposesis in the TMR, which can be tuned by tuning the gate voltage (Bernand-Mantel *et al.*, 2006).

9.2.1. *Spin transport through the double barrier tunneling junctions with ferromagnetic leads*

Very recently, Jung *et al.* measured the electron transport through single self-assembled InAs QDs grown on GaAs surfaces with metallic leads. The results show clear Coulomb staircases and Coulomb gaps which were observed at 4.2 K (Jung *et al.*, 2005a,b; Shibata *et al.*, 2007). By using ferromagnetic leads, one can inject spin-polarized electrons into the QD. If the SRT is long enough, spin accumulation inside the QDs can be detected experimentally by measuring the TMR. The spin accumulation was indeed detected in the TMR through semiconductor double barrier junctions (DBJs) [(Ga,Mn)As/AlAs/GaAs/AlAs/(Ga,Mn)As].

A lateral double barrier tunneling junction was recently used to measure spin transport through single semiconductor QD (Hamaya *et al.*, 2007a). This DBJ consists of a single self-assemble InAs QD and ferromagnetic Co leads. The transport through the MTJ shows clearly the Coulomb blockade and TMR. The TMR can be changed significantly by tuning the gate voltage, which means that one can control spin transport by using external electric fields. Self-assembled InAs QDs were grown by MBE on a (100)-oriented n^+-GaAs substrate. After successively growing a 100 nm-thick AlGaAs barrier layer and a 200 nm-thick undoped GaAs buffer layer, the (001) n^+-GaAs layer was used as a backgate electrode. Using standard electron beam lithography and lift-off method, metallic electrodes with 30–50 nm nanogap can be fabricated, as shown in Fig. 9.10(a). A single InAs QD with 50 nm diameter was contained between the two 40 nm-thick Co leads. The two Co leads have different widths, one Co lead with 1 μm width and 20 μm length, another Co lead with 200 nm width and 20 μm length; therefore, each lead displays different switching fields. The oxide layer between InAs QD and Co leads forms a tunneling barrier. The designed

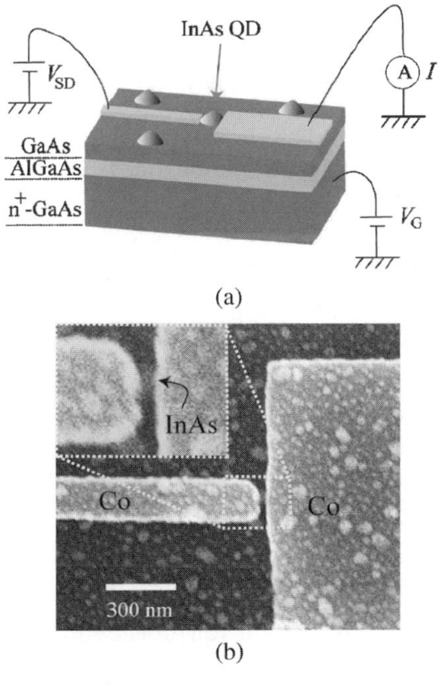

Fig. 9.10. (a) Schematic of Co/InAs/Co DBJ and (b) the SEM image of the DBJ (Hamaya *et al.*, 2007a).

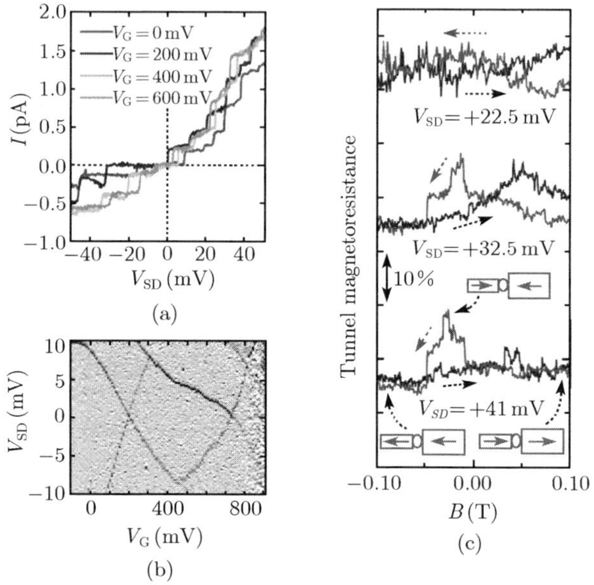

Fig. 9.11. (a) $I-V_{\mathrm{SD}}$ curve of a Co/InAs/Co MTJ at $T = 50\,\mathrm{mK}$ for various backgate voltages V_{G}. (b) Differential conductance $\mathrm{d}I/\mathrm{d}V_{\mathrm{SD}}$ as a function of the source-to-drain voltage V_{SD} and V_{G}. (c) Magnetic field dependence of the TMR at $V_{\mathrm{G}} = 200\,\mathrm{mV}$ and $T = 50\,\mathrm{mK}$, for various V_{SD} (Hamaya *et al.*, 2007a).

device structure behaves like a double barrier tunneling junction structure. Figure 9.10(b) is the SEM image of Co/InAs/Co DBJ, in which a single InAs QD is located in between two Co leads. The transport measurement was carried out at 50 mK using the dc method.

Under different backgate voltages, the $I-V_{\mathrm{SD}}$ curve of Co/InAs/Co MTJ is shown in Fig. 9.11(a), where V_{SD} is the source-to-drain voltage. An external magnetic field ($\approx 0.2\,\mathrm{T}$) is parallel to the Co leads. From the $I-V_{\mathrm{SD}}$ curve, one can see that the conductance near $V_{\mathrm{SD}} = 0$ vanishes, indicating the presence of Coulomb blockade, and the staircase of the $I-V_{\mathrm{SD}}$ curve, the so-called "Coulomb staircase." Notice that the resistance jumps of approximately gigaohm are much larger than the quantum resistance of $h/e^2 \sim 25.8\,\mathrm{k\Omega}$, and the junction resistance is very large. Therefore, the coupling between the Co leads and junction becomes very weak, and the tunneling processes play a dominant role in the sample. Further, the $I-V_{\mathrm{SD}}$ curve is asymmetric respect to V_{SD}, which means the double barrier in the device is asymmetric. In Fig. 9.11(b), the differential conductance $\mathrm{d}I/\mathrm{d}V_{\mathrm{SD}}$ is plotted as a function of V_{SD} and V_{G}. Coulomb blockade takes place in the dark-colored diamond-shaped areas, each of which is associated with a well defined number N of confined electrons. These results demonstrate that Coulomb blockade in a single semiconductor QD with ferromagnetic leads and the possibility for realizing single-electron transistor using this Co/InAs/Co MTJ structure.

Figure 9.11(a) shows the $I-V_{\text{SD}}$ curve of a Co/InAs/Co MTJ at $T = 50\,\text{mK}$ for different backgate voltages V_{G}. From this figure, one can see that the current is not sensitive to the backgate voltages. The staircases of the current against V_{SD} indicates the variation of the tunneling electron number. Figure 9.11(c) describes the dependence of TMR on the magnetic fields. The external magnetic field (H) is parallel to the Co leads and sweeps from -0.2 to $0.2\,\text{T}$, or vice versa. The TMR ratio is defined as $(\%) = \{(I_H - I_{-0.1T})/I_{-0.1T}\} \times 100$, I_H and $I_{-0.1T}$ are the tunneling currents at H and $H = -0.1\,\text{T}$, respectively. When $V_{\text{SD}} = 41\,\text{mV}$, the magnetic fields sweep from $+0.2$ to $-0.2\,\text{T}$ (grey curves), and the magnetization configuration of two parallel Co leads is changed from parallel to antiparallel configurations (see the inset of Fig. 9.11(c)), and the TMR approaches the maximum at $H \approx -0.02\,\text{T}$. The similar behavior can be found when the magnetic field sweeps from -0.2 to $0.2\,\text{T}$ (black curves), the so-called positive TMR effect. When $V_{\text{SD}} = 32.5\,\text{mV}$, the typical hypnosis behavior can be seen clearly. These features of TMR indicate that spin is conserved in the tunneling processes. At small source-to-drain voltage $V_{\text{SD}} = 22.5\,\text{mV}$, the TMR signal becomes very weak.

Figures 9.12(a)–9.12(f) show the TMR curves at low temperature $T = 50\,\text{mK}$ and a fixed source-do-drain $V_{\text{SD}} = 41\,\text{mV}$ for various gate voltages. Note that the magnitude of the TMR and the curve shape can be manipulated by tuning the backgate voltages. Decreasing the voltage V_{G} from 600 to 100 mV [Figs. 9.12(a)–9.12(d)], the TMR changes slightly and the variation of the TMR curve shapes can be seen. Surprisingly, the shape and magnitude of TMR show significant changes at $V_{\text{G}} = 50$ and 0 mV [Figs. 9.12(e) and 9.12(f)].

One of the possible mechanisms of the spin transport is that spin accumulation inside InAs QD is caused by the spin injection from ferromagnetic Co leads. The theory predicts that the MR ratio depends on the gate voltage and the source-to-drain voltage. The SRT in InAs QDs is about 1 ns, which is long enough for nonequilibrium spin accumulation. However, the surprisingly high TMR ration in Figs. 9.12(e) and 9.12(f) cannot be explained based on the previous theoretical works. The underlying physics of the variation of the TMR curve is still not clear. Recently, the TMR in

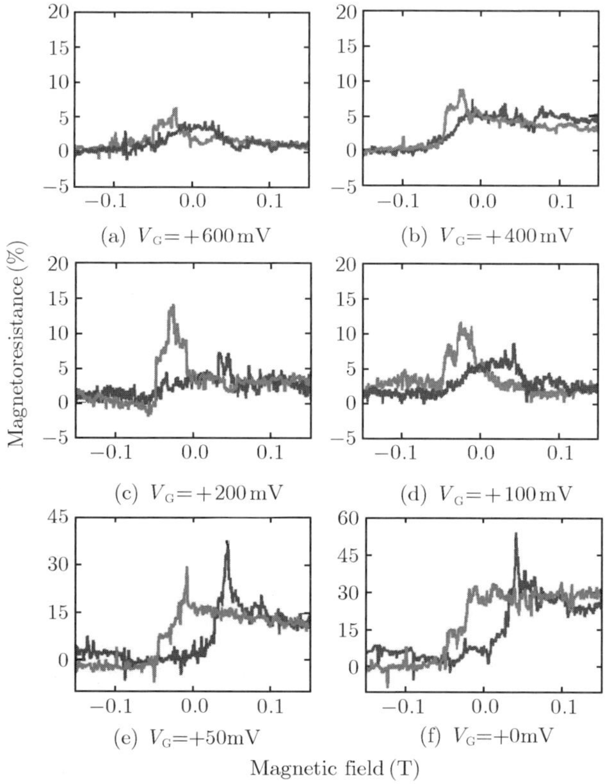

Fig. 9.12. The measured TMR at $V_{SD} = 41\,\text{mV}$ and $T = 50\,\text{mK}$ for different gate voltages V_G (Hamaya *et al.* 2007a).

PdNi/CNT/PdNi systems was reported. The gate voltage dependence of TMR could be caused by the discrete levels depending on the gate voltage and the spin orientation of the ferromagnetic leads. The influence of the discrete QD levels should be taken into account since the spin accumulation inside the QDs may induce spin splitting of the QD levels.

9.2.2. *The Kondo effect in semiconductor QDs with ferromagnetic leads*

Kondo effect, a many-body effect caused local spins screened by itinerant electrons in leads, has been investigated experimentally in semiconductor QDs with ferromagnetic leads. The zero-bias conductance is enhanced by spin-flip process below the Kondo temperature, the so-called Kondo resonance. In the presence of external magnetic fields, the Kondo effect in a

QD containing an odd number of electrons is suppressed and leads to the splitting of the Kondo resonance due to the spin splitting of the unpaired localized electron state in the QD.

The Kondo effect in a QD containing odd number of electrons was studied experimentally in the lateral transport through the QD spin device. Utilizing Ti/Au nanoelectrode technique, Jung *et al.* measured lateral spin transport through a InAs QD with ferromagnetic leads, and the charging energy of electron shell structure in the QD (Jung *et al.*, 2005a,b). Very recently, Shibata and Igarashi designed and fabricated a device with a strong coupling between single InAs QD and Ti/Au leads and observed the Kondo effect (Shibata *et al.*, 2007).

These experiments demonstrated clearly the TMR effect in the spin transport property of Co/InAs QD/Co and Ni/InAs QD/Ni spin devices and electrical control of the TMR (Hamaya *et al.*, 2007a). These QD spin devices could be used to construct new spintronic devices, e.g. the gate-controlled spin memory, etc. The experiment not only displays the suppresion of the Kondo effect, but also restores the Kondo effect at a finite magnetic field. In a parallel magnetic configuration of the Ni electrodes, a clear splitting at $N = 3$ can be seen without an external magnetic field, i.e. the zero-bias anomaly. The splitting gradually disappears with increasing magnetic fields, and the Kondo efefct almost restores at $B = 1.2\,\mathrm{T}$. This result indicates that an inverse effective magnetic field $B \sim 1.2\,\mathrm{T}$ is applied to the QD in the parallel magnetic configuration of ferromagnetic electrodes.

The detailed information about the sample is given as follow. The self-assembled InAs QDs was grown on the substrate consisting of 170 nm-thick GaAs buffer layer/90 nm-thick AlGaAs insulating layer/n^+-GaAs(001). The n^+-GaAs (001) layer is used as a back gate electrode. The wire-shaped Ni electrodes with a small gap (about 30 nm) were fabricated by using a conventional electron-beam lithography and a lift-off method. An amplified scanning electron micrograph of the QD spin valve is shown in Fig. 9.13(a). From this image, one can see a single InAs QD (120–140 nm) in contact with two 40 nm-thick Ni electrodes. The transport measurements were carried out at $T = 30\,\mathrm{mK}$. External magnetic fields were applied along the long axis of the wire-shaped Ni electrodes.

When an external magnetic field of $B = 1\,\mathrm{T}$ is applied to the wire-shaped Ni electrodes, a parallel magnetic domain configuration is formed. After the saturation of the magnetization, the magnetic configuration remains parallel even when the external magnetic field vanishes. Under

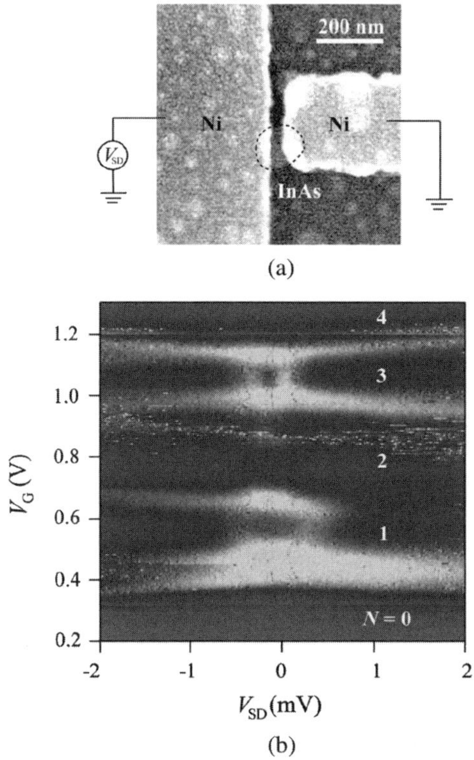

Fig. 9.13. (a) The amplified SEM image of InAs QD spin valve and (b) $B = 0\,\mathrm{T}$, the differential conductance $\mathrm{d}I/\mathrm{d}V_{\mathrm{SD}}$ as a function of the source-to-drain voltage V_{SD} and the gate voltage V_{G}. The light regions denote higher conductance regimes and the numbers indicate the number of electrons in the QD (Hamaya *et al.*, 2007b).

this condition, the differential conductance $\mathrm{d}I/\mathrm{d}V_{\mathrm{SD}}$ was measured as a function of V_{SD} and V_{G} (see Fig. 9.13(b)). The enhanced conductance was measured in the vicinity of zero bias $V_{\mathrm{SD}} \approx 0\,\mathrm{V}$ at $N = 1$ and 3 Coulomb valleys, which implies the $S = 1/2$ Kondo effect. A surprising splitting of the zero-bias anomaly can be observed at $N = 3$, i.e. suppression of the Kondo effect at $V_{\mathrm{SD}} = 0\,\mathrm{V}$, without an external magnetic field $B = 0\,\mathrm{T}$. The differential conductance $\mathrm{d}I/\mathrm{d}V_{\mathrm{SD}}$ versus V_{SD} was measured at $N = 3$ (see the red curve in Fig. 9.14). The peak splitting (about $0.74\,\mathrm{mV}$) shows an asymmetry to the polarity of V_{SD}. This feature may result from the asymmetric tunnel coupling between ferromagnetic electrodes and the QD or the difference in the spin polarization at the Ni/InAs interfaces.

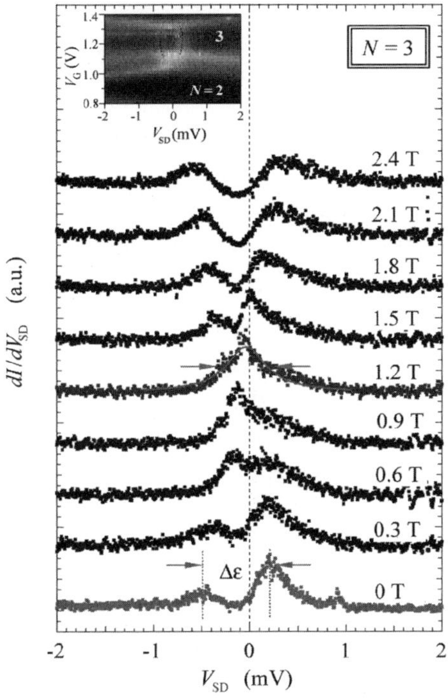

Fig. 9.14. The differential conductance dI/dV_{SD} as a function of V_{SD} for various external magnetic fields at $V_G = 1.25\,V$ ($N = 3$). The inset shows the corresponding Coulomb diamonds near $N = 2$ and 3 (Hamaya *et al.*, 2007b).

As magnetic fields parallel to the long axis of the wire-shaped electrodes increase, the splitting of the Kondo resonance is gradually reduced, and the observed double peaks merge into a single peak at $B = 1.2\,T$ (see the heavy grey curve in Fig. 9.14), which means that the Kondo effect is almost restored. From the V_{SD} of the Kondo resonance measured at $B = 1.2\,T$, the Kondo temperature $T_K = 5\,K$ can be roughly estimated. For further increase in applied magnetic fields, the Kondo effect is suppressed again due to the lifting of the spin degeneracy. Hamaya *et al.* (2007a) measured the magnetoconductance of the QDSV at $V_G = 1.25$ ($N = 3$) in Fig. 9.15, in order to examine the relationship between the zero-bias conductance and the magnetic configuration of Ni electrodes. The light and heavy grey curves correspond to the data for up-sweep and down-sweep measurements, respectively. At $B = \pm 0.015\,T$, we can see current changes due to the switching of the magnetic configuration of the Ni electrodes from parallel to antiparallel configuration. One can see an enhancement in the conductance

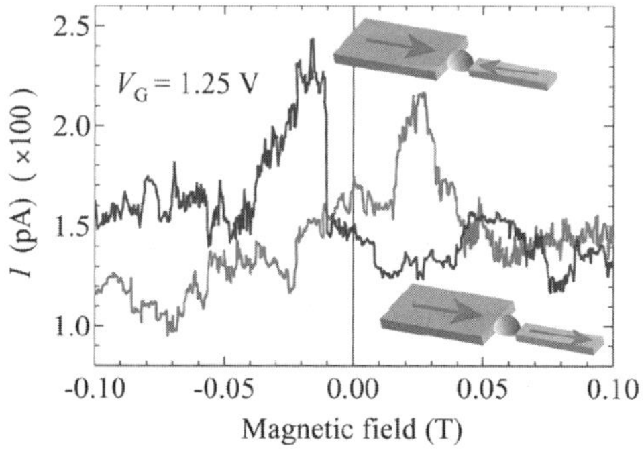

Fig. 9.15. The magnetoconductance versus magnetic fields at $V_G = 1.25\,\text{V}$ ($N = 3$) (Hamaya *et al.*, 2007b).

for antiparallel configurations, i.e. inverse TMR effect. This means that the Kondo effect is suppressed in a parallel magnetic configuration in the Ni electrodes and restored a little for the antiparallel configuration.

It was reported that the Kondo effect at $B = 0$ in Au(Co)/Au QD/Au(Co) metal tunneling junction is suppressed (Heersche *et al.*, 2006). The Kondo splitting arises from the competition between the Kondo effect and indirect RKKY interaction. However, it is well known that the RKKY interaction between the electron spin in the QD and static spins of impurities Co does not lead to an evident hysteretic TMR curves. Therefore, the underlying physical mechanism is not the RKKY interaction between the QD spin and the static spins of the Ni electrodes. When the magnetic configuration of the ferromagnetic electrodes is parallel, the local stray field orienting along the same direction of external magnetic fields can induce the Zeeman splitting. But an effective field deduced from Fig. 9.14 points along the opposite direction of external magnetic fields. Thus, the splitting of the Kondo resonance cannot be attributed to the local stray field. A possible mechanism for the suppression of the Kondo effect at $B = 0$ T may originate from a local exchange field. For parallel magnetic configurations of ferromagnetic electrodes in QD spin valve devices, the spin-dependent quantum charge fluctuations renormalize the QD's levels. The renormalization for the spin-down level is stronger than that for the spin-up level, so that a splitting between the spin-up and spin-down levels appears. Such an effective field is compensated by applying magnetic fields, being consistent with the present data in Fig. 9.15. For Ni/C_{60}/Ni spin valve (Sahoo *et al.*, 2005), since the exchange field probably acted as an effective field parallel to the magnetization orientation of the Ni electrodes, restoration of the

Kondo effect by applying external magnetic fields could not be detected. In the present case, the exchange field works as an effective field antiparallel to the magnetization orientation of the Ni electrodes, so that the restoration of the Kondo effect can be observed clearly by applying external magnetic fields. In summary, in the absence of magnetic fields, the Kondo effect is significant at $N = 3$ for the parallel magnetic configuration of the ferromagnetic electrodes. With increasing magnetic fields, the splitting of the Kondo resonance is suppressed, and disappears at a finite magnetic field, i.e. the restoration of the Kondo effect. This feature implies that an inverse effective magnetic field from ferromagnetic electrodes is applied on the QD.

9.2.3. *Theory of spin transport through magnetic semiconductor QDs*

The experiments mentioned above focussed on spin transport through non-magnetic semiconductor QDs. In the recent years high-quality magnetic semiconductor QDs can be fabricated using MBE. With the development of growth technique, one can fabricate semiconductor QDs containing a single magnetic ion. There have been many studies about the magneto-optical property of magnetic semiconductor QDs; the spin transport through magnetic semiconductor QDs is relatively unexplored. Next we will introduce a recent theoretical work (Qu and Vasilopoulos, 2006).

The Hamiltonian of the considered system reads

$$\hat{H} = \sum_{k,\sigma} \varepsilon_{k,\sigma} \hat{a}^{+}_{k,\sigma} \hat{a}_{k,\sigma} + \sum_{p,\sigma} \varepsilon_{p,\sigma} \hat{b}^{+}_{p,\sigma} \hat{b}_{p,\sigma} + \hat{H}_{\mathrm{D}} + \hat{H}_{\mathrm{T}}, \tag{9.6}$$

where the first and second terms denote the electron motion in the left (right) lead, H_{D} describes a few electron states in the QD, and H_{T} is the tunneling coupling between electrons states in the QD and leads. The Hamiltonian of the QD is (Qu and Vasilopoulos, 2006)

$$\begin{aligned} \hat{H}_{\mathrm{D}} = {} & \sum_{i,\sigma} E_{i,\sigma} c^{+}_{i,\sigma} c_{i,\sigma} + \frac{1}{2} \sum_{i,j,k,l}^{\sigma,\sigma'} \langle i,j | V_{\mathrm{ee}} | k,l \rangle \, c^{+}_{i,\sigma} c^{+}_{j,\sigma} c_{k,\sigma'} c_{l,\sigma'} \\ & - \frac{1}{2} \sum_{i,j} J_{i,j}(R) \left[(c^{+}_{i,\uparrow} c_{j,\uparrow} - c^{+}_{i,\downarrow} c_{j,\downarrow}) M_z + c^{+}_{i,\downarrow} c_{j,\uparrow} M^{+} + c^{+}_{i,\uparrow} c_{j,\downarrow} M^{-} \right] \\ & - \mu_{\mathrm{B}} B \left(g_{\mathrm{e}} \sum_{i} S^{(i)}_{z} + g_{\mathrm{Mn}} M_z \right). \end{aligned} \tag{9.7}$$

Here the first term is the single-particle energy levels of the QD, the second term denotes electron–electron Coulomb interaction, the third term describes the s–d exchange interaction between the electrons and magnetic ions, and the last term is the Zeeman term of electron and magnetic ions. The strength of the exchange interaction between electron and magnetic ion $J_{ij}(\boldsymbol{R})= J_{\rm C}^{2D}\varphi_i^*(\boldsymbol{R})\varphi_j(\boldsymbol{R})$ is determined by the magnitude of electron wavefunction at the site $\boldsymbol{R}$ of magnetic ion Mn, $J_{\rm C}^{2D} = 2J_{\rm C}/d$, d is the thickness of the QD.

The tunneling Hamiltonian between the QD and leads $H_{\rm T}$ is

$$\begin{aligned}\hat{H}_{\rm T} &= \sum_{k,i,\sigma} T^{\rm L}_{k,i,\sigma}(\hat{a}^+_{k,\sigma}\hat{c}_{i,\sigma} + \hat{c}^+_{i,\sigma}\hat{a}_{k,\sigma}) \\ &\quad + \sum_{p,i,\sigma} T^{\rm R}_{p,i,\sigma}(\hat{b}^+_{p,\sigma}\hat{c}_{i,\sigma} + \hat{c}^+_{i,\sigma}\hat{b}_{p,\sigma}),\end{aligned} \tag{9.8}$$

where T is the spin-dependent transmission, $t_\sigma^{\rm L/R} = 2\pi|T^{\rm L/R}_{k,i,\sigma}|2\rho_\sigma^{\rm L/R}\Omega\hbar$. The tunneling current is

$$I_\sigma = \sum_{i,j(j\neq i)} [\Gamma^{\rm L}_{i,j}(\sigma) - \Gamma^{\rm R}_{j,i}(\sigma)](N_j - N_i)P_j. \tag{9.9}$$

The TMR is defined as

$$\mathrm{TMR} = \frac{G_{\rm F} - G_{\rm AF}}{G_{\rm AF}}, \tag{9.10}$$

where $G_{\rm F(AF)}$ is the conductance for parallel magnetization (FM) and antiparallel magnetization (AFM) configurations, respectively.

Figure 9.16 describes how the TMR changes with increasing magnetic fields and source-to-drain voltage. The upper (lower) panel corresponds to the asymmetric (symmetric) QD-to-lead coupling case, i.e. $\alpha = 0.6$, $\Delta_{\rm L} = 0.9$, $\Delta_{\rm R} = 0.4$ ($\alpha = 1$, $\Delta_{\rm L} = \Delta_{\rm R} = 0.5$). At a weak magnetic field $\tilde{B} = \mu_{\rm B}B/J_{ss} < 1$, the ground state of the QD is $J = +3$, $m = +3$. As the magnetic field increases, the Zeeman splitting becomes larger than the splitting induced by the electron–magnetic ion exchange interaction. The interplay between magnetic field and source-to-drain bias leads to an oscillation of the TMR, which is caused by the opening of the spin-up and spin-down channels.

Fernández-Rossier and Aguado (2007) considered a simple constant interaction model, i.e. $H_{\rm C} = (Q + CV + C_{\rm L}\mu_{\rm L}/e + C_{\rm R}\mu_{\rm R}/e)^2/2C$, where $C = C_{\rm R} + C_{\rm L} + C_{\rm g}$ is the total capacitance to external circuit, $C_{\rm L}(C_{\rm R})$ is the capacitance of the left (right) junction, $V_{\rm B} = (\mu_{\rm L} - \mu_{\rm R})/e$ is the bias voltage, $C_{\rm g}$ is the capacitance to the gate, and Q is the extra charge in the QD. This simplified model neglects the correlation between electrons and the spatial distribution of electrons in the QD. This system behaves

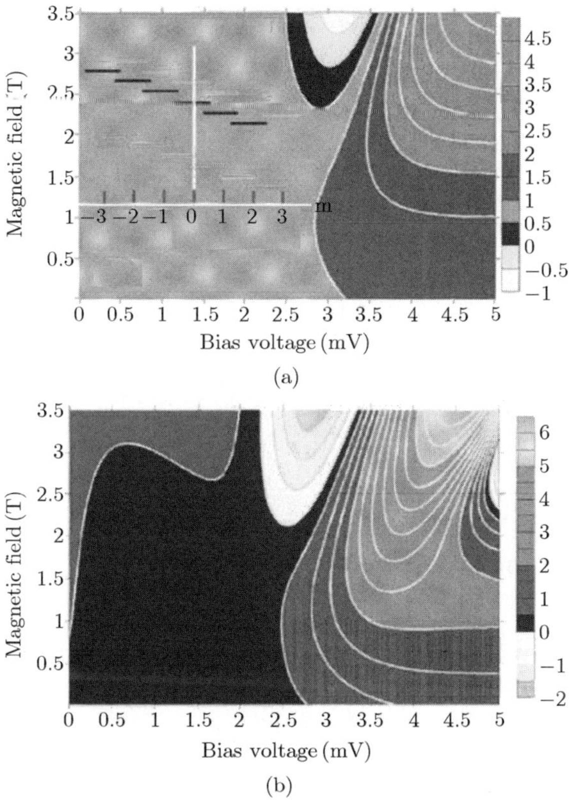

Fig. 9.16. Contour plot of the TMR through a QD as a function of magnetic field and bias voltage. The upper (lower) panel corresponds to the symmetric (asymmetric) QD-to-lead coupling case (Qu and Vasilopoulos, 2006).

like a quantum nanomagnet depending sensitively on the number of electrons and their orbital states. Assuming a weak coupling between the QD and reservoirs, the dispersive dynamics of the reduced density matrix is described by the quantum master equation. The tunneling current can be calculated from the density matrix:

$$\begin{aligned} I_{\mathrm{L/R}} = e\Gamma_{\mathrm{L/R}} \sum_{N\cdot N'} \sum_{M} \sum_{\alpha} \tilde{\rho}_{N\cdot N'} \{ & n^{+}_{\mathrm{L/R}}(E_N - E_{N'}) \\ & \times \langle N'|f_\alpha|M\rangle\langle M|f^{\dagger}_{\alpha}|N\rangle \\ & - n^{-}_{\mathrm{L/R}}(E_N - E_M)\langle N'|f^{\dagger}_{\alpha}|M\rangle\langle f_\alpha|N\rangle\}. \end{aligned} \quad (9.11)$$

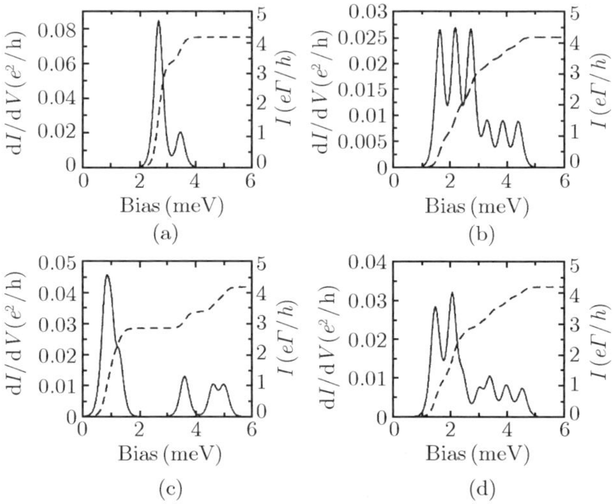

Fig. 9.17. Current and differential conductance against bias at different charge states Q (Fernández-Rossier and Aguado, 2007).

Notice that the equation includes the contributions from the diagonal and nondiagonal elements of the density matrix. In contrast to conventional semiconductor QDs, the spin tunneling through magnetic semiconductor QDs also depends on the spin states of the magnetic ion.

Figure 9.17 shows the current and differential conductance spectrum for different QDs. Figures 9.17(a)–9.17(c) correspond to the charge states $Q = 0$ and -1, 0 and $+1$, $+2$, and $+3$. Panel (d) is similar to panel (b), but for different QDs (sample B). From this figure, one can see that the current through the QD depends on the charge states in the QD. The differential conductance spectrum $\mathrm{d}I/\mathrm{d}V$ is closely related to the energy spectrum of the QD. The splitting of the current peaks in the differential conductance $\mathrm{d}I/\mathrm{d}V$ indicates the zero-field splitting of electron states in the QD (see Fig. 9.17(a)).

9.3. Magnetic Domain Transport in Magnetic Semiconductors

Magnetic domain in magnetic semiconductors is important for potential application of semiconductor information storing and spin transport devices (Yamanouchi *et al.*, 2004). The saturated magnetic moment in a p-type FMS GaMnAs is two orders of magnitude smaller than that in magnetic metals because of the low concentration of Mn magnetic ions. Threshold current of magnetic domain in GaMnAs is several orders of magnitude lower than that in magnetic metals. This is an important feature of DMS GaMnAs. Recent experiments demonstrated that the magnetic domains in GaMnAs can be driven by electric current and external fields (Tang *et al.*, 2006; Yamanouchi *et al.*, 2006). High-resolution electron holography technique provides direct images of domain wall (DW) magnetization profiles. For FMS GaMnAs, it becomes more difficult to obtain direct image of magnetic domains due to small magnetic moments. For perpendicular easy-axis GaMnAs samples, the spatial resolution of magneto-optical image should approach 1 μm (Yamanouchi *et al.*, 2004), and in the case of in-plane easy-axis GaMnAs samples, the spatial resolution needed to approach 10–100 nm.

In recent years, high-resolution electron holography technique was used to measure the magnetic domain structure in FMS GaMnAs samples (Sugawara *et al.*, 2008). This method is based on the transmission electron microscope (Hopster and Oepen, 2004) and holographic techniques. Utilizing this technique, one can measure the magnetic domain structures in an in-plane magnetized magnetic thin films. Usually, a general theory can only describe the magnetic domain qualitatively or semi-quantitatively due to disorder effect and uncertainty of the concentrations of Mn ions and holes. The theory of magnetic domain is based on the kinetic-exchange theory and the numerical simulation of Landau–Lifshitz–Gilbert equation (Dietl *et al.*, 2001a,b).

Figure 9.18 is the Lorentz micrographs of GaMnAs samples at various temperatures (Sugawara *et al.*, 2008). From this figure, one can see clearly that there are three types of magnetic DWs marked by white arrows: (i) a wall parallel to $(1\bar{1}0)$ that is near-180° type at high temperatures and near-90° type at low temperatures; (ii) a pair of DWs that is near-180° head-on type at high temperatures and near-90° type at low temperatures, and (iii) a near-90° type wall parallel to [110] that appeared at low temperatures. The local B directions are denoted by black arrows.

Due to the complex valence band structure of GaMnAs, the easy axis can be parallel or perpendicular to the thin film of GaMnAs, determined by strain and carrier concentration (Sawicki *et al.*, 2004, Thevenard *et al.*, 2005). Next we introduce a simplified model based on the micromagnetic theory. Within the framework of this model, uniaxial magnets is described by the magnetic stiffness A, uniaxial magnetic anisotropy energy K_{u}, and

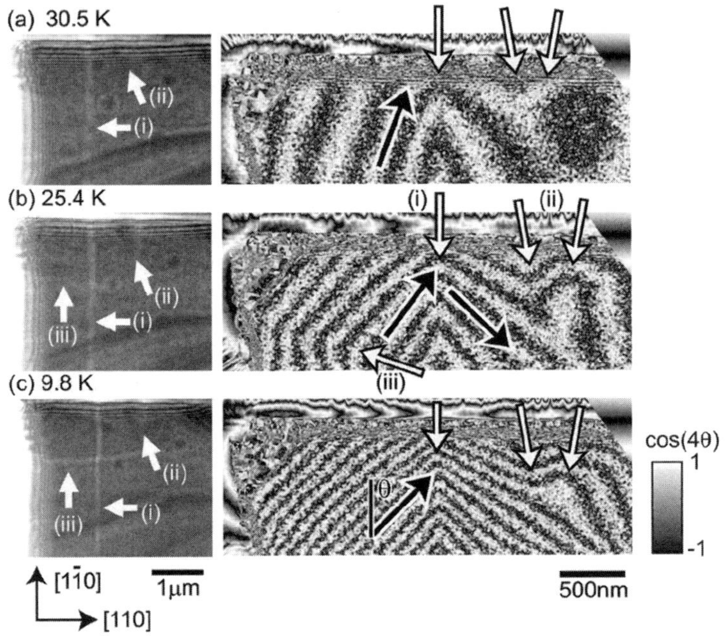

Fig. 9.18. Lorentz micrographs (left-hand column) and phase images amplified by a factor of 4 (right-hand column) acquired at different temperatures (a) 30.5 K, (b) 25.4 K, and (c) 9.8 K.

the saturated magnetization M_s. One can first calculate the temperature-dependent magnetization $M(T)$ and the exchange splitting $\Delta(T)$, then obtain the magnetic anisotropy energy $K_u(\Delta)$ and magnetic stiffness $A(\Delta)$. Notice that the fluctuation of magnetic moments is neglected in this model, and therefore this model is only valid at low temperatures.

It is more important to make sure that the continuous approximation in continuous-medium approximation underlying the micromagnetic theory is reasonable for FMS GaMnAs material. The shortest width of the Bloch DW δ_W can be estimated as $\delta_W = \pi/\sqrt{A/K_u}$, where A is the magnetic stiffness, and K_u is the anisotropic energy. In typical doped GaMnAs material ($p = 3\times10^{20}\,\mathrm{cm}^{-3}$), $\delta_W \gg r_m$ is required, where $r_m = (3/4\pi p)^{1/3}$ is the averaged spacing between holes. The magnetic domain can be controlled by a single parameter $\lambda_c = 4\sqrt{AK_u}/K_d d$, $K_d = \mu_0 M_s^2$ is stray-field energy, M_s is the saturation magnetization, and d is the thickness of samples. Figure 9.19 shows λ_c against the relative temperature T/T_C, where T_C is the Curie

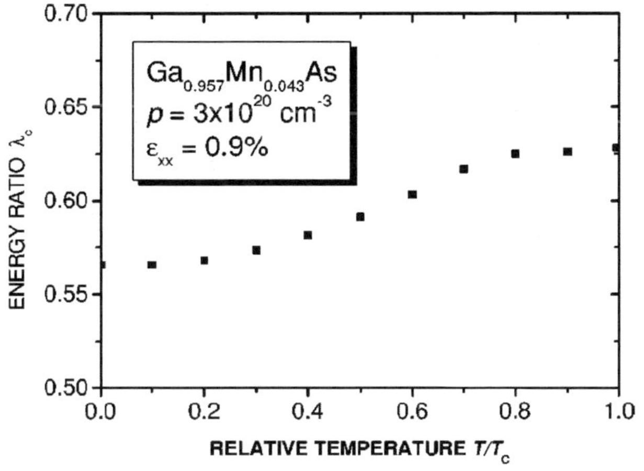

Fig. 9.19. The parameter λ_c of FMS GaMnAs versus the relative temperature T/T_C (Dietl *et al.*, 2001a,b).

temperature. From this figure, the parameter λ_c is not sensitive to the temperature.

A simple stripe domain structure is caused by the competition between energies associated with the stray fields and the formation of the Bloch DWs; therefore, the width of DW W is determined by

$$\lambda_c = \left(\frac{P^2}{\pi^2}\right) \sum_{n=1,3,5,\ldots} n^{-3}\left[1 - \left(1 + \frac{2\pi n}{P}\right) e^{-2\pi n/P}\right], \tag{9.12}$$

where $P = 2W/d$ is the normalized stripe period, and the parameter λ_c describes the ratio of the Bloch DW energy $\gamma_W = 4\sqrt{AK_u}$ and the stray-field energy K_d. In Fig. 9.20, the solid line is obtained from Eq. (9.12), the dashed line is the calculated results assuming that λ_c is by a factor of 1.8 greater (Dietl *et al.*, 2001a,b), and the full squares are experimental measurements (Shono *et al.*, 2000). From this figure, one can see that the observed width W is almost independent of the temperature; this behavior is consistent with the theory. At higher temperatures ($T/T_C > 0.8$), a strong temperature dependence of W is not explained by the simplified model. The break in the temperature dependence of the domain width W around 60 K indicates likely a new critical regime. In this regime, the model (Dietl *et al.*, 2001a,b) neglect the long-length-scale fluctuations in the magnetization M, which could be important for the stray-field energy K_d.

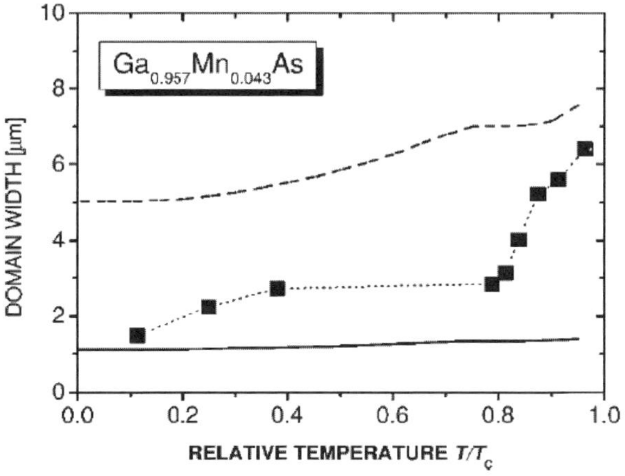

Fig. 9.20. The width of domain stripes in a $Ga_{0.957}Mn_{0.043}As$ film against the relative temperature T/T_C. The black squares are experimental measurements (Shono *et al.*, 2000), the solid and dashed lines are calculated domain widths assuming different λ_c (Dietl *et al.*, 2001a,b).

The motion of magnetic DW driven by electric current has been investigated extensively from both experimental (Saitoh *et al.*, 2004; Yamaguchi *et al.*, 2004; Yamanouchi *et al.*, 2004; Kläui *et al.*, 2005) and theoretical (Tatara and Kohno, 2004; Zhang and Li, 2004; Barnes and Maekawa, 2005) points of view. The DW can be driven by a pulse current with the density in ferromagnetic NiFe nanowires. While for GaMnAs, the pulse current density is 10^5 A/cm^2, which is two orders of magnitude smaller than that in ferromagnetic metals.

Figure 9.21 shows the experimental measurement of the DW motion driven by electric current (Yamanouchi *et al.*, 2006). A 30 nm $Ga_{0.955}Mn_{0.045}As$ sample was grown by MBE above a stack of (InAl)As buffer layers to make the magnetic easy axis of GaMnAs parallel to the growth direction utilizing the controlled strain. Stepped graded composition reduces the surface roughness due to crosshatch dislocations resulting from a lattice mismatch with the substrate, which may disturb the uniform DW motion. The Curie temperature T_C of the regions I and II in Fig. 9.21(a) is 112 and 115 K, respectively. Once DW is created in the channel, it can be moved back to the step boundary by the current.

Electric current generates two sources of the Oersted field driving DW motion. First, in a thin uniform conductor, the field generated by the current is concentrated on the two edges of the wire, reaching the magnitude

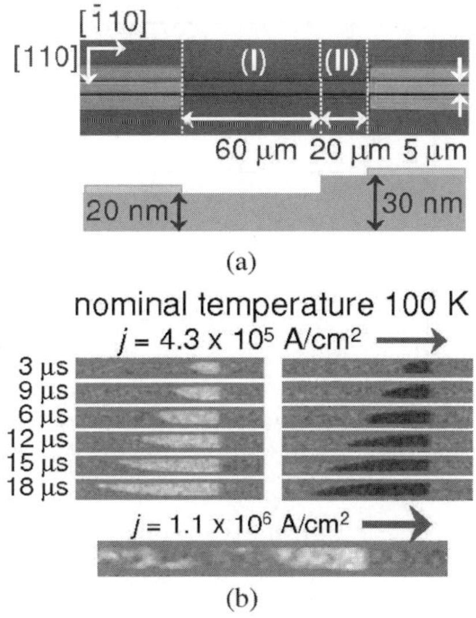

Fig. 9.21. (a) Layout of the device showing the 5 μm mesh and step for DW pinning in perpendicular magnetic anisotropy (Ga,Mn)As film. (b) Magneto-optical images showing that DW moves in the opposite direction to current (Yamanouchi *et al.*, 2006).

0.4 mT in the experiment. But the direction of DW motion caused by this field would depend on the initial M configuration, in contrast to the experimental observations. Second, the current and magnetization M produce a transverse (anomalous) Hall electric field that changes its sign on crossing DW. This generates an additional current that circulates around DW and induces a magnetic field reaching a maximum value in the DW center. But this hydromagnetic DW drag force moves DW in the direction of the current, again in contradiction to the experiment. The authors proposed to understand the experiment utilizing the concept of spin transfer. Since the sign of the p–d exchange integral between the hole carriers and localized Mn spins in (Ga,Mn)As is negative, a simple application of spin momentum conservation implies that the DW is expected to move in the opposite direction to the electric current, as observed (Yamanouchi *et al.*, 2006).

Theory of current-driven DW motion reveals that momentum and spin transfers between current and magnetic DW both exist, but play different roles in DW motion (Tatara and Kohno, 2004). In metallic wires, i.e. thick-wall case, the spin transfer dominates and the threshold current for wall motion is determined by the hard-axis magnetic anisotropy. In nanocontacts and magnetic semiconductors, i.e. thin-wall case, the

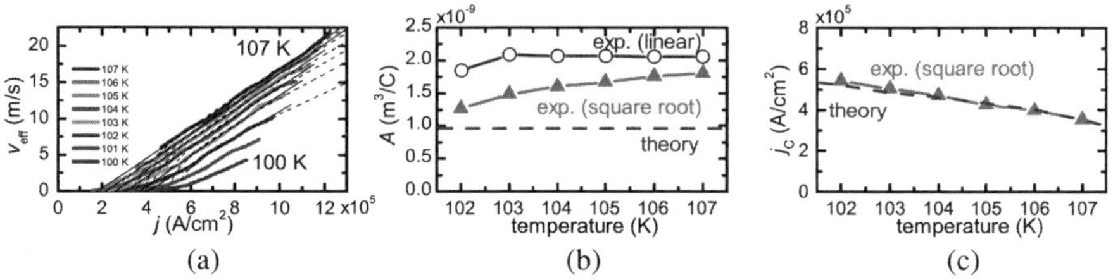

Fig. 9.22. (a) DW velocity as a function of current at various device temperatures. The thin line and broken thin line show fitted linear and square root dependencies of velocity on current, respectively. Efficiency factor A (b) and critical current density J_C (c) resulting from these two fits (open and solid symbols, respectively). Broken lines are theoretically calculated assuming that spin current polarization is equal to thermodynamic spin polarization. Adapted from Yamanouchi *et al.* (2006).

momentum-transfer process dominates. The velocity of DW from the theory of Tatara and Khono (2004) is $v_{\text{eff}} = A(j^2 - j_C^2)^{1/2}$, where the spin transfer efficiency factor $A = g\mu_B P/2eM$ (Yamanouchi *et al.*, 2006), the threshold current density is $J_C = 2eK\delta_W/\pi\hbar P$ neglecting damping and pinning, here P is the spin polarization of current, which can be estimated from the p–d Zener model. The motion of DW can be described by the Landau–Lifshitz–Gilbert equation. From the MOKE in FMS (Ga,Mn)As samples, Fig. 9.22(a) shows the DW velocity as a function of the current density, where the thin solid and dashed lines denote the linear and square-root dependencies of velocity on current, respectively. The efficiency factor A and threshold current J_C obtained from the fitting are shown as a function of temperature in Figs. 9.22(b) and 9.22(c), where the dashed line is the calculated assuming the spin polarization P of current is equal to the thermodynamic spin polarization P_s. From the p–d Zener model, $P_s = 6k_B T_C M/[(S+1)p\beta M_s]$, where p is the hole density and β is the p–d exchange integral between hole and local Mn spins. From Fig. 9.22(a), one can find that the maximum of DW velocity is about 22 m/s. The present phenomenal theory is based on the Landau–Lifshitz–Gilbert equation and spin transfer theory; the microscopic theory for the DW motion is still a challenging task.

References

Barnes SE, Maekawa S. (2005) *Phys Rev Lett* **95**: 107204.

Bernand-Mantel A, Seneor P, Lidgi N, Munoz M, Cros V, Fusil S, Bouzehouane K, Deranlot C, Vaures A, Petroff F, Fert A. (2006) *Appl Phys Lett* **89**: 062502.

Brataas A, Nazarov YV, Inoue J, Bauer GEW. (1999) *Phys Rev B* **59**: 93.

Dietl T, Ohno H, Matsukura F. (2001a) *Phys Rev B* **63**: 195205.

Dietl T, König J, MacDonald AH. (2001b) *Phys Rev B* **64**: 241201.

Fernández-Rossier J, Aguado R. (2007) *Phys Rev Lett* **98**: 106805.

Gui YS, Becker CR, Liu J, Daumer V, Hock V, Buhmann H, Molenkamp LW. (2004) *Europhys Lett* **65**: 393.

Hamaya K, Masubuchi S, Kawamura M, Machida T, Jung M, Shibata K, Hirakawa K, Taniyama T, Ishida S, Arakawa Y. (2007a). *Appl Phys Lett* **90**: 053108.

Hamaya K, Kitabatake M, Shibata K, Jung M, Kawamura M, Hirakawa K, Machida T, Taniyama T, Ishida S, Arakawa Y. (2007b) *Appl Phys Lett* **91**: 232105.

Heersche HB, Groot Z de, Folk JA, Kouwenhoven LP, van der Zant HSJ, Houck AA, Labaziewicz J, Chuang IL. (2006) *Phys Rev Lett* **96**: 017205.

Hikami S, Larkin AI, Nagaoka Y. (1980) *Prog Theor Phys* **63**: 707.

Hopster H, Oepen HP. (2004), *Magnetic Microscopy of Nanostructures*, Springer-Verlag, Berlin.

Imamura H, Takahashi S, Maekawa S. (1999) *Phys Rev B* **59**: 6017.

Jung M, Hirakawa K, Kawaguchi Y, Komiyama S, Ishida S, Arakawa Y. (2005a) *Appl Phys Lett* **86**: 033106.

Jung M, Machida T, Hirakawa K, Komiyama S, Nakaoka T, Ishida S, Arakawa Y. (2005b) *Appl Phys Lett* **87**: 203109.

Kato Y, Myers RC, Gossard AC, Awschalom DD. (2004) *Nature* (*Lond*) **427**: 50.

Kläui M, Vaz CAF, Bland JAC, Wernsdorfer W, Faini G, Cambril E, Heyderman LJ, Nolting F, Rüdiger U. (2005) *Phys Rev Lett* **94**: 106601.

Lee PA, Ramakrishnan TV. (1981) *Phys Rev B* **26**: 4009.

Lee PA, Ramakrishnan TV. (1985) *Rev Mod Phys* **57**: 287

Martinek J, Barnaś J, Michalek G, Bulka BR, Fert A. (1999) *J Magn Magn Mater* **207**: L1.

Martinek J, Barnaś J, Maekawa S, Schoeller H, Schön G. (2002) *Phys Rev B* **66**: 014402.

Ohno H, Chiba D, Matsukura F, Omiya T, Abe E, Dietl T, Ohno Y, Ohtani K. (2000) *Nature* (*Lond*) **408**: 944.

Qu F, Vasilopoulos P. (2006) *Appl Phys Lett* **89**: 122512.

Rudzinski W, Barnaś J. (2001) *Phys Rev B* **64**: 085318.

Sahoo S, Kontos T, Furer J, Hoffmann C, Gräber M, Cottet A, Schönenberger C. (2005) *Nat Phys* **1**: 99.

Saitoh E, Miyajima H, Yamaoka T, Tatara G. (2004), *Nature* (*Lond*), **432**: 203.

Sawicki M, Matsukura F, Idziaszek A, Dietl T, Schott GM, Ruester C, Gould C, Karczewski G, Schmidt G, Molenkamp LW. (2004) *Phys Rev B* **70**: 245325.

Shibata K, Jung M, Hirakawa K, Machida T, Ishida S, Arakawa Y, Sakaki H. (2007) *J Cryst Growth* **301–302**: 731.

Shono T, Hasegawa T, Fukumura T, Matsukura F, Ohno H. (2000) *Appl Phys Lett* **77**: 1363.

Smorchkova I, Samarth N. (1996), *Appl Phys Lett* **69**: 1640.

Smorchkova I, Samarth N, Kikkawa JM, Awschalom DD. (1997) *Phys Rev Lett* **78**: 3571.

Smorchkova IP, Samarth N, Kikkawa JM, Awschalom DD. (1998) *Phys Rev B* **58**: R4238.

Sugawara A, Kasai H, Tonomura A, Brown PD, Campion RP, Edmonds KW, Gallagher BL, Zemen J, Jungwirth T. (2008) *Phys Rev Lett* **100**: 047202.

Tang HX, Kawakami RK, Awschalom DD, Roukes ML. (2006) *Phys Rev B* **74**: 041310(R).

Tatara G, Kohno H. (2004) *Phys Rev Lett* **92**: 086601.

Teran FJ, Potemski M, Maude DK, Wilamowski Z, Hassan AK, Plantier D, Jaroszynski J, Wojtowicz T, Karczewski G. (2003) *Physica E* (*Amsterdam*) **17**: 335.

Thevenard L, Largeau L, Mauguin O, Lemaître A. (2005) *Appl Phys Lett* **87**: 182506.

Vasilopoulos P, Peeters FM (1989) *Phys Rev Lett* **63**: 2120.

Wetzels W, Bauer GEW, Grifoni M. (2005) *Phys Rev B* **72**: 020407(R).

Weymann I, König J, Martinek J, Barnas J, Schön G. (2005), *Phys Rev B* **72**: 115334.

Yamaguchi A, Ono T, Nasu S, Miyake K, Mibu K, Shinjo T. (2004) *Phys Rev Lett* **92**: 077205.

Yamanouchi M, Chiba D, Matsukura F, Ohno H. (2004) *Nature* (*Lond*), **428**: 539.

Yamanouchi M, Chiba D, Matsukura F, Dietl T, Ohno H. (2006) *Phys Rev Lett* **96**: 096601.

Yang W, Chang K. (2006) *Phys Rev B* **73**: 045303.

Yang W, Chang K, Peeters FM. (2005) *Appl Phys Lett* **86**: 192107.

Yang W, Chang K, Wu XG, Zheng HZ. (2006a), *Appl Phys Lett* **88**: 082107.

Yang W, Chang K, Wu XG, Zheng HZ, Peeters FM. (2006b) *Appl Phys Lett* **89**: 132112.

Zhang S, Li Z. (2004) *Phys Rev Lett* **93**: 127204.

Chapter 10

FUTURE QUANTUM DOT AND QUANTUM WIRE SPINTRONICS

Quantum computation and quantum communication need to design necessary hardware. Loss and DiVincenzo (1998) showed that the spin qubits in the quantum confinement systems, such as semiconductor QDs, atoms, or molecules satisfy all demands of scalable quantum computers. Besides, if the spin qubit is adhered at a movable electron, it can transport along the conductor. An interesting application of this possibility is the production (for example, in the coupled QDs or superconductor–normal metal interface) and transport of spin-entangled electrons, which is called Einstein–Podolsky–Rosen pair. The scheme of quantum computer based on the QD spins proposed by Loss and DiVincenzo (1998) is shown in Fig. 10.1. The system is composed of a magnetization layer or a high g layer, a modulation-doped heterojunction layer (2DEG), and surface and bottom gate arrays. The QDs can be formed by applying voltage on the gates, as shown by the circles in Fig. 10.1. Changing the gate voltage, the electrons can enter into the magnetization layer or the high g layer. A localized magnetic field can be produced by the electric current flowing in the left conductor. The electron spin in each QD can be set up individually through the ESR pulse produced by the alternating magnetic field in the plane. This mechanism can be used to produce initiation and rotation of spins. The exchange coupling between QDs is controlled by lowering the tunneling barrier between dots.

QDs are those systems where the space confinement scale is the electron Fermi wave length. For semiconductor QDs, the size is about 10 nm. It has several types: defined by the electrodes in 2DEG, self-assembled growth by MBE, and growth in solution by the chemical method. The applied magnetic field can obviously change the electron energy spectrum of QD, because the magnetic length of 1 T of magnetic field is comparable with the size of QD, i.e. the electron cyclotron radius equals the order of magnitude of the QD radius.

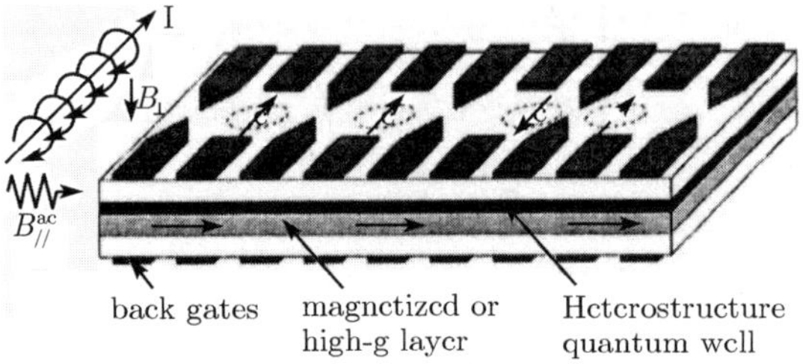

Fig. 10.1. Quantum computer scheme based on the QD spins.

Theoretically, the quantum two-energy-level system represents one qubit. To realize the quantum computation, it demands that the coupling between the system and the environment be very weak and the decoherence of spins be very slow. The SRT T_1 and the (transverse) decoherent time T_2 should be distinguished; generally, $T_2 \leq T_1$, thus T_2 is a more important quantity for quantum computation. The experiments showed that the spin coherent time is longer than the charge coherent time by several orders of magnitude. Khaetskii and Nazarov (2000) proved theoretically that in semiconductor QDs due to the discrete energy levels the momentum scattering is suppressed greatly and the spin relaxation mechanisms (DP and EY) related to the SOC are weakened, resulting in unusually low spin-flip rates and long spin coherent time.

In general 2DEG, the effective Hamiltonian for spin-flip is (see Eq. (5.87)),

$$H_{\text{so}} = \sum_{i,j=x,y} \beta_{ij}\sigma_i p_j, \tag{10.1}$$

where β_{ij} depends on the confinement intensity in the z-direction, for GaAs heterojunction, $\beta_{ij} \sim (1-3) \times 10^5$ cm/s. Because of SOC, a spin-up electron state is generally mixed with a small part of spin-down electron state, resulting in nonzero scattering matrix elements of reverse spin states. It is proved that in the 2D case $\beta \sim \Delta/E_{\text{g}}$, where Δ is the spin–orbit splitting of the valence bands of the bulk GaAs crystal and E_{g} the energy gap. Hence, the spin-flip rate is proportional to $(\Delta/E_{\text{g}})^2$. In QDs, it is proved theoretically that the terms which are linear in Δ/E_{g} canceled each other completely the spin flip rate $\propto (\Delta/E_{\text{g}})^6$ that stems from the third-order perturbation term of (Δ/E_{g}). Thus, in QDs, there will be very low spin-flip rate, i.e. very long spin coherent time.

A single qubit can be manipulated by using a g-factor-modulated material. The operation of quantum gate is described by the following effective Hamiltonian:

$$H(t) = \sum_{i,j} J_{ij}(t)\boldsymbol{S}_i \cdot \boldsymbol{S}_j + \sum_i \mu_B g_i(t)\boldsymbol{B}_i(t) \cdot \boldsymbol{S}_i. \tag{10.2}$$

This Hamiltonian depends on time through applied controlled fields $P(v(t))$, which include the exchange coupling $J_{ij}(t)$ or the Zeeman coupling $g_i(t)\mu_B\boldsymbol{B}_i(t)$. The exchange coupling is generally of short range and does not equal zero only between neighboring qubits; in some case, for example, in superconductor, it is long range. No matter the detail of the $P(v(t))$, only the integration $\int_0^\tau P(v(t))dt$ determines the quantum gate manipulation. The necessary condition is that the $P(v(t))$ is applied adiabatically, i.e. $|\dot{v}/v| \ll \delta\varepsilon/\hbar$, guaranteeing tenability of the effective Hamiltonian (10.2).

In the adiabatic approximation, if the effective g-factor changes Δg_{eff} in time τ, then a spin rotates relatively an angle $\phi = \Delta g_{\text{eff}}\mu_B B\tau/2\hbar$. In the magnetic field $B = 1\,\text{T}$, if the angle changes $\phi = 2\pi, \Delta g_{\text{eff}} = 1$, the needed time is $\tau \approx 30\,\text{ps}$. In semiconductor bulk materials, the g-factor is fixed and cannot be changed. While in the QDs, the g-factor not only depends on the size of the dot, but also can be changed by the applied electric field, so that the QDs provide the possibility of qubit manipulation. This chapter will discuss the spin-related properties of QDs and quantum wires.

10.1. Electron Structure and g-Factor of QDs

Pryor and Flatte (2006) used the $k \cdot p$ eight bands model, taking into account the strain, to calculate the g-factor as a function of the size of QDs for self-assembled growth InAs QDs. They found that the dot's discrete energy spectrum quenches the orbital angular momentum, pushing the electron g-factor toward 2, even when all the materials have negative bulk g-factors. Figure 10.2 shows the g-factors as functions of energy gap E_g. When the QD becomes smaller, E_g becomes larger, the g-factor increases from the bulk material value -14.5, becomes positive at $E_g \approx 0.7\,\text{eV}$, and approaches the g-factor of a free electron $+2$. The dashed line in Fig. 10.2 is the g-factor calculated from the Kane model,

$$g = 2 - \frac{2E_p\Delta}{3E_g(E_g + \Delta)}, \tag{10.3}$$

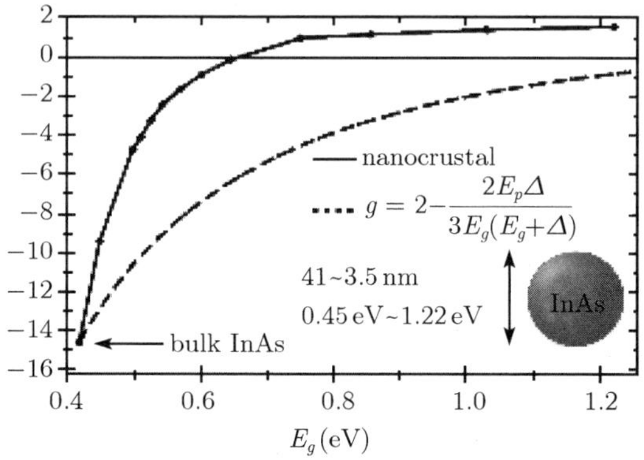

Fig. 10.2. g-Factors of self-assembled InAs QDs as functions of energy gap E_g. Dashed line is the g-factor calculated from the Kane model.

where Δ is the spin–orbit splitting energy, E_p is the coupling energy of the CB and the valence band. From Fig. 10.2, we see that according to the Kane formula (10.3), the g-factor is always negative. Actually, due to the quenching of the orbital angular momentum, the g-factor approaches the g-factor of a free electron in small QDs.

In this chapter, we will introduce some theoretical results of g-factor modulation in isolated QDs and quantum wires grown by the chemical colloidal method. First we write Luttinger's effective-mass Hamiltonian in the magnetic field (Luttinger, 1956) in the spherical coordinate. Because the QDs grown by the chemical method are generally spherical or ellipsoidal nanocrystals, it is convenient to calculate the electron states of QDs in the spherical coordinate. Luttinger's theory is for the zincblende crystal, we apply it to the case of the wurztite crystal. We take the c-axis of the wurztite crystal as the z-axis, and the basic functions of the valence band top,

$$|1,1\rangle = \frac{1}{\sqrt{2}}(X+iY), \quad |1,0\rangle = Z, \quad |1,-1\rangle = \frac{1}{\sqrt{2}}(X-iY). \tag{10.4}$$

The effective-mass Hamiltonian of the hole in the zero SOC and the zero magnetic field case is written as (Zhang and Xia, 2005),

$$H_{\mathrm{h}0} = -\frac{1}{2m_0}\begin{vmatrix} P_1 & S & T \\ S^* & P_3 & S \\ T^* & S^* & P_1 \end{vmatrix}, \tag{10.5}$$

where

$$P_1 = \gamma_1 p^2 - \sqrt{\frac{2}{3}}\gamma_2 P_0^{(2)},$$

$$P_3 = \gamma_1' p^2 + 2\sqrt{\frac{2}{3}}\gamma_2' P_0^{(2)} + 2m_0 E_\Delta, \tag{10.6}$$

$$S = AP_{-1}^{(1)} + \sqrt{2}\gamma_3 P_{-1}^{(2)},$$

$$T = \eta P_2^{(2)} + \delta P_2^{(2)},$$

$P^{(1)}$ and $P^{(2)}$ are the first- and second-order tensors of the momentum operator, respectively (Edmonds, 1957).

Assume that the external magnetic field is applied in the $x-z$ plane of the crystal structure, with an angle θ between the magnetic field and the z-axis. Choosing the symmetric gage, the vector potential of the magnetic field is written as

$$\boldsymbol{A} = \left(-\frac{1}{2}B_z y, \frac{1}{2}B_z x - \frac{1}{2}B_x z, \frac{1}{2}B_x y\right). \tag{10.7}$$

In the presence of a magnetic field, the momentum operator changes into $\boldsymbol{p} \Rightarrow \boldsymbol{p} + e\boldsymbol{A}/c$, whose components do not commute. Divide the Hamiltonian into two parts: the symmetric part is that the operator $p_\alpha p_\beta$ changes to its symmetrized product

$$\{p_\alpha p_\beta\} = \frac{1}{2}(p_\alpha p_\beta + p_\beta p_\alpha). \tag{10.8}$$

In the magnetic field, the components of the momentum operator in the spherical coordinate are given in Appendix A of Zhang and Xia (2005).

The antisymmetric part is simply written as

$$H_{\text{asym}} = K\mu_{\text{B}}\boldsymbol{I} \cdot \boldsymbol{B}, \tag{10.9}$$

where $\boldsymbol{I}$ is an angular momentum matrix (Luttinger, 1956). If the basic functions are $|1,1\rangle, |1,0\rangle$, and $|1,-1\rangle$, its components are

$$I_x = \frac{\sqrt{2}}{2}\begin{pmatrix} 0 & -1 & 0 \\ -1 & 0 & 1 \\ 0 & 1 & 0 \end{pmatrix}, \quad I_y = \frac{\sqrt{2}i}{2}\begin{pmatrix} 0 & 1 & 0 \\ -1 & 0 & -1 \\ 0 & 1 & 0 \end{pmatrix},$$

$$I_z = \begin{pmatrix} 1 & 0 & 0 \\ 0 & 0 & 0 \\ 0 & 0 & -1 \end{pmatrix}. \tag{10.10}$$

If the SOC is not zero, the SOC Hamiltonian is written as

$$H_{\mathrm{so}} = \begin{pmatrix} -\lambda & 0 & 0 & 0 & 0 & 0 \\ 0 & 0 & 0 & \sqrt{2}\lambda & 0 & 0 \\ 0 & 0 & \lambda & 0 & -\sqrt{2}\lambda & 0 \\ 0 & \sqrt{2}\lambda & 0 & \lambda & 0 & 0 \\ 0 & 0 & -\sqrt{2}\lambda & 0 & 0 & 0 \\ 0 & 0 & 0 & 0 & 0 & -\lambda \end{pmatrix}. \tag{10.11}$$

Here we take the basic functions as $|1,1\rangle\uparrow, |1,0\rangle\uparrow, |1,-1\rangle\uparrow, |1,1\rangle\downarrow, |1,0\rangle\downarrow$, and $|1,-1\rangle\downarrow, \lambda = \Delta/3$, and Δ is the spin–orbit splitting energy.

The electron Hamiltonian is written as

$$H_{\mathrm{e0}} = \frac{p^2}{2m_a} - \frac{1}{2m_b}\sqrt{\frac{2}{3}}P_0^{(2)}, \tag{10.12}$$

where

$$\frac{1}{m_a} = \frac{1}{3}\left(\frac{2}{m_x} + \frac{1}{m_z}\right), \quad \frac{1}{m_b} = \frac{1}{3}\left(\frac{1}{m_x} - \frac{1}{m_z}\right), \tag{10.13}$$

m_x and m_z are electron effective masses along the x, y, and z directions, respectively.

The total Hamiltonians of the electron and hole

$$H_{\mathrm{e}} = H_{\mathrm{e0}} + H_{\mathrm{mm\_e}} + H_{\mathrm{Zeeman\_e}}, \tag{10.14}$$

$$H_{\mathrm{h}} = H_{\mathrm{h0}} + H_{\mathrm{so}} + H_{\mathrm{mm\_h}} - H_{\mathrm{asym}} - H_{\mathrm{Zeeman\_h}}, \tag{10.15}$$

where H_{e0} and H_{h0} are Hamiltonians of electrons and holes in the zero magnetic field, respectively. H_{so} is the SOC Hamiltonian of holes (Eq. (10.11)), $H_{\mathrm{mm\_e}}$ and $H_{\mathrm{mm\_h}}$ are symmetric parts of Hamiltonians in the magnetic field subtracting H_{e0} and $H_{\mathrm{h0}}, H_{\mathrm{asym}}$ is the asymmetric part of the hole Hamiltonian in the magnetic field (Eq. (10.9)), and $H_{\mathrm{Zeeman\_e}}$ and $H_{\mathrm{Zeeman\_h}}$ are Zeeman terms of electrons and holes, respectively.

Assume that the potential outside the quantum sphere is infinite, thus the wave function at the spherical boundary ($r = R$) should be zero. The general form of the hole envelope function is

$$\Psi_{\mathrm{h}} = \sum_{l,n} \begin{pmatrix} a_{l,n\uparrow}C_{l,n}j_l(k_n^l r)Y_{l,m-1}(\theta,\phi) \\ b_{l,n\uparrow}C_{l,n}j_l(k_n^l r)Y_{l,m}(\theta,\phi) \\ d_{l,n\uparrow}C_{l,n}j_l(k_n^l r)Y_{l,m+1}(\theta,\phi) \\ a_{l,n\downarrow}C_{l,n}j_l(k_n^l r)Y_{l,m}(\theta,\phi) \\ b_{l,n\downarrow}C_{l,n}j_l(k_n^l r)Y_{l,m+1}(\theta,\phi) \\ d_{l,n\downarrow}C_{l,n}j_l(k_n^l r)Y_{l,m+2}(\theta,\phi) \end{pmatrix}. \tag{10.16}$$

$C_{l,n}$ is the normalization constant of the spherical Bessel function $j_l(k_n^l r)$, $k_n^l R = \alpha_n^l$ is the nth zero point of the $j_l(k_n^l r)$, and $m + 1/2$ is the z component of the total angular momentum. Because the hole Hamiltonians (10.5) and (10.6) consist of second-order tensor of the momentum operator, the summation of the wave function (10.16) includes summation over terms with angular momentum $l, l \pm 2, l \pm 4, \ldots$ and different n. If the magnetic field is zero or along the z-direction, the component of the total angular momentum $M = m + 1/2$ is a good quantum number. If the magnetic field is along an arbitrary direction, the summation in Eq. (10.16) should include terms with different components m of the angular momentum. The electron wave function can be similarly expanded as,

$$\Psi_{\mathrm{e}} = \sum_{l,n} \begin{pmatrix} e_{l,n\uparrow} C_{l,n} j_l(k_n^l r) Y_{l\cdot m}(\theta, \phi) \\ e_{l,n\downarrow} C_{l,n} j_l(k_n^l r) Y_{l\cdot m}(\theta, \phi) \end{pmatrix}. \tag{10.17}$$

Figure 10.3 shows the energies of hole states ε of CdSe QD ($R = 2.1\,\mathrm{nm}$) as functions of magnetic field b (Zhang and Xia, 2005), where ε and b are all dimensionless quantities, and the energy and magnetic field units are

$$\varepsilon_0 = \frac{1}{2m_0}\left(\frac{\hbar}{R}\right)^2, \quad b_0 = \frac{\hbar e B}{m_0 \varepsilon_0}. \tag{10.18}$$

The magnetic field is along the z-direction, figures 10.3(a)–10.3(d) correspond to $M = -3/2, M = 3/2, M = -1/2$, and $M = 1/2$. From Fig. 10.3, we see that in the presence of a magnetic field, the originally degenerate energy levels, for example, the energy levels with lowest energy $P_{\mathrm{X}-1}\downarrow$ and $P_{\mathrm{X}+1}\uparrow$ states and $S_{\mathrm{X}-1}\downarrow$ and $S_{\mathrm{X}+1}\uparrow$ states, split. The symbol at each level represents the main components of its wave function, for example, the $P_{\mathrm{X}-1}\downarrow$ means that the envelope function (Eq. (10.6)) consists of mainly $l = 1, n = 1$ state, multiplied with the $|1 - 1\rangle$ Bloch state of the valence band top and the spin-down state, thus mainly the sixth component. In the small magnetic field case, the splitting energy is proportional to the magnetic field, from which the g-factor of the hole state can be determined. The energy splitting is caused not only by the Zeeman term in Hamiltonian (10.15), but also by its orbit angular momentum part (symmetric and antisymmetric parts). When the quantum sphere is very small, its radius is smaller than the cyclotron radius; the contribution of the orbit angular momentum will be quenched, and the g-factor is determined mainly by the Zeeman term, approaching the g-factor of a free electron or a free hole.

It is interesting to notice from Fig. 10.3 that when the magnetic field is very small, the hole ground state is $P_{\mathrm{X}1}\uparrow$, i.e. optically passive. When

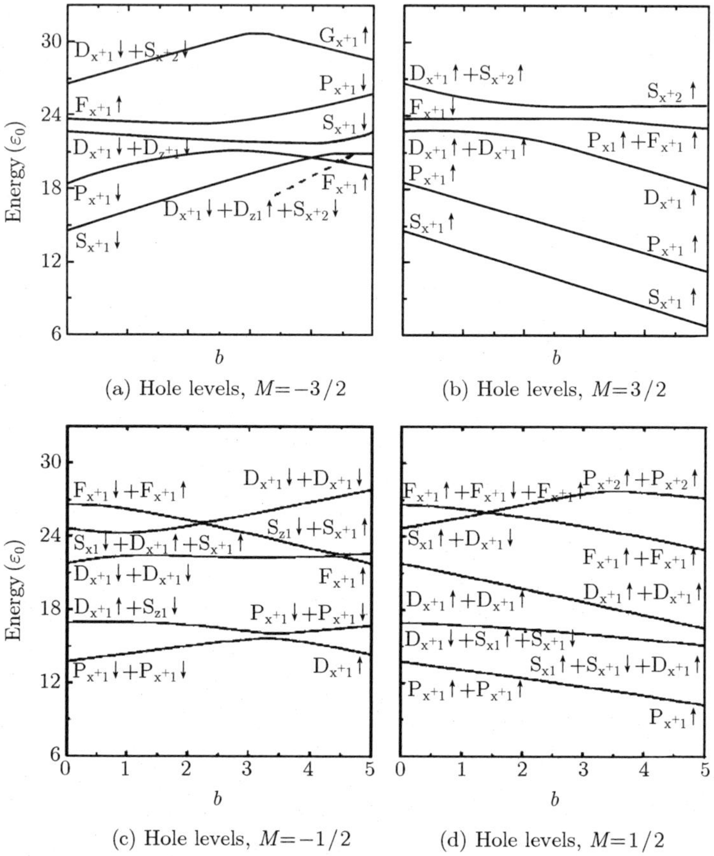

Fig. 10.3. Energies of hole states ε of CdSe QD ($R = 2.1\,\text{nm}$) as functions of magnetic field b: (a) $M = -3/2$, (b) $M = 3/2$, (c) $M = -1/2$, and (d) $M = 1/2$.

$b \geq 1$ ($B \geq 74\,\text{T}$ for $R = 2.1\,\text{nm}$ CdSe quantum sphere), the ground state becomes $S_{X1} \uparrow$, i.e. the transition changes from indirect transition (dark exciton) to direct transition (bright exciton). The magnetic field at which the change occurs is called the critical magnetic field B_c. If the radius of the quantum sphere is larger, B_c is smaller. For example, when $R = 2.5\,\text{nm}$, $B_c = 7.4\,\text{T}$. It means that through the magnetic field modulation, the indirect transition in the quantum sphere can be changed to the direct transition.

The experiment (Johnston-Halperin *et al.*, 2001) discovered that the randomly oriented dots of the wurtzite structure in the magnetic field gives off light with definite circular polarization. Define the circular polarization factor,

$$P = \frac{I_{\sigma} \quad - I_{\sigma^+}}{I_{\sigma^-} + I_{\sigma^+}}, \tag{10.19}$$

where I_{σ^-} and I_{σ^+} the are intensities of left and right CP lights, respectively. The experiment discovered that the circular polarization factor of the light from the randomly oriented CdSe dots of $R = 2.85\,\text{nm}$ increases with increasing magnetic field, approaches 0.8, not 1 in the large magnetic field. Figure 10.4(a) shows the theoretically calculated normalized intensities of σ^- and σ^+ transitions of CdSe dot ($R = 2.85\,\text{nm}$) at $T = 10\,\text{K}$ as functions of magnetic field b, and Fig. 10.4(b) shows circular polarization factors P as functions of b at different temperatures (Zhang and Xia, 2005). From Fig. 10.4(b), we see that P approaches 0.8 in the large magnetic field, in agreement with the experimental results. At low temperatures, P approaches saturation more quickly.

The QDs of zincblende structure are different from those of the wurztite structure. The zincblende QD has the cubic symmetry, and the x-axis is equivalent to the z-axis. The circular polarization factor P of light from the randomly oriented QDs approaches 1 in a large magnetic field. While the wurztite QD has a hexagonal symmetry, the c-axis and the direction perpendicular to the c-axis are not equivalent. Figure 10.5 shows the circular polarization factors P of single CdSe and GaAs dots as functions of

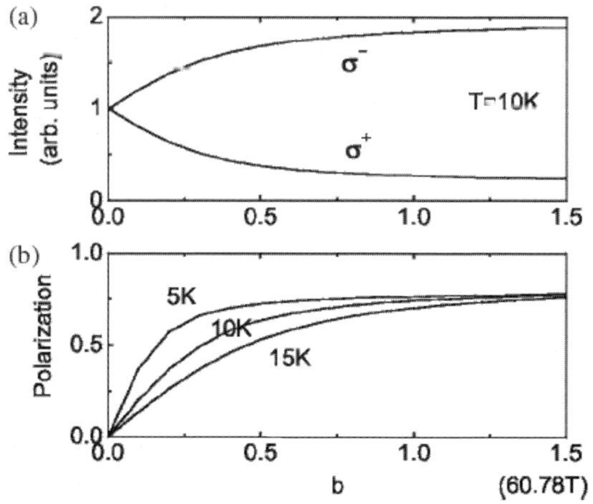

Fig. 10.4. (a) Normalized intensities of σ^- and σ^+ transitions of CdSe dot ($R = 2.85\,\text{nm}$) at T =10K as functions of magnetic field b. (b) Circular polarization factors P as functions of b at different temperatures.

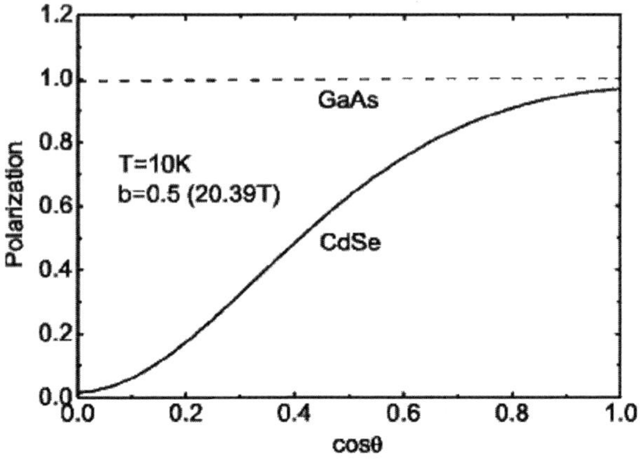

Fig. 10.5. Circular polarization factors P of single CdSe and GaAs dots as functions of $\cos\theta$ at $T = 10\,\text{K}$, $b = 0.5$ (20.39 T).

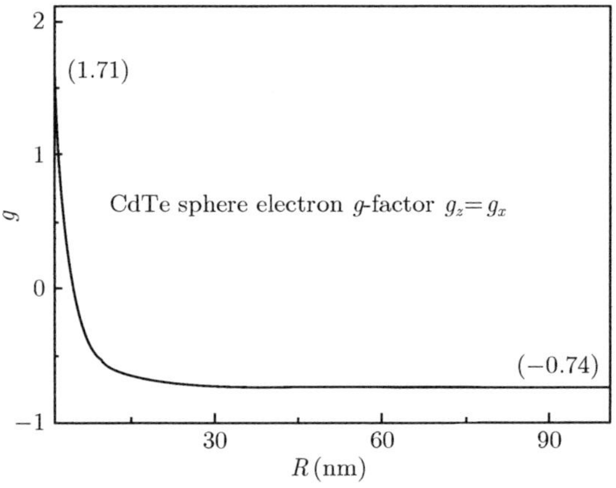

Fig. 10.6. Electron g-factor of CdTe quantum sphere as a function of radius R.

$\cos\theta$ at $T = 10\,\text{K}$, $b = 0.5\,(20.39\,\text{T})$, θ is the angle between the magnetic field and the z-axis (Zhang and Xia, 2005). From Fig. 10.5, we see that for GaAs P is independent of θ and always equals 1. While for CdSe the P varies with $\cos\theta$, changes from 0 to nearly 1, the average value over each θ equals 0.8.

Figure 10.6 shows the electron g-factor of CdTe quantum sphere as a function of radius R (Zhang and Xia, 2006a). The above theoretical model of the wurztite QDs is easily applied to the zincblende QDs. We only need that in hole Hamiltonians (10.5) and (10.6), let $\gamma_1' = \gamma_1$, $\gamma_2' = \gamma_2$, $E_\Delta = 0$,

$A = 0$, and $\delta = 0$. In order to study generally QDs of wide and narrow energy gaps, we accept $k \cdot p$ eight bands model (Zhu *et al.*, 2006a), taking into account CB and valence bands altogether. The basic functions are CB state $|S\rangle$ and the valence band states $|11\rangle, |10\rangle, |1-1\rangle$ with the spin-up and spin-down states, the $|S\rangle$ state is coupled with $|11\rangle, |10\rangle, |1-1\rangle$ by $-ip_0 P_1^{(1)}$, $ip_0 P_0^{(1)}$, and $ip_0 P_{-1}^{(1)}$, respectively, where $p_0 = \sqrt{2m_0 E_p}$, E_p is the Kane matrix element. The original effective-mass parameters calculated for CB and valence band individually should subtract the contribution from the interactions between the CB and the valence band.

From Fig. 10.6, we see that because CdTe is the zincblende structure, the three axes of the crystal structure are equivalent, and the g-factor is isotropic. Thus, $g_z = g_x$, g_z (g_x) is the g-factor for the magnetic field along the $z(x)$ direction. When the radius is large, the g-factor equals the value of the bulk material -0.74. When the radius decreases gradually, the g-factor increases, approaches 1.71, close to the value of the free electron. This indicates the quenching effect of the orbit angular momentum.

Not only the size of QD influences the g-factor, but also the shape of QD influences the g-factor. Figure 10.7(a) shows the electron g-factors g_z and g_x of CdTe ellipsoids with transverse radius $R = 2.0$ nm as functions of the aspect ratio e. When $e = 1$, in the sphere case, $g_z = g_x = 1.03$. As e increases for fixed transverse radius R, the ellipsoid enlarges along the c-axis direction, g_z and g_x both decrease, but $g_z > g_x$, and approach 0.78 and 0.69, respectively. $g_z(g_x)$ is the g-factor for the magnetic field along the $z(x)$-direction. When e increases for fixed R, the electron cyclotron movement in the x–y plane around the magnetic field in the z-direction does not change, so that the g_z factor changes a little. The electron cyclotron movement in the y–z plane around the magnetic field in the x-direction is relaxed in the z-direction, so that g_x decreases more quickly, $g_x < g_z$. Figure 10.7(b) shows the g_z and g_x of CdTe ellipsoids with longitudinal length $L = 4.0$ nm as functions of transverse radius R. The results are just reverse to (a). When R increases for fixed L, g_x and g_z both decrease, but $g_x > g_z$, and approach 0.26 and 0.015, respectively. The results can also be explained by the confinement of the electron cyclotron movement. It is concluded that as the size in one or two dimensions for electron cyclotron movement increases, the electron g-factor decreases.

Figure 10.8 shows the electron g-factor of InSb quantum sphere as a function of radius R (Zhu *et al.*, 2006b). InSb is a narrow energy gap semiconductor, and the strong interaction between the CB and valence band results in strong correlation between the electron g-factor and the

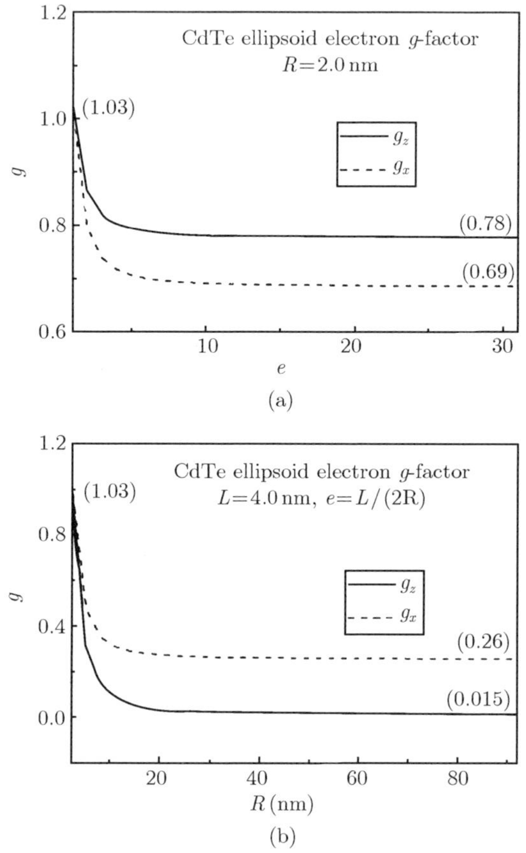

Fig. 10.7. (a) Electron g-factors g_z and g_x of CdTe ellipsoids with transverse radius $R = 2.0\,\text{nm}$ as functions of the aspect ratio e. (b) g_z and g_x of CdTe ellipsoids with longitudinal length $L = 4.0\,\text{nm}$ as functions of transverse radius R.

energy gap, as shown in Eq. (10.3). From Fig. 10.8, we see that when R is very large, the g-factor equals its value of bulk material -47.2. When R decreases, the g-factor increases, equals zero at $R = 2.25\,\text{nm}$. When R decreases furthermore, the g-factor becomes positive and approaches the free electron value 2. Therefore, for narrow gap semiconductors, one can modulate their g-factors in a large range.

The shape of QD can also change the g-factor, same as the case of CdTe. Figure 10.9(a) shows the electron g-factors g_z and g_x of InSb ellipsoids with transverse radius $R = 2.0\,\text{nm}$ as functions of the aspect ratio e. $g_z > g_x$ and approach -0.399 and -0.704, respectively. Figure 10.9(b) shows the g_z and g_x of InSb ellipsoids with longitudinal length $L = 4.0\,\text{nm}$ as functions of

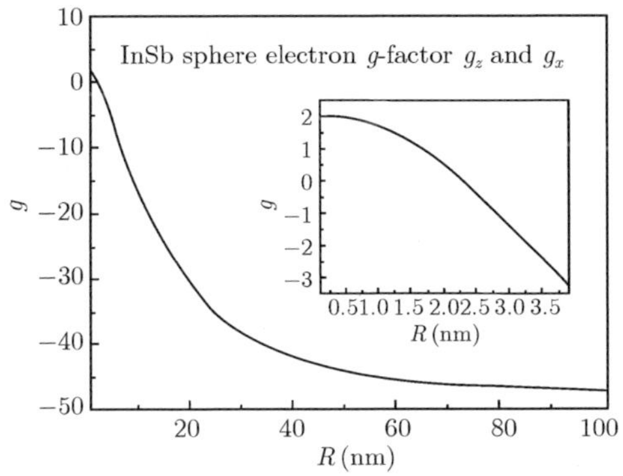

Fig. 10.8. Electron g-factor of InSb quantum sphere as a function of radius R. The inset is the detail for small R.

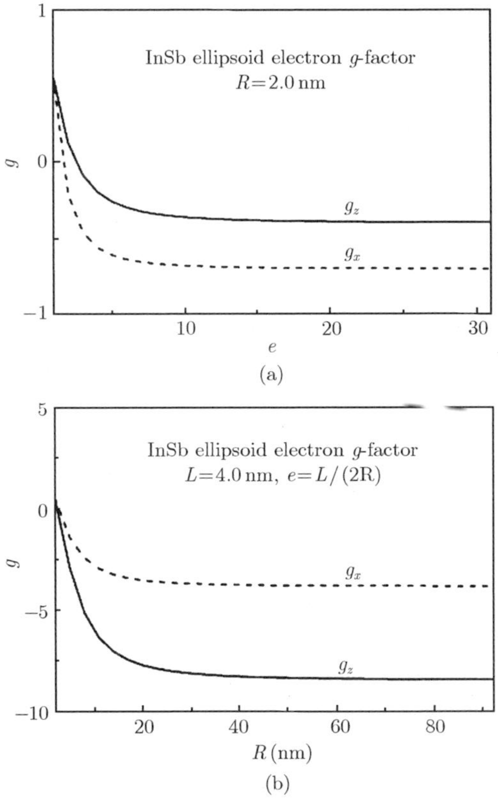

Fig. 10.9. (a) Electron g-factors g_z and g_x of InSb ellipsoids with transverse radius $R = 2.0\,\text{nm}$ as functions of the aspect ratio e. (b) g_z and g_x of InSb ellipsoids with longitudinal length $L = 4.0\,\text{nm}$ as functions of transverse radius R.

transverse radius R. $g_z < g_x$ and approach -8.455 and -3.836, respectively (Zhu *et al.*, 2006b).

10.2. Electron Structure and g-Factor of Quantum Wires

Except quantum spheres, the chemical colloid method can also grow quantum ellipsoids, quantum wires and quantum rods. This section studies the electron structures and g-factors of quantum wires and rods. Because the quantum wires or rods have confinement effect for the electron cyclotron movement, their radius also influences the g-factor.

In order to calculate the electron states of quantum wires (or rods) in the magnetic field, we write the effective-mass Hamiltonian in the cylindrical coordinate. In the case of zero magnetic field, the eight bands effective-mass Hamiltonian including the CB and valence band is written as (Zhang *et al.*, 2006a),

$$H_{\text{eh}} = \begin{pmatrix} H_{\text{int}} & 0 \\ 0 & H_{\text{int}} \end{pmatrix} + H_{\text{so}}, \tag{10.20}$$

where H_{so} is the SOC Hamiltonian (10.12), and

$$H_{\text{int}} = \frac{1}{2m_0}\begin{vmatrix} \varepsilon_g + P_e & \frac{i}{\sqrt{2}}p_0p_+ & ip_0p_z & \frac{i}{\sqrt{2}}p_0p_- \\ -\frac{i}{\sqrt{2}}p_0p_- & -P_1 & -S & -T \\ -ip_0p_z & -S^* & -P_3 & -S \\ -\frac{i}{\sqrt{2}}p_0p_+ & -T^* & S^* & -P_1 \end{vmatrix}, \tag{10.21}$$

$$\begin{aligned} P_e &= \alpha p_-p_+ + \alpha p_z^2, \\ P_1 &= \frac{L+M}{2}p_-p_+ + Mp_z^2, \\ P_3 &= Mp_-p_+ + Lp_z^2, \\ S &= \frac{1}{\sqrt{2}}Np_-p_z, \\ T &= \frac{L-M-N}{4}p_+^2 + \frac{L-M+N}{4}p_-^2, \\ p_\pm &= p_x \pm ip_y. \end{aligned} \tag{10.22}$$

$\varepsilon_g = 2m_0 E_g$, E_g is the band gap of the bulk material. $p_0 = \sqrt{2m_0 E_p}$, E_p is the Kane matrix element.

When the magnetic field is applied in the z-direction (wire direction), the vector potential is taken as,

$$\boldsymbol{A} = \left(-\frac{1}{2} B_z y, \frac{1}{2} B_z x, 0 \right). \tag{10.23}$$

Divide the effective-mass Hamiltonian into symmetric and antisymmetric parts. The product of the momentum operators in the symmetric part is

$$\{P_+ P_-\} = \frac{1}{2}(P_+ P_- + P_- P_+), \tag{10.24}$$

where

$$\begin{aligned} P_\pm &= \left(p_x - \frac{eB}{2c} y \right) \pm i \left(p_y + \frac{eB}{2c} x \right) = p_\pm \pm i \frac{eB}{2c} r_\pm, \quad r_\pm = x \pm iy, \\ \{P_+ P_-\} &= p_+ p_- + \beta^2 r^2 + i\beta(r_+ p_- - r_- p_+), \quad \beta = \frac{eB}{2c}. \end{aligned} \tag{10.25}$$

The matrix elements of the Hamiltonian in the magnetic field are given in the appendix of Zhang *et al.* (2006a).

Figure 10.10 shows the electron g-factors g_z and g_x of GaAs rods with $L = 2\,\text{nm}$ as a function of radius R (Zhang *et al.*, 2006b). When R increases, g_z and g_x both decrease and g_z decreases more quickly. When $R > 32\,\text{nm}$,

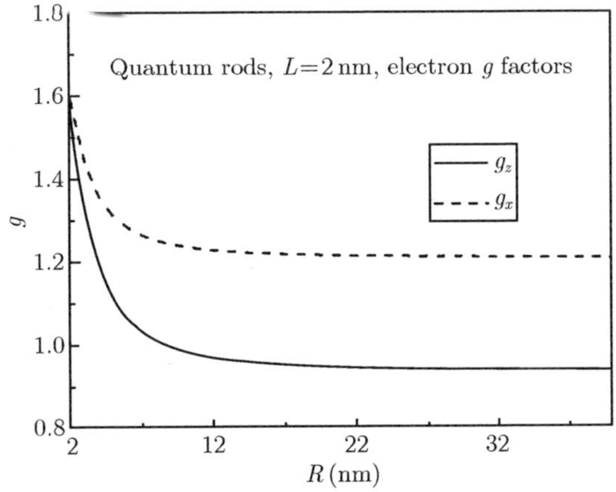

Fig. 10.10. Electron g-factors g_z and g_x of GaAs rods with $L = 2\,\text{nm}$ as a function of radius R.

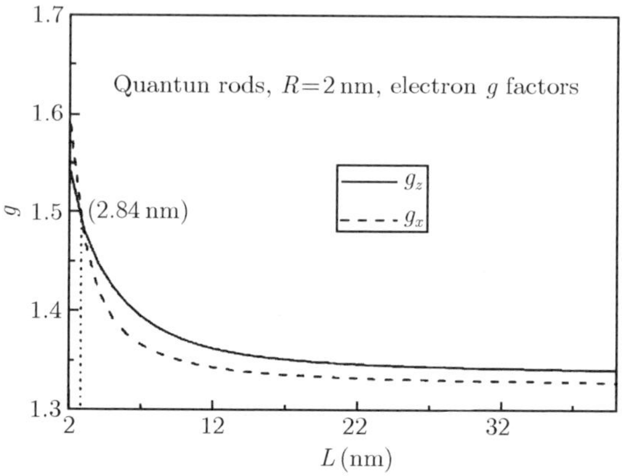

Fig. 10.11. Electron g-factors g_z and g_x of GaAs rods with $R = 2\,\text{nm}$ as function of radius L.

the g-factors approach the g-values of the GaAs quantum well of width 2 nm, similar to the case of ellipsoid in the last section.

Figure 10.11 shows the electron g-factors g_z and g_x of GaAs rods with $R = 2\,\text{nm}$ as function of length L (Zhang *et al.*, 2006b). When L increases, g_z and g_x both decrease and g_x decreases more quickly. When $L > 32\,\text{nm}$, they approach the g values of the quantum wire of $R = 2\,\text{nm}$.

In Sec. 4.2.3, we introduced the splitting of electron spin states of InSb QW in the transverse electric field (Zhang and Xia, 2006b), i.e. the RSS. Here we introduce the hole Rashba effect and g-factor of InP quantum wires (Zhang and Xia, 2007). Before we have only studied the g-factors and Rashba effect of electron, the calculation discovered that the hole has very different properties from those of electron.

Figure 10.12 shows (a) electron and (b) hole spin-splitting energy ΔE of the ground state as functions of k_z for an InP nanowire with $R = 15\,\text{nm}$ in transverse electric field $F = 5\,\text{mV/nm}$ (Zhang and Xia, 2007). From Fig. 10.12, we see that ΔE are basically proportional to k_z, but that of hole ΔE_h is larger than that of electron ΔE_e by two orders of magnitude. From ΔE_e and ΔE_h as functions of k_z, we can define the Rashba coefficients of electron and hole α and β,

$$\begin{aligned} \Delta E_\text{e} &= 2\alpha k_z + 0(k_z), \\ \Delta E_\text{h} &= 2\beta k_z + 0(k_z). \end{aligned} \tag{10.26}$$

The Rashba coefficient β is larger than α by two orders of magnitude in the InP nanowire.

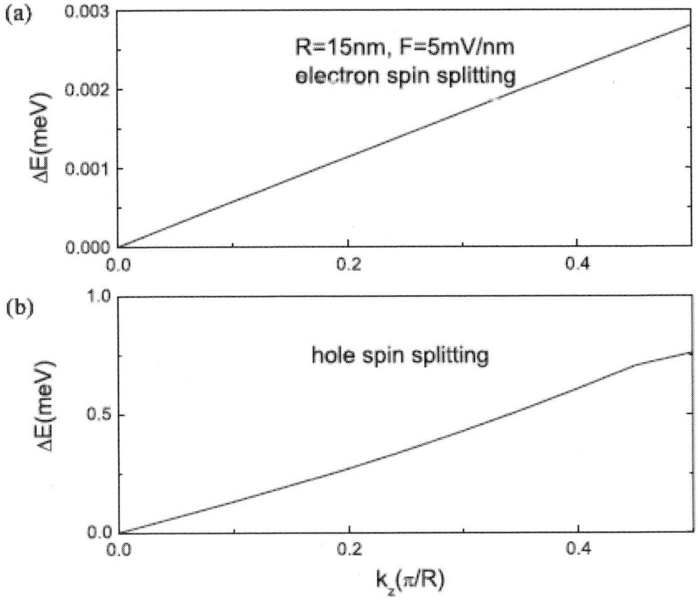

Fig. 10.12. (a) Electron and (b) hole spin-splitting energy ΔE of the ground state as functions of k_z for an InP nanowire with $R = 15\,\text{nm}$ in transverse electric field $F = 5\,\text{mV/nm}$.

Figure 10.13 shows the hole Rashba coefficients β of InP nanowire with (a) $R = 15\,\text{nm}$ and (b) $R = 10\,\text{nm}$ as functions of electric field F (Zhang and Xia, 2007). From Fig. 10.13, we see that in the small electric field β is proportional to F. When the electric field increases, β reaches a maximum, then decreases, equals zero at an electric field. We call this electric field to make $\beta = 0$, the critical electric field F_c. Comparing Figs. 10.13(a) and 10.13(b), we found that F_c is larger for thinner quantum wire. These specific properties of hole Rashba coefficient are caused by the complicated valence band structure; in common actions of quantum confinement effect and applied electric field, different hole levels are mixed. In the large electric field, the mixing becomes larger, resulting in the changing sign of β. If the radius of the quantum wire is smaller, the spacing between hole levels is large and different states are hard to mix, resulting in large critical electric field F_c.

The electric field can induce a great mixture of the hole states and change the state components of the highest hole states obviously, so that it affects hole g-factor. Figure 10.14 shows the hole g-factor of an

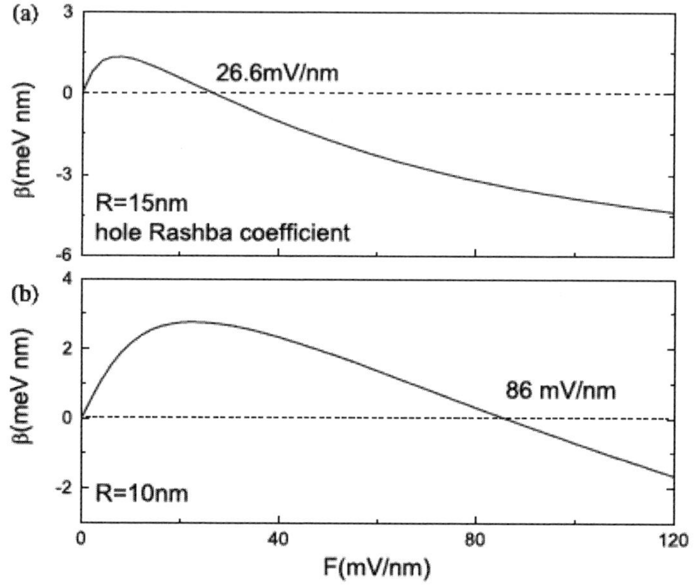

Fig. 10.13. Hole Rashba coefficients β of InP nanowire with (a) $R = 15\,\text{nm}$ and (b) $R = 10\,\text{nm}$ as functions of electric field F.

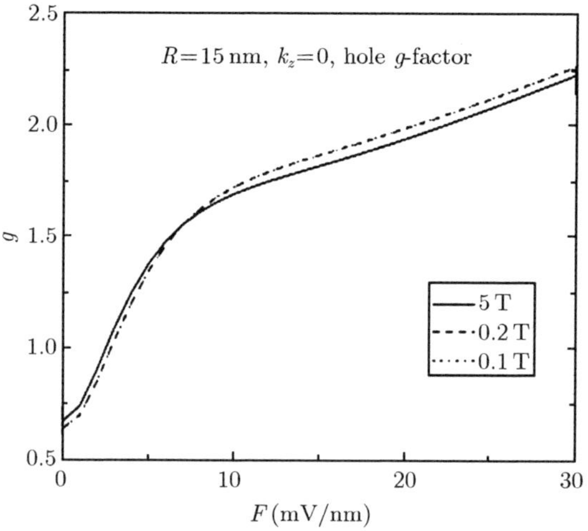

Fig. 10.14. Hole g-factor of an InP nanowire with $R = 15\,\text{nm}$ at $k_z = 0$ as functions of F. Three curves correspond to different magnetic field.

InP nanowire with $R = 15\,\text{nm}$ at $k_z = 0$ as functions of F (Zhang and Xia, 2007). From Fig. 10.14, we see that the hole g-factor increase with increasing electric field, quickly at the beginning, and slowly after $F > 10\,\text{mV/nm}$. Therefore, we can modulate the hole g-factor by applied electric field.

10.3. Electric Field Tunable g-Factor in QDs

Consider the electron states in QDs in the electric field. Due to the dielectric effect, the electric field in the QDs $\boldsymbol{F}$ does not equal the applied electric field $\boldsymbol{F}_{\text{ext}}$. Assume that the z-axis is the long axis of the ellipsoid, then the electric field inside the ellipsoid is $\boldsymbol{F} = (F_x, 0, F_z)$ (Li and Xia, 2003; Zhang *et al.*, 2007a),

$$\begin{aligned} F_z &= \frac{\varepsilon_0}{n_z\varepsilon_r + (1 - n_z)\varepsilon_0} F_{z,\text{ext}}, \\ F_x &= \frac{\varepsilon_0}{n_x\varepsilon_r + (1 - n_x)\varepsilon_0} F_{x,\text{ext}}, \end{aligned} \tag{10.27}$$

where ε_r and ε_0 are dielectric constants in and outside the ellipsoid, respectively.

$$\begin{aligned} n_z &= \frac{1 - e'^2}{2e'^2}\left(\ln\frac{1 + e'}{1 - e'} - 2e'\right), \quad e' = \sqrt{1 - \frac{1}{e^2}} \\ n_x &= \frac{1 - n_z}{2}, \end{aligned} \tag{10.28}$$

e is the aspect ratio of the length to diameter of the ellipsoid. The electric field potential,

$$V(\boldsymbol{r}) = -qeF_z r\left(\frac{4\pi}{3}\right)^{1/2} Y_{1,0} - qF_x r\left(\frac{2\pi}{3}\right)^{1/2}(Y_{1,-1} - Y_{1,1}). \tag{10.29}$$

Figure 10.15(a) shows the electron g-factor of InAs quantum ellipsoid with $R = 3\,\text{nm}$, at $F = 0$ as a function of e. (b) Electron g-factor of InAs ellipsoid with $R = 3\,\text{nm}$, $e = 3$ as a function of F (Zhang *et al.*, 2007b). From Fig. 10.15(a), we see that the shape can change the g-factor, the g-factor decreases as e increases and can be tuned to be zero by $e = 1.96$. From Fig. 10.15(b), we see that when $F \| x$, the g-factor is almost not affected, while it increases a lot by the electric field when $F \| z$. The g-factor can be tuned from negative to zero by the electric field $F = 12.3\,\text{mV/nm}$ and to positive under larger electric field. Quantum ellipsoid of $e = 3$ is a long ellipsoid, similar to the quantum rod. Thus, the transverse electric field affects the electron states less, while the longitudinal electric field

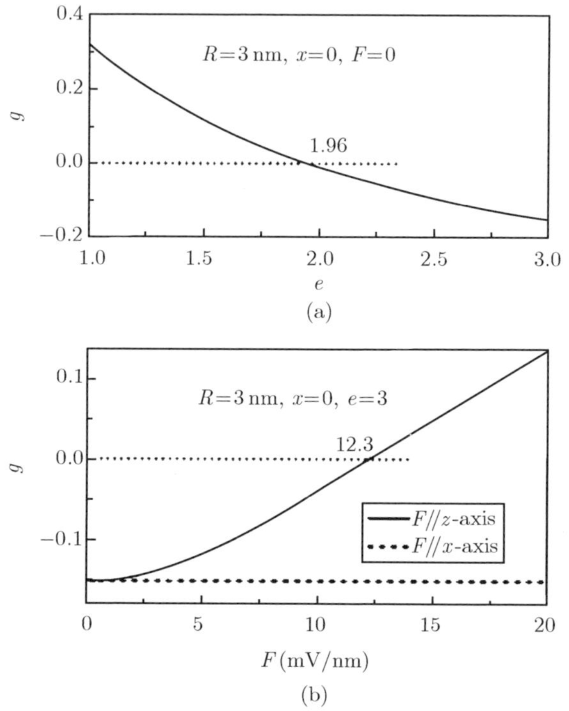

Fig. 10.15. (a) Electron g-factors of InAs quantum ellipsoid with $R = 3\,\text{nm}$, at $F = 0$ as a function of e. (b) Electron g-factors of InAs ellipsoid with $R = 3\,\text{nm}$, $e = 3$ as a function of F.

induces larger Stark effect, making the energy gap smaller. The smaller energy gap will enhance the interaction between the CB and valence band, resulting in larger mixing of electron states and hole states. The electron spin component changes, and the g-factor reverses the sign. Therefore, under a fixed magnetic field, we can use the electric field to tune the electron spin to be polarized, unpolarized, or antipolarized.

Figure 10.16 shows the (a) energy levels of $InAs_{0.95}N_{0.05}$ quantum ellipsoid with $R = 8\,\text{nm}$, $e = 4$ as functions of $F(F\|x)$, (b) $F\|z$, (c) $F = 15\,\text{mV/nm}$ as functions of the angle θ between the z-axis and the electric field (Zhang *et al.*, 2007a). The highly asymmetrical Stark effect is obviously seen. An electric field along the z-axis with intensity larger than $11\,\text{mV/nm}$ can make the CB and valence band overlapping, and the energy gap tuned to be zero, similar to the wire case. From Fig. 10.16(c), we see that due to the energy gap change the wave length of light emitted

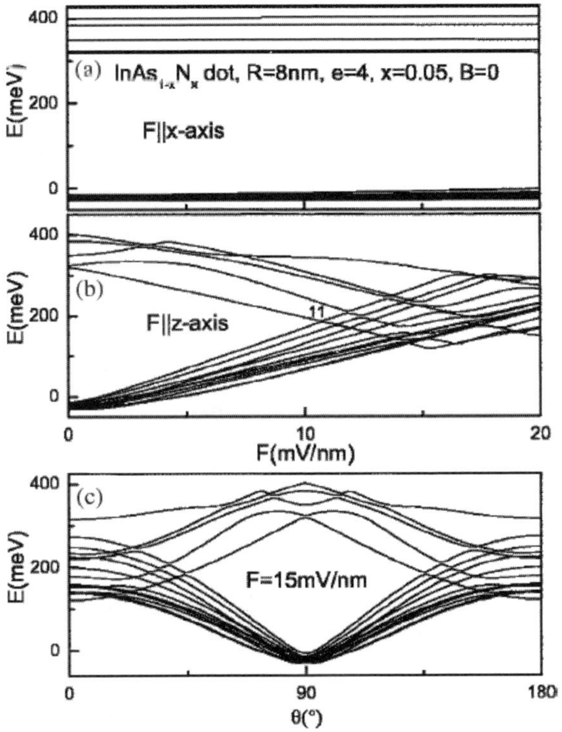

Fig. 10.16. (a) Energy levels of $InAs_{0.95}N_{0.05}$ quantum ellipsoid with $R = 8\,\text{nm}$, $e = 4$ as functions of $F(F\|x)$. (b) $F\|z$, (c) $F = 15\,\text{mV/nm}$ as functions of the angle θ between the z-axis and the electric field.

from the quantum ellipsoid can change in a large range when it rotates in a strong electric field.

Figure 10.17 shows the Stark shifts ΔE_S of InAs quantum ellipsoid with $R = 3\,\text{nm}$, $e = 3$ as functions of F; (a) electron and (b) hole (Zhang *et al.*, 2007a). When $F\|x$, the Stark shifts are nearly zero even if F is very large, while in the $F\|z$ case, the Stark shifts are as large as tens of meV when F is large. This highly asymmetrical Stark effect is due to the dielectric effect and the quantum confinement effect. Due to the dielectric effect, the electric field in the QD is smaller than the external electric field. When $e = 3$, from Eq. (10.28) we obtain $n_z = 0.102, n_x = 0.449$, so that the inner electric field when $F\|x$ is about three times smaller than that when $F\|z$. As the quantum confinement of the quantum ellipsoid along the x-axis is larger than that along the z-axis, the Stark shift in the

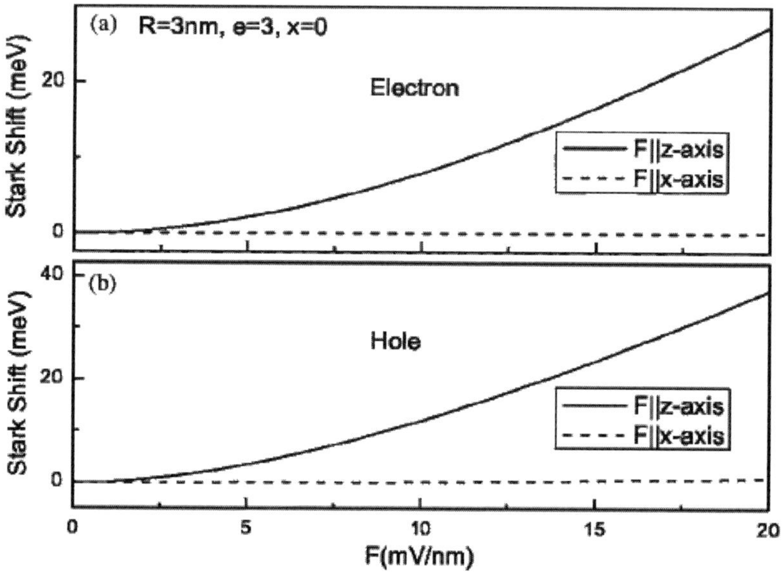

Fig. 10.17. Stark shifts ΔE_S of InAs quantum ellipsoid with $R = 3\,\text{nm}$, $e = 3$ as functions of F; (a) electron and (b) hole.

x-direction is smaller than that in the z-direction. Taking into account the dielectric effect and the quantum confinement effect simultaneously, the asymmetrical factor of the electron and hole Stark shifts are 319 and 40 from Figs. 10.17(a) and 10.17(b), respectively.

10.4. Influence of N Doping on the Rashba Coefficient and the g-Factor of Electrons

Most theoretical and experimental works have concentrated on the N-doped alloy III–V compounds, owing to its technological importance for fiber-optic communications at wavelengths of 1.3 and 1.55 μm. The energy level of N in the III–V compounds is in the CB, near the bottom of the CB. Due to the interaction of N and the CB bottom, the band gap of the N-doped compound is smaller than that of undoped compound. When the concentration of N is small, the decrease in the band gap is proportional to the concentration of N, resulting in the energy band bowing effect. For example, the band gap of $InSb_{1-x}N_x$ can be tuned to be near zero, even negative, by increasing the percentage of N.

In order to study the electron states in N-doped QDs or quantum wires, we used the 10 bands model including N, electron, and hole (Zhang *et al.*, 2007b). The Hamiltonian is written as

$$H_{\text{ten}} = \begin{pmatrix} H_{\text{five}} & 0 \\ 0 & H_{\text{five}} \end{pmatrix} + H_{\text{so}},$$

$$H_{\text{five}} = \frac{1}{2m_0} \begin{pmatrix} 2m_0 E_{\text{N}} & 2m_0 V_{\text{NC}} x^{1/2} & 0 & 0 & 0 \\ 2m_0 V_{\text{NC}} x^{1/2} & 2m_0 E_g + P_e & -ip_0 P_1^{(1)} & ip_0 P_0^{(1)} & ip_0 P_{-1}^{(1)} \\ 0 & -ip_0 P_{-1}^{(1)} & -P_1 & -S & -T \\ 0 & -ip_0 P_0^{(1)} & -S^* & -P_3 & -S \\ 0 & ip_0 P_1^{(1)} & -T^* & -S^* & -P_1 \end{pmatrix}, \tag{10.30}$$

where $E_N = 0.65\,\text{eV}$ is the N energy level in InSb relative to the valence band top, V_{NC} is the interaction energy of N and CB, and x is the composition of N.

Figure 10.18 shows the Rashba coefficients of $\text{InSb}_{1-x}\text{N}_x$ nanowires at $F = 0.5\,\text{mV/nm}$ as functions of radius R and N concentration x (Zhang *et al.*, 2007b). When $x = 0$, the Rashba coefficient α increases with the radius R and saturates, which is in agreement with Eq. (10.31). When $x \neq 0$, α increases very quickly especially for quantum wires with large

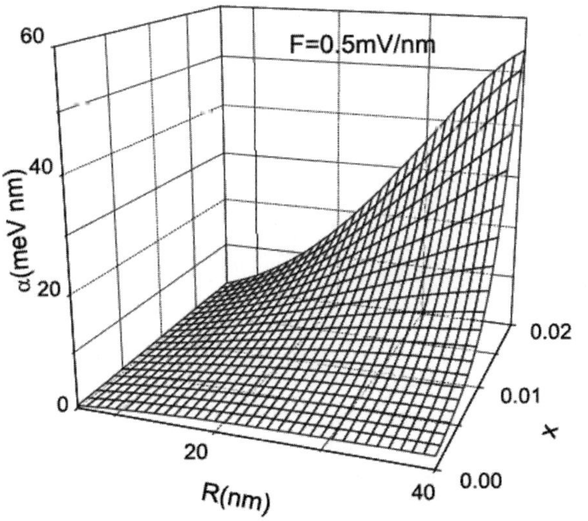

Fig. 10.18. Rashba coefficients of $\text{InSb}_{1-x}\text{N}_x$ nanowires at $F = 0.5\,\text{mV/nm}$ as functions of radius R and N concentration x.

radius. When $x = 0.02, R = 40\,\text{nm}$, $\alpha \approx 55\,\text{meV nm}$, about 20 times for the $x = 0$ case. The Rashba coefficient is given by

$$\alpha = \frac{\hbar^2}{2m_0} E_p \frac{\lambda(2E_g + 3\lambda)}{E_g^2(E_g + 3\lambda)^2} eF, \tag{10.31}$$

where E_g is the band gap of the quantum wires. When the N concentration x increases, E_g decreases, so that the coupling between the CB and valence band becomes strong. When R is larger, the band gap of InSb nanowires E_{g0} is smaller, and the relative decrease in the band gap $(E_{g0} - E_g)/E_{g0}$ is larger for a given x, resulting in larger increase in the Rashba coefficient.

Figure 10.19 shows the electron g-factor of $InSb_{1-x}N_x$ nanowires as functions of R and x (Zhang *et al.*, 2007c). When $x = 0$, the g-factor

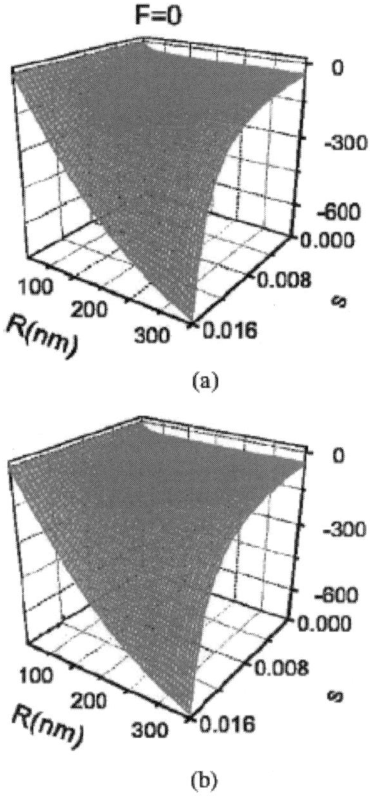

Fig. 10.19. Electron g-factor of $InSb_{1-x}N_x$ nanowires as functions of R and x; (a) g_z and (b) g_x.

decreases with increasing R and approaches the bulk material value -44.5. When $x \neq 0$, the g-factor decreases a lot. For nanowires with large radius (300 nm) the g-factor can decrease to -730. According to Eq. (10.3), these results are all caused by decrease of E_g due to N doping. The calculation indicated that the E_g of the $R = 300$ nm nanowire decreases from 230 meV to near zero as x increases from 0 to 0.016, while the E_g of the $R = 5$ nm nanowire decreases from 500 to 300 meV, the decreasing times differ a lot.

The spin properties of electron in semiconductors are decided not only by its magnetic moment ($g = 2$), but also are associated with its orbital movement, resulting in different g-factors in different semiconductors. In the narrow band gap semiconductors, due to a strong interaction between the CB and valence band the electron g-factor differs a lot from that of free electron. Besides, the Rashba splitting and Dresselhaus splitting effects caused by the SOC are most obvious in the narrow band gap semiconductors. The semiconductor QD has long spin coherent time; thus, it is one of the possible choices for realizing qubit. The quantum confinement effect in QDs and quantum wires limits, even quenches the orbital movement, certainly influences the spin states. From the theoretical results of this chapter, it is concluded that the size, shape of the dots and wires, the external electric field, stress field, magnetic field, and the N doping all can affect the g-factor, the Rashba coefficient, and other spin-related properties, providing varied means to manipulate the electron spin properties.

References

Edmonds AR. (1957) *Angular Momentum in Quantum Mechanics*, Princeton, New Jersey.

Johnston-Halperin E, Awschalom DD, Crooker SA, Efros AlL, Rosen M, Peng X, Alivisatos AP. (2001) *Phys Rev B* **63**: 205309.

Khaetskii AV, Nazarov YV. (2000) *Phys Rev B* **61**: 12639.

Li XZ, Xia JB. (2003) *Phys Rev B* **69**: 165316.

Loss D, DiVincenzo DP. (1998) *Phys Rev A* **57**: 20.

Luttinger JM. (1956) *Phys Rev* **102**: 1030.

Pryor CE, Flatte ME. (2006) *Phys Rev Lett* **96**: 026804.

Zhang XW, Xia JB. (2005) *Phys Rev B* **72**: 075363.

Zhang XW, Xia JB. (2006a) *Physica E* **35**: 69.

Zhang XW, Xia JB. (2006b) *Phys Rev B* **74**: 075304.

Zhang XW, Xia JB. (2007) *J Phys D* **40**: 541.

Zhang XW, Zhu YH, Xia JB. (2006a) *Euro Phys J B* **52**: 133.

Zhang XW, Zhu YH, Xia JB. (2006b) *Physica E* **33**: 376.

Zhang XW, Fan WJ, Li SS, Xia JB. (2007a) *Appl Phys Lett* **90**: 153103.
Zhang XW, Fan WJ, Li SS, Xia JB. (2007b) *Phys Rev B* **75**: 205331.
Zhang XW, Fan WJ, Li SS, Xia JB. (2007c) *Appl Phys Lett* **90**: 193111.
Zhu YH, Zhang XW, Xia JB. (2006a) *Phys Rev B* **73**: 165326.
Zhu YH, Zhang XW, Xia JB. (2006b) *Physica E* **35**: 57.

INDEX